DATE DUE

As a well balanced and fully illustrated introductory text, this book provides a comprehensive overview of the physical, technological and social components of natural disaster. The main disaster-producing agents are reviewed systematically in terms of geophysical processes and effects, monitoring, mitigation and warning. The relationship between disasters and society is examined with respect to a wide variety of themes, including damage assessment and prevention, hazard mapping, emergency preparedness, the provision of shelter and the nature of reconstruction. Medical emergencies and the epidemiology of disasters are described, and refugee management and aid to the Third World are discussed. A chapter is devoted to the sociology, psychology, economics and history of disasters.

In many parts of the world the toll of death, injury, damage and deprivation caused by natural disasters is becoming increasingly serious. Major earthquakes, volcanic eruptions, droughts, floods and other similar catastrophes are often followed by large relief operations characterized by substantial involvement of the international community. The years 1990–2000 have therefore been designated by the United Nations as the International Decade for Natural Disaster Reduction.

The book goes beyond mere description and elevates the field of natural catastrophes to a serious academic level. The author's insights and perspectives are also informed by his practical experience of being a disaster victim and survivor, and hence the unique perspective of a participant observer. Only by surmounting the boundaries between disciplines can natural catastrophe be understood and mitigation efforts made effective. Thus, this book is perhaps the first completely interdisciplinary, fully comprehensive survey of natural hazards and disasters. It has a clear theoretical basis and it recognizes the importance of six fundamental approaches to the field, which it blends carefully in the text in order to avoid the partiality of previous works. No other book covers the earth and social sciences, as well as engineering, architecture and development studies. This extraordinary breadth is made possible by virtue of a strong emphasis on simple principles of the interaction of geophysical agents with human vulnerability and response.

All students of environmental sciences/studies and geography will find this book essential. It is an introductory text which treats this fascinating and dramatic subject area as something demanding serious academic treatment and not just as an assemblage of horror stories.

David Alexander is an Associate Professor in the Department of Geology and Geography at the University of Massachusetts, Amherst.

NATURAL DISASTERS

NATURAL DISASTERS

David Alexander

Department of Geology and Geography
University of Massachusetts

First published in 1990 as *Calamità Naturali: Lineamenti di Geologia Ambientale e Studio dei Disastri* by Pitagora Editrice, Italy.

This enlarged and revised edition published in 1993 by UCL Press

Reprinted in 2001
by Routledge
11 New Fetter Lane
London EC4P 4EE

Routledge is an imprint of the Taylor & Francis Group

ISBNs: 1–85728–093–8 HB
1–85728–094–6 PB

Front cover: The Great Day of His Wrath by J. Martin
(courtesy of The Tate Gallery, London)

British Library Cataloguing-in-Publication Data
A catalogue record for this book is available
from the British Library.

Typeset by Columns Design and Production Services Limited.
Printed and bound by Biddles Ltd, King's Lynn and Guildford, England

for Lorenzo

Contents

Contents

Contents

List of tables

List of tables

Preface

This book evolved from a course of lectures which I have given on various occasions at the University of Massachusetts in the USA and the Universities of Siena and Urbino in Italy. During my research and teaching I found that, although there is a large body of literature on natural disasters, there is no general work that synthesizes it in an accessible and sufficiently comprehensive form. In part the book is simply a series of observations and interpretations of the many phenomena which together make up natural catastrophes. But my intention has also been to transcend the boundaries of disciplines, which, although they encourage profound analysis, so often restrict what the researcher can say about phenomena such as these, whose very essence is multidisciplinary.

Natural disasters, I believe, should be studied as complete entities. Over-emphasizing restricted aspects will not help us to design good mitigation strategies, for there is a strong chance of ignoring vital factors that defy classification within traditional disciplinary systems or that transcend their boundaries. Although it is fashionable to talk of inter-disciplinary studies, given the magnitude of current scientific endeavours there are remarkably few of them. Undoubtedly, over-specialization and fragmentation of effort have inhibited the growth of disaster research as an autonomous field and have restricted the development of theory. For example, only now in the 1990s are experts beginning to come together to investigate the relationship between building design, structural collapse, human behaviour and injury causation in natural disasters: their work looks likely to open a rich new field of enquiry.

The distinction between natural hazards or disasters and their man-made (or technological) counterparts is often difficult to sustain. Many researchers have called the "naturalness" of natural disasters into question, though the circumstances of purely anthropogenic hazards, such as oil spills and chemical explosions, are usually very different from those of, for example, earthquakes and floods. Nevertheless, in terms of the consequences there is a sizeable overlap. Hence, much that is of benefit to the study of natural catastrophe can be learned from technological risks and disasters, and from the manner in which these interact with their natural counterparts. For this reason the book contains sections on urban fires, dam disasters and the risks inherent in high-rise buildings. Man-made risks are important here not merely because technology has created new sources of risk, and increased old ones, but also because many aspects of disaster management and risk reduction, which are applicable to natural disasters, were developed in response to technological impacts. Hence the approach to these aspects must be wide-ranging and eclectic; it must take account of the cross-fertilization that has occurred between the two branches of hazards studies.

If disaster researchers and managers are to be trained effectively, it is vital that they gain a fully rounded appreciation of the phenomena in question – earthquakes, eruptions, floods, and so on. A broad perspective on these entities should encompass their physical manifestations and problems of identifying, classifying and predicting them, as well as questions of understanding and mitigating their impact on humanity. Hence the study of disasters must involve the physical, technological, economic and social (not to forget the *perceptual*) realities, and the absence of any one of these may compromise the level of understanding achieved in the other categories. For instance, to manage floods effectively, the physical risk and impact of high water levels must be understood together with the technological possibilities for monitoring the hazard, the nature of warning systems, the degree of public perception of warning messages and how the public reacts.

One of the principal failings of the all-embracing approach to natural hazards and disasters, as it has been applied in the past, has been the lack of a theoretical basis of general principles and models. Reviews of the subject have all too often degenerated to an anecdotal level in which each disaster is treated as unique and little connection is made between the events that are successively described. Although the field is still short enough of theory to be a long way from becoming a fully fledged discipline in its own right, common threads underlie the succession of events. For example, there is always a relationship between, on the one hand, the impact of disasters and, on the other, human vulnerability, the strength of geophysical hazards and the effectiveness of mitigation measures. The duality between physical impact and human risks and responses is one of the principal underlying themes in this book.

While it is important not to neglect the uniqueness of a particular disaster, as it will always have something distinctive to contribute to our knowledge of such phenomena, I have sought to emphasize the generalities inherent in each natural catastrophe. For I hope that one day there will be a sufficiently large body of theory to permit us to inaugurate a new "interdisciplinary discipline" dedicated to the understanding of disastrous natural phenomena and their effects, and hence to the service of humanity. Such a discipline must combine theoretical insight with the practical, applied dimension. Neglect of either can be construed as inefficient response to disasters.

Acknowledgements

The author and publishers thank the following organizations and individuals for permission to reproduce or redraw tables and figures used in this book:

Academic Press (Tables 3.7, 7.2, Fig. 7.8); American Association for the Advancement of Science (Figs 2.5, 2.8, 3.13, 3.14, 3.15, 3.17, 9.3); American Geophysical Union (Table 1.10, Figs 2.19, 2.23, 2.24, 4.5, 7.8); American Meteorological Society (Table 3.3, Fig. 3.16); American Public Health Association (Table 7.1); Edward Arnold (Fig. 4.16); Dr Peter Baxter (Table 7.1); Basil Blackwell (Table 1.1, Figs 5.2, 5.10, 6.11, 8.6); Prof. Salvatore Belardo (Fig. 6.8); California Department of Boating and Waterways (Fig. 4.25); Cambridge University Press (Fig. 8.4); Dr Andrew Coburn (Figs 5.4, 7.5, 7.6, 9.7); Prof. Ronald U. Cooke (Table 4.8, Fig. 6.2); Dr Doak C. Cox (Table 2.6a, Figs 2.14, 7.7); Croom Helm (Table 4.9); Dr Ian Davis (Table 6.2, Fig. 9.10); Earthscan Publications (Table 3.2); Elsevier Science Publishers (Fig. 2.24); Prof. Harold D. Foster (Fig. 6.5); Dr Peter W. Francis (Fig. 2.16); W.H. Freeman (Figs 2.1, 2.7, 5.1, 9.8); Prof. Hugh M. French (Fig. 4.23); Prof. T. Theodore Fujita (Table 3.4, Fig. 3.22, 3.23); Geological Society of America (Table 4.11); Dr Gary B. Griggs (Fig. 4.25); Harwood Academic Publishers (Fig. 4.9); Hawaii Institute of Geophysics (Table 2.6a, Figs 2.14, 7.7); HMSA (British Crown and Parliamentary Copyrights) (Fig. 4.17); Institute of Management Sciences (Fig. 6.8); International Association of Hydrological Sciences (Table 3.8, Fig. 3.26); International Civil Defence Organization (Fig. 4.28); International Glaciological Society (Fig. 3.29); Dr David K. Keefer (Table 4.11); Kluwer Academic Publishers (Tables 2.4, 2.6b, 8.4, Fig. 9.5); Prof. Edward R. LaChapelle (Table 3.8); Macmillan Press (Figs 3.4, 3.6); Methuen (Table 3.9, Figs 3.1, 3.21); Mettag (Fig. 7.11); MIT Press (Tables 1.2, 4.14, Figs 3.10, 4.24, 6.3, 6.5); Prof. J. Kenneth Mitchell (Table 5.2); National Academy Press (Table 4.10, Figs 2.21, 5.11); Natural Hazards Research and Applications Information Center, University of Colorado, Boulder (Figs 2.10, 3.30, 6.6); New York Public Library, Astor, Lennox & Tilden Foundations (Fig. 9.10a); Octopus Publishing Group Australia (Fig. 4.26); Prof. Cliff D. Ollier (Fig. 2.16); Oxford Polytechnic Press (Fig. 9.10); Oxford

University Press (Tables 3.1, 4.5, 9.2, Figs 1.5, 3.11, 3.27, 4.4, 4.6, 9.2); Pan American Health Organization (Fig. 6.10, 7.1); Penguin Books (Fig. 4.22); Dr Ronald I. Perla (Table 3.7); Prentice-Hall (Table 2.8); Royal Geographical Society (Fig. 8.11); Scandinavian University Press (Fig. 5.8); Scientific American (Table 4.12, Fig. 4.21); Skandia America Group (Fig. 6.1); Society for Computer Simulation (Fig. 6.9); Soil and Water Conservation Society of America (Table 5.5); Springer-Verlag New York (Fig. 6.5, 7.10); Dr Chauncy Starr (Fig. 9.3); Dr Karl V. Steinbrugge (Fig. 6.1); Transportation Research Board, National Research Council, Washington, DC. (Table 4.6, Figs 4.12, 4.13); UNESCO Press (Tables 1.9, 2.3, 3.6, Fig. 2.9); USA. Federal Centers for Disease Control (Tables 7.3, 7.4, 8.1, Fig. 8.5); USA Geological Survey (Table 2.5, Figs 1.1, 2.1, 2.3, 2.13, 2.22, 4.1, 6.4); USA Government Office of Emergency Preparedness (Figs 2.15, 3.24, 7.9); USA National Institute of Mental Health (Table 9.1); University of California Marine Science Institute (Fig. 4.25); University of Wisconsin Press (Fig. 9.6); Unwin-Hyman (Fig. 3.8, Fig. 6.2); Prof. Barry Voight (Fig. 2.24); Prof. Andrew Warren (Tables 4.3, 4.4); Dr Richard A. Warrick (Fig. 3.8); Dr John Whittow (Fig. 4.22); John Wiley (Table 2.1, Fig. 4.19); World Health Organization (Table 8.3)

Particular thanks are also due to Maureen Fordham and Ian Burton for reviewing a draft of this book.

Frontispiece: Effects of the 1982 landslide at Ancona, Italy.

CHAPTER ONE

Introduction

The global problem

Globally, it appears that the toll of death and damage in natural disasters is increasing (Fig. 1.1), although there is no international databank of sufficient comprehensiveness to verify this supposition. However, it seems that the frequency of the most severe impacts on the socio-economic system is decreasing, thanks to improvements in prediction, warning, mitigation and international aid. The cost to the global economy now exceeds US$50,000 million per year, of which a third represents the cost of predicting, preventing and mitigating disasters and the other two thirds represent the direct costs of the damage. Death tolls vary from year to year around a global mean of about 250,000, while major disasters kill an average of 140,000 people a year. About 95 per cent of the deaths occur in the Third World, where more than 4,200 million people live (Table 1.1).

Natural catastrophe can also have a severe impact on developed countries, notably Japan, Italy and the USA. In the territories of the Commonwealth of Independent States (the former USSR), natural disasters kill 150–200 people each year and injure several thousand (Porfiriev 1992), while in the USA, about 30 disasters, including 15 "states of emergency", are declared in an average year. For example, in 1983, 34 "major disasters" and four other states of emergency were declared federally, and 31 states and one overseas territory were affected. Floods account for about 40 per cent of the damage caused by natural disasters, while hurricanes and other tropical storms produce the largest number of fatalities, about 20 per cent of the total for all natural hazards. Overall, the number of deaths seems to show little increase (it averaged 332 per disaster over the period 1971–80), but damages and the costs of mitigation arising from natural hazards have been increasing for coastal

1

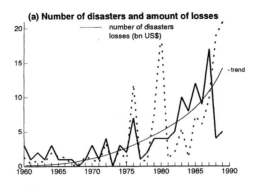

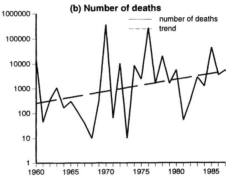

Figure 1.1 Number of disasters and average losses, 1960–88 (data from Berz 1988).

Table 1.1 Loss of life by disaster type and by continent, 1947–80 (after Shah 1983, p. 206).

Agent	No. of events	Asia	Oceania	Africa	Europe	South America	Caribbean & Central America	North America
Earthquake	180	354,521	18	18,232	7,750	38,837	30,613	77
Tsunami	7	4,459	—	—	—	—	—	60
Volcanic eruption	18	2,805	4,000	—	2,000	440	151	34
Flood	333	170,664	77	3,891	11,199	4,396	2,575	1,633
Hurricane	210	478,574	290	864	250	—	16,541	1,997
Tornado	119	4,308	—	548	39	—	26	2,727
Severe storm	73	22,008	—	5	146	205	310	303
Fog	3	—	—	—	3,550	—	—	—
Heatwave	25	4,705	100	—	340	135	—	2,190
Avalanche	12	335	—	—	340	4,350	—	—
Snowfall & extreme cold	46	7,690	17	—	2,780	—	200	2,510
Landslide	33	4,021	—	—	300	912	260	—
Total		1,054,090	4,502	23,540	28,694	49,275	50,676	11,531

erosion, earthquakes, floods, hurricanes, landslides and tornadoes. Mitigation costs have been relatively stable for drought, lightning, tsunamis, volcanic eruptions, avalanches, frost, hailstorms, urban snow and windstorms, but damages have been rising for the last five of these (White & Haas 1975).

Natural catastrophes thus have the power to exert a substantial and consistent influence on modern society. But there are several additional reasons why people should be encouraged to take an interest in and seek to understand them. First, in professional, administrative and political rôles, many people will have to deal with and face the effects of disaster in their own communities and spheres of influence. Secondly, the more technologically complex our society becomes, the more diverse are the risks and effects of disaster; the more people travel and relocate, the more they put themselves at risk. Thirdly, it is vital to dispel popular myths and misconceptions about the occurrence and impact of disasters, for their effects can only be combatted effectively if we have a rational and objective understanding of them. Finally, in keeping with the new spirit of global internationalism, an interest in problems of the Third World is to be encouraged. Natural disasters are a very important component of such questions.

Objective analysis requires that the sensational aspects of disasters be played down and that we assume a sober, responsible attitude to them. This also means that anecdotal and purely descriptive approaches should be avoided: case studies should be used only where they genuinely demonstrate or illustrate fundamental principles. Hence a rigorous approach to natural disasters requires that we look for the common regularities in each event, however unique it may at first seem. In this way will it be possible to improve our understanding of the phenomenon of natural catastrophe.

In this context there is good reason to ask whether the newly emerging field of disaster studies merits the status of a discipline. The answer may be "yes", in that it does have some general principles. On the other hand, it may be "no", in that the field constitutes, not so much a corpus of information as a very heterogeneous mixture of observations, regularities, hypotheses and techniques derived from a wide variety of other disciplines. It is also a field in which practical experience is as useful as academic study, especially as social and ethical problems must be tackled that cannot be resolved by science alone. Thus disaster studies are highly distinctive: they consist not merely of geology, geography, sociology, and other well-defined academic disciplines, but also of preparedness training and, quite simply, the application of sound judgement to practical problems. One may argue, indeed, that every citizen should receive basic training in disaster preparedness, as already is the case in Japan (Arnold 1984).

Having briefly reviewed the problem of natural catastrophe in its world-wide context, we will now establish a theoretical basis for the practical analysis of disasters.

Definitions and basic concepts

Initial definitions

The term **natural hazard** has been defined in four ways. It is:

(a) "A naturally occurring or man-made geologic condition or phenomenon that presents a risk or is a potential danger to life or property" (American Geological Institute 1984);
(b) "An interaction of people and nature governed by the co-existent state of adjustment of the human use system and the state of nature in the natural events system" (White 1973);
(c) "Those elements in the physical environment [which are] harmful to man and caused by forces extraneous to him" (Burton & Kates 1964);
(d) "The probability of occurrence within a specified period of time and within a given area of a potentially damaging phenomenon" (UNDRO 1982).

From this it is clear that we are dealing with a physical event which makes an impact on human beings and their environment and, unless this conjunction occurs, there will be no hazard or disaster. The hazard involves the human population placing itself at risk from geophysical events.

A **natural disaster** can be defined as some rapid, instantaneous or profound impact of the natural environment upon the socio-economic system. Turner specified the phenomenon more completely as "an event, concentrated in time and space, which threatens a society or a relatively self-sufficient subdivision of a society with major unwanted consequences as a result of the collapse of precautions which had hitherto been culturally accepted as adequate" (Turner 1976: 755–6). The concentration of effects must be emphasized: disease and malnutrition kill some 15 million children a year worldwide, yet this is not regarded as a disaster in the same sense.

One convenient operational definition of "disaster" postulated that an individual event must cause more than US$1 million in damage, or the death or injury of more than 100 people (see Burton et al. 1978). But this masks considerable potential for variation, especially as small damages can be very costly in highly developed societies, while catastrophic ones may appear cheap in very poor societies with few valuables. In fact, disasters seem to have a disproportionately large impact on very poor and very rich societies: the former present the highest casualty totals and the latter the highest property damages. Impacts can also be high in societies that are industrializing and have lost much of their traditional resistance but not yet developed sophisticated mitigation measures.

In general terms we are not only dealing with phenomena of high magnitude. In fact, we can define an **extreme event** as any manifestation in a geophysical system (lithosphere, hydrosphere, biosphere or atmosphere) which differs substantially or significantly from the mean. If human socio-economic and physiological systems do not have the capacity sufficiently to reflect, absorb or buffer the impact, then disaster may occur.

Adjustment and adaptation to risk
The degree to which a society remains unaffected by natural extremes reflects its ability to adapt to hazards (i.e. its absorptive capacity). **Adaptation** involves awareness of both hazards and the means by which they can be avoided. It depends on the available technology, on the economic viability of alternative strategies for avoiding disaster and on social processes which may be slow and complex.

Alexander (1991a) defined four forms, or levels, of adaptation to natural hazard risk. The first involves persistent occupation of the hazard zone despite the presence of danger. This may involve comprehensive measures for hazard mitigation and abatement. Alternatively, no measures other than warning and evacuation may be used, or there may be no protection measures at all, thus creating a state of maximum vulnerability. The second level of adaptation involves cohabitation with the damage caused by past disasters in a state of **maximum geographical inertia**. It is complemented by a third state, in which damaged or destroyed structures are abandoned, but populations relocate within the risk zone, hence creating **secondary geographical inertia**. Fourthly, there may instead be planned or unplanned migration to safer zones.

Collectively, efforts to reduce the negative impact of natural events are termed **adjustment** (examples are given in Table 1.2). Adjustment of human activity is often less expensive than control of physical forces (if the latter is in any case practicable). For example, it is often less costly to use floodplain land-use regulation to reduce flood risk than it is to build a substantial river-bank levée, which is unlikely to provide complete protection against extreme floods. There are, however, hidden costs associated with setting up bureaucracies and inspectorates, introducing legislation and ensuring compliance: these can vastly increase the costs of

Table 1.2 Examples of adjustment to potential disasters (after White & Haas 1975: 58). Copyright 1975 by MIT Press.

Type of hazard	Modify the event	Modify human vulnerability	Distribute the losses
Avalanche	artificial triggering	snow tunnels and barriers	emergency relief
Coastal erosion	beach nourishment programmes	breakwaters and groynes	insurance against flood inundation
Drought	cloud seeding	variation in crop planting sequence	insurance of crops
Earthquake	regulation of interstitial fluid pressure	anti-seismic construction	emergency relief
Flood	upstream water impoundment	flood-proofing	loans and insurance
Frost	heating system for orchards	warning network	insurance of crops
Hail	cloud seeding	crop selection	crop insurance
Hurricane	cloud seeding	land-use modification	emergency relief
Landslide	drainage of the headscarp area	land-use regulation	legal proceedings against developers
Lightning	cloud seeding?	lightning conductors	emergency relief
Snowstorm in urban area	—	snow clearing operations	taxes levied to finance snow clearance
Tornado	cloud seeding?	warning network	emergency relief
Tsunami	—	warning network	emergency relief
Volcanic eruption	—	land-use regulation	emergency relief
Windstorm	—	improved anchoring of light structures and mobile homes	insurance of property

non-structural mitigation. In fact, the degree of adjustment that is possible in a given society is a function of several other variables. These include the proportion of expected losses to present reserves, the availability of relief and aid, the degree of choice between alternative mitigation strategies, the prevailing type of government and the degree to which it is willing to involve itself in hazard mitigation.

According to Burton et al. (1978: 203–6), in using physical resources people behave in a way that combines adaptation to extreme events with both deliberate and spontaneous adjustment. In order to cope, they employ on the one hand social and cultural adaptation and on the other purposeful and unpremeditated forms of adjustment. The various ways of coping can be grouped as a whole into four main strategies – loss absorption, loss acceptance, loss reduction, and radical change – separated respectively by thresholds of awareness, action and intolerance. These four adjustments to hazard appear universal, but such are the varying fortunes of society that there will not necessarily be a steady progression through them (see Whyte 1982: 256).

At the simplest level, humanity has the option either to live in harmony with the natural environment, by a **symbiosis** in which ecological life-support systems are enhanced, or to exploit resources by a form of **parasitism**, in which hazards are ignored until they strike. The former provides opportunities for sustainable development and protection against environmental extremes, while the latter relies on exploitation of non-renewable resources, and places little emphasis on hazard abatement. In reality these are the end members of a continuum of possible development and mitigation strategies.

Conceptual equations of risk, impact and vulnerability
A **hazard** may be regarded as the predisaster situation, in which some **risk** of disaster exists, principally because the human population has placed itself in a situation of **vulnerability**. According to the Office of the United Nations Disaster Relief Co-ordinator (UNDRO 1982) this can be defined as "the degree of loss to a given element or set of elements at risk resulting from the occurrence of a natural phenomenon of a given magnitude. It is expressed on a scale from 0 (no damage) to 1 (total loss)." When the risk becomes tangible and impending, there is a distinct **threat** of disaster. Hence, the sequence of states pertaining to disaster is as follows:

$$\text{hazard} \rightarrow \text{risk} \rightarrow \text{threat} \rightarrow \text{disaster (impact)} \rightarrow \text{aftermath}$$

Alternatively, Harriss et al. (1978: 379) defined hazards as "threats to humans and what they value" and risks as quantified, conditional probabilities that the consequences of hazards will be harmful.

UNDRO (1982) offered a wider definition, in which the concept of **risk** can be considered in the light of three components:

(a) the **elements at risk (E)** comprise the population, properties, economic activities, public services, and so on, which are under the threat of disaster in a given area;
(b) **specific risk (R_s)** is the degree of loss likely to be caused by a particular natural phenomenon. It may be expressed as the product of the natural hazard, **H**, times the vulnerability, **V**;
(c) the **total risk (R_t)** consists of the number of lives likely to be lost, the persons injured, damage to property and disruption to activities caused by a particular natural phenomenon. It is the product of the specific risk (R_s) and the elements at risk (E):

$$R_t = (E)(R_s) = (E)(H.V)$$

Looked at another way, human vulnerability is a function of the costs and benefits of inhabiting areas at risk from natural disaster (Alexander 1991a):

$$\begin{array}{ccc} \text{total} & \text{risk} & \text{risk} & \text{risk} \\ \text{vulnerability} = \text{amplification} & - \text{mitigation} & \pm \text{perception} \\ \text{measures} & \text{measures} & \text{factors} \end{array}$$

Risk amplification occurs as a result of the continued development of past and future disaster areas, but it can be reduced by mitigation efforts. High levels of risk perception can provide the motivation for hazard mitigation, while low levels encourage *laissez faire.*

Mitigation can be divided broadly into structural and non-structural forms (see Table 1.3). However, Micklin (1973) preferred to divide the

Table 1.3 Structural and non-structural methods of disaster mitigation.

- *Structural methods:*
  Retrofitting of existing structures
  Reinforcement of new structures: design features
       overdesign
  Safety features: structural safeguards
       failsafe design
  Engineering phenomenology
  Probablistic prediction of impact strength

- *Non-structural methods:*
  (a) short-term:
  Emergency plans: (civil):   co-ordinator(s)
       police and firemen
       Red Cross and charities
       volunteer groups
       medical services
       military forces
  Evacuation plans: routes and reception centres
       for the general public
       for vulnerable groups: the very young,
       elderly, sick or handicapped
  Prediction of impact: monitoring equipment
       forecasting methods and models
  Warning processes: general message
       specialized warning (e.g. ethnic)
  (b) long-term
  Building codes and construction norms
  Hazard microzonation: selected risks
       all risks
  Land-use control: regulations, prohibitions, moratoria,
       compulsory purchase
  Probablistic risk analysis
  Insurance
  Taxation
  Education and training

ways in which mitigation can be applied into four categories: **engineering mechanisms** include technological innovations and application; **symbolic mechanisms** involve culture and its constituent norms and rôles; **regulatory mechanisms** define public policy and social control; and **distributional mechanisms** specify the movement of people, activities and resources.

Once mitigation is applied in some degree, the net impact of disasters can be viewed as a simple conceptual equation:

$$
\begin{array}{cccc}
\text{Net impact} & \text{total benefits} & \text{total costs} & \text{costs of} \\
\text{of disasters} = & \text{of inhabiting} - & \text{of disaster} - & \text{adaptation} \\
& \text{hazard zone} & \text{impact} & \text{to hazard}
\end{array}
$$

Complicating factors
Both disasters and the agents that produce them are sufficiently complex to defy easy classification. However, one basic distinction is between **sudden impact** and **slow onset (creeping)** disasters. The former may occur in a matter of seconds (earthquakes), minutes (tornadoes) or hours (flash floods); while the latter may take months (certain types of volcanic eruption), years (types of subsidence of the ground) or centuries (various forms of land degradation and erosion). There is in fact a continuum in both lengths of forewarning and speeds of onset (Table 1.4), with no set definition of what constitutes "abrupt". Moreover, one cannot make general rules about the magnitude at which a high-intensity event turns into a fully fledged disaster, as a small-scale geophysical impact may have very serious human implications if it occurs at a locality of high vulnerability. Nevertheless, the severity of sudden-impact disasters is usually judged according to casualty and damage figures, while that of creeping disasters is generally related to the size of the affected population and the magnitude of the threat which they face.

A further source of complexity lies in the fact that many disasters are composite. Thus, a single earthquake may have the potential to cause tsunami waves at sea, landslides or avalanches on slopes, dam failure at reservoirs, and building damage and fires in urban areas. The relationship between geophysical impact and human vulnerability – which forms the guiding principle of disaster studies – means that naturally induced effects are difficult to separate from anthropogenic ones. For example, disastrous flooding may depend as much upon the level of human settlement of a floodplain as it does upon the hydrological behaviour of a river. Conversely, although they usually have purely anthropogenic causes, serious episodes of pollution could be viewed as at least partially "natural" disasters, because of their ramified impact on natural ecosystems.

No geophysical event is inherently catastrophic. For instance, the 1964 Alaska earthquake caused the Sherman debris avalanche, in which 30

9

Table 1.4 Classification of disasters by duration of impact and length of forewarning.

Type of disaster	Duration of impact	Length of forewarning (if any)
Lightning	instant	seconds–hours
Avalanche	seconds–minutes	seconds–hours
Earthquake	seconds–minutes	minutes-years
Tornado	seconds–hours	minutes
Landslide	seconds–decades	seconds–years
Intense rainstorm	minutes	seconds–hours
Hail	minutes	minutes–hours
Tsunami	minutes–hours	minutes–hours
Flood	minutes–days	minutes–days
Subsidence	minutes–decades	seconds–years
Windstorm	hours	hours
Frost or ice storm	hours	hours
Hurricane	hours	hours
Snowstorm	hours	hours
Environmental fire	hours–days	seconds–days
Insect infestation	hours–days	seconds–days
Fog	hours–days	minutes–hours
Volcanic eruption	hours–years	minutes–weeks
Coastal erosion	hours–years	hours–decades
Accelerated erosion	hours–millennia	
Drought	days–months	days–weeks
Crop blight	weeks–months	days–months
Expansive soil	months–years	months–years
Desertification	years–decades	months–years

million m³ of rock were mobilized at an average speed of 30.8 m/sec (108 km/hr), but this material flowed and slid into an uninhabited valley (Shreve 1966). Two years later at Aberfan in South Wales, a tip of mining spoil failed under conditions of extreme saturation, and a tongue of debris flowed at walking pace into a school, engulfing 144 people, many of whom were children (Miller 1974; see Ch. 4). Despite being much smaller and slower, only the latter event can be viewed as a disaster, because, of course, it involved human impact and vulnerability of a kind that was totally lacking in the valleys of central Alaska.

The hazardousness of geographical locations
The late twentieth century has witnessed a substantial increase in the risk of natural disaster, which has been offset only partially by better preparation and mitigation. There are four main reasons for this. First, geographical inertia and the economic advantages of specific locations have led to a continued inhabitance of past and potential disaster zones (Fig. 1.2). Secondly, particular risks or dangerous conditions have

Figure 1.2 The cone of Mount Vesuvius (southern Italy) and the ramparts of the Somma, an earlier volcanic edifice destroyed in the eruption of AD 79. In the foreground are the harbour and railway station of Portici (pop. 83,000), Europe's most densely populated municipality. There are few better examples of how the risk of natural events interacts with human vulnerability in terms of the *hazardousness of place*.

sometimes been increased by neglect or environmental malpractice. For instance, flood-prevention schemes have increased the discharge of the River Danube, thus elevating the flood risk down stream of the protected area. Thirdly, the impact of hazards has become more profound as the complexity of society has increased. The accumulation of physical capital has been accompanied by its concentration in vulnerable locations or its distribution throughout potential disaster areas. As world population levels rise, such areas are simply becoming more populous. Citizens of the richer nations are becoming wealthier, are intent on socio-economic gain, are engaging in more activities, often in hazard zones, and thus are expecting more protection. Overall, to live in areas of the world known to be susceptible to disaster is a luxury for the rich and a constraint upon the poor.

Lastly, until recent times there has been a lack of knowledge, research and protective regulations concerning disasters and their impacts. In part this reflects the difficulty of studying events, recurrent on the geological timescale ($10^3 - 10^9$ years), which may manifest themselves only in the long term. Nevertheless, only in recent decades has the global community woken up to the need for a concerted strategy dedicated to understanding and tackling the problems of extreme natural events.

There are several reasons why groups of people continue to live in hazard zones, despite the existence of measurable and known levels of risk. Some of these areas offer their inhabitants superior economic opportunities, such as those given by fertile volcanic or floodplain soils. But for many groups there is a lack of satisfying alternatives, especially if the resources are not available to allow them to adjust to the hazard or migrate to more promising lands. Some groups instead are dedicated to short-term objectives. These may range from the realization of a quick profit on activities (tourism, agriculture, industry, and others) that locally may be threatened infrequently by natural hazards to strategies for avoiding the destitution likely to occur eventually at particular locations (shifting cultivation, nomadism, and so on). Groups and individuals may remain in hazard zones because they have generated high ratios of reserves to potential losses (including the use of insurance as a means of providing for losses that are occasionally extreme). Or, more simply, such people may have become very resilient, being able to bear losses or adapt to risk in whatever manner is necessary.

Schools of thought
The observations reported above have been subject to various distinct interpretations. In the English-speaking world, the field of natural hazards and disasters has grown in close association with the applied natural and social sciences, but has suffered from problems of fragmentation and over-specialization, as well as from the insularity that disciplinary studies tend to foster. To some extent this has prevented it from acquiring a separate identity.

During the past 125 years there has been a gradual shift of emphasis in the natural and social sciences from the impact of environment upon mankind to humanity's impact on environment (Goudie 1983). In one sense, disasters represent an extreme class of human-environmental or ecological phenomena. They cannot be considered truly natural, in that human vulnerability seldom results from purely natural states (rather than locational decisions based on socio-economic criteria) and human intervention often results in aggravated risk of geophysical impact. Hence, because they represent an environmental imperative to society, natural disasters have been studied using the tenets of **human ecology**, which Mileti (1980) defined as "seeking the determinants of human behaviour in the natural environment, and the processes that facilitate human adjustment to the physical world through social organization".

Although an all-embracing concept such as human ecology is capable of unifying disparate subject matter, it has not been adopted universally. For example, Hewitt (1983) argued that the direction of causality has been mistaken by the majority of scholars in the field. In catastrophe, nature is seen to decide which social conditions or responses will be significant. He has argued that vulnerability and human social organization are, instead, the critical determinants of both risk and impact. This view is reinforced by Kreps (1989) and Drabek (1989), who both redefine disasters as "non-routine *social* problems" (my emphasis).

The increasing division of knowledge into disciplines, which have in turn spawned subdisciplines, has meant that natural hazards and disaster studies have had to struggle for identity against a wide range of groups and subgroups, many of which have substantial overlap and common ground but often very different viewpoints, approaches or objectives. Regrettably, the field has responded to increasing disciplinary specialization by becoming ever more fragmented. Hence, at least six schools of thought and expertise can at present be identified. First, the **geographical approach** to natural hazards stems from Harland Barrows's work in the 1920s on human ecological adaptation to environment (Barrows 1923) and Gilbert F. White's seminal monograph on human adjustment to floods (White 1945). Social science methods are widely used, and emphasis is given to the spatio-temporal distribution of hazard, impacts and vulnerability. Geographers have also given particular thought to the question of how choices are made between different types of adjustment to natural hazards.

Secondly, the **anthropological approach** has focused on the rôle of disasters in guiding the socio-economic evolution of populations, in dispersing them and in causing the destruction of civilizations (although Torry (1979) noted that the last of these is highly debatable). A strong concern for the Third World has led anthropologists to search for the threshold points beyond which local communities can no longer provide the basic requirements for survival of their members. They have also studied the "marginalization syndrome" arising from the impoverishment of disadvantaged groups in underdeveloped societies (Oliver-Smith 1979).

Thirdly, the **sociological approach** stems from the work of Russell R. Dynes, Enrico L. Quarantelli and others. Vulnerability and impacts are considered in terms of patterns of human behaviour and the effects of disasters upon community functions and organization (Quarantelli 1978; Dynes 1970). In addition, psychologists have studied disaster in relation to factors such as stress (Glass 1970), bereavement (Church 1974) and the "disaster syndrome", a psychologically determined defensive reaction pattern (Wallace 1956; see Ch. 8).

Fourthly, the **development studies approach** considers problems of providing aid and relief to Third World countries, and addresses questions of refugee management, health care and the avoidance of starvation. Logistical aspects are emphasized (Knott 1987), as are nutritional studies (Chen et al. 1980). Davis (1978) stated that over 80 per cent of disaster impacts occur in developing countries, and it is clear that the epiphenomenon of poverty increases human vulnerability to natural hazards: locational constraints tend to place the poor more firmly in the path of impacts, while credit, savings, capital and alternative options that normally cushion the impact are lacking.

Fifthly, a new field of **disaster medicine and epidemiology** has recently been founded. It focuses on the management of mass casualties, the treatment of severe physical trauma and the epidemiological surveillance of communicable diseases whose incidence rates may increase during the disruption of public health measures following a disaster (Beinin 1985; PAHO 1981). Lastly, the **technical approach** prevails among natural and physical scientists. Emphasis is given to seismology, volcanology, geomorphology and other predominantly geophysical approaches to disasters (Bolt et al. 1977, El-Sabh & Murty 1988) and to engineering solutions.

Clearly, the prevailing theories and models in hazard and disaster studies have not succeeded in preventing the emergence of quite separate schools of thought that deal with what are essentially the same phenomena. Dichotomies exist between the **technocratic** school, for whom the solution to the disaster problem lies in the application of measuring and monitoring techniques and sophisticated managerial strategies, and the **development** school, who point out that, as a result of poverty, such luxuries are denied to the majority of disaster victims. The view that technology will eventually triumph over catastrophe is still fashionable, but is increasingly subject to criticism and doubt. Economic power, the force behind technocracy, is related to the seriousness of disaster impacts in a nonlinear way that is tempered by political factors and the tendency to accumulate physical capital and thus increase vulnerability. Concomitantly, most of the theory used by hazards specialists has been formulated for the developed world (especially the United States of America) and is often of doubtful validity elsewhere.

As Parker & Harding (1979) have noted, by emphasizing adjustment, adaptation and perception, the study of natural disasters draws attention to the dynamic relationship between humanity and environment and can be an excellent means of enhancing environmental awareness. The field can also be broadened to embrace ecological destruction (Ball 1979). In fact, Bunin (1989) argued that the International Decade for Natural Disaster Reduction (1990–2000) should take place explicitly within the compass of efforts to maintain ecological sustainability, as the main scope for disaster mitigation in the less developed nations lies in protecting their natural resources. Disasters must, however, be presented as holistic, interdisciplinary phenomena, as the boundaries between disciplines tend to impede understanding and restrict the creation of theory. This is especially true, given that almost all aspects of disasters occur on continua, rather than as discrete entities (Table 1.5).

Table 1.5 Continua, dichotomies and polychotomies in disaster studies (Alexander 1991b: 214).

CONTINUA IN HAZARDS AND DISASTERS

events:

anthropogenic disaster ↔ natural disaster

sudden impact disaster ↔ creeping disaster

short-term aftermath ↔ long-term aftermath
(restoration)   (reconstruction)

scientific organization:

technocentric approach ↔ ecocentric approach

natural hazards ↔ environmental geology
(social science)   (natural science)

attitudes and approaches:

symbiosis with environment ↔ parasitism (exploitation)

risk amplification ↔ risk reduction

optimizer ↔ satisficer

mitigation ↔ laissez faire

fatalism ↔ activism

environmental determinism ↔ probablism ↔possiblism

DICHOTOMIES AND POLYCHOTOMIES

recurrence interval | time-scale of
for most disasters | geological events
$[10^{-1}-10^2$ years] | $[10^3-10^9$ years]

prediction | warning

simple impact | composite disaster | secondary disaster
[earthquake] | [earthquake & tsunami] | [post-earthquake fire]

adaptation | adjustment

costs | benefits

structural mitigation | non-structural mitigation
[retrofitting of buildings] | [insurance]

Although theory is essential to their study, it is vital to consider disasters in the light of practical problems, one of the most important groups of which concerns the provision of aid and relief after disaster has struck.

Problems of providing disaster relief
The relief operations that usually follow major international disasters must take account of some salient problems. For example, Third World countries often have difficulty in assimilating aid supplied by the developed nations (see Ch. 8). Secondly, the evacuation of risk zones or

disaster areas may create a serious and long-term problem of homelessness. Thirdly, poverty tends to aggravate the various difficulties caused by natural disaster. The poorest parts of the world have a building stock which is particularly vulnerable to damage and destruction in disasters, but they lack the resources to upgrade it. Thus, in both geographical and social terms, natural disasters tend to accentuate the plight of the poorest and most disadvantaged sectors of society.

For the providers of disaster relief it is all too easy to confuse pre-existing problems with those caused directly by the catastrophe. Even if they are not affected by disaster, the victims of deprivation are likely to throw in their lot with disadvantaged survivors in the clamour for relief and aid – understandably, of course. In such situations, where resources are often severely limited, there is therefore a great dilemma as to whether relief should concentrate on remedying the effects of the disaster or should instead tackle the wider problem of underdevelopment, from which vulnerability to natural catastrophe eventually stems.

The difficulties of providing disaster relief contribute to the wider problem of misconception about natural catastrophe, which will now be considered in detail.

Misconceptions about natural disaster

Although having antecedents in the work of the Canadian minister, Samuel Henry Prince in 1920 (Scanlon 1988), disaster studies constitute a young field of enquiry that has had little time to render its conclusions well known outside its strict circle of academic and professional workers. One sad consequence of this is that ignorance has allowed misconceptions to proliferate, not merely among the general public, but among people whose business it is to involve themselves with disaster, such as emergency planners, journalists and local politicians. Accordingly, it is useful to examine the fallacies and misunderstandings that commonly arise, and to attempt to lay them to rest.

Misapprehension of the fundamental facts
The misapprehension of disasters begins with the assumption that the natural environment need not be treated with much respect. Such an attitude has been termed **technocentric**, and tends to assume that mankind's ingenuity is sufficient to overcome a particular hazard, either by modifying it or by making the environment safe. But human vulnerability has often been a result of ignorance not merely about the scope of natural disaster impacts but also about the limited degree to which a given technology is capable of mitigating them. Thus settlers in the Australian outback colonized infertile land with a poor weather

record and were forced to abandon their farms at the first serious water shortage, even though drought is not an uncommon phenomenon there.

Another common misconception is that disasters are unique and exceptional events. On the contrary, many extreme geophysical events occur with a periodicity only slightly longer than the timespan of human memory (perhaps 30–100 years). The tendency to forget during the lull between such impacts means that, unless a community is very well prepared against the disasters that threaten it, definition of the event and prediction of its consequences may be no more than afterthoughts, too late to affect the level of vulnerability. The result has often been a prevailing aura of surprise and confusion in the face of events that are essentially predictable and about which much has been written.

In a classic paper that referred specifically to the controversy over drought, desertification and international aid in the Sahel, Glantz (1977) listed a series of common fallacies of natural disaster. He demonstrated that politicians cannot be relied upon to give a straight and accurate answer to questions of relief and development, technology will not necessarily provide the answers to disasters and unsystematic solutions are not likely to bring change for the better. He also showed that, in the politics of foreign disaster aid, what is considered to be interdependence among nations is often dependence; and, far from encouraging people to train and think for themselves, education has frequently contributed to dependence by imposing extraneous ideas upon the stricken populations. Nor should much be assumed about traditional societies, many of whom are more than willing to abandon their ways of life in favour of something safer and more prosperous. But perhaps the greatest fallacy is that change for the better is inevitable, and "normality" will return when the disaster is over. Glantz argued that change must be made to occur, and that for the people most vulnerable to disaster there is no such thing as "normal", so hard is everyday life.

Popular misconceptions abound concerning specific aspects of natural disasters and can be dispelled only by better public education and the encouragement of a "culture of civil defence", in which each individual is held responsible for many aspects of his or her own safety and his or her knowledge of the risks being run. It would be impossible to describe all the particular misconceptions which disasters commonly engender, but a few examples will suffice. Although fissuring may occur in soft sediments, earthquakes have hardly ever caused the ground to open up and engulf people or objects. Although there are sizeable regions of the world where low-strength housing coincides with occasional high-magnitude earthquakes, contrary to popular belief, in many earthquakes houses do not collapse *en masse*, and hence it is probably often true that running outside during the tremors engenders a greater risk than remaining indoors, as minor damage may include the collapse of masonry into the street. Fire is actually not a common consequence of seismic events; yet it is not necessarily absent during water-related disasters, for example when a tsunami wave ruptures oil tanks situated on the coast and spreads burning fuel.

Exaggeration and rumour
Disasters are by their very nature chaotic, and those who are closely involved in them are often able to gain only a limited perception of events as they take place. On many occasions the tolls of death and injury, and the extent and value of damage, become exaggerated. Often, official tallies of the death toll caused by a sudden impact disaster rise as information is collected and search-and-rescue takes place. Meanwhile, unofficial estimates based on the perception of those who participate in the rescue operation tend to exaggerate the figures, if these individuals allow their judgement to be overwhelmed by the enormity of the event. The figures eventually stabilize, but long after they have ceased to be news (Fig. 1.3).

Another common phenomenon during the immediate aftermath of disasters is **rumour**. This is usually related to one or more of the following: the probability that the disaster will strike again in the immediate future; the likelihood that a secondary disaster such as a disease epidemic will occur; or the possibility that the government or other authority is keeping the true facts about the catastrophe from the people (for example, concerning the extent of the disaster area, the number of casualties or the degree of risk). According to Scanlon (1977), rumours tend to be spread between people of the same sex and similar socio-economic status, who speak the same language. Scanlon added that

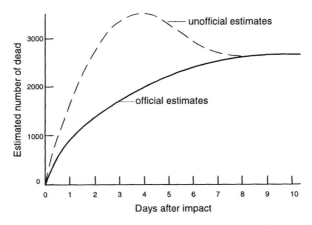

Figure 1.3 Estimates and official tallies of the death toll in a sudden-impact disaster.

the process of passing the rumour on simplifies it by eliminating some details, focuses it by concentrating on some particular aspect which interests the communicator, and re-embelishes it by adding new detail. None of this is likely to be productive: for instance, after the 1976 Guatemala earthquake, foreign epidemiological teams investigated more than 30 instances of rumour concerning the possibility of disease outbreaks or other potential hazards; they found all of them to have no basis in fact (Spencer et al. 1977). Such cases illustrate the need to collect accurate information, and to disseminate it carefully and fully to the press, survivors and the public at large.

It is also clear that officials must be prevented from exaggerating the facts which they communicate to the public and news media, preferably by supplying spokespeople with full and accurate information. The media need to be told a categorical "no" if they are to be prevented from prophesying an epidemic after natural disaster has struck. But health officials who are unsure of the likelihood of disease outbreaks may be reluctant to rule them out, even though the probability may be very small (see Ch. 7). Again, correct information is the key to the problem.

Misdirection and mismanagement
Preparedness and training are necessary if a disaster is to be managed well, but so are experience and level-headedness. Cool, detached judgement is difficult during the aftermath of a crisis that has convulsed the normal social institutions of a city or region: an official may find his or her own life, or the lives of family members, threatened by continuing danger; bereavement may have sapped the will to cope; survivors may be making clamorous accusations and demands; problems may seem insurmountable; and the situation may seem strange and incomprehensible.

Unfortunately, through lack of experience and training the mistakes made on previous occasions are often repeated. But some important lessons can be learned from the mistakes made during past catastrophes. For example, presumed looters should not be shot indiscriminately as they may merely be survivors trying to salvage personal belongings from among the wreckage of their homes. Although looting is rare in disasters, the widespread belief that it will occur often results in massive deployment of police or army personnel who would be best engaged on tasks other than guarding property. In the few cases in which looting does occur during the aftermath of natural catastrophe, it appears to be a fairly spontaneous act with little forethought or prior organization (Quarantelli & Dynes 1969).

Another common misconception is that unburied bodies often cause a health hazard and should be disposed of with great haste. They should instead be dealt with according to the proper procedures. The formalities associated with death include embalming, storage, identification, death certification (with autopsy, if necessary), funerary rights and burial or cremation (Blanshan 1977).

The legacy of unwanted aid suggests that measures should be taken to avoid the provision of unsuitable food, clothing and medicines to groups of survivors. Misplaced or excessive aid may be useful only if it can be sold by the survivors to whom it is distributed (except that the black market in such goods may soon reach a state of glut). Additionally, when providing relief after disasters, the social and biophysical customs of survivors should be respected: they should preferably not be offered supplies with which they are completely unfamiliar. Moreover, people will assist themselves as much as they can, and hence it will not be necessary to cater for all their needs. Aid that is geared to this will be at least partially redundant. Nevertheless, it can easily be predicted that hospitals which must cope with large number of casualties will need splints, bandages, pain killers, anaesthetics and radiographic film. At the same time, search-and-rescue activities will need not only personnel but also arc lamps and lifting gear.

A history of confusion in past relief operations suggests that the geography of the disaster area should be sorted out immediately after the impact, boundaries should be established and relief workers deployed where they are most needed, rather than being left to their own devices. Lastly, in most cases visits to the disaster area by VIPs use up resources of transportation and security, do little good for the morale of survivors and should be discouraged.

These observations will be explained in greater detail in the appropriate chapters below. Meanwhile, it is clear that one way of avoiding misconceptions is to analyze the phenomenon of disaster on the basis of its principal dimensions: time, space and intensity.

Time and space in disaster

Although not in itself a process, cause or prime mover, time is one of the unifying phenomena in disaster studies. As a linear measure, it is essentially the backbone of most models of how disasters occur or are managed. As an underlying factor, time can be examined in both a physical and a human context. Space is another common feature of all disasters, for hazards, vulnerability and impacts all have distinct geographical distributions; patterns, moreover, which may change dynamically as time progresses.

Timescales of geophysical phenomena
The **magnitude–frequency principle** states that the amount of work done by events of a given size is a product of their magnitude times their frequency of occurrence (Wolman & Miller 1960). In general, large

events are too infrequent to have the greatest overall significance and frequent events are too small to have a significant aggregate impact. The average effect is obtained by multiplying the magnitude of events by the time units in which they occur.

Frequency can be defined variously as the number of events of a given size in a given unit of time; the number of events, and their sizes, in each of a succession of time units; and how often a given magnitude of event occurs. The average length of time between events of a given size is known as the **recurrence interval** or **return period**, and is the reciprocal of their frequency. These factors vary considerably between different types of geophysical phenomenon (Table 1.6), but they usually bear a strong nonlinear relationship to magnitude, as shown in Figure 1.4.

It is very hard to define a particular **magnitude** at which geophysical events commonly become disastrous. This, however, may be known intuitively and may give rise to a change in terminology for example, from "whirlwind" to "tornado", or from "tropical storm" to "hurricane". In physical terms a magnitude of at least 6.0 is usually the minimum necessary for an earthquake disaster to occur (but note that the 1972 Nicaraguan earthquake involved $M_L = 5.6$), while severe damage tends to be caused by tsunamis only if the wave height exceeds 2.0 m. The official organizations responsible for prediction and warning are usually the first to try to quantify a particular event. In order to clarify the concept of magnitude semantically, arbitrary thresholds may be set, such as the definition of the hurricane on the basis of a minimum wind speed of 65 knots (see Ch. 3). But other aspects of the same phenomenon may be less easily classified: for example, no threshold can be defined for the diameter of the hurricane storm, or for its forward tracking speed.

Studies of the frequency of physical events have generally identified a strong correlation between increases in magnitude and decreases in frequency. When the logarithm of frequency is taken, a linear relationship usually appears. Problems occur when the magnitude–frequency curve is projected forward into time periods for which there are no data. Geophysical monitoring tends to extend back in time relatively few decades or, at the most, a couple of centuries, as the earth, atmospheric and hydrological sciences have no very long tradition of precise observation. Thus, it is extremely difficult to predict long return periods and therefore how often very large events will recur. Under these circumstances, the further into the future that projections are made, the more inaccurate they tend to be. This is true because the relationship between magnitude and frequency is statistical rather than precise, and because it is unlikely that accurate data on very infrequent events will ever be obtained.

One consequence of this uncertainty, or indeed of the relative infrequency of extreme events in general, is that it may not make

Table 1.6 Classification of disasters by frequency or type of occurrence.

Type of disaster	Frequency or type of occurrence†
Lightning	Random
Avalanche	Seasonal/diurnal; random
Earthquake	Log-normal
Landslide	Seasonal-irregular
Tornado	Seasonal; negative binomial
Intense rainstorm	Seasonal/diurnal; Poisson
Hail	Seasonal/diurnal; Poisson, gamma, negative binomial
Tsunami	Random
Subsidence	Sudden or progressive
Windstorm	Seasonal/exponential
Frost or ice storm	Seasonal/diurnal; Markovian, binomial
Hurricane	Seasonal/irregular
Snowstorm	Seasonal; modified Poisson‡
Environomental fire	Seasonal; random
Volcanic eruption	Irregular
Insect infestation	Seasonal; random
Fog	Seasonal/diurnal
Flood	Seasonal; Markovian, gamma, log-normal
Coastal erosion	Seasonal/irregular; exponential, gamma
Drought	Seasonal/irregular; binomial, gamma
Crop blight	Seasonal/irregular
Expansive soil	Seasonal or irregular
Accelerated erosion	Progressive (threshold may be crossed)
Desertification	Progressive (threshold may be crossed)

† Frequency distributions adapted from Hewitt (1970: 333–4).
‡ Eggenberger and Polya modification: see Hewitt (1970).

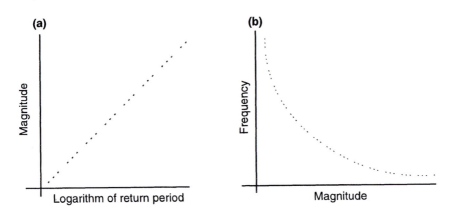

Figure 1.4 Relationship between magnitude, frequency and return period for natural events.

economic sense to invest resources in preparing for the very long-term event. This is especially true if such investment prevents the resources in question from being used in the meantime and unless the cost of doing nothing will be overwhelmingly great when the large event finally occurs. Of course, the political will must exist if infrequent events are to be tackled, and this may be difficult to stimulate in cases where the risk is perceived to be low, for example, when impacts occur only once or twice a century.

Many geophysical relationships involve a **lag** between cause and effect, or between input and output, and a **relaxation time** that elapses between a shock to the system and the completion by its components of a compensating response that restores stability (Chorley & Kennedy 1971). Viewed in the long term, systems that succeed in compensating for shocks to their stability by self-regulation through **negative feedback** are considered to be in **dynamic equilibrium**. Lags in the system often result from differences in the location of inputs and responses; for example, the temporal lag in flooding results from the time taken to translate snowmelt or rainfall into peak river flow. Understanding the physical **process–response system** requires us to obtain sound knowledge of the time-histories of lags and variations in system input values and known responses. In this context the rate of energy supply to the system varies simultaneously over many timescales, a number of which are periodic. Sudden and extreme concentrations of energy or input often account for the geophysical events that cause disasters.

Timescales of the human crisis

Disasters tend to be classifiable into distinct time phases. The first of these is the impact phase, in which survivors can do little except hold on until the worst has passed. As noted above, the speed of onset varies with the type of geophysical agent and impact. The emergency or crisis that is thus caused can usually be divided into phases of **isolation**, **rescue** and **remedy**. Isolation may last from a few hours to three days, depending on how accessible the disaster area is and how well organized the emergency services are. In historical times the isolation of communities stricken by disaster often lasted months or even years, particularly when communications were poor and surpluses insufficient for relief to be supplied. Even in the present day the survivors of disaster are often critical of delays in response on the part of the authorities.

The remedy to the crisis involves supplying food, shelter, medical care and organized assistance in order to make the environment safe and habitable. The trend is towards ever more rapid and sophisticated involvement. In major international disasters, specialized experts and foreign aid supplies may arrive in the disaster area within hours of the impact. Hence, organized, well-prepared developed societies can generally cope well: in fact, the arrival of supplies and volunteers is often too overwhelming and rapid to be assimilated into the local relief structure. On the other hand, the international aid community withdraws assistance quickly, as it is usually faced with many demands upon its resources. However, this is partly compensated by the fact that, throughout the post-impact phases, survivors will spontaneously organize themselves and help each other in a limited way.

During the initial phase of recovery, victims' lives are organized in artificial patterns. Temporary shelter dominates, and eating arrangements may be communal instead of private. Social cohesion tends to be reinforced under the duress of the disaster, which represents a common enemy to be faced by all survivors. However, as long-term effects begin to predominate, old inequalities reassert themselves and often become accentuated. The long term may be a period in which the authorities, faced with many other problems, lose interest in the disaster area; and this can lead to a lack of funds for recovery and reconstruction. In general, the strength of recovery and the time taken for it to be achieved are closely related to the socio-economic magnitude of the initial disaster impact. The complexity and variety of society mean that there is no standard timescale for reconstruction: the length of time it takes to overcome the effects caused by disaster may vary by as much as four times, depending on the size of the affected population, the availability of resources and the level of organization.

Natural catastrophe may have social, economic and medical consequences. For instance, when destruction is widespread, disasters create a sudden and massive demand for shelter, and the need to put reconstruction into effect as rapidly as possible may stimulate rebuilding to old, inadequate standards. Until the rebuilding is complete, some survivors will be constrained to move away from the area. Migration will probably first involve those residents who already have strong contacts outside the disaster zone and hence have a clear destination. Return migration may occur as rebuilding provides new sources of accommodation.

Although relatively few kinds of disaster destroy crops, major droughts, floods and hurricanes can lead to severe local shortages of food. At the same time, damage to water supplies may be severe and long-lasting, such that a black market in bottled water develops. The increased risk of disease transmission associated with unsafe water and damaged sewerage may put undue strain on medical and public hygiene facilities. However, this is unlikely unless these services were already inadequate.

Loss of livelihood is one potential economic effect. If this occurs, it would indicate a severe erosion of social organization, and hence is an improbable consequence, as most disasters do not overturn the structure of society. Likewise, unemployment is not a natural and immediate result of disaster. The construction industry, for example, may go through a "boom-and-bust" cycle, which may last 10–15 years and create a sizeable demand for manpower. But at the same time other sectors of the economy may be neglected.

Lastly, timescales of human perception and memory should be borne in mind. Disasters provide a "window of opportunity", in which public opinion may be strongly in favour of mitigating future impacts. But although extreme events tend to be periodic, the community rarely achieves total protection during the times when it is recovering from impact. In the lulls between catastrophes the hazard may persist in the minds of residents, be turned into fable or legend or simply be disregarded. On occasion, ostensible disregard of the risk may mask private anxiety (hazard perception is dealt with in Ch. 9).

The physical and human timescales in disaster are summarized for three hypothetical examples in Table 1.7. However, the time periods suggested here may vary considerably according to the physical, political, social and economic peculiarities of any given disaster situation. With regard to the human dimension, Burton et al. (1978) proposed a classification of timescales based on the process of coping with disaster. As noted above, this can be divided into adaptation and adjustment. The former may be separated into physiological and cultural components, while the latter consists of unpremeditated and deliberate forms. The timescale is different for each of these, as shown in Figure 1.5.

Spatial aspects of disaster
The spatial dimension of natural disasters has received remarkably little theoretical treatment, yet it is fundamental to the understanding of how emergencies evolve and how they can be managed. In 1956 A. F. C.

Table 1.7 Summary of time periods in disaster (with examples).

Time period	Purely hypothetical examples:		
	Earthquake	Tornado	River flood
"Incubation" or return period	150 years	5 years	100 years
Immediate precursor period	none	20 mins	15 hours
Impact	100 secs	5 mins	36 hours
Aftermath or crisis period:			
Isolation	8–48 hrs†	2 hours	2 hours
Search and rescue	2–7 days	12 hours	3 days
Repair of basic services	4 weeks	3 weeks	5 weeks
The long term:			
Restoration–reconstruction	12 years	2 years	4 years
Developmental reconstruction	25 years	3 years	12 years

† Time lapse represents centre-periphery dichotomy.

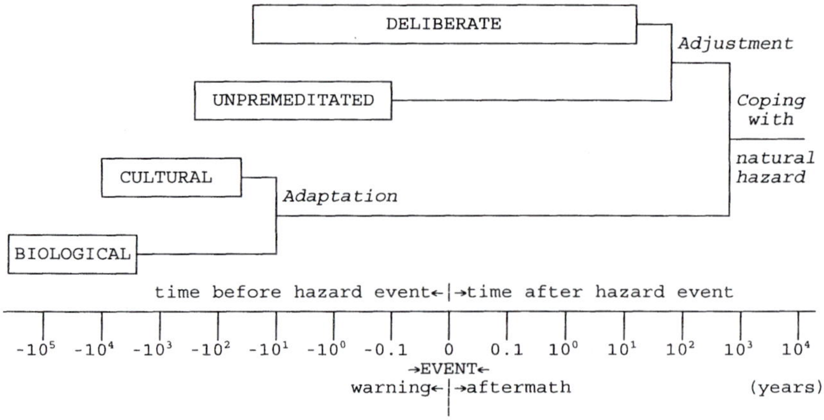

Figure 1.5 Timescale for coping with natural hazards (Burton et al. 1978: 38).

Wallace proposed a simple conceptual model of spatial relations in disaster consisting of four concentric areas (Fig. 1.6). At the centre is the **zone of total impact**, in which most structures are destroyed or severely damaged. This is surrounded by the **zone of marginal impact**, in which damage is less serious. Relief workers' perception in the "marginal impact" zone may tend to minimize the magnitude of the disaster, while the survivors and relief workers in the "total impact" area may find their perception overwhelmed.

Beyond the impact areas is a **zone of filtration**, which shows no physical damage but is where refugees may arrive in large numbers, over-taxing facilities such as shelter and medical aid. The outer ring is the **zone of national and international aid**, in which supplies are gathered and from which materials and personnel are sent into the disaster area. The magnitude of the response depends upon the nature of the relief appeal, the degree of interest on the part of national and foreign governments and the strength of public opinion.

Although widely reported in the literature, this model has seldom been put to the test. The concentric nature of the zones is, of course, little more than a theoretical construct, as is their relative size. Using data from the 1985 Val di Stava mudflow disaster in the Dolomite Mountains of northern Italy (in which there were 264 deaths), I found that the distance relationship between the zones is logarithmic – it increases sharply as one moves outwards from the total impact area (Alexander 1986a).

Formulating intensity measures

Having reviewed some fundamental questions of magnitude, frequency and time periods, the picture will be completed by examining another

26

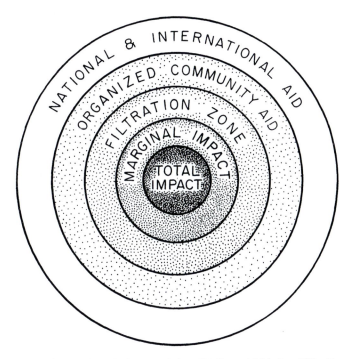

Figure 1.6 Spatial model of disaster (after Wallace 1956, De Ville De Goyet & Lechat 1976: 152).

common theme, that of intensity. The intensity of extreme events can be viewed in terms both of the concentration of energy in geophysical phenomena and the seriousness of their impact on human lives and on the built or natural environment. In terms of the former, here are five examples of meteorological extremes:

(a) maximum recorded precipitation per minute: 16.5 mm (California);
(b) maximum recorded precipitation per hour: 335 mm (Jamaica);
(c) maximum recorded precipitation per 24 hours: 1870 mm (Réunion Island);
(d) maximum recorded precipitation in 8 days: 3430 mm (Cherrapunji, India);
(e) upon rapid warming (with or without associated precipitation) snowmelt may reach 80 mm per 24 hour period.

Not all natural hazards are amenable to classification of the intensity of their physical impact or human effects. Floods, for example, have received little attention from this perspective. Of those hazards which have been subjected to intensity classification the most substantial efforts have been dedicated to formulating and refining the scales for earthquakes. These will therefore serve as an illustration of the general principles involved in the taxonomy of impacts.

Earthquake intensity scales
The first systematic classification of earthquake intensity is probably that invented in the 1780s by Domenico Pignataro, an Italian physician. He reviewed all available accounts of 1,181 earthquakes and classified them as slight, moderate, strong and very strong. He denoted the Calabrian earthquakes of 1783, which killed at least 29,500 people, as "violent". At the same time a priest, Father Elisio della Concezione, produced the first known damage map (at a scale of 1:145,600), in which he used astronomy to determine the co-ordinates of towns and a "star rating" to indicate the level of damage to each town or village.

The scale proposed in 1828 by P. N. G. Egen contained six classes of intensity, which were rather more sharply defined than those developed by Pignataro. The next advance was that of the Irish engineer Robert Mallet, who was one of the principal founders of modern seismology. Mallet combined theoretical and archival scholarship with a rigorous field investigation of southern Italy directly after the earthquake disaster of December 1857, which killed nearly 9,750 inhabitants of the Kingdom of Naples (Guidoboni & Ferrari 1987). His methods of recording earthquake damage were highly systematic and analytical, and new scales could thus be propounded on the basis of information that was much more comprehensive and less subjective than had previously been the case. However, the scales remained inductive, that is, they were derived directly from observed phenomena, with only limited input of explanatory or predictive theory.

The foundations of modern scales were laid in 1879 by De Rossi, an Italian, and were further developed in 1883 by Forel, a Swiss. Their scale of 10 classes was broadened to 12 by Giuseppe Mercalli, who in 1902 proposed what has proved to be the most durable system for classifying perceptible seismic phenomena. In 1931 the Mercalli–Wood–Neumann Scale was introduced in order to take account of the effect of earthquakes on motorized vehicles, tall buildings and other modern inventions that were not significant in Mercalli's time (Wood & Neumann 1931). The principal modern variants of Mercalli's original scale are the Modified Mercalli (MM) Scale of 1956 (Table 1.8), which is standard in the Americas, and the Mercalli–Cancani–Sieberg (MCS) Scale, which is used widely in Europe. Other scales include the Medvedev–Sponheuer–Kárník (MSK) Scale (Kárník et al. 1984), the Japan Meteorological Association (JMA) Scale (which has seven classes, rather than the 12 utilized in Western scales), and the GOST Scale, which is standard in the Commonwealth of Independent States (Bolt et al. 1977).

Seismic intensity scales constitute an attempt to classify the observed

Table 1.8 Modified Mercalli earthquake intensity scale (Wood & Neumann 1931).

Grade	Effects
I	Not felt, except by a very few people under especially favourable circumstances.
II	Felt only by a few persons at rest, especially on upper floors of buildings. Delicately suspended objects may swing.
III	Felt quite noticeably indoors, especially on upper floors, but many people do not recognize it as an earthquake. Standing motor cars may rock slightly. Vibration like that caused by a passing truck.
IV	During the day, felt indoors by many people, outdoors by few. At night some people are awakened. Crockery, windows, doors are disturbed; walls make a creaking sound. Sensation like that caused by a heavy truck striking the building. Standing motor cars rocked noticeably.
V	Felt by nearly everyone; many awakened. Some crockery, windows, etc., broken; a few instances of cracked plaster; unstable objects overturned. Disturbance of trees, poles and other tall objects sometimes noticed. Pendulum clocks may stop.
VI	Felt by all, many people are frightened and run outdoors. Some heavy furniture moved; a few instances of fallen plaster or damaged chimneys. Damage is slight.
VII	Everybody runs outdoors. Damage negligible in buildings of good design and construction; slight to moderate in well-built ordinary structures; considerable in poorly-built or badly designed structures; some chimneys broken. Noticed by persons driving motor cars.
VIII	Damage is slight in specially designed structures; considerable in ordinary, substantial buildings, with partial collapse; great in poorly built structures. Panel walls thrown out of frame structures. Fall of chimneys, factory stacks, columns, monuments, walls. Heavy furniture overturned. Sand and mud ejected from the ground in small amounts. Changes in water levels in wells. Earthquake disturbs persons who are driving motor cars.
IX	Damage considerable in specially designed structures; well-designed frame structures thrown out of plumb; damage great in substantial buildings, with partial collapse. Buildings shifted off their foundations. Ground cracks conspicuously. Underground pipes are broken.
X	Some well-built wooden structures are destroyed; most masonry and frame structures destroyed with foundations; ground badly cracked. Rails bent. Landslides considerable in the vicinity of river banks and steep slopes. Sand and mud deposits are shifted. Water splashed over banks.
XI	Few, if any, masonry structures remain standing. Bridges destroyed. Broad fissures in the ground. Underground pipelines completely out of service. Earth slumps and landslips occur in soft ground. Rails bent greatly.
XII	Damage is total. Waves seen on ground surface. Lines of sight and level distorted. Objects thrown upwards into the air.

and observable effects of earthquakes in order to give some idea of their relative distribution in time and across the land (i.e. spatially). The smallest effects (such as very light shaking that is incapable of permanently displacing objects) can be detected only by sensitive instruments and hence they form the lowest categories of the scales. Low to moderate categories are made up of effects that can be perceived or felt by the general public and by damage to furniture (or other artefacts). In higher categories damage to buildings and structures is more or less widespread and, at the highest end of the scales, destruction of urban landscapes and modification of terrain and natural drainage systems are widespread. The principal reasons why more than one scale is required are that the built environment differs from place to place and has changed over time. The diffusion of steel-frame and reinforced-concrete construction, for example, has changed the response of large buildings to seismic forces, necessitating a re-evaluation of their vulnerability to damage.

Under isotropic (spatially uniform) conditions, the effects of seismic shaking will diminish regularly with distance from the epicentre, because that is how the energy of shaking diminishes (Howell & Schultz 1975). However, irregularities may occur in the pattern. Thus the distribution of geological units is important, for the number, thickness and hardness or softness of soil, sediment or rock layers affect the frequency and other characteristics of the seismic waves which they transmit. In particular, soft sediments may lengthen the frequency of the waves and cause them to interfere with one another destructively (see Ch. 2), whereas hard rock is more strongly resonant. Additionally, hills and exposed slopes such as canyon walls may "radiate" and refract the waves.

Besides geology, any variation in the anthropogenic landscape, such as local variation in building materials or clustering of towns, will influence the distribution of damage. Moreover, the state of maintenance and conservation of buildings is critical to their survival during earthquakes (see Ch. 5). Dilapidated or aseismic buildings (those which are not designed to accommodate seismic forces) may lack resistance to the horizontal acceleration of the ground. Concentrations of such buildings will therefore represent nodes of vulnerability to damage or destruction. Finally, as earthquake intensity scales include the perceived and felt effects of the tremors, they are dependent to some extent on what people who are interviewed after the event claim to have experienced. An element of exaggeration may be present here.

The use of recorded earthquake source mechanisms and intensities to create a map of future earthquake hazard is known as **seismic zoning**. **Macroseismic intensity** is expressed as a number scaling the effects of each individual earthquake on humans, the built environment and the surface of the Earth; these effects are described in the text that defines the scale, and the mean of them is computed on the basis of a local **macroseismic survey**. Intensity values are assigned to all areas affected significantly by the earthquake and lines are drawn to separate intensities, e.g. III from IV, IV from V; these are known as **isoseismals**. The resulting map is affected by the level of generalization inherent in the macroseismic survey. It is unlikely in any case to express intensities lower than about IV (on the MM, MCS or MSK Scales). As mapped, seismic intensity also shows the size of the **macroseismic field** (the area affected by strong motions of the ground) and local irregularities in the radiation of seismic waves.

Expected intensity maps define the source regions of earthquakes and either the maximum intensity that can be expected at each location or the maximum intensity to be expected in a given time interval (i.e. with a given return period). They are thus capable of answering such questions as: "What intensity of earthquake will affect a given point about once every century?" Some expected intensity maps incorporate explicit assumptions about the quality of construction at affected localities, answering, for example, the question: "What level of damage (or what defined intensity) is to be expected at a given location?"

The spatial probability of a given intensity or magnitude over a given time interval can be expressed by a **seismic risk map**. In this case the risk can be expressed in purely seismic form as the return period of a damaging earthquake, in engineering terms as the probability of widespread structural failure or in insurance terms according to the number and size of claims lodged. These variables are usually expressed as decimal or percentage probabilities.

Earthquakes that occurred prior to the onset of instrumental monitoring, which began around 1900, can be expressed only in terms of intensity, and not with respect to their physical magnitude. Their observed effects may be reported in contemporary documents, which give a basis for the application of intensity scales. Thus, countries such as China, Iran and Italy, that have extremely rich historical records, have been able to reconstruct long sequences of earthquake impacts. However, although intensity is a useful measure which gives an indication of the strength of the earthquake at each location, it is not to be over-used. The following comparison indicates that physical magnitude and the severity of earthquake effects are not simply related:

Earthquake	Magnitude	Depth of focus	Deaths	Damage ($ million)
California, 17 October 1989	7.1	18.5	62	6,000
Roumania, 4 March 1977	7.2	94	1,500	800
Chile, 3 March 1985	7.8	33	200	1,200
Japan, 26 May 1983	7.7	33	104	600

Other intensity scales
As we have seen, measures of earthquake intensity refer to the direct effects of the shaking but are only indirectly related to the physical forces at work. The relationship, moreover, is often very approximate or indistinct. Certain other intensity scales are more closely related to physical variables, or include more precise measurements. Tsunami intensity, for example, is estimated on the basis of the height of the wave at landfall as well as the level of damage (Table 1.9). The scale encompasses other

Table 1.9 Scale of tsunami intensity (Soloviev 1978: 131). Copyright 1978 by UNESCO Press.

Intensity	Run-up height (m)	Description of tsunami	Frequency in Pacific Ocean
I	0.5	*Very slight.* Wave so weak as to be perceptible only on tide gauge records.	
II	1	*Slight.* Waves noticed by people living along the shore and familiar with the sea. On very flat shores waves generally noticed.	one per four months
III	1	*Rather large.* Generally noticed. Flooding of gently sloping coasts. Light sailing vessels carried away on shore. Slight damage to light structures situated near the coast. In estuaries, reversal of river flow for some distance upstream.	one per eight months
IV	4	*Large.* Flooding of the shore to some depth. Light scouring on made ground. Embankments and dykes damaged. Light structures near the coast damaged. Solid structures on the coast lightly damaged. Large sailing vessels and small ships swept inland or carried out to sea. Coasts littered with floating debris.	one per year
V	8	*Very large.* General flooding of the shore to some depth. Quays and other heavy structures near the sea damaged. Light structures destroyed. Severe scouring of cultivated land and littering of the coast with floating objects, fish and other sea animals. With the exception of large ships, all vessels carried inland or out to sea. Large bores in estuaries. Harbour works damaged. People drowned, waves accompanied by a strong roar.	once in three years
≥VI	16	*Disastrous.* Partial or complete destruction of man-made structures for some distance from the shore. Flooding of coasts to great depths. Large ships severely damaged. Trees uprooted or broken by the waves. Many casualties.	once in ten years

phenomena, such as the level of flooding, the hydrodynamic force of water, wave forces, scour effects, the destruction of buildings or damage to structures and shipping, and the occurrence of drownings (thus mixing physical and human factors). Continual use of this scale enables the recurrence intervals to be calculated for tsunamis of different sizes (Soloviev 1978).

Volcanic eruptions have so far been classified only on the basis of their physical manifestations. The two scales in use are the Tsuya classification (points I to IX) and the Volcanic Explosivity Index (VEI), which runs from 0 to 8 (Tables 1.10a & b). The former is an estimate of the size of eruption, and hence is more properly described as a scale of magnitude, and is based on the volume of ejecta, treating this for comparative

Table 1.10a Volcanic eruption scales (Tsuya 1955, Newhall & Self 1982, Fedetov 1985).

Volcanic explosivity index, VEI (Newell & Self 1982)	Volcanic intensity (Fedetov 1985)	Tsuya scale (Tsuya 1955)	Eruption rate (kg/sec)	Volume of ejecta (m^3)	Eruption column height (km)	Thermal power output (log. kW)	Duration (hours of continuous blast)
0	V	I	10^2–10^3	$<10^4$	0.8–1.5	5–6	<1
1	VI	II–III	10^3–10^4	10^4–10^6	1.5–2.8	6–7	<1
2	VII	IV	10^4–10^5	10^6–10^7	2.8–5.5	7–8	1–6
3	VIII	V	10^5–10^6	10^7–10^8	5.5–10.5	8–9	1–12
4	IX	VI	10^6–10^7	10^8–10^9	10.5–17.0	9–10	1–>12
5	X	VII	10^7–10^8	10^9–10^{10}	17.0–28.0	10–11	6–>12
6	XI	VIII	10^8–10^9	10^{10}–10^{11}	28.0–47.0	11–12	>12
7	XII	IX	$>10^9$	10^{11}–10^{12}	>47.0	>12	>12
8	—	—	—	$>10^{12}$	—	—	>12

Table 1.10b Other determinants of Volcanic Explosivity Index (VEI) (Newhall & Self 1982: 1,232)

VEI	Explosivity	Qualitative description	Classification	Tropospheric injection	Stratospheric injection
0	non-explosive	effusive	Hawaiian	negligible	none
1	small	gentle	Hawaiian, Strombolian	minor	none
2	moderate	explosive	Strombolian	moderate	none
3	moderate–large	severe	Strombolian, Vulcanian	great	possible
4	large	violent	Vulcanian, Plinian	great	definite
5	very large	cataclysmic	Plinian, Ultraplinian	great	significant
6	very large	paroxysmal	Ultraplinian	great	significant
7	very large	colossal	Ultraplinian	great	significant
8	very large	terrific	Ultraplinian	great	significant

purposes as if it were entirely pyroclastic (see Ch. 2). The latter is more comprehensively based and can be related approximately to the type and violence of the eruption, its magmatic products, its duration and the subsidiary phenomena (such as mudflows) which it creates. The intensity of the human impact of eruptions has not been classified.

Wind speeds can be classified from 0 to 12 on the Beaufort Scale, which determines wind forces according to speed and effects observed at sea and on land. Even though hurricanes involve strong winds, they have had their intensity classified only with respect to the phase of landfall in coastal areas (Table 1.11). The scale is based on wind speeds in km/h, depth of coastal flooding, damage and destruction and speed with which escape routes are cut off by the advancing storm surge (*Weatherwise* 1974). Tornadoes, instead, are classified according to three different

Table 1.11 Hurricane disaster potential scale (the Saffir-Simpson scale; after *Weatherwise* 1974: 169, 186).

1	*Winds 120–50 km/hr; central pressure >980 mb.* Damage to shrubbery, trees, foliage and poorly anchored mobile homes. Some damage to signs. Storm surge 1.2–1.5 m above normal. Some low-lying coastal roads flooded. Limited damage to piers and exposed small craft. Example: Agnes, 1972.
2	*Winds 151–75 km/hr; central pressure 965–79 mb.* Trees stripped of foliage and some of them broken down. Exposed mobile homes suffer major damage. Poorly constructed signs are severely damaged. Some roofing material ripped off; windows and doors might be affected. Storm surge may be 1.6–2.4 m above normal. Coastal roads and escape routes flooded 2–4 hours before hurricane centre arrives. Piers suffer extensive damage and small unprotected craft are torn loose. Some evacuation of coastal areas is necessary. Example: Cleo, 1964.
3	*Winds 175–210 km/hr; central pressure 945–64 mb.* Foliage stripped from trees and many blown down. Great damage to roofing material, doors and windows. Some small buildings are structurally damaged. Storm surge may be 2.5–3.6 m above normal. Serious coastal flooding and some coastal buildings may be damaged. Battering of waves might affect large buildings, but not severely. Coastal escape routes cut off 3–5 hours before hurricane centre arrives. Flat terrain 1.5 m or less above sea level is flooded as far inland as 13 km. Evacuation of coastal residents for several blocks inland may be necessary. Example: Betsy, 1965.
4	*Winds 211–50 km/hr; central pressure 920–44 mb.* Shrubs, trees and signs are all blown down. Extensive damage to roofing materials, doors and windows. Many roofs on smaller buildings may be ripped off. Mobile homes destroyed. Storm surge may be 3.7–5.5 m above normal. Flat land up to 3 m above sea level might be flooded to 10 km inland. Extensive damage to the lower floors of buildings near the coast. Escape routes cut 3–5 hours before hurricane centre passes. Beaches suffer major erosion, and evacuation of homes within 500 m of coast may be necessary. Example: Carla, 1961.
5	*Winds greater than 250 km/hr; central pressure <920 mb.* Increase on the extensive damage of the previous level. Glass in windows shattered and many structures blown over. Storm surge may be more than 5.5 m above normal. Lower floors of structures within 500 m of coast extensively damaged. Escape routes cut off 3–5 hours before hurricane centre arrives. Evacuation of low lying areas within 8–16 km of coast may be necessary. Example: Gilbert, 1988.

scales (see Ch. 3), relating to the strength of the phenomenon, which is a function of rotational wind speed (the Fujita Scale), and the length and width of paths after "touchdown" (the Pearson Scales). However, tornado damage has not been subjected to intensity classification.

Finally, the effects (but not the physical dimensions) of landslides have been classified by intensity (Alexander 1989). The aim here was to create a series of interconnected scales to express the damage caused by landslides to built-up environments at various levels of urbanization. Hence, the scales relate to degrees of cracking, tilting, collapse and extension or compression in architectural or engineering structures (Table 1.12; Fig. 1.7). The use of complementary scales for roads, utility lines, buildings, bridges, and so on, enables an intensity map to be formulated for overall patterns of damage, which can then be related to the physical characteristics of the movement, including its speed and pattern of evolution.

Table 1.12 Landslide damage intensity scale (after Alexander 1986b).

Grade	Description of damage
0	*None.* Building is intact
1	*Negligible:* Hairline cracks in walls or structural members; no distortion of structure or detachment of external architectural details.
2	*Light:* Building continues to be habitable; repair not urgent. Settlement of foundations, distortion of structure and inclination of walls are not sufficient to compromise overall stability.
3	*Moderate:* Walls out of perpendicular by 1–2 degrees, or substantial cracking has occurred to structural members, or foundations have settled during differential subsidence of at least 15 cm; building requires evacuation and rapid attention to ensure its continued life.
4	*Serious:* Walls out of perpendicular by several degrees; open cracks in walls; fracture of structural members; fragmentation of masonry; differential settlement of at least 25 cm compromises foundations; floors may be inclined by 1–2 degrees, or ruined by soil heave; internal partition walls will need to be replaced; door and window frames too distorted to use; occupants must be evacuated and major repairs carried out.
5	*Very serious:* Walls out of plumb by 5–6 degrees; structure grossly distorted and differential settlement will have seriously cracked floors and walls or caused major rotation or slewing of the building (wooden buildings may have detached completely from their foundations). Partition walls and brick infill will have at least partly collapsed; roof may have partially collapsed; outhouses, porches and patios may have been damaged more seriously than the principal structure itself. Occupants will need to be rehoused on a long-term basis, and rehabilitation of the building will probably not be feasible.
6	*Partial collapse:* Requires immediate evacuation of the occupants and cordoning off the site to prevent accidents with falling masonry.
7	*Total collapse:* Requires clearance of the site.

Summary

Essentially, time defines the linear direction of natural catastrophes and space is their medium of expression. The question of time in disasters can be considered in terms both of the frequency of occurrence and the phases of an individual event. Space can be considered on the one hand in terms of global, national and local events and, on the other, as the geographical manifestation of any particular impact or response. Distinctions are less clear regarding magnitude and intensity: ideally, the former would refer to the power or energy of a physical event and the latter to the strength of its consequences for humanity and the environment. The wide variety of existing usages, however, has made the two terms somewhat interchangeable. Rather clearer is the distinction – and also the link – between physical impact and human vulnerability, which constitutes the very essence of disaster. As shown above, natural catastrophe can be analyzed from a variety of disciplinary viewpoints in the physical, natural, technical and behavioural sciences. But full understanding requires that geophysical causes and impacts be viewed in the light of their human consequences, including adaptation, adjustment and mitigation efforts. No one discipline is uniquely qualified to achieve this.

Figure 1.7 Urban effects corresponding to category V on the landslide damage intensity scale (Table 1.12). This building in the town of Montelupone, in the central Italian region of Marche, is situated astride the headscarp of a rotational slumping movement which is slowly pulling it apart.

References

Alexander, D. E. 1986a. Northern Italian dam failure and mudflow, July 1985. *Disasters* **10**(1), 3–7.

Alexander, D. E. 1986b. Landslide damage to buildings. *Environmental Geology and Water Science* **8**, 147–51.

Alexander, D. E. 1989. Urban landslides. *Progress in Physical Geography* **13**, 157–91.

Alexander, D. E. 1991b. Natural disasters: a framework for teaching and research. *Disasters* **15**: 209–26.

American Geological Institute 1984. *Glossary of geology*. Falls Church, Virginia: American Geological Institute.

Arnold, C. 1984. Planning against earthquakes in the United States and Japan. *Earthquake Spectra* **1**, 75–88.

Ball, N. 1979. Some notes on defining disasters: suggestions for a disaster continuum. *Disasters* **3**, 3–7.

Barrows, H. H. 1923. Geography as human ecology. *Annals of the Association of American Geographers* **13**, 1–14.

Beinin, L. 1985. *Medical consequences of natural disasters*. New York: Springer.

References

Berz, G. 1988. List of major natural disasters, 1960–87. *Earthquakes and Volcanoes* **20**, 226–8.

Blanshan, S. A. 1977. Disaster body handling. *Mass Emergencies* **2**, 249–58.

Bolt, B. A., W. L. Horn, G. A. MacDonald, R. F. Scott 1977. *Geological hazards: earthquakes, tsunamis, volcanoes, avalanches, landslides, floods*, 2nd edn. New York: Springer.

Bunin, J. 1989. Incorporating ecological concerns into the IDNDR. *Natural Hazards Observer* **14**, 4–5.

Burton, I. & R. W. Kates 1964. The perception of natural hazards in resource management. *Natural Resources Journal 3*, 412–41.

Burton, I., R. W. Kates, G. F. White 1978. *The environment as hazard*. New York: Oxford University Press.

Chen, L. C., A. K. M. Chowdhury, S. L. Hoffman 1980. Anthropometric assessment of energy-protein malnutrition and subsequent risk of mortality among pre-school age children. *American Journal of Clinical Nutrition* **33**, 1,836–45.

Chorley, R. J. & B. A. Kennedy 1971. *Physical geography: a systems approach*. Englewood Cliffs, New Jersey: Prentice-Hall.

Church, J. S. 1974. The Buffalo Creek disaster: the extent and range of emotional problems. *Omega* **5**, 61–3.

Davis, I. 1978. *Shelter after disaster*. Headington, Oxford: Oxford Polytechnic Press.

De Ville De Goyet, C. & M. F. Lechat 1976. Health aspects in natural disasters. *Tropical Doctor* **6**, 152–7.

Drabek, T. E. 1989. Disasters as nonroutine social problems. *International Journal of Mass Emergencies and Disasters* **7**, 253–64.

Dynes, R. R. 1970. *Organized behaviour in disaster*. Lexington, Mass.: D. C. Heath (Lexington Books).

El-Sabh, M. I. & T. S. Murty (eds) 1988. *Natural and man-made hazards*. Dordrecht: Kluwer.

Fedetov, S. A. 1985. Estimates of heat and pyroclast discharge by volcanic eruptions based upon the eruption cloud and steady plume observations. *Journal of Geodynamics 3*, 275–302.

Glantz, M .H. 1977. Nine fallacies of natural disaster. *Climatic Change* **1**, 69–84.

Glass, A. J. 1970. The psychological aspects of emergency situations. In *Psychological aspects of stress*, H. S. Abram (ed.), 62–9. Springfield, Illinois: Charles C. Thomas.

Goudie, A. S. 1983. *The human impact*. Cambridge, Mass.: MIT Press.

Guidoboni, E. & G. Ferrari (eds) 1987. *Mallet's macroseismic survey of the Neapolitan earthquake of 16th December 1857* (4 vols). Bologna: Società-Geofisica-Ambiente.

Harriss, R. C., C. Hohenemser, R. W. Kates 1978. Our hazardous environment. *Environment* **20**, 6–15, 38–40.

Hewitt, K. 1970. Probablistic approaches to discrete natural events: a review and theoretical discussion. *Economic Geography Supplement* **46**, 332–49.

Hewitt, K. 1983. The idea of calamity in a technocratic age. In *Interpretations of calamity*, K. Hewitt (ed.), 3–32. London: Allen & Unwin.

Howell, B. F., Jr. & T. R. Schultz 1975. Attenuation of Modified Mercalli intensity with distance from the epicentre. *Seismological Society of America, Bulletin* **65**, 651–66.

Kárník, V., Z. Schenková, V. Schenk 1984. Vulnerability and the MSK scale. *Engineering Geology* **20**, 161–8.

Knott, R. 1987. The logistics of bulk relief supplies. *Disasters* **11**, 113–6.

Kreps, G. A. 1989. Future directions in disaster research: the rôle of taxonomy. *International Journal of Mass Emergencies and Disasters* **7**, 215–41.

Micklin, M. 1973. *Population, environment and social organization.* Hinsdale, Illinois: Dryden Press.

Mileti, D. S. 1980. Human adjustment to the risk of environmental extremes. *Sociology and Social Research* **64**, 327–47.

Miller, J. 1974. *Aberfan: a disaster and its aftermath.* London: Constable.

Newhall, C. G. & S. Self 1982. The Volcanic Explosivity Index (VEI): an estimate of explosive magnitude for historical volcanism. *Journal of Geophysical Research* **87**, 1,231–8.

Oliver-Smith, A. 1979. Post disaster consensus and conflict in a traditional society: the 1970 avalanche of Yungay, Perú. *Mass Emergencies* **4**, 39–52.

PAHO 1981. *Emergency health management after natural disaster.* Washington, DC: Pan American Health Organization.

Parker, D. J. & D. M. Harding 1979. Natural hazard evaluation, perception and adjustment. *Geography* **64**, 307–16.

Porfiriev, B. N. 1992. The environmental dimension of national security: a test of systems analysis methods. *Environmental Management* **16**, 735–42.

Quarantelli, E. L. (ed.) 1978. *Disasters: theory and research.* Beverly Hills: Sage.

Quarantelli, E. L. & R. R. Dynes 1969. Dissensus and consensus in community emergencies: patterns of looting and property norms. *Il Politico* **34**, 276–91.

Scanlon, J. T. 1977. Post disaster rumour chains: a case study. *Mass Emergencies* **2**, 121–6.

Scanlon, J. T. 1988. Disaster's little known pioneer: Canada's Samuel Henry Prince. *International Journal of Mass Emergencies and Disasters* **6**, 213–32.

Shah, B. V. 1983. Is the environment becoming more hazardous? A global survey 1947 to 1980. *Disasters* **7**, 202–9.

Shreve, R. L. 1966. Sherman landslide, Alaska. *Science* **154**, 1,639–43.

Soloviev, V. 1978. Tsunamis. In *The assessment and mitigation of earthquake risk.* Paris: UNESCO Press.

Spencer, H. C., C. C. Campbell, A. Romero et al. 1977. Disease surveillance and decision-making after the 1976 Guatemala earthquake. *Lancet II* (23 July 1977), 181–4.

Torry, W. I. 1979. Anthropological studies in hazardous environments: past trends and new horizons. *Current Anthropology* **20**, 517–40.

Tsuya, H. 1955. Geological and petrological studies of volcano Fuji. *Tokyo Daigaka Jishin Kenkyusho Iho* **33**, 341–2.

Turner, B. A. 1976. The development of disasters: a sequence model for the analysis of the origin of disasters. *Sociological Review* **24**, 753–74.

UNDRO 1982. *Natural disasters and vulnerability analysis.* Geneva: Office of the United Nations Disaster Relief Co-ordinator.

Wallace, A. F. C. 1956. *Human behaviour in extreme situations.* Washington, DC: Disaster Research Group, National Academy of Sciences.

Weatherwise 1974. The hurricane disaster-potential scale. *Weatherwise* **27**, 169, 186.

White, G. F. 1945. *Human adjustment to floods: a geographical approach to the flood problem in the United States.* Chicago: Department of Geography, University of Chicago.

White, G. F. 1973. Natural hazards research. In *Directions in geography*, R. J. Chorley (ed.), 193–216. London: Methuen.

White, G. F. & J. E. Haas 1975. *Assessment of research on natural hazards.* Cambridge, Mass.: MIT Press.

Whyte, A. V. T. 1986. From hazard perception to human ecology. In *Themes from the work of Gilbert F. White: geography, resources and environment*, Vol. 2, R. W. Kates & I. Burton (eds), 240–71. Chicago: University of Chicago Press.

Wolman, M. G. & J. P. Miller 1960. Magnitude and frequency of forces in geomorphic processes. *Journal of Geology* **68**, 54–74.

Wood, H. O. & F. Neumann 1931. Modified Mercalli intensity scale of 1931. *Seismological Society of America, Bulletin* **21**, 277–83.

Select bibliography

Bailey, K. D. 1989. Taxonomy and disaster: prospects and problems. *International Journal of Mass Emergencies and Disasters* **7**, 419–32.

Beatley, T. 1989. Towards a moral philosophy of natural disaster mitigation. *International Journal of Mass Emergencies and Disasters* **7**, 5–32.

Bryant, E. A. 1991. *Natural hazards*. Cambridge: Cambridge University Press.

Burton, I., R. W. Kates, G. F. White 1968. *The human ecology of extreme geophysical events*. Boulder, Colorado: Natural Hazards Research and Applications Information Center.

Clarke, J. I., P. Curson, S. L. Kayastha, P. Nag (eds) 1989. *Population and disaster*. Oxford: Basil Blackwell.

Costa, J. E. & V. R. Baker 1981. *Surficial geology: building with the earth*. New York: John Wiley.

Doornkamp, J. C. & R. U. Cooke 1989. *Geomorphology in environmental management*, 2nd edn. Oxford: Oxford University Press.

Drimmel, J. 1984. A theoretical basis for macroseismic scales and some implications for practical work. *Engineering Geology* **20**, 99–104.

Foster, H. D. 1976. Assessing disaster magnitude: a social science approach. *Professional Geographer* **28**, 241–7.

Glantz, M. H. 1976. Nine fallacies of natural disaster: the case of the Sahel. In *The politics of natural disaster: the case of the Sahel drought*, M. H. Glantz (ed.), 3–24. New York: Praeger. (Reprinted 1977 in *Climatic Change* **1**, 69–84.)

Graves, P. E. & A. E. Bresnock 1985. Are natural hazards temporally random? *Applied Geography* **5**, 5–12.

Hays, W. W. (ed.) 1981. *Facing geologic and hydrologic hazards: earth science perspectives*. US Geological Survey Professional Paper 1240B.

Kates, R. W. 1971. Natural hazards in human ecological perspective: hypothesis and model. *Economic Geography* **47**, 438–51.

Kroll-Smith, J. & S. R. Couch 1991. What is disaster? An ecological–symbolic approach to resolving the definitional debate. *International Journal of Mass Emergencies and Disasters* **9**, 355–66.

Maybury, R. (ed.) 1986. *Violent forces of nature*. Mount Airy, Maryland: Lomond.

Mitchell, J. K. 1974. Natural hazards research. In *Perspectives on environment*, I. Manners & M. Mikesell (eds), 311–41. Washington DC: Association of American Geographers.

O'Keefe, P., K. Westgate, B. Wisner 1976. Taking the naturalness out of natural disasters. *Nature* **260**, 566–7.

Palm, R. I. 1990. *Natural hazards: an integrated framework for research and planning*. Baltimore: The Johns Hopkins University Press.

Ponting, J. R. 1973. Rumour control centres: their emergence and operations. *American Behavioural Scientist* **16**, 391–401.

Smith, K. 1992. *Environmental hazards: assessing the risk and reducing disaster.* London: Routledge.

Taylor, A. J. 1990. A pattern of disasters and victims. *Disasters* **14**, 291–300.

White, G. F. (ed.) 1974. *Natural hazards: local, national, global.* New York: Oxford University Press.

The Geophysical Agents

CHAPTER TWO

Earthquakes and volcanoes

In many respects, an earthquake may cause the archetypal sudden-impact disaster, especially as the opportunities for immediate forewarning are at present severely limited. Most catastrophic volcanic eruptions are also sudden and dramatic events that, like earthquakes, can take a very high toll of human life and result in very widespread destruction. The two are linked, not merely by volcano-seismicity (the earth tremors of minor to moderate power which eruptions frequently cause), but also by having a series of monitoring techniques in common. They are also both capable of causing tsunamis.

As this chapter will demonstrate, seismic and volcanic disasters have provided some of the fundamental physical, technological and social research in the field of natural catastrophes – work that has often been a model for studies of other hazardous natural agents and their effects. In this respect earthquakes usually have the shorter recurrence interval and hence demand more comprehensive mitigation. The dilemma posed by many volcanoes is that there may be very long intervals between eruptions, and many preparedness techniques may thus be uneconomic. This is less likely for earthquakes.

The physical nature of earthquakes

More than 3,000 perceptible earthquakes occur each year. Perhaps only 7 to 11 of these will involve significant loss of life, but the combined average annual death toll may exceed 10,000. In fact, the loss of life in extreme earthquakes may be more than ten times that caused by extreme volcanic eruptions. Such events may result not only in high death tolls, but also in overwhelming numbers of casualties requiring immediate medical treatment, and in serious long-term difficulties over housing and reconstruction. However, they seldom cause major problems of lack of food and drink and rarely result in large-scale population migrations.

Using a combination of instrumental monitoring and popular observation of natural phenomena the Chinese were able to predict the earthquake which devastated the city of Haicheng in Liaoning Province on 4 February 1976 (Adams 1975). Prompt evacuation almost certainly saved many thousands of lives (see below). However, the following year it proved impossible to predict the tremors which levelled Tangshan, a city of 1 million inhabitants, 230,000 of whom died in the catastrophe. More than a decade and a half later, the Haicheng prediction remains the only large-scale success of its kind, and routine forecasting of strong motions of the ground is unlikely to be achieved in the near future, although the Chinese do claim to have successfully predicted at least ten earthquakes, including some with magnitudes in the range 6.7–7.2 (Blundell 1977).

Seismology

The science of elastic, or seismic, waves is called **seismology**. It deals with their origin, propagation across and through the Earth, and how they are recorded and interpreted. Just as all other waves, **seismic tremors** can be characterized in terms of wavelength (the distance between two consecutive peaks or troughs), frequency (the number of waves passing a point in a given unit of time, measured in Hertz, Hz), period (the length of time taken for a given wave to pass a point), and amplitude (the difference in elevation between an adjacent peak and trough). In addition, when waves are superimposed, the irregularities produced in the distance between peaks or troughs of uniform frequency is known as 'phase difference' or 'phase shift'. In all cases, it is the energy of the wave, rather than the medium in which it occurs, that travels: the medium itself undergoes an oscillatory motion.

Severe shaking of the ground is known as **strong motion**, and is characterized by its duration (in seconds), the frequencies that are present, the maximum wave amplitude, attenuation with distance from the fault, the maximum velocity in metres per second and the maximum acceleration (as a percentage of gravitational acceleration, which is 9.81 m/sec^2). The larger the earthquake magnitude (see below) and the greater the area of the fault plane that ruptures, the more extensive will be the zone from which seismic waves progressively emanate and hence the longer that strong motion will last.

The duration of strong motion may be the single most important factor in producing damage to buildings (although the latter is often easier to correlate in the field with maximum accelerations). We can also define **bracketed duration** as the length of time over which ground shaking exceeds a defined level: for instance, 10 km from the epicentre of a

43

shallow-focus earthquake of Richter magnitude $M_L = 7.0$ the bracketed duration of shaking in excess of 0.05g may be 25–30 seconds. Duration correlates with earthquake magnitude better than it does with acceleration. Peak vertical acceleration is usually about half the horizontal acceleration of the ground, and both quantities decrease with distance from the epicentre and hypocentre of the earthquake. During a highly powerful shallow focus earthquake horizontal accelerations may momentarily exceed 0.6 g or locally even twice that (see Ch. 5).

There are six kinds of earthquake wave (the motion of the four main ones is shown in Fig. 2.1). **Body waves** comprise longitudinal movements

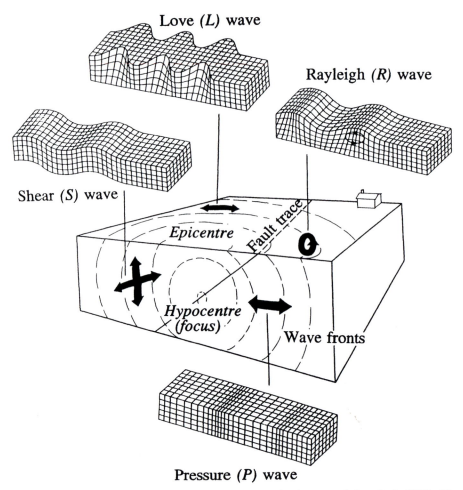

Figure 2.1 Generation of different types of seismic wave (after Bolt 1988: 29; Hays 1981: 29).

known as **P** (for pressure or "primus") waves, and transverse motions known as **S** (for shear or "secundus") waves. These movements are transmitted through the body of geological media such as rock units. **Surface waves** consist of Love (**L**), Rayleigh (**R**), Stoneley (**S**) and Channel (**C**) waves, of which the first three are named after their discoverers. L and R waves, can only travel along the surfaces of certain geological structures, while S waves travel only along interior crustal discontinuities and C waves travel in structural "channels" or low-velocity layers in the crust. In order to propagate themselves, L waves need a layered surface and R waves require a homogeneous one.

The velocity of seismic waves is a function of the elastic properties of the materials through which they pass. Hence the speed of P and S waves in the crust and upper mantle is:

$$V_p = \sqrt{\frac{k + 4/3\mu}{\rho}}$$

$$V_s = \sqrt{\frac{\mu}{\rho}}$$

where k = incompressibility modulus (bulk modulus)
      μ = modulus of rigidity (shear modulus)
and   ρ = density of material

Wave velocity (in km/sec) differs between continental and oceanic crust and between sedimentary and igneous rocks.

Seismic waves normally travel faster and generate more consistent resonances in hard, uniform rock formations than in soft sediments, in which they have a tendency to lose their energy and interfere with one another. As with all other forms of wave, seismic frequencies that become superimposed lead to changes in amplitude which are additive if two peaks coincide or subtractive if a peak and a trough are superimposed. Interferences are more numerous and irregular in waves that pass through bodies of soft sediment or soil.

The principal causes of strong motion are volcanic activity, ground subsidence and tectonic stress caused by the mobility of the Earth's crust. Tectonics are responsible for almost all of the world's earthquake disasters and for much **geological faulting**, which involves the movement of adjacent blocks of the Earth's rigid crust (Fig. 2.2). Extension produces vertical slippage in the downward direction which is termed **normal faulting**, while compression leads to upward thrusting or over-thrusting of one block onto another, which is called **reverse faulting**. Lateral movements are called **transcurrent** or **strike-slip faulting**, which is **left-lateral** if the left-hand block moves towards the observer and **right-lateral** if it is instead the right-hand block. **Oblique faulting** occurs when

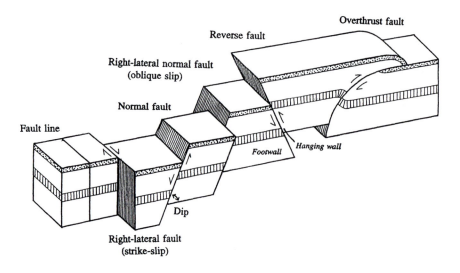

Figure 2.2 Types of faulting.

the movement of blocks is diagonal, rather than vertical or horizontal. The San Andreas Fault in California is perhaps the world's most famous strike-slip fault. It has a mean slip rate of 2–3 cm per year, and in some places the strain is released by infrequent large earthquakes. In general terms, an earthquake of magnitude 4.5 requires a fault plane approximately 5 km long and about 2 cm of slip.

Fault plane locations and orientations are usually determined by analysis of the distribution of first motion (compression or dilation) of the longitudinal P wave. As this method cannot separate the main slip plane from subsidiary ones at right-angles to it, the radiation pattern of S waves must also be studied, and field investigation is often necessary.

Seismogenic faulting is common at **subduction zones**, where one crustal plate descends below another to be consumed in the hotter material of the Earth's upper mantle. It is also widespread at the point of collision between adjacent plates and in areas where compressional strains accumulate. Hence, some **seismogenic areas** respond directly to global or regional forces, and others to purely local stresses. Earthquake source regions can be defined with regard to the manifestations of recent tectonics. These include differences in the amplitude of vertical movements of the ground surface over the past 25 million years, as deduced by structural geological analysis and absolute dating. The most detailed information usually pertains to the amplitude and direction of historical fault movements, and for these it is useful to reconstruct the pattern of seismic activity over the past 2,000 years. Significant geological phenomena also include longitudinal deep faults that have been active in recent times, transverse deep faults and the pattern of fault intersections. Finally, the presence of young volcanism is often indicative of local seismogenesis: the crust may become highly stressed in the vicinity of volcanic eruptions, leading to **volcano-seismic activity** of magnitude generally inferior to $M_L = 5$ (see below).

Observation of strike-slip faulting on the San Andreas Fault led the seismologist H. F. Reid to propound the **elastic rebound theory** (Fig. 2.3), in which stress causes the rock to deform on either side of the fault until such strain can no longer absorb the forces applied to it. At this point friction is overcome and the strain is released in an abrupt movement along the fault plane, which is where the two crustal blocks abut. The point at which movement begins is known as the **hypocentre** or **focus**, and is situated at some depth below the ground surface (Fig. 2.1). In the Pacific Basin earthquake foci occur down to 719 km below the surface, although with few exceptions foci in the Asiatic and Mediterranean regions do not exceed 300 km in depth. This defines the maximum depth of rigid crust, below which the Earth's mantle is situated and rocks are too plastic to slip in abrupt seismogenic movements.

Although a strong earthquake will send waves to the other side of the globe, it cannot do so through the Earth's core and mantle, which do not consist of rigid media, and hence there is a "seismic shadow" in the mode

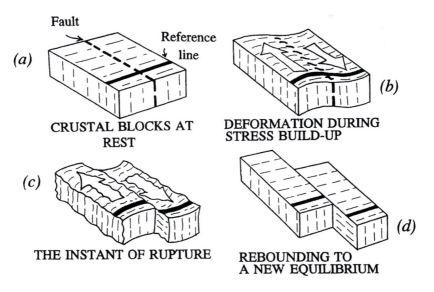

Figure 2.3 Elastic rebound theory (Plafker & Galloway 1989: 5).

in which tremors are picked up at stations located far away from their origin. In fact, seismic waves are refracted around the interior plastic and molten layers, or transmitted only weakly through some of them. Most damaging earthquakes have shallow foci situated less than 30 km below the surface: intermediate depth foci (50–80 km deep) also produce damaging tremors, while few deep-focus (>80 km) earthquakes are ruinous. As a general rule, the shallower the hypocentre, the greater the magnitude at the epicentre. However, although seismic shaking may be traced back to a point, the actual fault movement will involve the frictional slippage of a plane which may be tens of square kilometers in size. Geophysicists define the **seismic moment**, M_o, according to the rigidity of the faulted material, μ, the average fault dislocation caused by the earthquake, L, and the fault area, S:

$$M_o = \mu LS$$

The point on the Earth's surface which is vertically above the hypocentre is called the **epicentre** (Fig. 2.1). **Epicentral determination** makes use of the fact that different kinds of earthquake wave travel at different speeds: P waves travel faster than S waves, which travel faster than L waves, which are more rapid than R waves (in fact, the P waves travel at 5.95–6.75 km/sec in the crust and 8.0–8.5 km/sec in the mantle beneath it, while S waves move at 2.9–4.0 km/sec in the crust and 4.3–4.5 km/sec in the mantle). The relationships between wave velocities are generally constant, hence for any given geological material, $V_p/V_s = \sqrt{3/1} = 1.73$ and $V_r = 0.92V_s$. At a particular measuring station situated, say, 500 km from the epicentre, P waves will arrive before S waves of various frequencies, and the other wave types will follow. The difference (in fractions of a second) between the arrival times enables the distance from the station to the point of origin of the waves to be calculated. The epicentre must lie somewhere on a circle with radius corresponding to the distance thus determined, but its direction is not known. When the calculation is made at two stations, it is clear that the epicentre must lie at one of the two points of intersection of the radii which they produce. The addition of a third station enables one single point of intersection to be determined, and this is where the epicentre must lie.

Although it is normally considered to be a point, the epicentre should, strictly speaking, be plotted as a probability ellipse, as for the most part there are no absolutely precise means of locating it. If the seismometric monitoring network is dense, the epicentre can usually be located with an accuracy of ±10–20 km, while the hypocentre can be determined to within ±1–5 km if it is shallow and ±10–20 km if it is deep (>70–80 km).

Seismic waves are registered by an instrument called a **seismometer**, which traces the oscillation caused by ground shaking on drum, paper or magnetic tape in the form of a **seismograph**. There are various types of seismometer, which use a pendulum, light source or electromagnetic galvanometer to register the tremors; and a seismic station may require an array of instruments, each of which is sensitive to a particular frequency range. The ideal seismometer combines sensitivity to vibration with robustness in the face of strong ground motions. As this is a combination which is very difficult to achieve, for the most part few useful data can be gained from seismic stations which are very close to the epicentre of major strong motions, as the instruments tend to be overwhelmed by the magnitude of the tremors. Given that the earthquake may occur at any time of the day or night, stations must be designed to record continuously and hence they require much maintenance and supervision. Thus, an efficient seismological network requires very substantial investment.

Deficiencies can, however, be made up at lesser cost by using **accelerometers**. These are instruments that function only when set in motion by strong tremors (i.e. by waves of sufficient velocity, acceleration and amplitude) and are often situated in places such as electricity substations. The lack of need for regular maintenance allows a larger network to be maintained than the seismometric one. The **accelerographs** thus produced give no information on continuous small-scale seismic activity (known as **background seismicity**) but can be useful records of major tremors.

Nowadays, earthquake activity is monitored in a co-ordinated way across the globe. The World-Wide Network of Seismic Stations (WWNSS) has more than 120 continuously recording stations, each with six short-period and six long-period seismometers, which give a standardized response to seismic events. The distribution of stations reflects a compromise between the need for regular spacing and the desire to cluster the measuring points around the world's major seismic belts. Eventually, the Global Telemetered Seismograph Network (GTSN) will bring seismic data in real-time via satellite to points of collection and analysis in New Mexico and Colorado from four stations in Africa, four in South America and one in Antarctica. After primary analysis, the data will be encoded on CD-ROM (which is capable of storing large quantities of information on a single disc) and distributed to users throughout the world. The value of real-time seismographic data transmission cannot be over-emphasized: for example, it enabled the epicentre of the 1985 Mexico earthquake to be determined only 17 minutes after the event.

The initial measure of earthquake **magnitude** was devised by the Californian seismologist Charles F. Richter (1958), and describes the total energy of seismic waves radiated from the hypocentre (or, by approximation, the epicentre). It is based on the amplitude of the largest trace written by a standard (Wood-Anderson torsion) seismometer located

100 km from the epicentre. When calculating magnitude, interpolation is used to correct deficiencies in the location of seismometers. The measure consists of an open-ended scale from <1 to >8 in which the progression of values is logarithmic between each unit. Moreover, the increment in energy expended during the tremors increases 31.6 times as one moves up through each unit in the scale. This means that high magnitude earthquakes are thousands of times larger than low magnitude ones (see Table 2.1).

Nowadays, the Richter magnitude of earthquakes is known as "local magnitude", M_L, and is calculated according to the following formula:

$$M_L = \log(a/T) + f(\delta,h) + C_s + C_r$$

where: a = amplitude of ground waves (μm)
T = wave period (seconds)
δ = distance from epicentre (degrees lat./long.)
h = focal depth (km)
C_s = correction for site conditions at station
C_r = regional correction

(f is an empirically and theoretically determined function.)

But this is not the only equation to be used in calculating magnitude. For surface (R) waves, which have a period of about 20 seconds (and a wavelength of approximately 60 km), the Gutenberg formula can be used:

$$M_S = \log a + c_1.\log \delta + c_2$$

where c_1 and c_2 are constants, and a is the amplitude of the horizontal component of Rayleigh waves.

The magnitude of surface waves can also be derived from the Moscow-Prague formula of 1962:

$$M_S = \log(a/T) + 1.66 \log \delta° + 3.3$$

where a is the horizontal component of Rayleigh waves, T should be in

Table 2.1 Earthquake magnitude (M_L) in relation to other variables (after Costa & Baker 1981: 61). Copyright © 1981 by John Wiley & Sons, Inc.

Magnitude M_L	Energy (Joules)	Acceleration (g)	MM intensity	Expected annual incidence	Felt area (km²)	Distance felt (km)
3.0–3.9			≈II–III	49,000	1,940	25
4.0–4.9			≈IV–V	6,200	8,850	50
5.0–5.9	2.75×10^{12}	0.006	≈VI–VII	800	38,850	110
6.0–6.9	7.59×10^{13}	0.15	≈VII–VIII	120	165,350	200
7.0–7.9	2.09×10^{15}	0.50	≈IX–X	18	518,000	400
8.0–8.9	5.75×10^{16}	0.60	≈XI–XII	1	2,072,000	725

the period range 10–30 seconds, and station and regional corrections should be made. Finally, there is an empirical formula for the determination of body-wave magnitudes, which have periods of 1–10 seconds:

$$M_b = 0.56 \ M_S + 2.9$$

As each of the scales relates only to certain wave frequencies, whose composition changes with the energy radiated by the earthquake, none provides a satisfactory index of the largest seismic events. For these, a new scale, the **moment magnitude, M**, is now used. This is based on the surface area of the fault displaced during the earthquake, its average length of movement and the rigidity of the rocks involved. Whereas no earthquake of $M_L \geq 9$ has ever been registered, some of the largest have been upgraded to **M** values ≥ 9.

In general, the standard error of magnitude determinations is about ± 0.3 of a scale unit. The relationship between earthquake magnitude and frequency of occurrence is semi-log linear:

$$\log N \ (M) \ dM = (a + bM) \ dM$$

where a and b are constants, and N is the number of earthquakes per unit time, per unit area, corresponding to a given magnitude, M. This relationship can be simplified by approximation to:

$$\log N = a + bM$$

For convenience of units, seismic energy is usually expressed as the logarithm of measured energy in ergs or Joules.[1] It is related to the magnitude of surface waves (M_S) and body waves (m_b) using simple parametric equations:

$$\log E = 12.24 + 1.44 \ M_S \ (\text{for } M_S > 5)$$

and $\quad \log E = 4.78 + 2.57 \ m_b$

where energy E is given in ergs (see Table 2.1). The energy released by a magnitude 8.0 earthquake is approximately 12,000 times greater than that liberated by the nuclear bomb dropped on Hiroshima in 1945. The energy release of a magnitude 6.8 earthquake would supply a medium-sized town with electricity for a year, but that of a magnitude 8.75 earthquake would provide 670 times as much power. Hence, despite their relative rareness, earthquakes of magnitude greater than 6.1 account for over 90 per cent of total seismic energy expenditure.

---

[1] The Joule (J) is a unit of energy: 1 J is equivalent to the work done by 1 Newton acting through a distance of 1 m. The Newton (N) is a unit of force: 1 N will impart an acceleration of 1 m/sec² to a mass of 1 kg. One J is equal to 10^7 ergs, and 1 N is equivalent to 10^5 dynes.

Attenuation of seismic intensity with distance from epicentre

As shown in Chapter 1, seismic intensity is a descriptive measure of earthquake effects (Fig. 2.4). Although not truly empirical, it can be related by statistical approximation to more quantitative variables. Thus, the decrease in intensity with distance from the epicentre is approximately as follows:

$$I_o - I_n = 3 \log \frac{r^2 + h^2}{h^2}$$

where I_o = epicentral intensity

I_n = intensity at a given distance, n, from epicentre

r = distance from point n to the epicentre, o (km)

and h = focal depth (km).

This being the case, the decrease in intensity is related as well to distance from the hypocentre, D:

$$I_o - I_n = a + b \log D + c \log D$$

where a, b and c are empirical constants.

Figure 2.4 Damage to Intensity X on the Mercalli–Cancani–Sieberg (European) scale. This dwelling in the southern Italian village of Conza di Campania was destroyed by the 23 November 1980 earthquake (magnitude M_B = 6.8).

Attenuation curves are used to overcome the paucity of macroseismic data at distances less than 300 km from epicentre, where detailed information on earthquake variables is badly needed for hazard zoning purposes, but where the force of ground motion is strong enough to distort local instrumental readings. Attenuation (x) occurs as the waves lose their seismic energy through travel, and it can be related to magnitude, M_L, and hypocentral distance, D, as follows:

$$x = c_1 \exp c_2\, M_L D^{-c_3}$$

where the c values are constants.

At hypocentral distances of less than 500 km, a magnitude calibration curve is used to estimate the maximum displacement that a building or other structure is likely to experience at a given site:

$$\log A_m = M_L - 1.73 \log D - 3.17$$

where A_m is the maximum acceleration experienced at that point. Epicentral intensity is also related parametrically to the maximum acceleration at the epicentre (A_o, in cm/sec^2):

$$I_o = 3 \log A_o + 1.5$$

Attenuation curves are thus simple empirical relationships that give the largest amplitudes as a function of earthquake magnitude and distance from the fault. At distances of less than about 150 km from the hypocentre, direct, refracted and reflected P and S waves predominate on seismograms, while at greater distances surface waves begin to dominate the records. Moreover, as a result of the differing rheologies of the media through which they pass, deep-focus earthquakes tend to show a different pattern of attenuation to shallow-focus ones.

Earthquake control and man-made earthquakes
Many small to moderate earthquakes have been provoked by the superpowers' programmes of nuclear testing, and hence the size of the international seismic network also reflects investment in intelligence work designed to monitor such trials. The manufacture of weapons also led to the first inadvertent experiments in earthquake control.

In 1961 at Denver, Colorado, the US army drilled a well through 3,638 m of nearly horizontal sedimentary formations into Precambrian crystalline rocks below. The aim was to dispose of contaminated waste water left over after chemical weapons manufacturing at the army's Rocky Mountain arsenal. Fluid pressure during injection into the well reduced the frictional resistance of fractured rock and stimulated adjustment by slip faulting. This induced 710 perceptible earthquakes at up to 25 per day and magnitudes that reached 4.3 on the Richter Scale, all within a NW–SE ellipse extending no more than 8 km from the arsenal (Healy et al. 1968). Increased fluid pressure in the voids, or spaces between rock fragments, was responsible for the faulting, which was copious when 17–23 million litres of fluid were injected per month. Decreasing the fluid pressure involved increasing rock strength by sealing the cracks, and increasing it reopened them. The rate of percolation of the fluid along discontinuities in the rock was so slow that there was a lag time of 1–4 months between maximum pressure of injection and maximum seismic activity. Fluid injection was continued until February 1966, when the controversy surrounding it led to the decision to halt it.

The discovery at Denver raised speculation that it would be possible to control earthquakes in areas of favourable geology using pressurized fluid injection. Hence, accumulated strain could be released selectively and without harm in areas of seismic gaps. Accordingly, between 1969 and 1970 the United States Geological Survey carried out experiments at Rangely, Colorado (Fig. 2.5; Raleigh et al. 1976). After background seismicity had been monitored, four redundant oil wells 1,800 m deep were pressurized with water. When the pressure was raised from 235 to 275 bar about 900 small earthquakes occurred, 367 of them within a 1 km radius of the bottom of the wells. Over the following three months, pressure was reduced to 203 bars and only three earthquakes occurred. Further experiments confirmed that the resulting sequence of earthquakes could be enhanced or inhibited by altering the fluid pressure across a threshold value of 264 kg cm^{-2} per month. Furthermore, the technique can help locate active faults.

In one control scheme that has been proposed, wells are drilled to a depth of 5,000 m and are 5 km apart. Fluid pressure is reduced sufficiently, and over a large enough area, for fault slippage to cease temporarily. New wells are drilled between the initial boreholes and injected at high pressure with fluids in order to stimulate earthquakes. In this way, stress is relieved in the centre of the control zone and concentrated on the margins, where pressure has been lowered. For the next six months, the initial wells are pressurized and the relief wells are depressurized, and vice versa, until the strain which might generate a major earthquake has been dispersed. However, although it was once hoped that such techniques could be used to lower the probability of strong motion in areas of high seismicity, there is too great a risk of provoking a large, uncontrolled earthquake for this to be feasible.

The central mathematical concept of earthquake control is the **Hubbert-Rubey failure criterion**. In this, the critical shear strength of rocks is:

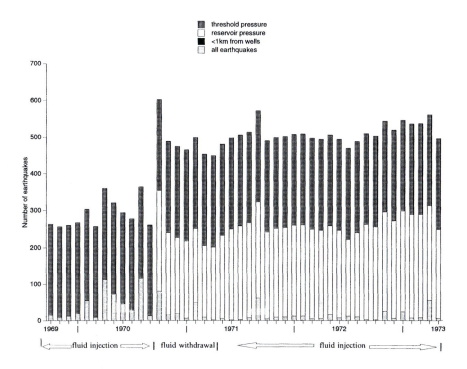

Figure 2.5 Induced earthquake experiments at Rangely, Colorado (Raleigh et al. 1976: 1,236). Copyright 1976 by the AAAS.

$$\tau_c = (\sigma_n - P_c)\,\mu$$

where τ_c = critical shear resistance (strength)
σ_n = normal stress across the failure plane
P_c = fluid pressure necessary to cause earthquakes
μ = coefficient of static friction

The shear resistance of rocks at failure is:

$$\tau = \tau_o + \mu.\sigma_n$$

where τ_o = intrinsic strength

Fluid pressure (P) tends to reduce the effective normal stress, and hence the strength drops:

$$\tau = \tau_o + \mu \, (\sigma_n - P)$$

Experiments in the Weber Sandstone Formation of Colorado (Raleigh et al. 1976) gave the following values:

$$\mu = 0.81$$
$$P_c = 257 \text{ bars}$$
$$\sigma_n = 342 \text{ bars}$$

Therefore: τ_c = 68 bars (where 1 bar = 1,000,000 dynes/cm^2 ≈ 1 atmosphere, and 1 dyne is the force necessary to accelerate 1 gram 1 cm/sec^2).

Some induced seismicity is, however, inadvertent (see Judd 1974). The phenomenon was first noted in Greece in 1929, and on the filling of the Hoover Dam on the Colorado River, which provoked about 600 earthquakes, the largest of which reached a magnitude of 5.0. In 1965 in Greece the filling of the man-made Lake Kremasta caused an earthquake of magnitude 6.3, which killed one person, injured 60 and damaged 1,680 houses. Two years later in west-central India, the Koyna Reservoir, which has a capacity of 2,780 million m^3 of water, caused a magnitude 6.5 earthquake which killed 177 people, injured 2,200 and left thousands homeless (Gupta & Combs 1976). The local geology consists of old, stable rocks and was considered one of very low seismicity. Worldwide, at least 90 other cases of reservoir-induced seismicity have been documented (see Table 2.2).

Reservoir-induced seismicity can be a problem when hydraulic conductivity is strong all the way down to highly fractured rocks situated deep beneath the impoundment, and where water pressure is high in the saturated clefts between rock blocks. However, percolation is often so slow that the seismic reaction does not necessarily follow directly upon the filling of the reservoir (in fact the Koyna Reservoir had been full for about five years before the earthquake occurred).

It is likely that if the water in the reservoir is deeper than about 100 m its weight causes stresses to which the underlying rocks must adjust, and this may cause earthquakes with very shallow focus to occur immediately on filling the impoundment. For example, the Oroville Reservoir in California impounds 4,365 km^3 of water behind a dam that is 236 m high. The pressure on the lake bed is 21 kg/cm^2, or 22 atmospheres, and earthquakes of magnitudes up to 5.7 were caused when it was filled. However, although it tends to be somewhat of an unpredictable phenomenon, not all such seismic activity is hazardous, and it tends to diminish after the reservoir has been full for some years.

Since the 1960s the underground testing of nuclear bombs has often

Table 2.2 Selected cases of documented reservoir-induced seismicity (after Simpson 1976: 125–6).

Reservoir	Country	Height of dam (m)	Volume of storage million m³	Year of impound- ment	Year of largest earthquake	Maximum magnitude earthquake
Koyna	India	103	2,708	1964	1967	6.5*
Kremasta	Greece	165	4,750	1965	1966	6.3†
Xingfengjiang	China	105	10,500	1959	1962	6.1
Oroville	California, USA	236	4,295	1968	1975	5.9‡
Kariba	Zimbabwe	128	160,368	1959	1963	5.8
Hoover	Arizona, USA	221	36.703	1936	1939	5.0
Marathon	Greece	63	41	1930	1938	5.0
Benmore	New Zealand	118	2,100	1965	1966	5.0
Monteynard	France	155	240	1962	1963	4.9
Kurobe	Japan	186	199	1960	1961	4.9
Bajina-Basta	Jugoslavia	89	340	1966	1967	4.5–5.0
Nurek	former USSR	317	10,400	1969	1972	4.5
Vouglans	France	130	605	1968	1971	4.5
Clark Hill	France	67	2,500	1952	1974	4.3

* Extensive damage, 177 deaths, 2,200 injuries.
† 1,680 houses damaged, 1 death, 60 injuries.
‡ Accompanied by surface rupture.

stimulated faults to move. For instance, up to 1 m of displacement has occurred along 600 m of the Yucca Fault near the central Nevada test site, where a blast which was set off in 1968 created a new fault with a maximum displacement of 4.5 m. The effects tend to increase substantially with increasing megatonnage of the bomb detonated, but in most cases they can be estimated with a fair degree of accuracy. Nuclear testing has not so far created new areas of natural seismicity.

Predicting earthquakes

In China, as mentioned above, the authorities ordered the evacuation of three large cities 48 hours before the Haicheng-Yingkou earthquake of 4 February 1975. Foreshocks constituted the main precursor, and over the three days before the main earthquake they reached $M_L = 4.7$. Hours after an official warning had been issued, the earthquake struck with a magnitude of 7.3, damaging 22 million m² of property. Some 1,328 people were killed and 16,980 injured (0.02 per cent of the population living and working within the intensity VII isoseismal), which is an exceptionally low casualty toll for such a large event (Haicheng Earthquake Study Delegation 1977). Yet subsequently there have been very few successful earthquake forecasts, and none that merits the fame of the Haicheng prediction. In order to understand why, the nature of earthquake faulting will be examined and the relative predictability of the various physical phenomena that accompany it.

The scope of prediction
The main purpose of earthquake forecasting is to reduce loss of life and damage to property by allowing time for preparation and by furnishing data needed for mitigation efforts. The secondary purpose is to improve the scientific understanding of earthquake source mechanisms. In 1976 the Seismological Committee of the US National Research Council convened a panel on earthquake prediction (US NRC 1976), which reported that "The apparent public impression that routine prediction of earthquakes is imminent is not warranted by the present level of seismic understanding." The same holds true more than a decade and a half later, for although substantial advances in the understanding of seismic forces continue to be made, the phenomenon is sufficiently complex at the local level to defy exact characterization using general models.

A comprehensive earthquake prediction should specify: the location of the event, and the geographical area likely to be affected; the time interval during which the event is expected to occur; the expected magnitude range of the tremors, and other physical parameters, and the effects likely to be provoked at the Earth's surface, including the intensity and the probable distribution of damage. As it is unlikely that complete precision will ever be achieved, the prediction should be assigned a confidence level, representing the percentage likelihood that it will come true during the period of its validity. Low percentages may be more realistic than the false precision which could result from high ones, but they are less likely to generate credibility and hence stimulate a response from the public and civil authorities.

One should distinguish between several types of prediction. In the loosest sense, statements about the general seismicity of an area are predictions, and they are useful for establishing norms for the quality of construction and plans for emergency preparedness. Long-term earthquake prediction usually involves using studies of historical seismicity to establish the probability that a given magnitude of earthquake or intensity of damage will recur during a given period of time (such as 50 or 100 years). This will help encourage gradual adjustment to a specific form of risk (see Table 2.3). If they are turned into a warning, short-term predictions (valid for hours, days or weeks) will motivate temporary or transient responses, such as evacuation, and will also initiate a scientific response in the form of intensive monitoring of seismic and geological phenomena. Thus, for a given area, earthquake alerts can be couched in terms of a gradation of prediction, from the long-term to the immediate (Table 2.4).

Earthquake hazard zoning is a form of broad prediction of seismic risk.

Table 2.3 Safety measures based on earthquake predictions (Savarenskij & Neresov 1978: 86). Copyright 1978 by UNESCO Press.

Period of prediction	Buildings	Material assets	Safeguards for human life	Special measures
Operative (a few hours to one or two days)	Evacuate dangerous buildings; cease activities in places of public assembly	Evacuate the most important material assets	Allocate emergency equipment in the danger area; prepare medical establishments	Cut off electricity and gas mains; shut down nuclear reactors and dangerous chemical plants
In the short term (from 2 to 4 months)	Estimate probable damages; prepare public evacuation plans	Preserve major assets	Prepare emergency measures and medical establishments	Remove or safeguard hazardous substances; lower reservoir levels, etc.
In the long term (12 months)	Strengthen buildings of particular vulnerability to earthquakes		Plan emergency food stores; plan the use to be made of medical establishments	Transfer of hazardous substances to other places of storage

Table 2.4 Gradation of earthquake predictions (Wyss 1981: 112). Reprinted by permission of Kluwer Academic Publishers

Earthquake alert	Conditions and observations necessary
Stage 1	An approximately defined area is estimated to be more likely than surrounding seismic areas to experience a future earthquake (e.g seismic gap or occurrence of at least one geophysical, geological or geodetic anomalous observation).
Stage 2	One or several crustal parameters show the beginnings of a long- to medium-term pattern of change known to have occurred before some other earthquakes. At least one of the prediction elements (location, size or time) is still poorly defined (e.g. occurrence time uncertainty is approximately equal to 50 per cent of precursor time).
Stage 3	Changes in crustal parameters are observed which can be interpreted as indicating that the end of the long-term preparatory process is near (e.g. the anomalies return to normal). The three prediction elements are fairly well defined (e.g. occurrence time uncertainty is less than about 20 per cent of precursor time).
Stage 4	In addition to the conditions of stage 3, an anomaly is measured which can be interpreted as a short-term precursor. Occurrence time uncertainty may range from hours to weeks.

It involves several stages, beginning with the mapping of recent crustal movements and active faults, including the length and age of fractures, their patterns of intersection and contrasts in the movement of different structural blocks (known as **structural units**). Next, seismic phenomena can be mapped at the regional level, especially by isoseismals (lines of equal earthquake intensity; see example shown in Fig. 2.6) and by plotting the locations of epicentres of recorded earthquakes of different magnitudes. Recurrence intervals can be calculated on the basis of past frequencies and magnitudes. A seismic risk map can then be drawn up by extrapolating and interpolating the other maps. Maximum risk levels are based on the assumption that future earthquakes will occur to the maximum magnitudes and intensities observed at each given location in the past. Such is the importance – and measurability – of acceleration as a cause of damage to buildings and structures, that the most general forms of risk map can usually be based on the spatial distribution of maximum seismic acceleration values in past earthquakes (see Fig. 2.7).

Earthquake prediction is also intended to improve the physical model of seismic source mechanisms and determine the best precursors of strong motion. This means integrating the monitoring and interpretation of all known phenomena that precede earthquakes and investigating the links between them. It also requires better geophysical data collection and the

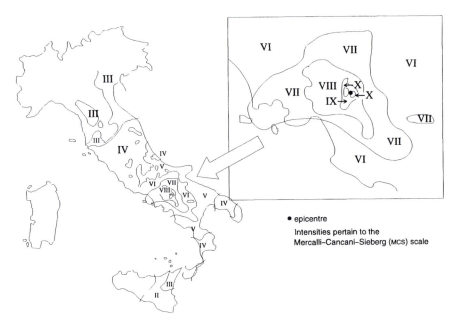

Figure 2.6 Isoseismal (macroseismic) map of the 23 November 1980 southern Italian earthquake (after Postpischl 1985).

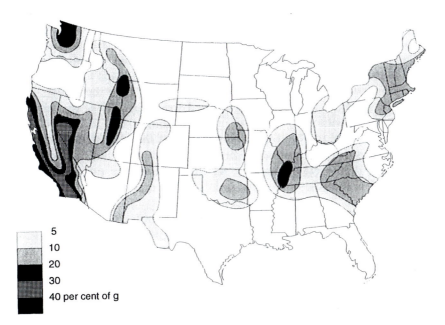

Figure 2.7 Map of peak accelerations in hard rock in the USA (Bolt 1988: 205).

intensification of field, laboratory and theoretical research on the geodynamics of seismicity. Thus, dense networks of monitoring and observation points should be established in many parts of the world, and the instruments which they contain must be robust, reliable and sensitive, such as seismometers that cover a broad frequency range while maintaining a good dynamic response to ground shaking. In this respect, data collection, transmission and processing should be automated, as computer methods allow large quantities of data to be handled and transformed into an intelligible form. Computers also permit digital simulation of forces which cannot be recreated in the laboratory or properly monitored in the field.

A model of earthquake generation
In order to understand how earthquake precursors work and fit together, and how prediction of strong motions will one day be possible, it is necessary to understand how earthquakes occur. Two models have been proposed, one in the USA and one by the Russian Academy of Sciences, and they are broadly similar in conception. This account will be restricted to the Russian **slip faulting model** of earthquake event generation (also known as the "dilation-instability" model), in which there are four phases (see Mjachkin et al. 1984).

In the first stage, prolonged stress in the fault system and its surrounds causes complex patterns of microfissuring, in which tiny cracks in the rock, which is saturated with groundwater, open and close during fault creep. As the fissuring is spasmodic, it can be recorded in the form of continual micro-earthquakes, known as **background seismicity**. The second phase is marked by accelerated creep, which leads to the birth of the earthquake focus as fissures open more rapidly than they close. This may be the result of an increase in applied stress (usually of the tectonic variety) with or without a decrease in rock strength, but in any case the stress level is equal to about half the strength of the saturated rock mass. Open fractures develop parallel to the principal direction of compression. The smallest of the cracks may be no longer than one μm (micron, or millionth of a metre), while the largest may reach several metres in length.

In the third phase, the fissures coalesce, larger fractures develop and slip faulting occurs around the nascent hypocentre. This relieves tension in the centre of the focal zone, where water gradually fills up the new fractures, but causes it to concentrate around the edge of this area. Finally, the growth of fractures and alteration of stress patterns cause the focal zone to be transformed into a series of rock blocks. Stress continues to build up until one or more of the blocks yields and the movement spreads dynamically with extreme rapidity. At this point the earthquake occurs.

Dilatancy preceding brittle fracture has been successfully studied in laboratory experiments at high temperature and pressure. In fact, in the USA about one third of laboratory capacity for high-temperature and high-pressure rock deformation experiments is used in earthquake prediction research. Hence, much is now known about crack propagation, the physical properties of rocks and the flow of fluids in porous media. As a result, models such as the one outlined here are very helpful in explaining the phenomena which precede major ground tremors. The various manifestations will now be reviewed, since they form the basis of current earthquake monitoring efforts with respect to prediction.

Earthquake precursors
Seven groups of phenomena have been studied in terms of how anomalous values may herald the arrival of earthquakes (see Fig. 2.8; an indication of how physical precursors relate to the general elements of earthquake prediction is given in Fig. 2.9). Each phenomenon can be related to the model of slip faulting outlined above.

(a) **Velocity ratio.** During periods of background seismicity the ratio of the velocities of pressure and shear waves, V_p/V_s, is normally about 1.73. Hours or days before a major earthquake it may diminish by 6–15 per cent, largely as a result of reductions in V_p. The duration, but not the magnitude, of the decrease appears to be proportional to

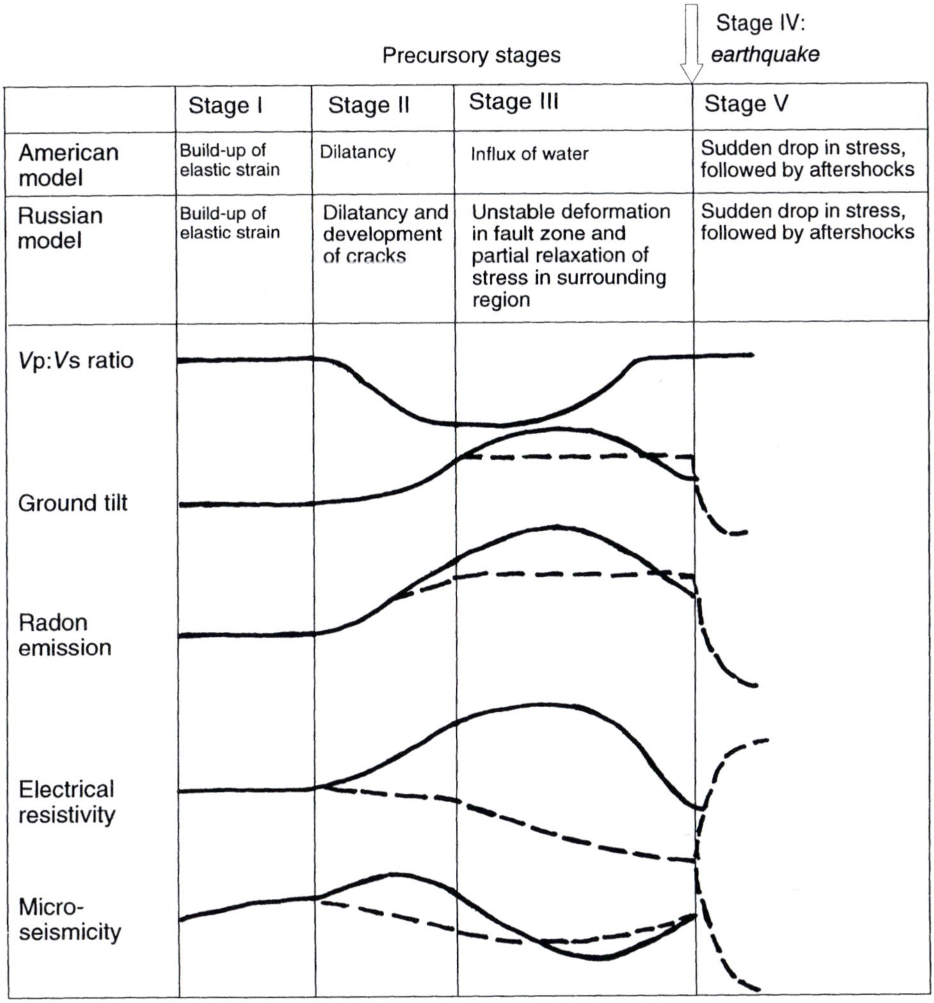

	Stage I	Stage II	Stage III	Stage V
American model	Build-up of elastic strain	Dilatancy	Influx of water	Sudden drop in stress, followed by aftershocks
Russian model	Build-up of elastic strain	Dilatancy and development of cracks	Unstable deformation in fault zone and partial relaxation of stress in surrounding region	Sudden drop in stress, followed by aftershocks

Precursory stages

Stage IV: *earthquake*

Vp:Vs ratio

Ground tilt

Radon emission

Electrical resistivity

Micro-seismicity

Figure 2.8 Expected changes in physical variables before an earthquake (Scholz et al. 1973: 806). Copyright 1976 by the AAAS.

the earthquake magnitude. The ratio will return to normal at the start of the "critical" period just before the earthquake.

(b) **Crustal deformation.** The earthquake may be preceded by ground lengthening, shortening, tilting, uplift or subsidence. This is a manifestation of crustal adjustment to stresses and is likely to increase during the phase of enhanced slip faulting. It can be measured by very precise **tiltmeters** and **extensiometers**, which are capable of registering geodetic changes of fractions of a millimetre over distances of tens of kilometres (Agnew 1986). Extensiometers

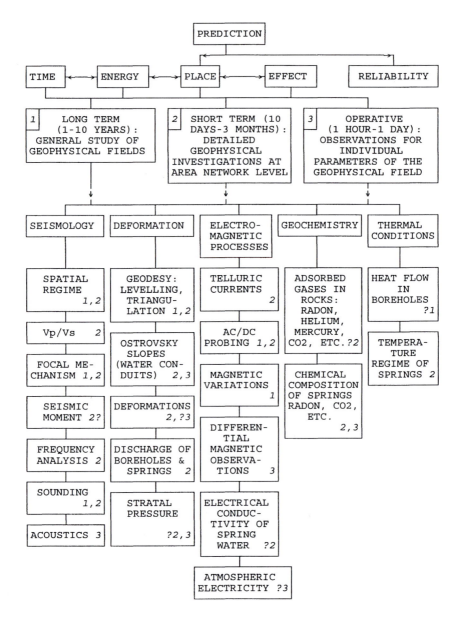

Figure 2.9 Factors in earthquake prediction (Savarenskij & Nersesov 1978: 80). Copyright 1978 by UNESCO Press.

must be made of stable metals or other materials (such as quartz) and located in controlled environments in order to minimize error resulting from thermal expansion of the instrument. Modern tiltmeters are often laser devices of very high precision. However, the point at which movements become critical is not always easy to detect. For example, uplift of 15–25 cm has been registered since 1960 at Palmdale, 65 km north of Los Angeles in California (Langbein et al. 1982), but the strain on this locked part of the San Andreas Fault system was subsequently not released in more than twenty years.

(c) **Geochemistry and groundwater hydrology**. The movement and chemical content of groundwater may change before earthquakes. Stresses in saturated strata may cause spring discharge to increase, or the water in wells to alter its level or become turbid. For example, the 1980 earthquake in the Sele Valley of southern Italy (magnitude 6.8) caused the discharge of hot, sulphurous springs to increase from 300 to 7,200 litres per day. At Tashkent in the former USSR an earthquake in 1966 caused a 200 per cent increase in the radon content of groundwater. In fact, the inert gases of low molecular weight (argon, helium, neon, radon and xenon) are the elements that appear to be most indicative of earthquake activity. During micro-fracturing, migrating groundwater comes into contact with increased surface areas of rocks and dissolves the halogens in trace form. Stress in the rocks may speed up the movement of water in the phreatic zone and hence the transmission of these elements to the surface, where they can be monitored. In addition, many active fault traces are also zones of geothermal energy release, where it may be possible to correlate the temperature of hot springs with incipient seismic activity. Thus it is important to make regular measurements of stream and spring discharge and water quality and temperature.

(d) **Tidal stresses**. Attempts to correlate tides with the occurrence of earthquakes have produced equivocal results, but it does appear that, through altering the local position of large bodies of water, tides may help overstress rock formations which are already highly stressed.

(e) **Magnetic susceptibility and electrical resistivity**. Stresses also alter the magnetic and electrical properties of rock. Hence, periods of high stress can often be detected by monitoring changes in the magnetic susceptibility and electrical resistivity of the rock formations in which they are taking place. Electrical resistivity differs in fractured and unfractured rocks, while the local magnetic field becomes stronger shortly before earthquakes.

(f) **Atmospheric effects**. Meteorological phenomena often accompany earthquakes. Elastic (piezometric) deformation of quartz particles may cause surface electrical charges and hence lightning bolts or glows in the sky, the so-called "earthquake lights". Unusual colours in the sky (such as violets, blues or oranges) may be caused by the release of hydrocarbons from rock formations under stress. Chinese seismologists have observed fireballs in the sky, which may be the result of piezometric ignition of methane released from seismically stressed rocks. Because of their unpredictability, it is difficult to monitor such atmospheric phenomena systematically.

(g) **Animal behaviour**. Unusual behaviour has been observed in many kinds of animal prior to earthquakes (with a mean precursor time of 0.56 days, according to Rikitake 1984). It may be a reaction to high-frequency noises emitted by fracturing rocks, or perhaps represents the effect on their brains of variations in magnetic strength. Moreover, hibernating animals, such as snakes, may wake up and emerge from their hiding places. However, attempts to study these phenomena scientifically have not yielded reliable results, perhaps because little is known about the psychology of wild animals, or indeed of animals in general. Thus the intense public interest in animal behaviour as an earthquake precursor is probably not warranted.

Rikitake (1979: 10) examined 391 reports of earthquakes with precursors and placed them in the following order with regard to the number of times they appeared: ground tilt, strain and deformation (119); foreshocks, microseismicity and anomalous seismicity (101); V_P, V_S and their ratio (69); electrical conductivity, resistivity, telluric currents and geomagnetism (58); radon (12); other (32).

China, Japan and the Commonwealth of Independent States have devoted much attention to field observations designed to help predict earthquakes. The USA has implemented monitoring schemes in Alaska, Missouri, Nevada, New York State, Utah, Washington State, and, above all, in California. The Californian network involves 300 seismic stations, about half of which are directly involved in prediction and the rest of which are designed primarily to locate earthquake epicentres. Crustal distortion is measured by 1,200 laser devices, with an average range of 20 km. There are 20 sites for measuring electrical resistivity along the San Andreas fault in central California, where there are also 30 sites for weekly monitoring of radon emanation (which is similarly monitored at Blue Mountain Lake in New York State). On the San Andreas Fault seven magnetometers record continuously (Raleigh et al. 1982).

Much attention is concentrated on the "seismic gap" at Parkfield in California, where strain has built up to the extent that a sizeable earthquake is expected soon. Within a 25 km radius of Parkfield the US Geological Survey and California Division of Mines and Geology have installed more than 130 seismometers and accelerometers, 80 geodolite lines, 19 alignment arrays, 18 water well sampling sites, 7 dilatometers, 6 soil hydrogen meters, 4 tiltmeters, 2 radon-monitoring boreholes, 2 two-colour laser geodimeter networks (one portable and one permanent), 1 electrical resistivity meter, 1 radio frequency meter and numerous engineering experiments (such as liquefaction arrays and a pipeline deformation experiment). There are 18 seismometers connected to a real-time processor capable of determining locations and magnitudes of quakes within 3–5 minutes of occurrence. Waterwell sites transmit data via satellite for analysis every 15 minutes (Bakun 1988). But elsewhere in California the instruments are widely distributed and clear precursor signals are seldom detected at more than one or two stations at a time. Yet if all monitoring facilities were as concentrated as they are at Parkfield, critical points would be left unguarded.

In general, much more research is needed in order to make earthquake prediction routine and to demonstrate the full meaning of observed precursors.

The relationship between prediction, warning and public policy

A prediction involves forecasting that event X will occur at location Y on day Z. Warning (see Ch. 6) is a separate process, which urges people to modify their behaviour for a period of time on the basis of the predicted hazard. As a general rule, prediction should be carried out by scientists (in the USA, for example, earthquake forecasting is the preserve of the nation's Geological Survey), while public administrators and elected representatives should be responsible for warning (see US National Research Council 1975). Before being publicized, predictions should be properly evaluated as to their validity and probable impact on the recipients. Warnings should be devised not merely by the public administrators who will have to issue them, but should have a significant input from earth scientists, and hence will require collaboration between both groups.

The principal priority in predicting earthquakes is, of course, to save lives, and this goal should not be neglected in order to minimize the social and economic disruption that adjustment to an impending earthquake will inevitably bring (see Fig. 2.10). In this context, the legal implications of earthquake prediction and warning should be considered carefully by the authorities, as there are likely to be important questions of responsibility for public safety. In any event, earthquake prediction should be integrated into comprehensive plans for seismic mitigation and relief, which should be operationalized through existing public authorities, such as regional and local government (Fig. 2.11). Such a strategy must, of course, be preceded by an equally thorough evaluation of the geophysical risk and hazard to social, economic and structural systems (e.g. Figure 2.12).

Real-time earthquake response

If prediction cannot be achieved with adequate reliability, some of the most vulnerable pieces of modern technology can be protected by a very

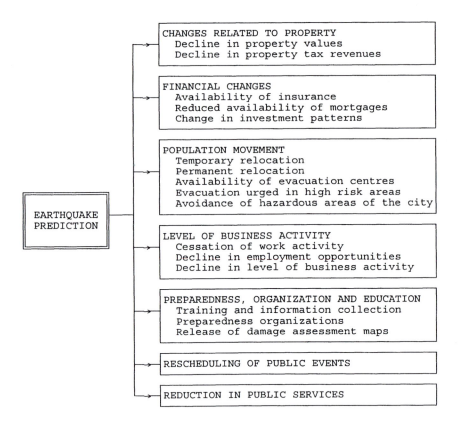

Figure 2.10 Types of socio-economic impact and adjustment to earthquake predictions (after Haas & Mileti 1976).

particular kind of earthquake response. If strong motions of the ground can be sensed as soon as they begin, it may be possible to safeguard certain equipment and processes by shutting them down immediately, as the following two examples illustrate.

In Japan, an earthquake alert system has been developed to protect the 240-km/hr Shinkansen railway network, which is used by an average of 340,000 passengers a day (see Fig. 2.13). Accelerometer stations are located every 20 km along the Tokaido line. If the accelerographs record a horizontal acceleration of more than 0.04 g (39.24 cm/sec^2), electrical power is automatically shut off for 10 km on either side of the seismic station and the train's brakes are applied. The track is then inspected with a thoroughness that is determined by the level of ground shaking. At maximum speed, the train will come to a stop in 70 seconds over no more than 2,300 m of track. However, it is difficult to ensure that the carriages

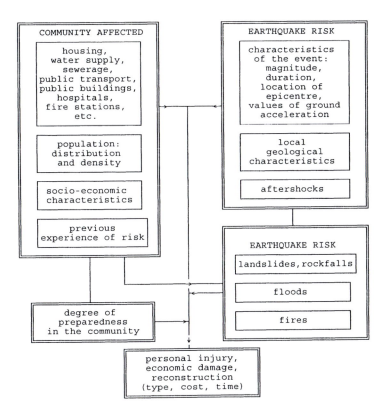

Figure 2.11 Civil defence strategy for earthquake hazards (Solbiati & Marcellini 1983: 175).

are not derailed if rolling motions occur with accelerations of ≥0.5 g and frequencies of ≈1 Hz. In 20 years of operation the system has stopped trains about 100 times, 5–10 of which were false alarms (Nakamura & Tucker 1988).

The relatively slow movement of P and S waves (≤3 m/sec) means that tens of seconds of warning can be given, provided that the onset of ground shaking can be detected promptly in the vicinity of the epicentre. According to the proposal for SCAN, a Seismic Computerized Alert Network (Heaton 1985), once the signature of strong motions is detected, alarm radio signals can be sent to technical processes which are controlled by computer, such as power grids, hazardous chemical systems, gas valves, nuclear plants, strategic defence systems and hospital generators. Emergency services can also be alerted and computers themselves isolated against power failure or surge. In order to give any warning at all, delay in data processing and telemetry must be minimal, which

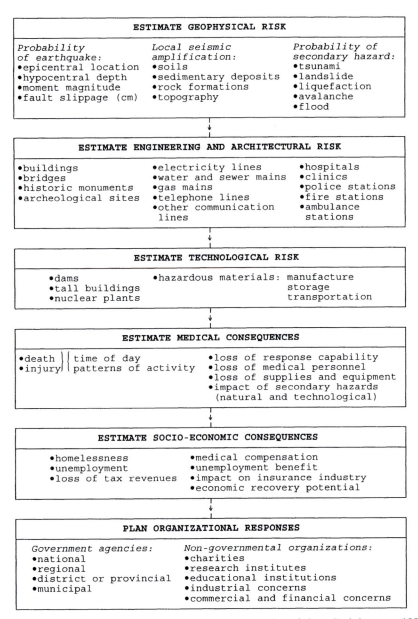

ESTIMATE GEOPHYSICAL RISK

Probability of earthquake:	*Local seismic amplification:*	*Probability of secondary hazard:*
•epicentral location	•soils	•tsunami
•hypocentral depth	•sedimentary deposits	•landslide
•moment magnitude	•rock formations	•liquefaction
•fault slippage (cm)	•topography	•avalanche
		•flood

ESTIMATE ENGINEERING AND ARCHITECTURAL RISK

•buildings	•electricity lines	•hospitals
•bridges	•water and sewer mains	•clinics
•historic monuments	•gas mains	•police stations
•archeological sites	•telephone lines	•fire stations
	•other communication lines	•ambulance stations

ESTIMATE TECHNOLOGICAL RISK

•dams	•hazardous materials: manufacture
•tall buildings	storage
•nuclear plants	transportation

ESTIMATE MEDICAL CONSEQUENCES

•death ⎫ time of day	•loss of response capability
•injury ⎭ patterns of activity	•loss of medical personnel
	•loss of supplies and equipment
	•impact of secondary hazards (natural and technological)

ESTIMATE SOCIO-ECONOMIC CONSEQUENCES

•homelessness	•medical compensation
•unemployment	•unemployment benefit
•loss of tax revenues	•impact on insurance industry
	•economic recovery potential

PLAN ORGANIZATIONAL RESPONSES

Government agencies:	*Non-governmental organizations:*
•national	•charities
•regional	•research institutes
•district or provincial	•educational institutions
•municipal	•industrial concerns
	•commercial and financial concerns

Figure 2.12 Pre-earthquake vulnerability estimation (after Steinbrugge 1989: 419).

requires a dense network of seismic stations and a simplified approach to calculating the variation in ground motions with magnitude and distance from epicentre. A similar system has been proposed in order to protect subway trains in Mexico City.

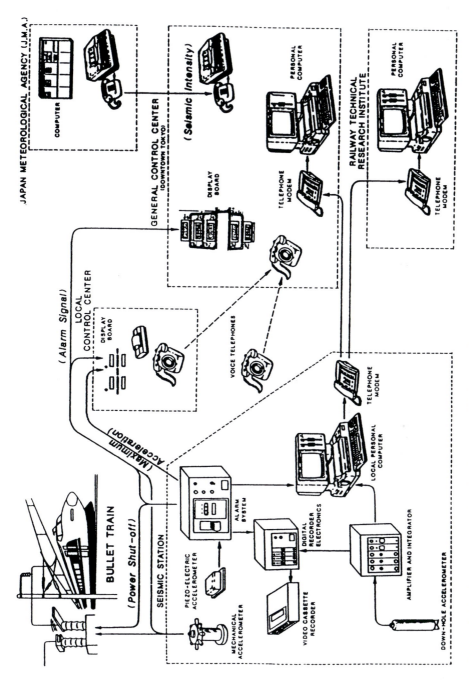

Figure 2.13 Alarm system for the Japanese Shinkansen Railway (Nakamura & Tucker 1988: 36).

71

The human impact of seismic hazards

The earthquake hazard is multiple: **primary effects** include strong motions of the ground and the collapse of buildings, while **secondary effects** include landslides, tsunamis, floods and fires. The principal cause of injury during seismic shaking is **construction failure**, but relatively little is yet known about the relationship between the total or partial collapse of buildings and the resulting patterns of death, injury and homelessness (see Chs 5 & 7).

The impacts that earthquakes cause can be summarized according to the following categories:

(a) loss of life;
(b) physical or psychological injury and bereavement;
(c) destruction of property;
(d) economic disruption and indirect losses (such as loss of employment, but note that rebuilding after earthquakes tends to create jobs in the construction industry);
(e) geological effects resulting in physical alterations of the ground, including cracking, subsidence or settlement, landslides, liquefaction phenomena, alteration of drainage patterns and the scouring effects of tsunamis;
(f) ecological losses (including loss of wildlife habitat and loss of farmland or natural vegetation) as a result of disruption of the land surface.

Loss of life and injury are known respectively as **mortality** and **morbidity**. They obviously depend on magnitude, intensity, focal depth, distance to epicentre and the other physical parameters of the earthquake. They can also be increased considerably by secondary hazards, such as landslide, fire or tsunami. But they respond to a number of other controls, such as the time of day at which the earthquake occurs, given that human activity is more or less regular and in the aggregate predictable with respect to each part of the day. Hence, the pattern of casualties will also be correlated with the social, cultural and economic activity patterns of the society which suffers the impact. Generally, the journey to or from work and the times at which people are gathered in factories, offices, places of worship or entertainment are the periods of greatest vulnerability. In areas with heavy-roofed, weak-walled housing, the night may be the time of highest risk, as sleep will reduce the residents' capacity to react to structural damage. Hence, the number and distribution of casualties also depend upon whether buildings collapse. This in turn may be partly a function of the level of enactment, implementation and enforcement of anti-seismic building codes.

Casualty patterns are often critically affected by what people do during the tremors or, in other words, whether behaviour is prudent and rational or unwise and irrational. Lastly, not all fatalities are immediate, and so much depends on whether there is sufficient – and sufficiently rapid – aid during the first 48 hours after the earthquake impact. The effectiveness of **search-and-rescue** operations is a vital determinant of the number of survivors who can be extracted from damaged structures before they die of their injuries. This operation requires lifting gear and floodlights for round-the-clock clearance of rubble, and prompt transport to hospital for medical aid and trauma surgery (Olson & Olson 1987). These sources of mortality in earthquakes will be considered in Chapter 7.

Vulnerability to earthquakes is not distributed equally around the world, or even with respect to the major seismogenic zones. Instead, it tends to be concentrated in areas of recurrent strong earthquakes, where population is highly nucleated and where the human environment is particularly lacking in resistance to seismic shaking. One of the principal concentrations of such vulnerability is the great seismic belt that runs from the Mediterranean Basin, through Asia Minor (Turkey) and the Middle East, to Iran and the northwest of the Indian subcontinent (Pakistan).

There are several reasons why this belt of land has been the scene of so many casualties in the earthquakes which periodically occur there. First, many parts of the belt have an arid or semi-arid climate in which the days are hot and the nights cold. Building styles have responded to this factor much more than to seismic risk: traditionally, walls tend to be massive but weak and roofs heavy and, although they insulate the occupants against extremes of temperature, they are extremely vulnerable to collapse during seismic acceleration of the ground. Secondly, in many regions of the belt there is a severe shortage of timber, and often not much good quality building stone, such that there are few materials which are capable of carrying tension adequately.

Thirdly, poverty and lack of accumulated expertise in many countries of the belt mean that dwellings tend to be erected by their future inhabitants, who may have no knowledge of anti-seismic methods, or by builders who are not trained in anti-seismic construction. Such buildings are therefore **aseismic**. Fourthly, the use of heavy rubble masonry or mud-brick imparts an inertia to walls during swaying motions that can eventually cause them to collapse. The variability and weakness of man-made structures in such zones is often a source of risk of far greater importance than any geological factor except distance from epicentre. Lastly, building density and population concentration are important influences upon vulnerability. In most parts of the Eurasian seismic belt buildings tend to be highly clustered, even in sparsely settled areas. This gives rise to population densities that are locally very high (perhaps in the range 5,000–10,000 persons per km^2), and hence to extremes of seismic vulnerability.

Some of the greatest seismic risk and vulnerability is found in Iran, where traditional vernacular housing is mainly single-storey with thick adobe walls, crossbeams of medium-size tree trunks and heavy roofs of mud, twigs and beams. Likewise, in some seismogenic areas of Turkey, random rubble, hand-hewn timber frames, adobe, tile brick and reinforced concrete are the main building types. In the 1970 Gediz earthquake most deaths occurred in houses which incorporated a mixture of these styles and performed with great lack of homogeneity (Mitchell 1976).

The lessons of recent events

Here are several case studies that illustrate some of the principles and problems associated with recent seismic disasters.

The Nicaragua earthquake of 1972

In 1972 Managua, the capital of Nicaragua, contained 405,000 inhabitants, comprising 20 per cent of the national population. It had grown tenfold after being devastated by the earthquake in 1931. The area of Lake Managua abounds in active faults (Brown et al. 1973). On 23 December 1972 one of these located to the northeast of the city under the lake produced a magnitude 5.6 earthquake at shallow focus. At the time the only preparedness measures in force were an anti-seismic construction law, which had not been implemented, a natural hazards insurance scheme, which covered only upper-income housing, and a spare radio frequency for emergency broadcasts.

Of the residents of Managua, 1 per cent died, 4 per cent were injured, 50 per cent lost their source of employment, 70 per cent were made homeless, and 75 per cent were affected by the disaster (Kates et al. 1973). Three quarters of the homeless sought shelter in the immediate vicinity of the wreckage of their homes, and there was thus a massive problem of relief, including the need to provide alternative accommodation on a very large scale.

The response involved a small amount of regional decentralization of the city and reduction in the density of buildings. Some progress was made in reconstruction to anti-seismic standards, but the pressures of a vast refugee problem did not help the process of reconstruction to be rational. Speculation in the building trade and corruption in government circles were both rife, and hence construction standards were difficult to enforce. There was an absence of critical information and a general sense of confusion and indecision. In other words, the earthquake caught Nicaragua quite unprepared, and the general consequences were heightened disaster and long-term misery.

The December 1988 Armenian earthquake

On 7 December 1988 an earthquake of magnitude $M_L = 6.9$ occurred at the southern end of the Lesser Caucasus Mountains in the Soviet Republic of Armenia. Northeast dipping thrust faulting was complemented by some right-lateral strike-slip motion (i.e. the movement was part compressional, part lateral). Although larger tremors have been recorded in the Caucasus, the local area had not experienced such a strong earthquake in historical times.

Damage and loss of life were severe in three settlements: Leninakan (pop. 290,000), Spitak (pop. 20,000) and Kirovakan (pop. not known). Fifty villages also suffered various levels of damage. Approximately 500,000–700,000 people were affected by the earthquake. Estimates of the death toll varied from 25,000 to 100,000; and 130,000 people were injured, of whom 18,000 required hospital care.

Search-and-rescue missions were initiated by untrained local people immediately after the disaster. These were supplemented within 48 hours by 1,000 Soviet specialist volunteers and 16 international teams. The bulk of this operation was concluded by day 10, but live victims were extracted as late as 19 days after the impact. Owing to the vulnerability of construction in the area (see below) the manpower and equipment needs for search-and-rescue were grossly underestimated. The equipment normally used thus proved inadequate. Precast concrete frame buildings tended to disintegrate, leaving little space in which occupants could survive until rescued. Hence crush injuries (often fatal) were extremely widespread. Freezing conditions caused many trapped survivors to die of hypothermia and exposure before they could be rescued. Initially, central control and co-ordination of the rescue effort were significantly lacking.

In the area of total impact (comprising Leninakan, Spitak and Kirovakan) hardly any medical facilities survived the earthquake, and many doctors and nurses were killed. Thus, injured victims had to be transported long distances in order to receive medical care, but their evacuation was hampered by bad weather and lack of adequate transportation facilities.

In Leninakan 80 per cent of structures collapsed or were severely damaged (representing 1.6 million m^3 of floor space), while in Spitak all buildings were damaged and in Kirovakan 40 buildings were destroyed and 450 damaged. As some 14 properly reinforced buildings in Leninakan survived the disaster with negligible damage, the principal cause of building collapse appears to have been aseismic (rather than **anti-seismic**) construction (i.e. not incorporating construction elements that would have allowed the buildings to perform homogeneously without disintegrating during the tremors).

At Kirovakan the worst damage occurred in an area about five blocks square where land had been reclaimed from marshes by backfilling (conditions are similar to this in San Francisco and Boston, where reclaimed land and soft sediments increase the probability of seismic damage to structures). Although soft sediments 300 m thick amplified ground motion under Leninakan, and similar conditions prevailed at Spitak, construction type and quality were much more important determinants of damage to buildings. Older buildings were mostly constructed of load-bearing stone masonry, often using soft volcanic tuff (i.e. welded or cemented volcanic tephra, which forms a soft, porous rock that splits or shears easily), and they had heavy reinforced concrete and hollow-brick floors. Lack of steel reinforcing ties tended to cause such buildings to crack at corners until the end wall fell out and the floor beams collapsed. Newer buildings were constructed of reinforced concrete columns and beams, or of precast concrete panels. Inadequate splices between columns and beams, or between these and shear walls, led to structural failure and building collapse. Precast concrete stairs tended to detach themselves during the tremors, making evacuation of such buildings difficult. Clearly, with regard to earthquake hazards, the system of prefabrication and site assembly of structural components in use in Armenia was deeply flawed.

Many industrial and commercial facilities in this heavily industrialized area were severely damaged, representing the largest portion of the estimated US$15,000 million of losses caused by the earthquake. Water supplies to Leninakan were interrupted by a rock avalanche caused by the tremors, while the railway line to Spitak was severed for 300 m by an embankment slump. A 110 kV electricity substation was damaged beyond repair by the tremors, but two 440 MW pressurized water nuclear reactors sustained no damage, as their site experienced only weak ground motion. Elsewhere, steel-framed facilities with equipment that was well anchored to the ground performed best.

Thousands of mass movements were caused by the earthquake, principally in the form of rockfalls, slumps and debris or soil slides. In the Caucasus Mountains, north of Nalband, several million m^3 of rock were mobilized in a single movement, and liquefaction phenomena occurred.

The principal investigation team sent to Armenia from the United States concluded that a similar disaster could happen in the USA, which is not sufficiently prepared for a major search-and-rescue effort. Although the Armenian case was an extreme one, it illustrated many problems of world-wide relevance, such as the nature of building collapse (Wyllie & Filson 1989), the significance of landslide damage (US Geological Survey 1989) and the difficulty of providing medical care rapidly when hospitals and lifelines have been destroyed (Noji 1989).

Earthquake hazard in the USA
Seismicity is one of the most serious natural hazards to threaten the United States, although the risk is more latent than manifest. Nevertheless, more than 500 potentially damaging earthquakes have occurred in California or near its borders in the twentieth century including those at San Francisco in 1906, Long Beach in 1933, Kern County in 1952, the San Fernando Valley in 1971, Mammoth Lakes in 1980, Coalinga in 1983, Morgan Hill in 1984 and Santa Cruz in 1989 (Wallace 1990). These events were responsible for the deaths of 1,000 people and billions of dollars in property damage (damage caused to San Francisco and Santa Cruz by the 17 October 1989 earthquake may have exceeded $6,000 million; Plafker & Galloway 1989). A magnitude 8.2 earthquake occurred on the Alaskan coast in 1964 and anomalous earthquakes have helped define poorly known seismic zones elsewhere in the country, including the 1959 tremors at Hebgen Lake, Montana (Eckel 1970; US National Academy of Sciences 1973; US Geological Survey 1964). Earthquakes at Cape Ann near Boston in 1755, New Madrid, Missouri, in 1811 and 1812, and Charleston, South Carolina, in 1886 all suggest that there is high potential in these places for renewed disaster (McKeown & Pakiser 1982; Nuttli et al. 1986). Moreover, the areas in question have become significantly more developed, and hence more vulnerable to damage, since the last major earth tremors there.

California is home to one of the most graphic expressions of seismicity in the world. The San Andreas Fault is 1,000 km long and 30 km deep and forms part of a complex fault system marking the boundary between the North American and Pacific crustal plates (Wallace 1990). Some segments of the system release the stress and strain which plate movement imposes on them by **fault creep** as a series of small earthquakes or micro-tremors. Parts of the Calveras and Hayward Faults, for instance, undergo 2.5 cm of creep each year through hundreds of small tremors. But in 1906 a locked-up part of the fault released its strain by producing 6.4 m of near instantaneous displacement (with a magnitude of 8.0–8.2, a Modified Mercalli intensity of XI and damage valued at $600 million). At Pallett Creek, 55 km northeast of Los Angeles on the San Andreas Fault, radiocarbon dating of deformed strata suggests that earthquakes of magnitudes ≥8 occur on average once every

160 years (see Table 2.5). This means that a major damaging tremor will probably take place in southern California during the next half century.

In the past 120 years there have been seven major earthquake disasters in California:

Date	Location	Magnitude	Death toll	Losses ($ million)
1872	Owens Valley	8.3	60	
1906	San Francisco (and fire)	8.2	700	1,600
1933	Long Beach	6.7	115	89
1940	Imperial Valley	7.1	9	12
1952	Kern County	7.7	12	63
1971	San Fernando Valley	6.6	65	439
1989	Loma Prieta	7.1	62	6,000

The last of these events displaced 12,053 people and damaged 18,306 homes and 2,575 businesses (McNutt & Sydnor 1990). Damage was severe in San Francisco and Oakland, more than 80 km north of the

Table 2.5 Probabilities of major earthquakes on primary fault segments in California (after USGS 1988)

Fault segment	Date of most recent event	Expected magnitude	Estimated recurrence interval (years)	Probability of recurrence 1988–2018 (per cent)	Level of reliability
San Andreas Fault					
North Coast	1906	8	303	<10	high
San Francisco Peninsula	1906	7	169	20	medium
Southern Santa Cruz Mts	1906	6.5	136	30	very low
Central creep zone	—	—	—	<10	very high
Parkfield	1966	6	21	>90	very high
Cholame	1857	7	159	30	very low
Carrizo	1857	8	296	10	high
Mojave	1957	7.5	162	30	high
San Bernadino Mts	1812(?)	7.5	198	20	very low
Coachella Valley	1680±20	7.5	256	40	medium
Hayward Fault					
Northern East Bay	1836(?)	7	209	20	low
Southern East Bay	1868	7	209	20	very low
San Jacinto Fault					
San Bernadino Valley	1890(?)	7	203	20	very low
San Jacinto Valley	1918	7	184	10	medium
Anza	1892(?)	7	142	30	low
Borrego Mts	1968	6.5	189	<10	high
Imperial Fault					
Imperial Valley	1979	6.5	44	50	medium

epicentre, yet the disaster was by no means the worst that can be expected in this area over the next few decades.

The occurrence of a magnitude 8.2 earthquake on the San Andreas Fault with approximately the same isoseismal distribution as the 1906 earthquake would generate damage with maximum intensities IX or X in some parts of San Francisco. It has been estimated that between 3,000–5,000 people would be killed if the earthquake occurred outside business hours and between 10,000 and 12,000 if it took place during the working day (Cochrane et al. 1974). The number of injured would also depend on the time of day but might reach 40,000, and 20,000 people would be homeless. Property damage would exceed $10,000 million, as would economic costs such as loss of earnings. Of the damage, 40 per cent would occur in the centre of San Francisco.

Similarly, it is estimated that a magnitude 7.5 earthquake on the Hayward Fault would result in up to 4,500 deaths, 135,000 injuries (again depending on timing and day of the week), damage valued at more than $40,000 million, the destruction of at least one hospital and damage to all major routeways.

The highest cost would be borne by those survivors who lost their homes or jobs, especially people with low incomes living in old, multi-storey apartment blocks that predated the 1933 anti-seismic construction laws and were never retrofitted. Through higher taxation the entire United States would provide money for compensation and reconstruction. In 1971 it was estimated that the disaster would cost one third of the total personal income of Californians and 3 per cent of the US Gross National Product (White & Haas 1975). Although the reconstruction of San Francisco after the 1906 earthquake and fire was principally financed from private sources and with insurance monies, modern trends are towards ever greater involvement of the Federal Government, and private funding has ceased to play a major rôle in reconstruction.

In the context of seismic hazards in California, it is worth remembering that the campus of Stanford University was almost destroyed by the 1906 San Francisco earthquake. Since then vast sums of money have been spent on hazard mitigation throughout the state, and Stanford has become one of the principal centres of expertise in seismic engineering. But like Berkeley, its campus remains vulnerable: the cost of seismically retrofitting the California State University system, estimated at $4,000 million, is prohibitive. The most vulnerable buildings are those constructed of unreinforced masonry before 1934, when the first comprehensive anti-seismic building codes were introduced. There are estimated to be 50,000 of these structures in California, of which 8,000 are in Los Angeles and 6,000 in San Francisco (Alesch & Petak 1986). A number of the latter caught fire or collapsed during the 1989 Loma Prieta earthquake (Plafker & Galloway 1989).

Tsunamis

The word **tsunami** is Japanese for "harbour wave". These fast-travelling long waves are often known as "seismic sea waves", although they are not always caused by earthquakes. The popular epithet "tidal waves" is, however, a complete misnomer, as they have nothing to do with tides. The danger posed by tsunamis rests in the fact that the energy of a wave is embodied in its wavelength (or period), amplitude and phase. A change in any one of these will cause a compensating adjustment to take place in the other two. Hence, if the speed of propagation decreases, as occurs in coastal areas, the amplitude will increase. Tsunamis travel fast in open ocean conditions, yet their wavelengths are broad (usually 150–250 km, but up to 1,000 km in open ocean waters), their amplitudes are small (not more than a few metres) and their periods are long (10–60 minutes). They are thus difficult to observe until they reach land and these characteristics start to change.

The mean global cost of tsunami damage is relatively low at $8 million per annum, and although casualty totals vary markedly, on average about 300 people are killed by the waves each year.

Stages in the existence of a tsunami

The four stages in the existence of a tsunami are generation, propagation in deep water, propagation in shallow water and landfall.

In order of importance, the **tsunamigenic forces** are:

(a) vertical displacements of a column of sea water caused by undersea earthquakes which involve abrupt movement of fault blocks on the sea bed;

(b) horizontal displacements of the water column as a result of earthquake movements (those which occur on land may generate a tsunami if coasts are impacted with sufficient violence);

(c) volcanic eruptions at sea (which must displace the surrounding water with considerable violence if a tsunami is to be generated);

(d) rapid mass movements (landslides) on the sea floor (these, however, are seldom responsible for tsunamis of any great size).

More than 90 cases of volcanigenic tsunamis have been recorded, roughly one quarter of which resulted from the impact on the sea of pyroclastic flows and another quarter volcanically generated submarine earthquakes (Blong 1984). One fifth resulted from submarine explosions, while 7 or 8 per cent resulted from landsliding into the sea, which caused pressure waves with effects similar to those of a true tsunami, as occurred during the 1792 eruption of Unzen in Japan, when some 15,000 people were drowned. But tsunamis principally result from tectonically generated earthquakes of magnitude greater than 6.5 and focal depth shallower than 50 km. They may be dangerous if the hypocentre is less than 25 km below the surface. However, not all undersea earthquakes are tsunami-genic, as this depends on the nature and degree of displacement of the sea-water column.

In tsunamigenesis, the rapid movement of the sea floor causes the water column above it to be displaced and, as water is not a compressible medium, this sets up oscillations at the sea surface. Under the influence of gravity the water surface returns to equilibrium after one or two oscillations, but waves radiate from the source in all directions (as ripples travel out from the point at which a stone is dropped into a tranquil pond). As with other forms of wave, in tsunamis, energy is transported rather than mass. Hence, during the passage of the wave, water particles move in a horizontally elongated ellipse, which diminishes in size at increasing depths below the surface. Oscillatory currents are set up with periods of 10–60 minutes, affecting the whole water column from surface to bottom and spreading outward from the source.

The generating mechanism differs between tsunamis, but is usually a deep ocean-floor cataclysm. A point source gives rise to smooth, highly dispersive waves, a source which is as long as the water is deep (such as a fault trace) gives beat-type oscillations, while when the source is a large area of sea floor the result may be a single large wave compounded by oscillations of shorter period.

Propagation in deep ocean water involves the movement of concentric rings of crests and troughs bounded by an intangible front that expands outward in all directions at the limiting velocity. All subsequent waves travel slower, and hence the separation between wave crests increases progressively outwards from the source. The speed of propagation of the tsunami is expressed by the celerity equation

$$v = \sqrt{(g.h)}$$

in which v is velocity, g is the acceleration due to gravity (9.81 m/sec^2) and h is the depth of water. The average water depth in the Pacific Basin is 5,500 m. Hence, this gives a velocity of $\sqrt{(9.81 \times 5500)}$, which is 230 m/sec or about 830 km/hr. The wavelength is

$$L = v.T$$

and, if the period T = 15 minutes, then 830 km/hr $\times$ 0.25 hrs gives a period of about 208 km.

Propagation in shallow water involves progressively greater interaction between the wave and the ocean depth, during which energy of movement is conserved. A wave of amplitude 5 m moving at 170 m/sec in deep water may be refracted to a 30 m-high wave travelling at 15 m/sec (which is still 54 km/hr) at the coastline. At this stage, the amplitude of waves is determined by four factors. The first of these is the nature of the source: earthquake, volcanic eruption or undersea landslide. The second is the distance of the wave from its source, given that it loses its energy as it travels. The third is the depth of water over the route that the wave has taken to reach the coastal area in question. And, finally, waves can be refracted in various ways by different shapes of coastline or offshore topography, such as a continental shelf (see Fig. 2.14).

Wave energy declines logarithmically during propagation, as the wave is not only interacting with its surrounds but also spreading itself over an increasingly wide area. In fact, energy is lost by reflection, scattering, refraction, absorption into the bed or body of the sea and leakage into other seas or oceans. Any change in the depth of water will affect the

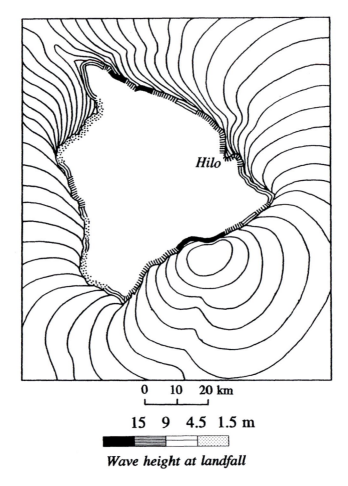

Figure 2.14   Hawaii Island tsunami hazard map (after Cox & Morgan 1984: 12, Hays 1981: B36).

speed and energy of the wave, but the scale of the tsunami phenomenon means that only marine topography of hundreds of square kilometers will have an appreciable effect, until, that is, the wave makes landfall.

Tsunamis become dangerous during **run-up** and **landfall**. About 40 per cent of the wave energy that remains after travel from the source is scattered back out to sea, while the rest is expended on the shoreline. There is usually a sequence of waves, in which the fourth or fifth may be the largest, and the first arrival may be a trough, or in other words the drawing back of water preceding a major surge. The interval between the arrival of each wave will be some fraction of an hour. The size of the wave, as it arrives at the shore will be a function of several variables, including the magnitude of the original water displacement at the source and the distance which the wave has travelled. The offshore topography is also important: broad, gently sloping continental shelves allow the tsunami to increase its amplitude without great loss of velocity and energy, while deep water close inshore minimizes bottom friction and hence restricts the growth of wave size. The growth of wave amplitude is favoured by narrow inlets and bays, which concentrate the energy of the tsunami. Finally, differential effects are exerted by irregular coasts, while the impact of wave run-up is generally more widespread on low-lying coasts.

Tsunami magnitude, M, is calculated at run-up and is a simple parametric function of wave height, h, in metres:

$$M = 3.32 \log h$$

In completely or partially land-locked bodies of water (such as lagoons and sounds) **seiches** may be generated. These are resonant oscillations of the water body caused by abrupt impact and the resulting inertial response. Periods tend to be shorter than those of tsunamis and may reach 15 minutes for a body of water a few tens of kilometers wide.

The destructive effects of tsunamis are of three kinds: **hydrostatic effects** involve the lifting and carrying of light structures, such as wooden buildings; **hydrodynamic effects** involve tearing buildings apart, washing away soil, and so on; and **shock effects** occur when objects are battered by debris or other objects, or carried away by the wave, which drags or floats them. Japanese research indicates that wooden buildings are likely to be partially damaged by a 1 m wave. A 1–2 m-high wave will float a poorly secured wooden building, while waves higher than 2 m may wash out the lower floor and cause the structure above it to collapse (Soloviev 1978). Death may occur by drowning, by building collapse or by impact when the victim is lifted and battered or hit by moving debris.

Of damaging tsunamis, 90 per cent occur in the Pacific Basin at an average of more than two each year. In the past 190 years the Hawaiian Islands have experienced more than 150 such events (Table 2.6a), and

Table 2.6a Origin of 159 tsunamis recorded in Hawaii over the period 1801–1975 (data from Cox & Morgan 1977).

Source location		Numer of tsunamis	Percentage
Central Pacific	Hawaii	30	18.9
	Tahiti	1	0.6
East Pacific	Japan	21	13.2
	Philippines	2	1.3
	Bonin Island	1	0.6
Southeast Pacific	Solomon Islands	6	3.8
	Tonga	3	1.9
	Indonesia	2	1.3
	New Guinea	2	1.3
	New Ireland	2	1.3
	Fiji	1	0.6
	Samoa	1	0.6
Former USSR	Kamchatka	10	6.3
	Kuril Islands	10	6.3
North America	Alaska	8	5.0
	Aleutian Islands	8	5.0
	British Columbia	1	0.6
Central America	Mexico	6	3.8
	Costa Rica	1	0.6
	Guatemala	1	0.6
South America	Ecuador-Colombia	1	0.6
	Perú-Chile	26	16.4
Unknown		14	8.8
Total		158	100.0

Table 2.6b Occurrence of known tsunamis in the Mediterranean Basin (Soloviev 1990). Reprinted by permission of Kluwer Academic Publishers.

Period	I	II	III	IV	V	≥VI	Unknown	Total
Before 1700	—	3	43	28	11	5	10	100
1701–1800	—	2	20	8	2	2	1	35
1801–1900	—	15	64	24	1	1	7	112
1901–1990	1	17	11	6	4	3	1	43
Total	1	37	138	66	18	11	19	290

destruction has often been widespread. The USA suffers more than $26 million in property damage each year, and 500,000 people live in areas of the country at risk from tsunami waves 15 m high, while 1.2 million are at risk from waves reaching 30 m (White & Haas 1975). The west coast and Hawaiian Islands run the greatest risk. The Atlantic and Mediterranean Basins are by no means immune from tsunamis (see Table 2.6b), but these are less common there and hence records are incomplete and return periods are largely unknown. Nevertheless, the eruption of Santorini in 1500 BC resulted in tsunamis which travelled all over the eastern

Mediterranean Sea, and the earthquake which occurred in 1755 near Lisbon, Portugal, is estimated to have caused 25,000 deaths by tsunami in that city and to have involved waves that arrived in the Caribbean (Table 2.7).

Tsunami monitoring and warning systems

Various instruments are regularly used to detect and monitor the passage of tsunami waves (Clark 1987). Seismographs have been devised which are very sensitive to long-period oscillations, such as those caused by a wave as it travels across the sea bed. Tide gauges consist primarily of a float in a vertical stilling well, which is connected to an instrument that records changes in water level. These need to be located in harbours in order to reduce the effect of ocean swell. They provide an excellent record of the tsunami when it arrives, but are of little use in predicting it. Bottom pressure recorders are located on the sea floor in shallow water and are connected to instruments on land. Changes in water pressure are

Table 2.7 Some of the largest tsunamis in history (Bolt 1988, and other sources).

Date	Zone of origin	Height	Damage to:-	Comments
1530 BC	Santorini (eruption)	?	Crete	Mediterranean coasts devastated
1755	E. Atlantic	5–10 m	Lisbon, Portugal	also felt in the Caribbean
1837	Chile	6 m	Hilo, Hawaii	
1867	Formosa Strait	3.5 m	China, Taiwan	500 deaths
1868	Perú-Chile	>18 m	Arica, Perú	also felt in New Zealand; damage in Hawaii
1883	Krakatoa (eruption)	12 m	Java	36,000 drowned
1896	Honshu	24 m	Sanriku, Japan	26,000 drowned
1908	Strait of	5 m	Messina, Reggio Calabria	>8,000 deaths
1923	Kamchatka	6 m	Waiakea, Hawaii	
1925	Pacific Ocean	11 m	Zihuatanejo, Mexico	
1932	Pacific Ocean	10 m	Cuyutlán-San Blas, Mexico	10–75 deaths (estimated)
1933	Honshu	>20 m	Japan	3,000 deaths
1946	Aleutian Islands, Alaska	17 m	Wainaku, Hawaii	173 killed, 163 injured
1957	Aleutian Islands	16 m	Hawaii	Heavy damage
1960	Chile	>10 m	Hawaii	61 killed, 282 injured
1964	Alaska	8.5 m	Crescent City, California	119 deaths; 200 injured; $104 million in damage
1975	Hawaii Island	8 m	Hilo, Hawaii	
1976	West Pacific	5 m	Philippines	3,000 drowned
1983	Hokkaido	6–14 m	Japan	100 deaths

registered by electrical transduction as the wave passes. Such instruments are expensive to maintain if they require situating in distant, widely dispersed or inaccessible locations. Eventually, networks of deep-sea pressure and long-wave transducing stations will be situated throughout the Pacific Basin and monitored remotely by telemetering. Finally, taut-wire buoy stations consist of a float anchored to the sea bed by a high-tensile wire. The measuring signal is stable because the buoyancy force is several orders of magnitude greater than the horizontal forces of waves. The buoys, which are expensive and essentially temporary measures, are used to measure changes in the sea surface level during the passage of the wave.

The location of run-up and inundation, and the damage caused by tsunamis, can be mapped after the event. In this way Japan kept records sporadically over the period 1933–45, and (like Hawaii) since 1946 it has accumulated large, systematic data sets concerning the impact of tsunamis on particular pieces of coastline. The location of future damage and flooding are now well known in both archipelagoes, and on that basis run-up risk maps have been constructed (e.g. Fig. 2.14). What cannot be predicted from past data can be simulated. Thus, laboratory models are particularly useful for the study of tsunami impact at landfall. Hydraulics can be studied using hardware models or digital computers, and engineering defences can be designed and tested in controlled conditions. As energy dissipation at boundaries is difficult to scale down, physical models must be large and may need to compensate by exaggerating the vertical dimension. Despite such limitations, hardware models can reproduce convincingly the processes of generation, deep-water propagation and run-up of tsunamis.

New analytical techniques are continually being developed to investigate the tsunamigenic mechanism, with the aim of discovering exactly what type of earthquake causes tsunamis. Probablistic computer models are used to express "tsunamicity", the likelihood that an earthquake of particular focal parameters and magnitude will cause a tsunami of a given magnitude. There is, however, no successful way of identifying tsunami waves in deep ocean water, where their periods are excessively long, unless they have already been detected at genesis (Kulikov et al. 1983).

Prediction and monitoring problems are greatest across the vast expanse of the Pacific Ocean Basin. According to Foster & Wuorinen (1976), over the period 1900–70 four of 138 tsunamis which occurred in the Pacific Basin were locally destructive and nine caused damage at great distances. Given the considerable risks caused by Pacific tsunamis, in 1946 the US National Oceanographic and Atmospheric Administration set up the Seismic Sea Wave Warning System (SSWWS) for the Circum-Pacific Zone, which is now known as the **Pacific Tsunami Warning System** (PTWS; Fig. 2.15). There are 23 member nations of the PTWS (Pararas-Carayannis

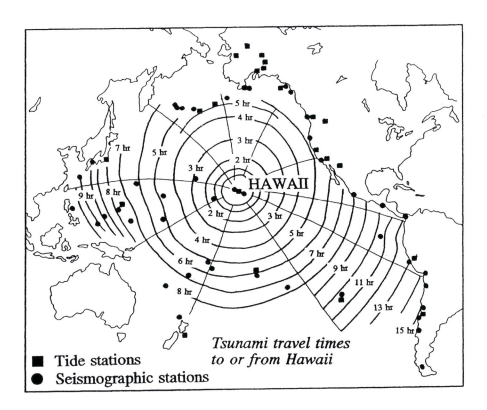

Tide stations
Seismographic stations

*Tsunami travel times
to or from Hawaii*

Figure 2.15 Pacific tsunami travel times and warning system (Office of Emergency Preparedness 1972, Vol. 1: 92).

1986). The system uses 69 seismic stations and 65 tide stations, which are located in all the major harbours of the Pacific Basin. Earthquakes of Richter magnitude ≥6.5 will cause alarms to sound and those ≥7.5 give rise to a round-the-clock tsunami watch, but it remains difficult to predict the size of waves and to avoid false alarms. The hub of the system, at Honolulu in Hawaii, is connected to the nations of the Pacific rim by a sophisticated communications system which functions in an anti-clockwise direction (Dohler 1988). Within one hour of the detection of a tsunami, 101 dissemination points are notified by teletype and telephone of earthquake location (±50 km), magnitude (±0.3) and the arrival time of waves (±20 minutes).

A large tsunami may traverse the Pacific Basin in 20–25 hr, which gives ample time to warn distant locations (Fig. 2.15). But 99 per cent of tsunami deaths occur within 400 km of the point of wave generation and, along with much of the damage, most occur within 100 km. As the waves

travel at speeds of 800–950 km/hr in deep ocean water, only 10 minutes'
alarm can be given regionally and it is therefore difficult to warn nearby
communities. Thus the US National Oceanographic and Atmospheric
Administration (NOAA) has developed an inexpensive, self-contained
system for local warning (THRUST: Tsunami Hazard Reduction Utilizing
Systems Technology). When an earthquake of magnitude ≥7.0 occurs
within 100 km of the coast, an accelerometer transmits a signal to the
geostationary GOES satellite positioned over the eastern Pacific Basin.
From there it is relayed to a standard microcomputer, which decodes the
signal and is programmed to begin monitoring water-level sensors and to
send a prerecorded telephone message to officials responsible for
evacuation plans. The success rate in tests was 84 per cent and the mean
response time was 88 seconds (Bernard 1991). THRUST technology has
already been integrated successfully into the Chilean national emergency
system based at Valparaiso (Lorca 1991).

However, technology such as THRUST and the PTWS cannot solve the
problem of warning, which is the preserve of the national governments
and the authorities which they appoint. One of the few studies carried out
on the social aspects of tsunami impact showed that in the United States
up to two thirds of people warned of an impending tsunami will respond
inappropriately, perhaps by ignoring the call to evacuate (Anderson
1970). Hence, a tsunami caused by the May 1960 Chilean earthquake
inundated the lower parts of Hilo, Hawaii, killed 61 people, injured
several thousand and destroyed 500 homes. A warning was given four
hours in advance (although the authorities actually had more than ten
hours' notice) and was heard by 95 per cent of residents (Lachman et al.
1961). Of these, 40 per cent evacuated their homes, 13 per cent
disregarded the call to leave home and 40 per cent failed to interpret the
warning correctly and therefore did nothing. The first wave struck while it
was still night-time: 43 per cent of the people who did not evacuate were
asleep, 48 per cent were at home but awake and at least 14 people had
gone down to the shore to watch the wave come in. Hindsight shows that
the warning was confusing and ambiguous.

It is obvious that the design and implementation of warning systems
depend upon the recognition of danger. But this is not always
forthcoming. For example, in 1963 Canada withdrew from the PTWS, as its
National Committee on Oceanography had decided that the west coast
ran no risk of tsunamis. Early on 28 March 1964 a tsunami which had
been generated by earthquake in the Gulf of Alaska overran the twin
communities of Alberni and Port Alberni on Vancouver Island. No
warning had been given to residents, but they were alerted by the first
wave and evacuated. Subsequent waves destroyed 58 buildings and
damaged 320. Here and elsewhere on the west coast of Canada losses
from the 1964 tsunami exceeded $10 million (Foster 1980).

Counter-tsunami measures

Various measures can be taken to avoid the worst forms of tsunami impact. First of all, warning should be accompanied by thorough evacuation of risk zones, and this requires that there be adequate escape roads to higher ground. It also requires that there be an effective system for contacting the public, an evacuation plan and a programme of popular education designed to publicize the risk.

Land-use zoning can be made a function of hazard maps which report the expected heights of tsunamis likely to recur at a particular location during intervals of 20, 50 or 100 years. Houses and other buildings can be removed to higher ground and new construction banned in the principal risk areas. In this way, after devastating tsunamis in 1946 and 1960, the central business district of Hilo, Hawaii, was removed to a safer area and the waterfront and estuarine areas made into public open space. Special effort must be made to ensure that urban uses do not surreptitiously reinvade land officially rezoned for low-intensity uses. This happened in Japan after urban areas devastated by tsunami in 1934 were converted to agricultural use only to be devastated again, after being successively invaded by new urban development, and overrun by the 1960 tsunami that originated in Chile.

Breakwaters can be constructed to weaken (though not stop) the approaching wave. Sea walls are more costly but can provide better protection. In Japan they have been built with a curved profile capable of reflecting and deflecting wave energy. Dense groves of trees can be planted at the shore, and these also serve to absorb wave energy that otherwise might destroy buildings and structures. Alternatively, where there are particular grounds for using the waterfront, as in ports, frontage sheds can be constructed at right angles to the line of approach of the waves and hence parallel to the coast. These are large buildings which are not likely to be intensively occupied and hence can be allowed to take some of the effect of the tsunami. They can also be reinforced against damage.

Other buildings can be built to resist the passage of the waves. Reinforced concrete frames and infills can be used, ground floors can be left open such that the waves pass between the concrete piles on which the building rests, and buildings can be situated perpendicular to the shore so that their smallest elevation bears the brunt of the waves. Special measures must be taken to protect power stations, which need much cooling water and therefore are often situated on the coast. Where there is a tsunami risk they must be built to resist impact and to keep sea water away from boilers in order to resist explosive cooling. They must also have water intake ducts which extend beyond the furthest likely point of water retreat during the arrival of the wave trough.

Clearly, the best strategy for protecting a coast against tsunamis is to use an appropriate mixture of these methods.

Volcanic eruptions

The nature of volcanism and volcanoes

About half of the 1,300 volcanoes that have erupted during the past 10,000 years can be considered active, and about 50 of them erupt each year, although cataclysmic eruptions happen only about once in a thousand years. During the past five centuries at least 200,000 people have been killed either directly by eruptions or by starvation when crops or livestock were affected. The worst example of a volcanic disaster in the twentieth century occurred when in 1902 Mount Pelée on the Island of Martinique sent a glowing ash cloud into the town of St Pierre (pop. 25,792) and killed 28,000 people, including all but two of its inhabitants (Boudon & Gourgaud 1989).

In volcanism, magma in subsurface chambers wells up under the influence of its confining pressure and superheated steam and gases and is emitted at the surface in gaseous, liquid or solid form. The causes of the world pattern of volcanic activity are connected with global tectonics and crustal dynamics (Simkin et al. 1981). The formation of new lithosphere at the mid-ocean rifts is a volcanic process of enormous importance. **Subduction** of one crustal plate under another causes escape of magma among the contorted rock formations buckled up by the descending plate. Finally, "hotspots" exist in the mantle at mid-plate locations, and as the crust spreads over them plumes of magma migrate to the surface and produce chains of volcanoes, as in the Hawaiian archipelago.

The oft-quoted classification of volcanoes into **active**, **dormant** and **extinct** is somewhat misleading, as there is often no reliable way of distinguishing between the categories over long time periods. However, volcanoes (such as Paricutin and El Chichón in Mexico) that erupt only once before the eruptive centre shifts elsewhere are known as **monogenetic**.

Eruptions can be classified as **explosive** or **effusive**, depending on the violence with which they occur. Easy-flowing or low-viscosity magmas escape in lava flows as effusive volcanism, which is of the Hawaiian or Icelandic type, depending on the viscosity (which is higher in the former case). High-viscosity magmas allow pressure to build up (perhaps in Domean extrusions) and the eruption to be explosive (Sparks 1986). Using the names of particular volcanoes (and of volcanism's first literary victim, Pliny the Elder), such eruptions are classified as Strombolian, Vulcanian, Vesuvian, Plinian or Krakatoan (ultra-Plinian) in increasing order of explosivity (Table 2.8; Fig. 2.16). The first two types are

Table 2.8 Types of volcanic eruptions (after Macdonald 1972: 211). Copyright ©
1972 Prentice-Hall, reprinted with permission.

Type	Characteristics
Icelandic	Fissure eruptions that quietly release large volumes of free flowing (fluid) basaltic magma which is poor in gas and which flows as sheets over large areas and builds up plateaux. Minor amounts of tephra.
Hawaiian	Fissure, caldera and pit crater eruptions; mobile lavas with some gas in thin, fluid flows; quiet to moderately active eruptions; occasional rapid emission of gassy lava in fountains; only minor amounts of ash; builds up lava domes.
Strombolian	Stratocones (summit craters); moderate, rhythmic to nearly continuous explosions, resulting from spasmodic gas escape; clots of lava ejected, as bombs and scoria; periodic, thicker outpourings of moderately fluid lava; light-coloured clouds (mostly steam) reach upwards only moderate heights.
Vulcanian	Stratocones (central vents); viscous lavas in short, thick flows; lavas crust over in vent between eruptions, allowing gas to build up below surface; eruptions increase in violence and are interspersed with longer quiescent periods until lava crust is broken up, clearing vent and ejecting bombs, pumice and ash; lava flows out of summit after main explosive eruption; dark, convoluted and cauliflower-shaped ash-laden clouds, rising to moderate heights more or less vertically, depositing ash along flanks of volcano. (N.B.: In *pseudo-Vulcanian* eruptions phreatic steam clouds vent fragmental matter explosively.)
Surtseyan	Violently explosive, phreatomagmatic eruption, ejecting fragmental material and glassy tephra.
Vesuvian	More paroxysmal than preceding types; extremely violent expulsion of gas-charged magma from stratocone vent; eruption occurs after long interval of quiescence or mild activity; vent tends to be emptied to considerable depth; lava ejects in explosive spray (glowing above vent) with repeated clouds that reach great heights and deposit ash.
Plinian	More violent form of Vesuvian eruption; sometimes associated with caldera collapse; last major phase is uprush of gas that carries clouds rapidly upward in vertical columns that are narrow at their bases but are kilometers high and expand outwards at upper altitudes; wide dispersal of tephra, including pumice; cloud is generally low in ash. The largest kinds of explosive eruption are known as *ultra-Plinian or Krakatoan*.
Peléan	Results from high viscosity lavas; delayed explosiveness; conduit of stratovolcano usually blocked by dome or plug; gas (some lava) escapes from flank openings or by destruction or uplift of plug; gas, ash and blocks move down slope in one or more blasts as *nuées ardentes* or other forms of pyroclastic flow, producing directed deposits.
Bandaian	Collapse of part of volcanic edifice, producing massive landslide or rock avalanche.
Katmaian	Voluminous production of ignimbrites (welded tuff deposits).

relatively gentle and small-scale, while the Vesuvian and Plinian varieties send a vertical column of ash into the air above the volcanic vent, and Krakatoan eruptions are extremely paroxysmal (Carey & Sigurdsson 1989). Eruptions can also be classified in terms of how much solid material is emitted, and in the largest events the totals can be expressed in terms of 10^0–10^2 km^3 (Table 1.10).

Lava that is dark and highly fluid is called **basalt** and is known by the

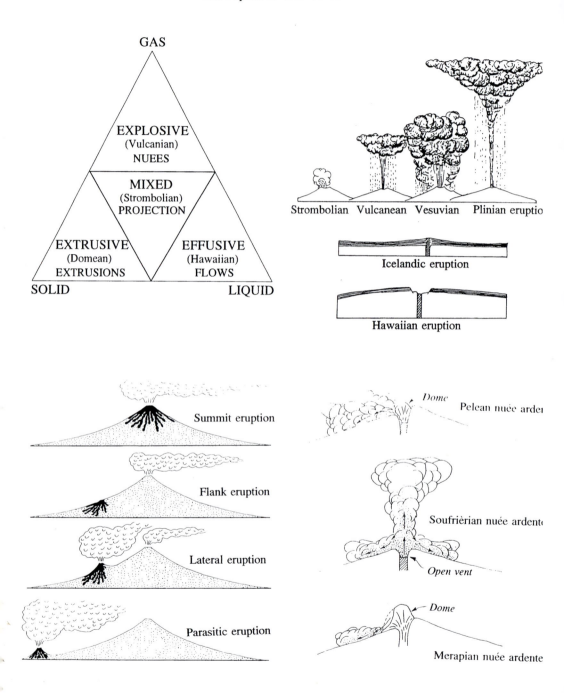

Figure 2.16 Types of volcanic eruption (after Ollier 1969, Francis 1976: 103).

Polynesian terms **a'a** if it cools in **brecciated** (broken up) form or **pahoehoe** if it becomes ropy and smooth (Fig. 2.17). When polyhedral boulders are extruded, the material is termed **block lava**. Magma which is light coloured and viscous is **rhyolite**, while the intermediate form is **andesite**. The solid, **pyroclastic** (literally "fire-broken") debris emitted during eruptions is known as **tephra**. Particles greater than 32 mm in diameter are known as **blocks** or volcanic **bombs**, those in the range 4–32 mm are called **lapilli**, and those which are smaller than 4 mm are classified as **ash**. Dust and ash solidify as **tuff**, which may be welded by high pressure and temperature into a layer of **ignimbrite**, while larger fragments may solidify into a volcanic breccia. Glassy solidified lava is known as **obsidian**, while frothy magmas laden with gases may solidify as porous **pumice** deposits.

Volcanic products can also be classified in terms of the silica content of the magma from which they are formed, according to the following scheme:

Magma type	*SiO$_2$ % by wt*	*Example of mineral*
acid	>68	rhyolite, obsidian
acid	63–68	dacite
acid/intermediate	57–68	trachyte
intermediate	57–63	andesite
intermediate	52–57	tholeiitic basalt
basic	45–52	alkali basalt
ultrabasic	<45	leucite

Volcanism produces a remarkable variety of constructional landforms or **volcanic edifices**. The smallest and frequently the most symmetrical forms are **cinder cones**, which usually result from gentle Strombolian or Vulcanian activity. In 1943, for example, steam began to rise from a field in rural Mexico and within 8 months a 410-m high cinder cone, christened Paricutin, was formed (Segerstrom 1966). This emitted 80 million m^3 of ash and other ejecta at a maximum rate of 6 million m^3 per day (activity continued until 1952). Similarly, Izalco, which was born in El Salvador in 1770, erupted almost continuously until 1958, by which time it was 900 m high.

Calderas are essentially very large craters, in which individual bursts of volcanic activity are localized and can produce smaller craters. Examples include Kilauea and Halemaumau on Hawaii Island, the Pozzuoli caldera near Naples in Italy, and Krakatoa at the tip of Java Island in Indonesia. In some cases, gravity causes the magma chamber to collapse, resulting in large **subsidence calderas** (Gudmundsson 1988). At a smaller scale, **explosion pits** are created by sudden blast leading to the collapse of an underlying magma chamber and a consequent depression in the surface.

There are continua of both size and steepness among different types of volcanic edifice. **Volcanic plains and plateaux** mainly result from gigantic

Figure 2.17   In this volcanic landscape at the southern end of Hawaii Island the dark, broken textured lava on the left is *a'a*, while the smooth, ropy textured lava on the right is *pahoehoe*.

outpourings of basalt (rhyolite is too viscous to engage in this sort of activity). Thus the Columbia–Snake River Plateau of Washington, Oregon and Idaho States is 520,000 km^2 in size and consists of lava 3,000 m thick in individual flows up to 30 m thick. The Deccan Plateau of northern India is even larger. **Shield volcanoes** are so called because of their gently sloping, convex shape. They include some of the world's largest volcanoes, such as Mauna Loa on Hawaii Island, and result from gentle, effusive outpouring of large quantities of basaltic lava, sometimes in lava fountains up to 400 m high. In contrast, **viscous lava cones** are usually made of rhyolite, are massy and have steep sides, but are not very large. **Strato-volcanoes** consist of alternate layers of lava and fallen tephra, as at Mount Fuji in Japan and Mount Shasta in northern California. They often form highly symmetrical cones of considerable size.

Volcanic hazards
Table 2.9 gives a classification of volcanic hazards with examples of some of the main events and associated death tolls.
   Lava flows are one of the most characteristic manifestations of

94

Table 2.9 Classification of volcanic hazards, with examples (after Blong 1984, Tilling 1989).

Type of hazard	Example	Death toll
	DIRECT HAZARDS	
Fall processes:		
• Ballistic projectiles	Soufrière (St Vincent), 1812	56
	Mount Etna (Sicily, Italy), 1979	9
• Tephra falls	Vesuvius (Italy), AD 79	≤16,000
	Santa María (Guatemala), 1902	≈2,000
	Vesuvius (Italy), 1906	250
	Rabaul (Papua New Guinea), 1937	375
	Agung (Indonesia), 1963	163
Flowage processes:		
• Debris avalanches	Unzen (Japan), 1792	≈10,000
• Lateral blasts	Mount St Helens (USA), 1980	51
• Lava flows	Vesuvius (Italy), 1630	700
	Nyiragongo (Zaire), 1977	72
• Primary lahars		
–crater lake eruption	Kelut (Indonesia), 1919	5,150
–Jökulhlaup	Öraefajökull (Iceland), 1727	3
–Pyroclastic flows		
and surges	Mount Pelée (Martinique), 1902	29,025
	El Chichón (Mexico), 1982	≥2,000
Other processes:		
• Phreatic explosions	Soufrière (Guadeloupe), 1976	none
• Volcanic gases	Dieng Plateau (Indonesia), 1979	142
	Lake Nyos (Cameroon), 1986	1,887
• Acid rain	Masaya (Nicaragua), 1979 *et seq.*	none
	INDIRECT HAZARDS	
• Atmospheric effects	Mayon, 1814	≤1,200
• Earthquakes and		
ground movements	Colima (Mexico), 1806	2,000
	Campi Flegrei (Italy), 1983–5	2
• Post-eruption erosion		
and sedimentation	Irázu (Costa Rica), 1963–4	none
• Post-eruption famine		
and disease	Lakigigar (Iceland), 1783	9,340
	Tambora (Indonesia), 1815	82,000
• Secondary lahars:		
–Saturated tephra	Irazú (Costa Rica), 1963	30
–Melting ice or snow	Nevado del Rúiz (Colombia), 1985	≥22,000
–Crater lake breach	Ruapehu (New Zealand), 1953	151
	Mount St Helens (USA), 1980	6
–Seismic causes	Mauna Loa (Hawaii, USA), 1868	31
• Tsunamis	Unzen (Japan), 1792	15,190
	Krakatoa (Indonesia), 1883	32,800

volcanism. They are usually less than 5 m thick and confined to the lowest sections of valleys. Basaltic lava can flow at speeds varying from less than 1 m per day to 3 m per second, but the threat to human life is slight. Despite this, the 1977 eruption of Nyiragongo in Zaire emitted 0.02 km^3

of highly fluid material in less than an hour from a lava lake at the summit of the volcano and from radial fissures. The lava covered 20 km^2 and the largest flow travelled at 40 km/hr finally coming to a halt 10 km from its origin. It overran 400 houses and killed 72 people (UNDRO/UNESCO 1985).

Lava flows fastest near to its source and progressively slower at greater distances, as it cools in contact with the ground and atmosphere. Cooling can cause it to congeal and flow continues within **lava tubes** of solidified material. These can be disrupted with explosives in order to slow the lava down and enable it to cool and become solid before travelling much further. Hence, in 1937 and 1942 the US army used 120 kg bombs to disrupt lava flows that threatened the city of Hilo, on Hawaii Island. However, bombing tends to be inaccurate, has debatable results and cannot be carried out in the poor visibility that accompanies many eruptions (Lockwood & Torgerson 1980).

If it is fluid enough, lava will flow around obstructions and hence will tend to fill a well-constructed house rather than flatten it. As a result, stone walls can be an effective means of diverting flows, and generally 3 m thickness of wall is needed to deflect each 1 m depth of lava (walls should also have ample height). As, under the influence of gravity, lava follows the lowest course it can, channels can be excavated in order to alter its direction. In fact, the 1983 eruption of Mount Etna in Sicily was successfully mitigated by the use of walls, channels and explosives (Abersten 1984).

For several months in early 1973 Helgafell Volcano, on the Icelandic island of Heimaey, erupted lava at 1030–1055°C from a new vent which was christened Eldafell. In the town of Westmannaeyjar (pop. 5,300), 800 buildings were damaged or destroyed, although 20 per cent was added to the area of the island and its fishing port was spontaneously improved. By pumping seawater at up to 1,200 litres per second onto the lava, thus cooling it and arresting its movement, 400 buildings were saved (Williams & Moore 1973). Despite the smallness of the local landing-strip most of the 10,000 sheep on the island were evacuated by air during the emergency, which must have posed some very particular logistical problems!

Doming can be a serious problem in volcanoes composed of andesite and rhyolite. Material that has solidified in the volcanic vent is forced upwards and may eventually collapse onto the flanks of the volcano in the form of a hot or cold rock avalanche, or it may cause serious ground deformation and instability. Thus, before the 1902 eruption of Mount Pelée on Martinique Island, a 3,300 m-high spire consisting of 90 million m^3 of rock built up and then collapsed (Chrétien & Brousse 1989).

Explosively erupting volcanoes pose a serious hazard of **tephra ejection**. Blocks and bombs may fall in significant quantities within a 5 km radius of the eruptive centre. **Ashfall** is likely to be more widespread and can cover or bury farmland, ruin crops, clog sewers and machinery, cause excessive wear to mechanical parts, suffocate fauna and create a severe weight on shallow or flat roofs. One cm of ash may add 19 kg to the weight of each m^2 of roof, and the density of ash varies from 0.5 to 1.0 ton/m^3, depending on its wetness. Roofs need to be pitched at 20° so that the ash will slide off. Ash may kill herbivorous animals in several ways: by destroying their feedstock, poisoning them, clogging their intestines or wearing down their teeth. Such problems led to famine in eighteenth-century Iceland and nineteenth-century Indonesia. At an intercontinental scale the masking effect that ash has on sunlight (see example below) may reduce crop fertility and cause temporary changes in climate. Ashfall is measured either volumetrically or in terms of centimetre or metre of accumulation on the ground. The latter measure can be plotted as a map of isohyets, which are often related to the direction of wind or blast patterns. Deposited ash can suffer very rapid fluvial erosion, which may result in the clogging of water courses and which often coincides with intense rains provoked by volcanic particulates injected into the local atmosphere.

At Pompeii and nearby towns in Roman Campania, the eruptions of Mount Somma (the modern Vesuvius) in AD 79 killed about 16,000 people as a result of ash and gas emissions and blast and mudflow effects (Fig. 2.18; Sigurdsson et al. 1985).

Nuées ardentes ("glowing avalanches") involve the fluidization of a cloud of ash, dust and gas at 600°C. In the form of heavy **base surges**,

Figure 2.18 Plaster casts of some of the victims found outside the Nucerian Gate of Pompeii. Either they were asphyxiated or they suffered internal burns during the US 79 eruption of Vesuvius.

nuées may flow at up to 100 km/hr and for distances of up to 10 km. The Hopi type involves tephra collapsing onto the volcano's flanks, while the Peléan type involves explosive venting at low angle from under a collapsing volcanic dome. This violent and rapid phenomenon can pose a severe threat to life: 3,000 people died in a nuée at Mount Lamington, New Guinea, in 1951, and similar death tolls have been recorded in the Caribbean.

A larger, more directionally oriented form of nuée composed of suspended rock froth and gases is known as the **ash flow** or **pyroclastic flow**, and it can travel at 200 km/hr for distances of up to 25 km. Ash flows may also reach volumes of 10 km^3. Like nuées, they result from the expansion of trapped air and gases, which are released from hot pyroclastic debris, carrying suspended rock froth. Surprisingly, they are not necessarily connected with explosive volcanic activity. At Mount Katmai in Alaska in 1912 about 7 km^3 of tephra were erupted in pyroclastic flows, which covered some 750 km^2. However, the flows which occurred about 1,800 years ago at Lake Taupo in New Zealand were 20 times larger, burying 15,000 km^2 under an average of 1 m of deposits. Pyroclastic flows have travelled up to 35 km during historical eruptions and more than twice that during prehistoric ones.

Air pollution and associated risks caused by volcanic eruptions are not restricted to ash emissions. **Poisonous gases** may be emitted (for example, they are thought to have killed the elder Pliny on the beach at Pompeii in AD 79). The composition of gases emitted varies from one volcano – and from one eruption – to another. Water vapour emitted during eruptions may be accompanied by CO, CO_2, HF, H_2S, and SO which may be toxic to plants (including crops), animals and mankind. Thus in 1984 at Lake Manoun and in 1986 at the nearby Lake Nyos, both in Cameroon, large amounts of carbon dioxide were released, asphyxiating 37 and 1,887 people, respectively. The Nyos tragedy occurred at night and the gas collected in low-lying areas of ground as far as 20 km from the lake (Kling et al. 1987; see Ch. 8). Sulphur fumes emitted in 1946 by Masaya volcano in Nicaragua damaged 130 km^2 of coffee crops, killing about 6 million of the plants.

Large eruptions also cause **atmospheric modification** in the form of **volcanic aerosols** injected through the tropopause, which is situated at altitudes of 8–16 km. Thus the stratospheric cloud produced by the 1982 eruption of El Chichón in Mexico was rich in sulphur dioxide and was also one of the largest of the century (Hofmann 1987). In this context, suspended particulates can orbit the globe in the general atmospheric circulation for months or even years. An increase in the **stratospheric aerosol** reduces the amount of solar radiation reaching the ground by absorbing, reflecting or scattering some of it. Small average temperature decreases (e.g. 0.3–1.0°C) caused by **solar masking** are thus a likely consequence of volcanic air pollution. The intensity of solar masking depends on the height of the dust veil and size and chemical composition of both the dust particles and the aerosol in general. The latter may be a complex of several different layers.

Using a Javanese word, volcanic mudflows are called **lahars**, and are classified as **primary** when they result directly from eruption and **secondary** if there are other causes. They can occur at any time before, during or after an eruption and may consist of hot or cold material. Lahars can be both large and repetitive: for instance, at Mount Rainier in Washington State a mudflow 110 km long consists of a 500 year-old deposit 150 million m^3 in volume superimposed on an older deposit of similar volume (Crandell 1971). The rate of flow may reach 50 km/hr on relatively low slopes and with viscous material, or twice that down steep gradients with low-viscosity debris. In one extreme case, the 1877 eruption of Cotopaxi Volcano in Ecuador caused lahars that travelled 300 km in 17 hours. In New Zealand lahars transported a 37-tonne rock 57 km, while at Mount Kelut in Java in 1919 lahars killed 5,000 people. In fact, lahars are a serious and increasing problem throughout the volcanic zones of southeast Asia: 1,200 people were killed by them in the 1814 eruption of the Philippine volcano Mayon, but the population of the zone in question has now risen to 800,000, and therefore so has the risk of fatalities. A description of the lethal Nevado del Ruiz lahars of 1985 is given below.

The causes of volcanic mudflows generally involve the spontaneous release of a large amount of water during some cataclysmic event. During an eruption, for example, water may suddenly be ejected from a crater lake, or a nuée ardente may flow into a stream at the foot of the volcano. Alternatively, explosions may launch rock avalanches into streams, causing the water to be impounded and the debris dam eventually to breach. In similar fashion, brecciated lava may flow into streams, across wet ground or across snow and ice, or material extruded in blocks may become saturated with surface water. In rather looser association with eruptions, ash deposits can be saturated by torrential rain induced by volcanic particulates suspended in the atmosphere and functioning as condensation nucleii. Finally, several phenomena that induce lahars can occur either with or without eruption. Thus, water-saturated material may be induced to flow by earthquakes, geothermal heat may spontaneously melt ice and snow on the flanks of a high volcano, or water may suddenly be released as the wall of a crater lake collapses.

The lahar warning system installed at Mount St Helens by the US Geological Survey after the 1980 eruptions consists of one stream gauge and six lake gauges that telemeter an alert signal if they detect a rapid change in water level (such as the massive failure of a natural dam). Three seismometers can detect the vibrations caused by rockfalls and avalanches that are capable of initiating mudflows (Brantley & Power 1985).

As described in the previous section (and in Ch. 9 for the case of Krakatoa), eruptions can cause **tsunamis**. Finally, there is a danger in some volcanic areas of duplication of the hazard (or alteration of its geography) if the eruptive centre is in the habit of shifting, as tends to occur in large calderas. New courses are created for lava flows and lahars, and overall there is a changing balance of geographical and geomorphological problems. For example, the 1783 eruption of Laki, in Iceland, gradually emitted 12 km^3 of basalts from a fissure 25 km long, leading to complex invasions of 565 km^2 of the surrounding land (Thorarinsson 1970).

By 1980 records and dates of 5,564 Holocene eruptions had been accumulated with respect to 1,343 volcanoes, 529 of which erupted in historical times. Regarding the larger total, falls and explosions of tephra occurred in about 60 per cent of eruptions, seismic activity and ground deformation occurred in 50 per cent, lava flows in 24.1 per cent, 8.8 per cent were phreatomagmatic (explosive venting of steam and magma), 5.9 per cent occurred under the sea, 5.6 per cent generated lahars, and 5.1 per cent caused pyroclastic flows (Simkin et al. 1981).

Prediction of volcanic activity

Various geophysical precursors give grounds for arguing that there are excellent prospects for predicting eruptions when the activity of a volcano is intensively monitored (see Fig. 2.19). However, because of cost and accessibility problems, monitoring is intensive at only about a dozen of the world's volcanoes, which are thus the only ones for which full-scale volcanic forecasts are likely. Even there, it can be very difficult to distinguish between an incipient or future surface eruption and a magmatic intrusion (that is, an eruption that does not reach the surface): the premonitory signs are similar, yet the human impact is entirely different. But once an eruption becomes likely, it may be as important to predict its climax as to forecast its inception.

Upwelling magma that fills and stresses underground magma chambers may give rise to **earthquake swarms**. The earthquakes are called **A-type** if they have hypocentres 1–10 km deep, **B-type** if the foci are more than 10 km deep and are unclassified if they result from explosions in or just underneath the crater. They may reach magnitude 5 and frequencies of 700 per day, as detected by seismograph. One characteristic form of volcano-seismicity is **harmonic tremor**, a narrow band of nearly continuous seismic vibration dominated by a single frequency. This is usually associated with the rise within the Earth's outer crust of magma or volcanically heated fluids. The period of earthquake precursors is, however, difficult to determine, as it may last from a few days to a year or

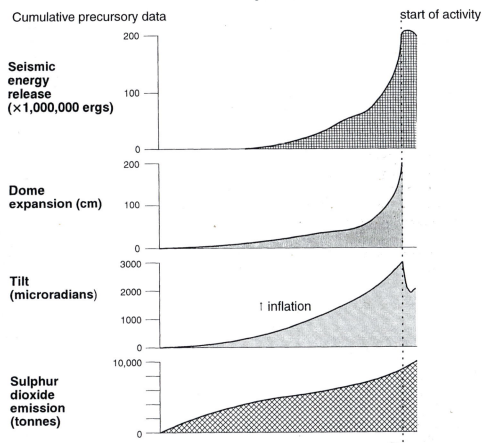

Figure 2.19 Increase in precursory activity before an eruption (Tilling 1989: 247). Copyright 1989 by the American Geophysical Union.

more. Moreover, eruption does not always follow enhanced volcano-seismic activity as the magma can remain underground.

Monitoring requires that seismic stations and geophones located on the flanks of the volcano are organized into networks. Thus the 14 December 1990 eruption of Redoubt Volcano, Alaska, was forecast when intensifying swarms of long-period (volcano)-seisms were detected. Five short-period seismometers were placed within a 21-km radius of the volcano and connected to a desktop computer, which was programmed to detect the pattern of seismicity associated with events. Real-time averages of the seismic signal were obtained from each seismometer every ten minutes. The largest volcano-seismic swarm, representing the main pre-eruptive phase, was detected and monitored 23 hours before the first large explosive eruption. Slow-scan TV systems were then used to monitor the eruption, but the short span of daylight and poor weather conditions prevented this method from reaching its full potential (Alaska Volcano Observatory Staff 1990).

During the eight months preceding the 1910 eruption of Mount Usu in Japan, 2 km² of land was elevated 150 m, and it went on to subside 36 m after the event. This rather extreme example illustrates that **tiltmeters** can be used to detect distortion of the ground caused by filling of the underlying chamber or vent with gas and magma, which can often reach 1–10 cm per day. At Rabaul Caldera in Papua, New Guinea, ground deformation is monitored using tide gauges, which measure change in the level of lake waters (McKee et al. 1984).

Gas is emitted from **fumaroles** on the flanks of the volcano (Fig. 2.20) and monitoring may detect increases in HCl, HF and SO_2 emissions, or ratios such as S/Cl. These may be correlated with impending eruption, as the proximity of molten magma tends to dissolve and emit more sulphur and fluorine. Moreover, increases in the dissolution of acid volcanic gases in the hydrothermal system often result in small decreases in pH level. In addition, lake water chemistry may alter in other ways: for example, at

Figure 2.20 Fumaroles in an active volcanic crater, the Solfatara at Pozzuoli, near Naples in southern Italy. As an eruption may be preceded by changes in heat and gas emissions, fumaroles such as these can be monitored for prediction purposes.

Ruapehu Volcano in New Zealand, the magnesium chlorine (Mg/Cl) ratio increased by 25–33 per cent before eruption.

As a volcano prepares to erupt, it emits more heat. **Thermal anomalies** can be detected in the ground and in hot springs, crater lakes and fumaroles (where gas temperatures may increase four to six fold). They can be detected by ground-based infra-red telescopes or airborne and satellite-based sensors, although ground instruments are limited in scope, satellite images tend to have relatively low resolution and aerial IR survey tends to be too expensive to carry out on a routine basis. Heating agitates rock to the point at which it loses its magnetic signature and its gravitational properties and electrical conductivity alter. Hence, **magnetic, gravitational and electrical anomalies** can be regarded as possible precursors of eruptions, and gravity metres, for instance, can be installed on the flanks of the volcano.

Finally, at Kilauea on Hawaii Island increases of 70 per cent have been detected in the Earth's weak natural electrical current, or **self-potential**. Using a controlled-source electromagnetic sensor or very-low frequency electrical sensor, magmatic intrusions can sometimes be detected as they begin.

For predictions to be successful they must allow sufficient lead time to complete evacuation and other safety measures. They must also be reliable and must define a short "time-window", or period of application, or else confidence in them will progressively erode.

Enhancing benefits and reducing drawbacks of volcanic zones
Of the world's population 10 per cent live on or near potentially active volcanoes, at least 91 of which are high-risk areas (42 in southeast Asia and the western Pacific, 42 in the Americas and 7 in Europe and Africa). The eruption, or precarious conditions which it creates, may continue for months, and therefore volcanic emergencies are often remarkably long-lasting in comparison with other sudden-impact natural disasters. Hence, during the 1980s Indonesia successfully evacuated more than 140,000 residents from the flanks of its volcanoes, but not without severe economic consequences (Suryo & Clarke 1985). With the rarity of eruptions of most of the world's volcanoes, it can be difficult to maintain a balanced perspective on volcanic risk.

Despite this, volcanic forecasting is likely to enjoy a bright future and, while the hazard should not be neglected, the benefits of volcanic zones are notable. Volcanic ash retains soil moisture in dry climates and releases nutrients to plants quickly and easily. Geothermal energy associated with volcanic activity is a cheap and useful source of electrical power. Water supplies are often good in permeable volcanic terrains and, lastly, there is usually abundant scope for tourism in picturesque or spectacular volcanic environments, which often have fine climates and high amenity value.

Losses caused by volcanic eruptions can be reduced by measures very similar to those employed for other forms of hazard, namely a combination of prediction, preparedness and land-use control. Appropriate structural measures for hazard reduction are not numerous, but include building steeply pitched, reinforced roofs that are unlikely to be damaged by ashfall, and constructing walls and channels to deflect lava flows. Hazard mapping and zoning, insurance, local taxation and evacuation plans are appropriate non-structural measures to put into effect. But the geological timescale on which volcanoes erupt may be very different from the human scale on which risk is conceived. Hence, unless people are shaken into awareness, a *laissez faire* attitude may prevail, especially as many volcanoes located in hazardous areas have no more than a 2 per cent chance of erupting in a 100-year period.

As with earthquakes, so for volcanic eruptions false predictions and unnecessary evacuations cannot always be avoided. Nevertheless, risk management depends critically on identifying the hazard zones and detecting incipient eruptions. In the long term, appropriate controls should be placed on land use and the location of settlements. The nature and location of volcanic risk must be assessed, and the deposits left by past eruptions must be dated and mapped precisely. This will show how active the volcano is and how its eruptions are likely to interact with the surrounding environment (for example, whether explosive blasts tend to be directional). Volcanic zoning tends to be inhibited by the fact that in general people do not want to make costly adjustments to a hazard that may have, say, only a 1 per cent chance of realizing itself in any 50-year period. Data on recurrence intervals are thus critical to zonation, and events which occurred in the early Quaternary and even the Tertiary periods are often significant to the understanding of current risk.

In the short term, most eruptions give some forewarning, although their more cataclysmic effects may occur with startling suddenness. Volcanologists and civil protection authorities must co-operate closely in the design of evacuation measures. The levels of risk must be defined and linked to appropriate social responses. Those who make only occasional use of volcanic hazard zones (such as loggers and holiday-makers) must be educated about the risks and public officials must be induced to respect the need to close certain zones to access during times of imminent danger. A quick-response plan for volcanic emergency management might involve three phases (UNDRO/UNESCO 1985):

(a) **alert**: civil protection services are mobilized 5–15 days ahead of the expected eruption;

(b) **readiness**: 2–5 days before the expected crisis, the sick, aged and very young are evacuated and the emergency services are put on stand-by for immediate action as required;

(c) **evacuation**: 1–2 days before the emergency a general evacuation is carried out (allowing sufficient time to complete it successfully).

The speed and efficiency with which phases (a) and (b) are carried out will affect the success of phase (c), especially as the most dangerous volcanic phenomena occur too quickly to be avoided if one is caught nearby.

Some case histories of recent volcanic eruptions will now be reviewed to illustrate the types of hazard and impact which may occur.

Mount St Helens volcanic eruption, 1980

Volcanism in the USA is restricted to the western part of the country, from New Mexico through Arizona and northeast California to Washington State, where several volcanoes of the Cascades Range are active, and finally through western Alaska. Crater Lake, Oregon, erupted 7,000 years ago and covered an area equivalent to four states with a 15 cm layer of ash. Lassen Peak in California erupted in 1914 and 1915, while Mount Baker in Washington State erupted in 1843, 1854, 1858 and 1870. In Alaska Mount Pavlov erupted in 1912, 1950 and 1973, Mount Katmai erupted in 1912 and Mount Redoubt erupted in 1989. Evidence of very powerful prehistoric eruptions is strong at Mount Hood in Oregon and Mount Shasta in northern California. In Washington, Mount Rainier shows evidence of 12 large ash eruptions and 55 mudflows: about 5,000 years ago one of the latter moved 2.1 km^3 of material, while a mudflow deposit dated to 500 years ago is located where 3,000 people now live. In the Sierra Nevada of California, Mammoth Lake has recently undergone apparent filling of its magma chamber, but this has not yet resulted in an eruption.

The eruption of Mount St Helens, in Washington State, on 18 May 1980 must surely rank as the most intensively studied volcanic event ever (Lipman & Mullineaux 1982). Fifty-seven people lost their lives, while damage to forests cost $450 million, to property $103 million and to agriculture $39 million. Clean-up costs were $363 million, but most of these sums were absorbed by the Federal Government without a significant impact on the national economy.

A minor eruption in March 1980 proved that Mount St Helens was active (it had last erupted in 1843) and this was succeeded by two months of earthquakes. A fracture 5 km long appeared in the surface of the volcano and the mountain side bulged at up to 2 m per day until it was displaced 100 m. Finally, at 8.32 hours on 18 May 1980, the northeast side of the volcano gave way in an avalanche of 2.7 km^3 of debris, which initiated a lateral blast that eventually extended 20 km outwards from the summit (Fig. 2.21). Pyroclastic flows reached 540 km/hr and were still at more than 700°C on the day after their occurrence. Avalanches breached a crater lake (Spirit Lake) and sent lahars of debris into the surrounding streams, the Toutle and Cowlitz Rivers, and eventually into the Columbia River (Cummans 1981).

The eruption was Plinian in style, and a vigorous ash column rose for nine hours 20 km into the stratosphere, reaching a height of 27,000m after only 30 minutes and ejecting more than 1.1 km³ of uncompacted tephra (0.25 km³ of magma-equivalent) into the atmosphere in only nine

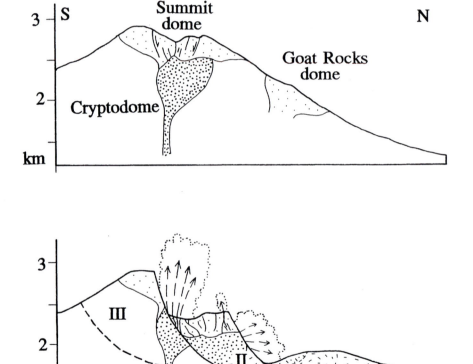

Figure 2.21 Mount St Helens eruption of 18 May 1980 (Moore & Rice 1984: 134). Reprinted with permission of the publishers, National Academy Press.

hours. The cloud travelled 400 km downwind during the first six hours, and eventually ash deposits were 7 cm thick at a distance of 40 km from the volcano. Ash fall-out during and after the eruption contained 3–7 per cent of free silica (SiO_2), which if inhaled over long periods is sufficient to cause silicosis of the lungs. At peaks of 13,800–35,800 $\mu g/m^3$, there were periods when total suspended particulates considerably exceeded danger levels (the warning level was set at 625 $\mu g/m^3$ and the damage level at 1,000 $\mu g/m^3$). Sulphurous steam emission rose 3,000–5,000 m above the surface and lasted for months.

During the eruption total darkness occurred within 390 km of the volcano, and migrating dust was later detected by radio balloon over Europe. Ash borne on dominant winds fell out over a large proportion of the Columbia River Basin, an area of important irrigated and dryland agriculture (Fig. 2.22). Ash fell both as dry powder and with rain as wet slurry. Fruit, hay, cereal grains and potatoes were adversely affected, the last two of these crops showing the worst impact (Cook et al. 1981). Some insects desiccate when their waxy layers are abraded, while others suffer scouring damage. Hence, many insects died when they came into contact with minute abrasive particles of ash: flies, cockroaches and bees were worst affected, moths and mealworms less so. Although crop pests such as the grasshopper and Colorado beetle were killed, bee populations were

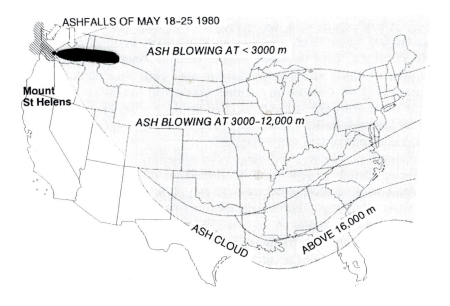

Figure 2.22 Distribution of ash erupted by Mount St Helens in May 1980 (Hays 1981: B86).

greatly reduced, honey production dropped and there were few pollenator bees. Plants with hairy leaves took a year to rid themselves of ash; the growth rate of plants was reduced in general and overall the ecological balance was altered by the eruption. Farming proved difficult where several centimetres of ash had accumulated on fields. The grain harvest was partially contaminated and abrasive ash caused several years' wear on agricultural machinery in only one season.

The governor of Washington State and the US Forest Service were both attentive to the results of volcanic monitoring carried out by the US Geological Survey. Continued seismic activity in March, April and early May 1980 showed that Mount St Helens remained active, but there was no way to predict the 18 May eruption to the day. Fortunately, the authorities did not give in to incessant demands from the public and logging industry to reopen the volcano to general access. Instead, they closed the central and northeastern sectors and restricted access to areas peripheral to these zones. This undoubtably reduced the eventual death toll, and most of the 62 people known to have died in the eruption did not have authorization to be in the area.

Nevado del Ruiz volcanic eruption, 1985
Nevado del Ruiz in Caldas State, Colombia, is the northernmost active volcano of the Andes of South America and rises to 5,389 m above sea level. It has experienced major eruptions at least ten times during the past 10,000 years, with an average recurrence interval in the range 160–400 years. One thousand people died in lahars produced by the eruption of 1845, and serious damage was done by eruption in 1916. Early in 1985 the Colombian National Geological Institute (INDEOMINAS) began monitoring the volcano, assisted by the US Geological Survey and the UN Disaster Relief Commission (UNDRO). Three geophones and four portable seismometers were installed.

Premonitory activity, including volcano-seismicity and increased gas emission, lasted a year and intensified in September 1985. These phenomena were monitored from 20 July 1985 until, on 11 September that year, a phreatic eruption occurred, accompanied by ground shaking, avalanches and rockfalls. A lake of 250,000 m^3 of water was impounded by the debris and was recognized as a hazard to the town of Armero, down stream. Electronic ground-tilt measurements began in October 1985 and hazard evaluation and mapping were completed by the second week of that month (Fig. 2.23). Upon advice from the volcanologists, the mayor of Armero decided to request facilities for evacuation of the inhabitants at this time. His demands went unheeded by the state authorities.

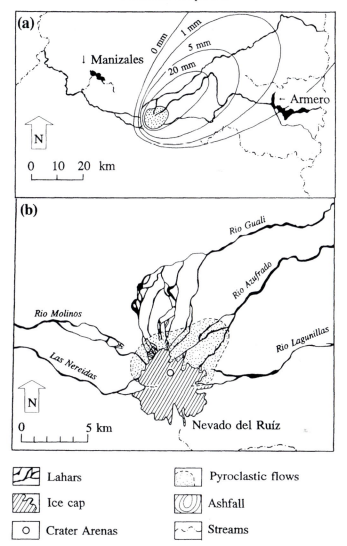

Figure 2.23 Volcanic hazards map for Nevado del Ruiz, Colombia (Herd et al. 1986: 459). Copyright 1986 by the American Geophysical Union.

The eruption of 13 November 1985 was the second worst volcanic disaster of the century (the worst being that of Mount Pelée, Martinique, in 1902, as mentioned above). It began in mid-evening with a strong phreatic eruption that gave little advance warning and culminated at 21.08 hours in the paroxysmal phase. Within 20 minutes a Plinian tephra column had started to drop flaming blocks over a radius of 11 km from the summit. Over the course of an hour andesite and dacite (58–65 per cent SiO_2) were erupted in pumiceous pyroclastic surges and flows, one of

109

which was welded. As it lasted only one tenth as long, the 1985 eruption of Ruiz ejected only about 3 per cent of the ash produced by Mount St Helens in May 1980 (0.039 km^3 against 1.1 km^3). But as about 10 per cent of the volcano's ice cap melted, this relatively small quantity of ejecta was able to generate 0.06 km^3 of lahar deposits from 0.01–0.02 km^3 of meltwater (Naranjo et al. 1986).

Figure 2.24 shows the distribution of pyroclastic flows and lahars caused by the eruption. The Rio Chichiná lahar flowed at more than 22 km/hr, that on the Rio Guali moved at 28 km/hr, while the Rio Azufrado-Rio Lagunillas one flowed at 38 km/hr. Peak flow in the last of these reached 48,000 m^3/sec, which is equivalent to the discharge of the Vajont Dam overflow (see Ch. 5). At about 23.30 hours the lahar disgorged from the canyon up stream of Armero as a wave 40 m high, which flooded through the town at about 30 km/hr subsiding to 2–5 m deep. Three major pulses and two minor ones occurred over a two-hour period. About 21,000 people were killed, of whom between 1,000 and

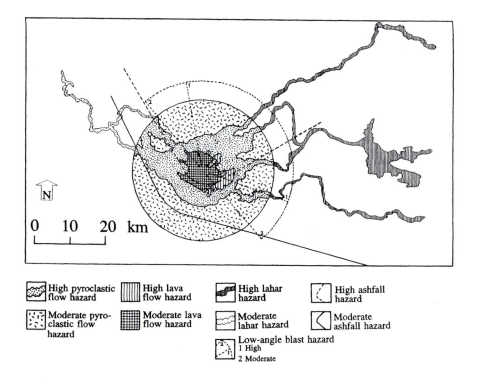

Figure 2.24 Lahars and pyroclastic flows produced by the 13 November 1985 eruption of Nevado del Ruiz (Herd et al. 1986: 460, copyright 1986 by the American Geophysical Union; Voight 1990: 170).

2,000 died slowly while trapped in the mud, from which only 65 were rescued alive. In addition, the Chinchiná lahar caused approximately 1,100 fatalities and destroyed 200 houses and 3 bridges. Homes were also destroyed near the villages of Mariquita by the Rio Guali lahar (Lowe et al. 1986).

In addition to more than 22,000 people killed, the eruption caused over $212 million of property damage. In all, the destruction encompassed 5,150 homes, 50 schools, 2 hospitals, 200 km of roads, 18 km of railway lines, 10 km of electricity transmission lines, 58 factories, 343 commercial enterprises, 3 oil pipelines, 5,000 ha of the best agricultural land, and water supplies to 50,000 people. About 7,700 people were left homeless.

From the human rather than the volcanic point of view, the disaster was effectively caused by weak leadership and lack of decisive action. Decisions were neither taken nor implemented rapidly enough and foresight was lacking. Although in general terms the risks were known in advance, in retrospect the decision not to evacuate until the moment that danger was fully recognized proved fatal. The authorities were unwilling to assume responsibility for the economic and political costs of premature or unnecessary evacuation.

It was widely believed on the basis of the October 1985 hazard map that Armero would have two hours to evacuate before dangerous lahars struck. Yet the map made it clear that no safe place existed within 1 km of the town. In fact, the experience of eruptions occurring after the 1985 disaster indicated that evacuation would not even *begin* until 4–7 hours after the danger had first been recognized. In the end, local civil defence officials were alerted by late afternoon on 13 November, but it seemed that abnormal conditions ended by 19.30 hours, thus reducing the impetus for action. Hence, evacuation, if it had ever been ordered, came too late, and attempts to warn Armero by civil defence radio 15 minutes after the eruption started were unsuccessful.

Emergency plans had been developed separately in the provinces of Caldas, Tolima, Risaralda and Quindo. Thus, the success of emergency responses varied between these jurisdictions and co-operation between them was inadequate. However, volcanic awareness programmes had been initiated in the schools of Caldas, and civil defence authorities were actively engaged in assessing the vulnerability of the populations of valleys leading from the volcano.

As usual, there was a broad differential between the costs of prior mitigation and those of subsequent relief. Had only part of the huge sums spent on helicopter operations after the event been spent on volcanological monitoring before it, evacuation would probably have been made more feasible. This conclusion is highlighted by the danger of helicopter missions during volcanic disturbances of the atmosphere: at Ruiz five people died and three helicopters were destroyed in crashes.

Earthquakes and volcanoes

Only slight increases in seismic energy release were observed before the 13 November magmatic eruption, in net contrast to the phreatic eruption of the previous September. But despite this, scientific data were capable of establishing the risk of a major volcanic event. However, they might not have been capable of convincing the civil defence authorities and vulnerable residents. In this respect, the phreatic eruption of September 1985 made future volcanic events credible.

Whatever the civil authorities believed, the scientific effort undoubtedly had its deficiencies, including inadequate instrumentation, the slowness with which seismic data were processed (for a while they had to be mailed to Bogotá) and rivalry, rather than collaboration, between scientific groups. Hence neither the Colombian government nor foreign agencies reacted sufficiently to calls by American and Italian volcanologists to install more instruments, especially those with semi-automatic data processing and data telemetering capabilities. In fact, the report of an Italian team in October concluded that monitoring was seriously inadequate and unable to provide rapid warning of changing conditions.

One year after the event, the Colombian government made the area the site of the best permanent volcanic observatory in South America. (For an exhaustive analysis of the 1985 eruptions see Williams 1989–90.)

References

Abersten, L. 1984. Diversion of a lava flow from its natural bed to an artificial channel with the aid of explosives, Etna, 1983. *Bulletin Volcanologique* **47**, 1,165–77.

Adams, R. 1975. The Haicheng earthquake of 4 February 1975: the first successfully predicted major earthquake. *Earthquake Engineering and Structural Dynamics* **4**, 423–37.

Agnew, D. C. 1986. Strainmeters and tiltmeters. *Reviews of Geophysics* **24**, 579–624.

Alaska Volcano Observatory Staff 1990. The 1989–1990 eruption of Redoubt Volcano. *EOS: American Geophysical Union, Transactions* **1**, 265–75.

Alesch, D. J. & W. J. Petak 1986. *The politics and economics of earthquake hazard mitigation*. Boulder, Colorado: Natural Hazards Research and Applications Information Center.

Anderson, W. A. 1970. Tsunami warning in Crescent City, California, and Hilo, Hawaii. In *The great Alaska earthquake of 1964: human ecology*, Committee on the Alaska Earthquake (ed.), 116–24. Washington, DC: National Academy of Sciences.

Bakun, W. H. et al. 1988. The Parkfield earthquake prediction experiment in central California. *Earthquakes and Volcanoes* **20**, 41–91.

Bernard, E. N. 1991. Assessment of Project THRUST: past, present, future. *Natural Hazards* **4**, 285–92.

Blong, R. J. 1984. *Volcanic hazards: a sourcebook on the effects of eruptions*. Orlando, Florida: Academic Press.

References

Blundell, D. J. 1977. Living with earthquakes. *Disasters* **1**, 41–6.

Bolt, B. A. 1988. *Earthquakes*. San Francisco: W. H. Freeman.

Boudon, G. & A. Gourgaud (eds) 1989. Mont Pelée. *Journal of Volcanology and Geothermal Research* **38**, 1–213.

Brantley, S. & J. Power 1985. Lahars (reports from the US Geological Survey's Cascades Volcano Observatory). *Earthquake Information Bulletin* **17**, 20–2.

Brown, R. D., Jr, P. L. Ward, G. Plafker 1973. *Geologic and seismologic aspects of the Managua, Nicaragua earthquakes of December 23, 1972*. USGS Professional Paper 838.

Carey, S. & H. Sigurdsson 1989. The intensity of Plinian eruptions. *Bulletin of Volcanology* **51**, 28–40.

Cheng Yong, Kam-ling Tsoi, Chen Feibi, Gao Zhenhuan, Zou Qijia, Chen Zhangli (eds) 1988. *The great Tangshan earthquake of 1976: an anatomy of disaster*. Oxford: Pergamon.

Chrétien, S. & R. Brousse 1989. Events preceding the great eruption of 8 May, 1902, at Mount Pelée, Martinique. *Journal of Volcanology and Geothermal Research* **38**, 77–95.

Clark, H. E., Jr. 1987. *Tsunami alerting systems*. Albuquerque, New Mexico: Albuquerque Seismological Laboratory.

Cochrane, H., J. E. Haas, M. Bowden, R. W. Kates 1974. *Social science perspectives on the coming San Francisco earthquake: economic impact, prediction, and construction*. Boulder, Colorado: Natural Hazards Research and Applications Information Centre.

Cook, R. J. et al. 1981. Impact on agriculture of the Mount St Helens eruption. *Science* **211**, 16–22.

Cooke, R. U. & J. C. Doornkamp 1974. *Geomorphology in environmental management*. Oxford: Clarendon Press.

Costa, J. E. & V. R. Baker 1981. *Surficial geology: building with the earth*. New York: John Wiley.

Cox, D. C. & J. Morgan 1977. *Local tsunamis and possible local tsunamis in Hawaii*. Honolulu, Hawaii: Hawaii Institute of Geophysics.

Cox, D. C. & J. Morgan 1984. *Local tsunamis in Hawaii: implications for warning*. Honolulu, Hawaii: Hawaii Institute of Geophysics.

Crandell, D. R. 1971. *Post-glacial lahars from Mount Rainier volcano, Washington*. USGS Professional Paper 677.

Cummans, J. 1981. *Mudflows resulting from the May 18, 1980, eruption of Mount St Helens, Washington*. USGS Circular 850B.

Dohler, G. C. 1988. A general outline of the ITSU Master Plan for the tsunami warning system in the Pacific. *Natural Hazards* **1**, 295–302.

Eckel, E. B. 1970. *The Alaskan earthquake, March 27, 1964: lessons and conclusions*. USGS Professional Paper 546.

Foster, H. D. 1980. *Disaster planning: the preservation of life and property*. New York: Springer.

Foster, H. D. & V. Wuorinen 1976. British Columbia's tsunami warning system: an evaluation. *Syesis* **9**, 113–22.

Francis, P. 1976. *Volcanoes*. Harmondsworth: Penguin.

Gudmundsson, A. 1988. Formation of collapse calderas. *Geology* **16**, 808–10.

Gupta, H. K. & J. Combs 1976. Continued seismic activity at the Koyna Reservoir site, India. *Engineering Geology* **10**, 307–13.

Haas, J. E. & D. S. Mileti 1976. *Socio-economic impact of earthquake prediction on government, business, and community*. Boulder, Colorado: Institute of Behavioural Science, University of Colorado.

Haicheng Earthquake Study Delegation 1977. Prediction of the Haicheng earthquake, China. *EOS: American Geophysical Union, Transactions* **58**, 236–72.

Hays, W. W. (ed.) 1981. Facing geologic and hydrologic hazards: earth science perspectives. USGS Professional Paper 1240B, 1–108.

Healy, J. H., W. W. Rubey, D. T. Griggs, C. B. Raleigh 1968. The Denver earthquakes. *Science* **161**, 1,301–10.

Heaton, T. H. 1985. A model for a seismic computerized alert network. *Science* **228**, 987–90.

Herd, D. G. & Comité de Estudios Vulcanológicos 1986. The 1985 Ruíz Volcano disaster. *EOS: American Geophysical Union, Transactions* **67**, 457–60.

Hofmann, D. J. 1987. Perturbations to the global atmosphere associated with the El Chichón volcanic eruption of 1982. *Reviews of Geophysics* **25**, 743–59.

Judd, W. R. (ed.) 1974. Seismic effects of reservoir impounding. *Engineering Geology* **8**, 1–212.

Kates, R. W., J. E. Haas, D. J. Amarel, R. A. Olson, R. Ramos, R. Olson 1973. Human impact of the Managua earthquake. *Science* **182**, 981–90.

Kling, G. W., M. A. Clark, H. R. Compton et al. 1987. The Lake Nyos gas disaster in Cameroon, West Africa. *Science* **236**, 169–75.

Kulikov, E. A., A. B. Rabinovich, A. I. Spirin, S. L. Poole, V. Soloviev 1983. Measurement of tsunamis in the open ocean. *Marine Geodesy* **6**, 311–29.

Lachman, R., M. Tasuoka, W. J. Bonk 1961. Human behaviour during the tsunami of May 1960. *Science* **133**, 1,405–9.

Langbein, J. O., M. F. Linker, A. McGarr, L. E. Slater 1982. Observations of strain accumulation across the San Andreas fault near Palmdale, California, using a two-colour geodimeter. *Science* **218**, 1,217–19.

Lipman, P. W. & D. R. Mullineaux 1982. *The 1980 eruptions of Mount St Helens, Washington.* USGS *Professional Paper* 1250.

Lockwood, J. P. & F. A. Torgerson 1980. Diversion of lava flows by aerial bombardment: lessons from Mauna Loa Volcano, Hawaii. *Bulletin Volcanologique* **43**, 727–41.

Lorca, E. 1991. Integration of the THRUST project into the Chile tsunami warning system. *Natural Hazards* **4**, 293–300.

Lowe, D. R., S. N. Williams, H. Leigh 1986. Lahars initiated by the 13 November 1985 eruption of Nevado del Ruiz, Colombia. *Nature* **324**, 51–3.

MacDonald, G. A. 1972. *Volcanoes.* Englewood Cliffs, New Jersey: Prentice-Hall.

McKee, C. O., P. L. Lowenstein, P. de St Ours et al. 1984. Seismic and ground deformation crises at Rabaul caldera: prelude to an eruption? *Bulletin Volcanologique* **47**, 397–411.

McKeown, F. A. & L. C. Pakiser (eds) 1982. *Investigations of the New Madrid, Missouri, earthquake zone.* USGS Professional Paper 1236.

McNutt, S. R. & R. H. Sydnor 1990. *The Loma Prieta (Santa Cruz Mountains), California, earthquake of 17 October 1989.* Sacramento, California: Department of Mines and Conservation.

Mitchell, W. A. 1976. Reconstruction after disaster: the Gediz earthquake of 1970. *Geographical Review* **66**, 296–313.

Mjachkin, V. I., B. V. Kostrov, G. A. Sobolev, O. G. Shamina 1984. The physics of rock failure and its links with earthquakes. In *Earthquake prediction*, T. Rikitake (ed.), 319–42. Tokyo: Terra Scientific Publishing & Paris: UNESCO Press.

Moore, J. G. & C. J. Rice 1984. Chronology and character of the May 18, 1980, explosive eruptions of Mount St Helens. In *Explosive volcanism: inception, evolution, and hazards*. Geophysics Study Committee, National Academy of Sciences (ed.), 133–42. Washington, DC: National Academy Press.

Nakamura, Y. & B. E. Tucker 1988. Earthquake warning system for Japan Railways' bullet train: implications for disaster prevention in California. *Earthquakes and Volcanoes* **20**, 140–55.

Naranjo, J. L., H. Sigurdsson, S. N. Carey, W. Fritz 1986. Eruption of the Nevado del Ruiz Volcano, Colombia, on 13 November 1985: tephra fall and lahars. *Science* **233**, 961–3.

Noji, E. K. 1989. The 1988 earthquake in Soviet Armenia: implications for earthquake preparedness. *Disasters* **13**, 255–62.

Nuttli, O. W., G. A. Bollinger, R. B., Herrmann 1986. The 1886 Charleston, South Carolina, earthquake: a 1986 perspective. *USGS Circular* 985.

Office of Emergency Preparedness 1972. *Disaster preparedness*, Vol. 1. Washington, DC: Executive Office of the President, US Government Printing Office.

Ollier, C. D. 1969. *Volcanoes*. Canberra: Australian National University Press.

Olson, R. S. & R. A. Olson 1987. Urban heavy rescue. *Earthquake Spectra* **3**, 645–58.

Pararas-Carayannis, G. 1986. The Pacific tsunami warning system. *Earthquakes and Volcanoes* **18**, 122–30.

Plafker, G. & J. P. Galloway (eds) 1989. *Lessons learned from the Loma Prieta, California, earthquake of October 17, 1989*. USGS Circular 1045.

Postpischl, D. 1985. *Atlas of isoseismal maps of Italian earthquakes*. Rome: Consiglio Nazionale delle Ricerche, Progetto Finalizzato "Geodinamica".

Raleigh, C. B., J. H. Healy, J. D. Bredehoeft 1976. An experiment in earthquake control at Rangely, Colorado. *Science* **191**, 1,230–7.

Raleigh, C. B., K. E. Sieh, L. R. Sykes, D. L. Anderson 1982. Forecasting southern California earthquakes. *Science* **217**, 1,097–104.

Richter, C. F. 1958. *Elementary seismology*. San Francisco: W. H. Freeman.

Rikitake, T. 1979. Classification of earthquake precursors. *Tectonophysics* **54**, 293–309.

Rikitake, T. 1984. Earthquake precursors. In *Earthquake prediction*. T. Rikitake (ed.), 3–20. Tokyo: Terra Scientific Publishing & Paris: UNESCO Press.

Savarenskij, E. F. & I. L. Neresov 1978. Earthquake prediction. In *The assessment and mitigation of earthquake risk*, 66–90. Paris: UNESCO Press.

Scholz, C. H., L. R. Sykes, Y. P. Aggarwal 1973. Earthquake prediction: a physical basis. *Science* **181**, 803–10.

Segerstrom, K. 1966. *Paricutin, 1965: aftermath of eruption*. USGS Professional Paper 550C.

Sigurdsson, H., S. Carey, W. Cornell, T. Pescatore 1985. The eruption of Vesuvius in AD 79. *National Geographic Research* **1**, 332–87.

Simkin, T., L. Siebert, L. McClelland, D. Bridge, C. Newhall, J. Latter 1981. *Volcanoes of the world: a regional directory, gazetteer, and chronology of volcanism during the last 10,000 years*. Stroudsburg, Pennsylvania: Hutchinson & Ross, for the Smithsonian Institution.

Simpson, D. W. 1976. Seismicity changes associated with reservoir loading. *Engineering Geology* **10**, 123–50.

Solbiati, R. & A. Marcellini 1983. *Terremoto e Società*. Milan: Garzanti.

Soloviev, S. L. 1978. Tsunamis. In *The assessment and mitigation of earthquake risk*. UNESCO, 91–143. Paris: UNESCO Press.

Soloviev, S. L. 1990. Tsunamigenic zones in the Mediterranean Sea. *Natural Hazards* **3**, 183–202.

Sparks, R. S. J. 1986. The dimensions and dynamics of volcanic eruption columns. *Bulletin of Volcanology* **48**, 3–15.

Steinbrugge, K. V. 1989. Reducing earthquake casualties: an overview. In *International workshop on earthquake injury epidemiology for mitigation and response*, 415–19. Baltimore: The Johns Hopkins University Press.

Suryo, I. & M. G. C. Clarke 1985. The occurrence and mitigation of volcanic hazards in Indonesia as exemplified at the Mount Merapi, Mount Kelut and Mount Galunggung volcanoes. *Quarterly Journal of Engineering Geology* **18**, 79–98.

Thorarinsson, S. 1970. The Lakigigar eruption of 1783. *Bulletin Volcanologique* **33**, 910–27.

Tilling, R. I. 1989. Volcanic hazards and their mitigation: progress and problems. *Reviews of Geophysics* **27**, 237–69.

undro/unesco 1985. *Volcanic emergency management.* New York: United Nations.

US Geological Survey (USGS) 1964. *The Hebgen Lake earthquake of August 17, 1959.* USGS Professional Paper 435.

US Geological Survey 1988. Probabilities of large earthquakes occurring in California on the San Andreas fault. *US Geological Survey Open File Report* 88–398, 162.

US Geological Survey 1989. Notes about the Armenia earthquake of 7 December 1988. *Earthquakes and Volcanoes* **21**, 68–78.

US National Academy of Sciences 1973. *The great Alaska earthquake of 1964* (8 vols). Washington, DC: National Academy of Sciences.

US National Research Council (NRC) 1975. *Earthquake prediction and public policy.* Washington, DC: National Academy Press.

US National Research Council 1976. *Predicting earthquakes: a scientific and technical evaluation with implications for society.* Washington, DC: National Academy Press.

Voight, B. 1990. The 1985 Nevado del Ruiz Volcano catastrophe: anatomy and retrospection. *Journal of Volcanology and Geothermal Research* **42**, 151–88.

Wallace, R. E. (ed.) 1990. *The San Andreas fault system.* US Geological Survey Professional Paper 1515.

White, G. F. & J. E. Haas 1975. *Assessment of research on natural hazards.* Cambridge, Mass.: MIT Press.

Williams, R. S. & J. G. Moore 1973. Iceland chills a lava flow. *Geotimes* **18**, 14–17.

Williams, S. N. (ed.) 1989–90. The November 13, 1985, eruption of Nevado del Ruiz Volcano, Colombia. *Journal of Volcanology and Geothermal Research* **40/ 41**.

Wyllie, L. A., Jr, & J. R. Filson 1989. Armenia earthquake reconnaissance report. *Earthquake Spectra* **5**.

Wyss, M. 1981. Recent earthquake prediction research in the United States. In *Current research in earthquake prediction*, T. Rikitake (ed.), 81–127. Dordrecht: D. Reidel.

Select bibliography

Adams, W. M. (ed.) 1970. *Tsunamis in the Pacific Ocean.* Honolulu, Hawaii: East–West Centre Press.

Select bibliography

Båth, M. 1973. *Introduction to seismology*. Berlin: Birkhäuser.

Berberian, M. 1978. Tabas-e-Golshan (Iran) catastrophic earthquake of 16 September 1978: a preliminary field report. *Disasters* **2**, 207–19.

Bolin, R. (ed.) 1990. *The Loma Prieta earthquake: studies of short-term impacts*. Boulder, Colorado: Programme on Environment and Behaviour, Institute of Behavioural Sciences, University of Colorado.

Boore, D. M. 1977. The motion of the ground during earthquakes. *Scientific American* **237**, 68–87.

Braddock, R. D. 1969. On tsunami propagation. *Journal of Geophysical Research* **74**, 1,952–7.

Bullard, F. M. 1976. *Volcanoes of the Earth*. Austin & London: University of Texas Press.

Casadevall, T. J. (ed.) 1991. First international symposium on volcanic ash and aviation safety. *US Geological Survey Circular* 1065.

Decker, R. W. 1986. Forecasting volcanic eruptions. *Annual Reviews of Earth and Planetary Science* **14**, 267–91.

Decker, R. W. & B. Decker 1981. *Volcanoes*. San Francisco: W. H. Freeman.

Decker, R. W., T. L. Wright, P. H. Stauffer (eds) 1987. *Volcanism in Hawaii*. US Geological Survey Professional Paper 1350.

Dudley, W. C. & M. Lee 1988. *Tsunami!* Honolulu: University of Hawaii Press.

Eiby, G. A. 1980. *Earthquakes*, 2nd edn. London: Heinemann.

Esteva, L. 1988. The Mexico City earthquake of September 19, 1985: consequences, lessons and impact on research and practice. *Earthquake Spectra* **4**, 413–26.

Fisher, R. V., A. L. Smith, M. J. Roobol 1980. Destruction of St Pierre, Martinique, by ash cloud surges, May 8 and 20, 1902. *Geology* **8**, 472–6.

Forrester, F. H. 1987. Tsunami. *Weatherwise* **40**, 84–90.

Fournier D'Albe, E. M. 1966. Earthquakes: avoidable disasters. *Impact of Science on Society* **16**, 189–202.

Geological Society of London 1979. Special issue on prediction of volcanic eruptions, *Geological Society of London, Journal* **136**, 321–60.

Geophysics Study Committee 1984. *Explosive volcanism: inception, evolution and hazards*. Washington, DC: National Research Council, National Academy Press.

Gere, J. M. & H. C. Shah 1984. *Terra non firma: understanding and preparing for earthquakes*. New York: W. H. Freeman.

Grandori, G., E. Guagenti, F. Perotti 1988. Alarm systems based on a pair of short-term earthquake precursors. *Seismological Society of America, Bulletin* **78**, 1,550–62.

Handler, P. 1989. The effect of volcanic aerosols on global climate. *Journal of Volcanology and Geothermal Research* **37**, 233–49.

Healy, J. H., W. W. Rubey, D. T. Griggs, C. B. Raleigh 1968. The Denver earthquakes. *Science* **161**, 1,301–10.

Hirose, H. 1985. Earthquake prediction in Japan and the United States. *International Journal of Mass Emergencies and Disasters* **3**, 51–66.

Hwang, L.-S. & D. Divoky 1970. Tsunami generation. *Journal of Geophysical Research* **75**(33), 6,802–17.

Iida, K. & T. Iwasaki (eds) 1983. *Tsunamis: their science and engineering*. Dordrecht: D. Reidel.

Isikara, A. M. & A. Vogel (eds) 1982. *Multidisciplinary approach to earthquake prediction*. Bristol: Adam Hilger.

Kasahara, K. 1981. *Earthquake mechanics*. Cambridge: Cambridge University Press.

Kowalik, Z. & T. S. Murty 1989. On some future tsunamis in the Pacific Ocean. *Natural Hazards* **1**, 349–70.

Krafft, K. & M. Krafft 1980. *Volcanoes: earth's awakening*. Maplewood, New Jersey: Hammond.

Kulhánek, O. 1990. *Anatomy of seismograms*. Amsterdam: Elsevier.

Lockridge, P. A. 1990. Nonseismic phenomena in the generation and augmentation of tsunamis. *Natural Hazards* **3**, 403–12.

Lomnitz, C. 1970. Casualties and behaviour of populations during earthquakes. *Seismological Society of America Bulletin* **60**, 1,309–13.

Lomnitz, C., 1974. *Global tectonics and earthquake risk*. Amsterdam: Elsevier.

Lomnitz, C. & E. Rosenblueth 1976. *Seismic risk and engineering decisions*. Amsterdam: Elsevier.

Loomis, H. G. 1978. Tsunamis. In *Geophysical predictions*, 155–65. Washington, DC: National Academy of Sciences.

Luhr, J. F. & J. C. Varekamp (eds) 1984. El Chichón volcano, Chiapas, Mexico. *Journal of Volcanology and Geothermal Research* **23**, 1–191.

Ma Zongjin, Fu Zhenxiang, Zhang Yingzhen, Wang Chengmin, Zhang Guomin, Liu Defu 1990. *Earthquake prediction*. New York: Springer.

Mileti, D. S., J. R. Hutton, J. H. Sorensen 1981. *Earthquake prediction response and options for public policy*. Boulder, Colorado: Natural Hazards Research and Applications Information Center.

Milne, W. G. (ed.) 1976. Induced seismicity. *Engineering Geology* **10**, 832–88.

Mogi, K. 1985. *Earthquake prediction*. Orlando, Florida: Academic Press.

Mooney, M. J. 1980. Tsunami. *Sea Frontiers* **26**, 130–7.

Ohta, Y. & H. Ohashi 1985. Field survey of occupant behaviour in an earthquake. *International Journal of Mass Emergencies and Disasters* **3**, 147–60.

Ollier, C. 1988. *Volcanoes*, 2nd edn. Oxford: Basil Blackwell.

Olson, R. S., B. Podesta, J. M. Nigg 1989. *The politics of earthquake prediction*. Princeton, New Jersey: Princeton University Press.

Pakiser, L. C., J. P. Eaton, J. H. Healy, C. B. Raleigh 1969. Earthquake prediction and control. *Science* **166**, 1,467–74.

Papoulia, J. E. & G. N. Stavrakakis 1990. Attenuation laws and seismic hazard assessment. *Natural Hazards* **3**, 49–58.

Pararas-Carayannis, G. 1980. The International Tsunami Warning System. *Sea Frontiers* **23**, 20–7.

Perry, R. W. & M. K. Lindell 1990. *Living with Mount St Helens: human adjustment to volcanic hazards*. Pullman, Washington: Washington State University Press.

Press, F. 1975. Earthquake prediction. *Scientific American* **232**, 14–23.

Rampino, M. R. & S. Self 1984. The atmospheric effects of El Chichón. *Scientific American* **253**, 48–57.

Reiter, L. 1991. *Earthquake hazard analysis*. Irvington, New York: Columbia University Press.

Rikitake, T. 1976. *Earthquake prediction*. Amsterdam: Elsevier.

Rikitake, T. 1981. *Current research in earthquake prediction*. Dordrecht: D. Reidel.

Rikitake, T. (ed.) 1983. *Earthquake forecasting and warning*. Dordrecht: D. Reidel.

Schaal, R. B. 1988. An evaluation of the animal-behaviour theory for earthquake prediction. *California Geology* **41**, 41–5.

Schick, R. 1988. Volcanic tremor-source mechanisms and correlation with eruptive activity. *Natural Hazards* **1**, 125–44.

Shah, H. C. 1984. Glossary of terms for probablistic seismic-risk and hazard analysis. *Earthquake Spectra* **1**, 33–40.

Sheets, P. D. & D. K. Grayson (eds) 1979. *Volcanic activity and human ecology*, Orlando, Florida: Academic Press.

Shuto, N. 1991. Numerical simulation of tsunamis: its present and near future. *Natural Hazards* **4**, 171–91.

Simkin, T. & R. S. Fiske 1983. Krakatau, 1883: the volcanic eruption and its effects. Washington, DC: Smithsonian Institution Press.

Simon, R. 1981. *Earthquake interpretations: a manual for reading seismograms.* Los Altos, California: William Kaufmann.

Simpson, D. W. & P. G. Richards (eds) 1981. *Earthquake predictions: an international review*. Washington, DC: American Geophysical Union.

Spence, W., S. A. Sipkin, G. L. Choy 1989. Measuring the size of an earthquake. *Earthquakes and Volcanoes* **21**, 58–63.

Stothers, R. B. 1984. The great eruption of Tambora and its aftermath. *Science* **224**, 1191–8.

Tazieff, H. & J-C. Sabroux (eds) 1983. *Forecasting volcanic events*. Amsterdam: Elsevier.

UNDRO/UNESCO 1976. *Disaster prevention and mitigation*. Vol. 1, *volcanological aspects*. New York: United Nations.

UNESCO, 1978. *The assessment and mitigation of earthquake risk*. Paris: UNESCO Press.

US Geological Survey 1982. Magnitude and intensity: measures of earthquake size and severity. *Earthquake Information Bulletin* **14**, 209–19.

US National Research Council 1991. *Real-time earthquake monitoring*. Washington, DC: National Academy Press.

Van Doorn, W. G. 1965. Tsunamis. *Advances in Hydroscience* **2**, 1–48.

Wesson, R. L. & R. E. Wallace 1985. Predicting the next great earthquake in California. *Scientific American* **252**, 35–43.

Williams, H. & A. R. McBirney 1979. *Volcanology*. San Francisco: Freeman, Cooper.

Wright, T. L. & T. C. Pierson 1992. *Living with volcanoes*. US Geological Survey Circular 1073, 1–57.

CHAPTER THREE

Atmospheric and hydrological hazards

As the atmospheric system provides the components of the terrestrial phase of the hydrological cycle, it is hardly surprising that there are strong linkages between the various natural hazards caused by extremes of cold, heat, wind, precipitation and dryness. Hence, in regions with extreme climates, floods may be interspersed with droughts. In coastal areas of the tropics hurricanes often cause flooding and may also be accompanied by tornadoes. Lightning and hailstorms may be associated with tornadoes or flash floods, and periods of winter snowfall, which give rise to an avalanche hazard, may later swell rivers with spring meltwater, causing a seasonal flood risk. For these reasons, the following chapter is devoted to the hazards of both atmospheric and hydrological extremes.

Floods

Basic definitions
A flood can be defined as the height, or **stage**, of water above some given point, such as the banks of a river channel. The **flood hazard** consists of the threat to life or property posed by rising or spilling water. There are four principal types of flood. **Riverine** floods can be divided into slow-rising kinds resulting from rainfall or snowmelt and the more abrupt flashfloods caused mainly by intense thunderstorms. **Estuarine** floods usually result from the combination of a tidal surge at sea, caused by storm-force winds, and river flooding caused by rainstorms inland. **Coastal** floods can be caused by hurricanes and other severe sea storms, or by tsunami waves. Finally, various catastrophic causes can be identified, such as dam-burst or the effects of earthquake or volcanic eruption. River floods resulting from heavy rainfall or intense snowmelt are the most common of these phenomena and often result in the heaviest damage.

The area around a river that is periodically covered by water spillage in excess of **channel capacity** is known as the **floodplain**. It is usually covered by relatively level deposits of sand, silt and clay brought by overflowing water. Floodplains occur principally on low-lying coasts and in the lower reaches of major alluvial rivers. The flood hazard is often related to the **100-year floodplain**, which is the area covered by a flood with an average **return period**, or **recurrence interval**, of once in a century (see below). This is thus an area which per century has a 1 per cent chance of inundation, while the 50-year floodplain has a 2 per cent chance. More generally, the larger the recurrence interval, the longer the return period and the greater the magnitude of flood flow.

River flooding occurs in the context of the **drainage basin** (also known as the **catchment**, or **watershed**). This consists of the network of channels, together with the slopes which they drain, bounded by the **drainage divide** which separates the basin from its neighbours. Drainage basins vary in size from a few hectares to the thousands of km^2 occupied by the world's largest river systems. They can, however, be considered as a **nested hierarchy**, in which the smaller catchments form an integral part of the larger ones. The length of channels draining a particular basin can be expressed in km/km^2 as the **drainage density**, and is a function of rainfall intensity and the potential for erosion of the rock or sediments in which the basin occurs. Generally, the larger the basin the greater its propensity to store water (on the valley floor, in stream channels and lakes, etc.). **Flood magnitude**, which can be expressed in m^3/sec/km^2, increases with increasing basin size, but at a declining rate, as increased storage results in slower increases in water yield (Costa & Baker 1981: 351).

In summary, various phenomena control the propensity of a river to flood (see Fig. 3.1). **Transient controls** consist of the climatic factors (precipitation, evaporation, soil-moisture storage capacity, etc.) that vary seasonally and daily. The intensity and amount of rainfall are important, but so also are the size, track and rate of movement of the storm. **Permanent controls** comprise the characteristics of the drainage basin, such as its size, shape, slope, drainage density and permeability. These factors are considered permanent in the short term, but in the long run of geological time they will undergo geomorphological change. Finally, **land use and occupance** are neither necessarily transient nor necessarily permanent controls.

The scientific study of surface water flows and the physical characteristics of flooding is called **hydrology**. The circulation of water through the atmosphere and across the Earth's surface is known as the **hydrological cycle**, and flooding falls within its **land phase**. For simplicity, the following sections will be concerned with riverine flooding (that which occurs in the **fluvial system**), although many of the concepts expressed below can be adapted to coastal or other flood situations. River hydrology is divided

121

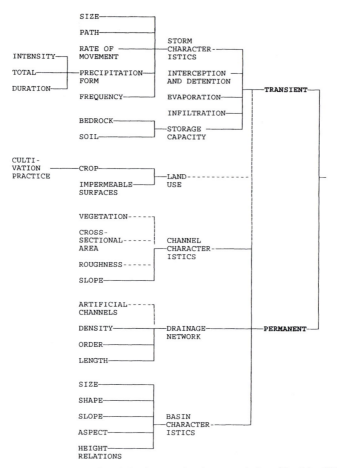

Figure 3.1 Controls on flood hydrograph characteristics (Rodda 1969: 164).

into **downstream** and **at-a-station** components. The former deals with spatial or spatio-temporal changes in water flow, while the latter refers to temporal changes at one or more fixed locations.

Discharge

River flow is measured as **discharge**, the volume of water passing a point per unit time (m^3 per second, or "cumecs"). The height or stage of water in the channel depends on discharge and on the capacity and shape of the channel. Channel capacity is related to **bankfull discharge** (Fig. 3.2a), the maximum throughput of water at a given velocity of flow which the river channel will sustain without **overbank spillage**. Spilled water will eventually drain back into the channel, or it will pond on the surface of the floodplain, infiltrate it or evaporate. The largest flow, or **peak discharge**, Q, can be related to catchment area, A, by a nonlinear parametric equation:

$$Q = CA^n$$

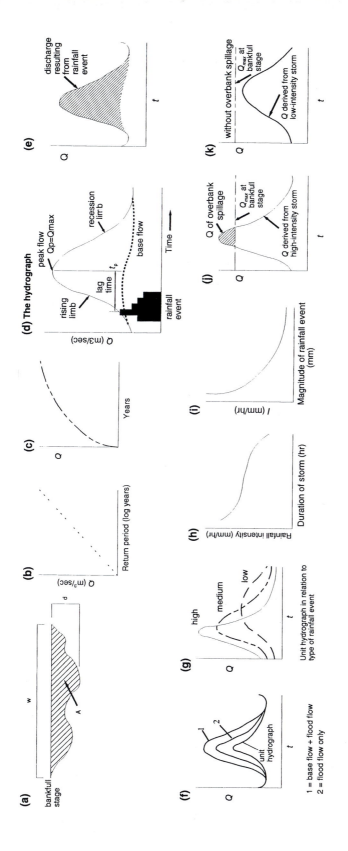

Figure 3.2 Some graphical methods for analyzing flood flow.

where the parameters are functions of climate and the other characteristics of the drainage basin. Peak discharge is calculated with respect to a given recurrence interval. Alternatively, it can be expressed in terms of rainfall intensity (in mm/hr), I, by the so-called 'rational equation':

$$Q = CIA$$

where C is a constant related to basin characteristics.

It is also possible to predict the mean annual flood, Q_m, in terms of basin area and mean basin slope, S:

$$Q_m = 0.07A^{0.74}S$$

where the parameters (here given for humid temperate conditions and average basin permeability) will vary with climate. In a study of 57 British drainage basins (NERC 1975), slope and area were found to predict 86 per cent of the variance in Q_m, but there is still a need to include climatic factors, such as the mean annual rainfall, R, which is an index of the basin's flood susceptibility:

$$Q_m = 0.009A^{0.85}R^{2.2}$$

(where the parameters are supplied by the British example). To improve the precision of estimates, the logarithm of discharge can be related by multiple regression to the logarithms of average annual maximum daily rainfall, P, and drainage density, D:

$$\log Q = c_1 + c_2\log A + c_3\log P + c_4\log D$$

Stepwise multiple regression equations that are rather more comprehensive can be formulated to include explanatory variables such as area, topography (basin relief, circularity, number of streams per unit area, or stream frequency), average channel slope, drainage density, rainfall intensity, frequency of flow-generating precipitation, daily rainfall maxima, percentage of rainfall stored at the surface and percentage of the surface storing rainwater.

Recurrence intervals
The recurrence interval or return period (Figure 3.2b&c) of a series of events of different sizes is defined as

$$RI = (n + 1)/r$$

where n is the number of items in the series and r is their rank. The probability, p, that a given magnitude of flow will be equalled or exceeded in a particular period of time, such as a year, is the reciprocal of the recurrence interval (p=1/RI). The probability, q, of a flood of recurrence interval t years being equalled or exceeded in the next n years is

$$q = 1 - (1 - 1/t)^n$$

124

Recurrence interval is an imprecise, empirical measure and cannot be predicted beyond the existing record without probable errors of estimation (Dalrymple 1960).

The design of flood prevention measures requires some estimate of the worst flooding to be expected during the life of the structure, and may utilize the concept of the **probable maximum precipitation** (Paulhus & Gilman 1953). This estimates the most severe flood resulting from the most severe rainfall and most severe combination of meteorological and hydrological conditions likely to be experienced by the region in question. It assumes a natural upper limit to rainfall intensity and duration variables and usually postulates maximum discharges that are five to six times the 100-year flood.

The flood hydrograph

The **stream hydrograph** (Fig. 3.2d) is a graphical measure of changes in discharge over time, as recorded or predicted for a given measuring station in the drainage basin (Rodda 1969). It can be drawn (as the **flood hydrograph**) for individual flood events. Essentially, the hydrograph converts rainfall into channelled runoff, or streamflow. It begins with the **rising limb**, in which discharge increases rapidly with time. This is followed by **peak discharge**, representing the maximum flow for the particular flood event (Q_p). There is a measurable **lag time** (t_p, or time-to-peak) between peak rainfall and peak streamflow, which is related to the time it takes for rainfall to collect on surfaces or in soils, find its way into channels and flow to the particular measuring station. Then, during the **recession limb**, flood discharge subsides, usually at a slower rate than it builds up, such that the slope of the graph is shallower here than during the rising limb. Beneath the storm hydrograph is the component of discharge, known as **base flow**, which is given by springs, soil moisture contributions and other sources that are at least partially independent of storms. Once the recession limb has ended, discharge returns to the level of base flow. Rivers that flow constantly are termed **perennial streams**, while those that lack discharge during dry spells (such as the North African wadis or southwest American arroyos) are called **ephemeral streams**. For the latter there is little or no base flow.

The stream hydrograph represents all flow occurring in the channel during a given period. If base flow is subtracted from the discharge caused by a particular rainfall event, the resulting lower curve is the **stormflow hydrograph** (Fig. 3.2e). One can also define a **unit hydrograph**, which is the pattern of discharge produced by 1 cm of rainfall (Fig. 3.2f&g). The shape of the unit hydrograph changes with storms of differing effective duration, being flatter (i.e. with a lower peak discharge) and longer drawn out for longer-lasting storms. Hence, changing the intensity (in mm/hr) of precipitation changes the response of the basin, even though the volume of discharge may be the same. A low-intensity storm (of, say, 6 mm/hr) is thus less likely to produce a peak discharge that exceeds channel capacity than a high-intensity storm (of, say, 70 mm/hr).

Water yield from the drainage basin
As Figure 3.2h&i show, the magnitude and duration of precipitation tend to be inversely related to storm intensity. If the basin has already been saturated by previous storms, low intensity can be compensated for by high magnitude: hence the importance of **antecedent moisture**, which is the water accumulation during a previous period usually defined as between 5 and 14 days. The rôle of previous rainfalls is expressed by the **antecedent precipitation index**, which for a ten-day period is:

$$\text{API}^{10} = \sum_{t=0}^{t=10} P_t K^{-t}$$

where P_t is the precipitation on day t, K is a constant and the exponent $-t$ expresses the negative exponential decay rate of the significance of precipitation on that day. API has the following rôle in flood discharge determination:

$$Q_s = -c_1 + c_2 P_s + c_3 \text{API}$$

where Q_s is storm runoff in mm and P_s is storm rainfall in mm. In such a calculation, the lag time in translating rainfall to stream discharge should not be neglected.

Antecedent moisture influences discharge by affecting the amount of water stored in the basin at any point in time. The **water balance** of the basin can be expressed as

precipitation = evapotranspiration + runoff ± change in storage

in which evaporation from the land surface and transpiration from plants are combined into one measure. Water storage at the land surface is altered by percolation or infiltration into soils, sediments or rocks and throughflow within them, surface detention (interception, ponding or lake impoundment) and overland flow on slopes or in channels. A saturated drainage basin will absorb and detain less moisture than one that is dry when a given precipitation event occurs. Forest cover tends to retard overland flow and, through **interception** of falling rain by the vegetation, reduce the speed and amount of the translation of precipitation into discharge. Clay soils tend to be sufficiently impermeable to generate high

126

water yields after storm precipitation, but they manifest a **desiccation factor**, in that they absorb much moisture through surface cracks when rain falls onto dry surfaces.

Water yield from a drainage basin is critically influenced by its average **infiltration capacity**. This is the rate of absorption of moisture into the ground (in mm/hr). It can be related to surface runoff rates by comparing it to rainfall intensity. For example, if the infiltration capacity of a clay loam is 25 mm/hr, and rainfall intensity is 50 mm/hr, runoff will be 25 mm/hr, which over 4,000 m^2 is about 0.03 m^3/sec. Over time, infiltration tends to follow a negative exponential curve towards a stable final value, which is low in impermeable terrains and high in permeable ones.

Hydraulic geometry

The study of channel cross section and how it transmits water flow is termed **hydraulic geometry**, which is defined as "the graphical analysis of the hydraulic characteristics of the stream channel" (Leopold et al. 1964). It can be used to relate the hydrograph to the flood hazard: clearly, unless the cross section is altered by erosion, a given channel capacity represents a fixed maximum discharge. If the flood peak exceeds this, there will be overbank spillage that will last until discharge abates to the level which the channel can transmit within its banks (Fig. 3.2j&k). Discharge is the only fully independent variable in hydraulic geometry, and the **dependent variables** include channel width, slope, velocity, bed material, sediment load and roughness.

It should be noted here that in many large river systems that are subject to regular flooding, such as the River Senegal in West Africa, a major part of the annual water discharge moves down the floodplain rather than along the channel. This would require modification of the concept of hydraulic geometry to include a much wider cross-section than that of the channel alone.

For a fixed cross-sectional area, A, discharge is related to the velocity of flow, V, by the **continuity equation**:

$$Q = VA \equiv V.w.d \text{ (because } A = w.d)$$

where w is channel width and d is its mean depth. The **hydraulic radius** of the channel, R, is defined as its cross-sectional area, A, divided by its wetted perimeter, w + 2d:

$$R = A/(w + 2d)$$

R differs with the shape of the channel, as this governs the degree of contact between the water body and its confining perimeter. Together with channel slope, this can be related to flow velocity by the **Manning equation**, which in metric units is as follows:

$$V = 1/n \ (R^{0.67}S^{0.5}) \equiv 1/n \ (R^{2/3}\sqrt{S})$$

The equation includes **Manning's n**, which is a surrogate for channel roughness and which varies from 0.01 in smooth, concrete-lined canals to 0.05 in boulder-strewn natural channels. Flood risk (i.e. the magnitude of peak discharges) can sometimes be reduced by increasing the area of channel cross section (channel capacity) and reducing the roughness, thus increasing velocity. But increasing discharge down stream may augment the flood hazard there.

Velocity can also be expressed using **Chezy's equation**:

$$V = C(RS)^{0.5} \equiv C\sqrt{(RS)}$$

where **Chezy's C** is a function of hydraulic roughness and hence of hydraulic radius. It can therefore be related through R to Manning's n (in fps units):

$$C = 1.5(R^{0.133}n^{-1})$$

Thus, the roughness of the channel is related to its wetted perimeter and shape, and hence the design of man-made channels will govern their ability to transmit flood flows efficiently.

Some problems with empirical measures

Estimates of stream–channel flow variables and areal values of water yield usually assume that the characteristics of the drainage basin are fixed. This is usually reasonable in the short timespans represented by a few years or decades. But effects such as the transition from slow to accelerated erosion, or urbanization of the catchment, will mean that the equations need to be recalculated with new parameters. Bankfull flow and hydraulic geometry, moreover, may alter during a particular flow event if the channel is highly mobile or erodible.

There are also grounds for considering the hydrograph to be a much more dynamic phenomenon than simple analysis would imply. Baseflow may alter if the perennial moisture supply to the stream is unstable. Alternatively, if a second precipitation event follows closely upon the first, the second storm hydrograph may "ride" upon the recession limb of the first one. This complicates the process of hydrological analysis.

Dating flood deposits

Estimates of the magnitude and frequency of floods can be improved by dating the sedimentary deposits which they leave behind on the floodplain (Lucchitta & Suneson 1981). Flood deposit dating has been carried out for a variety of reasons. For example, it has been used to construct a better magnitude–frequency curve by improving long-term estimates of the recurrence intervals of floods of different sizes. It has also been employed to reconstruct past changes in climate, and thus to ascertain whether at the location in question floods were more or less frequent during the past. Finally, it has been used to clarify the relationship between humans and the natural environment. Archaeologists may need to know whether primitive people were forced to migrate as a result of catastrophic floods, while Quaternary scientists may be concerned to find out whether environmental changes wrought by mankind have increased or decreased flood propensity.

Several dating methods can be applied to the alluvial stratigraphy created by floods (Costa 1978). Relative dating can be applied to the position, thickness and extensiveness of flood deposits, under the assumptions that the stratigraphy is regular and undisturbed (the lowest deposits being the oldest) and that thicker, more widespread deposits are the effects of larger floods. The interpretation of flood stratigraphy is generally facilitated when there are buried soil horizons and sedimentary structures whose origin can be reconstructed. Palaeo-hydraulic analysis is aided by investigation of forest beds, sand and clay lenses, transported boulders and buried flood channels. Flood sizes can also to some extent be estimated from relict fluvial landforms (Jarret 1990).

But as such methods tend to be imprecise, absolute dates are often sought. Flood deposits may include datable material, such as plant remains that contain organic carbon which is susceptible to absolute dating of the isotope ^{14}C. Dating is also possible for human artefacts created by extinct cultures known to have occupied the floodplain, and the date is often determined by measuring thermoluminescence. Vegetation damage may provide a clue to the age of flood deposits. On occasion trees are knocked over by floods but regenerate themselves: the physiological damage can help differentiate the variety of flood events. The analysis of tree ring growth, termed **dendrochronology**, can help date floods which have damaged individual growth rings (Sigafoos & Sigafoos 1966), although relatively recent events are the only ones that can be studied by this method.

Flash floods

Flash floods are an extreme, though short-lived, form of inundation. They usually occur under stationary or slowly moving clusters of thunderstorms, or result from motionless or slowly progressing storms which have an unabated inflow of air that has a high moisture content. The storm usually lasts less than 24 hours, but the resulting rainfall intensity greatly exceeds infiltration capacity (and will probably peak in the range 50–100 mm/hr). Runoff is generated very rapidly and hence streamflow is nearly instantaneous. As a result lag time is low, the rising limb of the hydrograph is steep and the peak discharge is high (Vishnevskiy &

Shcherbak 1969). Moreover, flash floods are often very destructive as the high energy of flow may carry much mud and sediment and many boulders (Fig. 3.3; Clarke 1991).

The US National Weather Service issues two kinds of flash flood alert. A **flash flood watch** signifies that it is possible that rains will cause flash flooding in a specified area, while a **flash flood warning** means that flash flooding is occurring or is imminent in the specified area and people who are at risk should evacuate to safe ground immediately. These warnings are broadcast on television and radio and, if appropriate, by loudspeaker from touring emergency vehicles (Mogil et al. 1978).

Although a large river such as the Mississippi may permit up to a week's warning of an impending flood, in small catchments and flash flood conditions there may be only a few hours between the identification of the event and the arrival of the flood peak down stream. However, warning methods have been designed, such as the Automated Local Evaluation in Real Time (ALERT) system which employs raingauges and stream sensors equipped with self-activating radio transmitters that communicate with a central microcomputer. Using software calibrated to the conditions in the local drainage basin, the microcomputer can process

Figure 3.3 In 1983 and 1984 the Rio Cuyocuyo flooded this small settlement in a remote part of the Cordillera Oriental of southern Perú, killing two people and burying the upstream part of town under 2 m of fluvial sediment.

data from up to 100 sensors and provide summary data and visual or auditory warning.

By 1983, 797 local flood warning systems and 42 ALERT systems were in use in the USA. Two years later, 1 in 20 communities at risk from flooding had a local warning system (Platt & Cahail 1987). But the price, function and scale of different systems varied widely: at opposite ends of the scale Lycoming County, Pennsylvania, protects 118,000 residents for an initial investment of $500 using a network of unpaid volunteer spotters, while Santee Cooper North Dam in South Carolina uses water sensors and seismometers to protect fewer than 3,000 people for $4 million (Gruntfest & Huber 1989). ALERT systems vary in cost from $35,000, for three stream gauges, two precipitation recorders and a microcomputer, to over $500,000 for more than 100 gauges. Generally, however, the system must be economical and simple to operate if it is to gain acceptance locally.

The human use of floodplains

There are several reasons why floodplains have enjoyed a long history of settlement. First, they often carry deposits of fertile alluvium that has been revitalized by continual flooding. Hence, the Rivers Euphrates, Ganges, Indus, Nile, Tigris and Yangtse have all supported major civilizations as a result of the fertility of their floodplains and the use of river water for irrigating crops. Secondly, activities besides agriculture that depend on the river may be located there or in the adjacent floodplain: these include fishing, log transport, barge traffic and power generating stations (which require supplies of cooling water). Thirdly, many floodplains and the rivers they contain are important corridors for rail, road and water transport. Moreover, as they offer level land for construction, floodplains impart a certain freedom of locational choice to the makers of the human environment. Lastly, river valleys often have particular scenic value, which complements the economic advantages of locating in them.

The particular social and economic benefits of inhabiting and using floodplains are sometimes outweighed by the negative effect of flood disasters. Nevertheless, it is clear that living on floodplains is second nature to humanity, and hence so is cohabiting with the flood risk. This has created a distinctive pattern of flood perception, which has been thoroughly investigated in many parts of the developed world (Kates 1962, Parker & Harding 1979), and which furnishes us with a considerable number of behavioural regularities (see also Ch. 9). First, accurate knowledge of the flood hazard does not necessarily prevent people from inhabiting floodplains. People tend not to be convinced by hazard maps, although they sometimes respond to more sophisticated information and to a clear presentation of the precise risk which they are running. In fact, people who are aware of the risk are often very sensitive to the depths and locations of past and probable future floods. On the other hand, those who are unaware of the risk are not necessarily convinced by technical information, or moved by officious floodplain managers. In general, the more recently a flood has affected an area, the greater the awareness of risk manifested by local people (White 1973).

Flood reduction requires the support of floodplain residents in order to succeed, but people who are disaffected by government rarely co-operate with such schemes. Others may become disaffected if there are substantial delays in implementing the schemes. With respect to floodplain users and residents, mitigation has to face particular problems: members of the public may be unaware of the risk or intransigent about it and, despite the obvious legal and practical dangers, there is a tendency to speculate in property menaced by floods. Finally, mitigation strategies and other hydrological modifications carried out upstream are often blamed by floodplain users down stream, who may claim – rightly or wrongly – to have suffered increased damages as a result (Liebman 1973).

Flood mitigation strategies

Adjustments to floods can broadly be classified into structural and non-structural, according to whether they use engineering or administrative methods (Thampapillai & Musgrave 1985). Non-structural approaches involve adjustment of human activity to accommodate the flood hazard (James 1975), while structural methods are based on flood abatement or the protection of human settlement and activities against the ravages of inundation (Fig. 3.4) The range of possible options is as follows.

Structural change involves modification to the built environment in order to mitigate flood damage directly. It includes the construction of levées, floodwalls and landfills and the modification of building design. Temporary measures include lining vulnerable areas with sandbags, while permanent measures include building floodwalls. The latter should be constructed before the protected area is urbanized, as they will usually be prohibitively expensive afterwards. Another structural method, flood control, consists of the channel phase (known as **flood protection**) and the land phase (known as **flood abatement**). The two approaches should complement each other, rather than be used separately. Protection measures in the channel phase include the construction of dykes, dams, reservoirs, flow retarding structures, and measures for accelerating or retarding flow, reducing bed roughness and deepening, widening or straightening channels. These latter responses are collectively termed **channelization**. Abatement measures include gully control, modified cropping practice, soil conservation, revegetation and stabilization of banks and floodplains (especially as silt and sand transport are usually important components of floods).

Structural measures are expensive and require major government

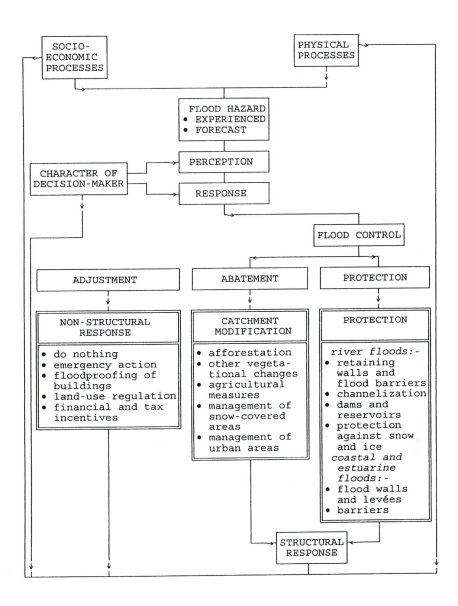

Figure 3.4 Response to the flood hazard (Ward 1978: 115).

participation and financing. Where there is a need for extra protection it may be necessary to design for more than the 100-year flood, however costly it is to do so. But the measures seldom ever give complete security: thus flood control which does not work may do more harm than good by encouraging the further development of the floodplain under the misapprehension that the protection is absolute. As a result of these considerations, the trend is away from complete reliance on structural measures towards combining them with a package of non-structural approaches.

Semi-structural measures are a secondary form of response to the hazard. Areas of low-intensity usage (such as farmland, water meadows or wasteland) are set aside for floodwater storage, with the intention of reducing peak discharge in channels that pass through more vulnerable areas. This method is seldom useful on its own, but can complement other measures. Another hybrid measure, **flood proofing**, involves combining structural change with emergency action. Hence, evacuation plans can be combined with physical protection measures. The drawbacks to this approach are that it tends to foster persistent occupation of the flood hazard area and, again, the structural component of it may give a false sense of security without providing absolute protection.

The simplest **non-structural measure** is to accept the loss. This results in no mitigation of future flood hazard and presupposes that society either has sufficient resources to absorb the impact or is too poor to be able to mitigate the hazard. A slightly less simple, but still very basic, non-structural method is to provide public relief. The aid provided by the Red Cross and voluntary organizations falls into this category, as do loans, deferment of payments, grants and government subsidies. Financial assistance of this kind is sometimes referred to as "forgiveness money", as it tends to pardon floodplain users for running the risk and not to encourage them to seek mitigation.

Emergency action and rescheduling involve plans for evacuation and mounting temporary flood defences. The effectiveness of such measures is strongly influenced by the ability to predict and capacity to warn against the flood. Hence, the brunt of the responsibility will be borne by local and regional government, whose degree of organization will be a vital determinant of the effectiveness of response. Much also depends upon public perception; for example, it seems that the public consistently underestimates the power and speed of onset of flash floods.

Land-use control can be one of the most effective ways of reducing flood hazards. Statutes, ordinances, regulations and compulsory purchases can be employed, and relocation can be subsidized. However, enactment of such measures must lead to their enforcement, which requires full co-operation between local and central government. Through land-use regulation it may be possible to create a **floodway** through a city, in which urban uses are subtracted from the floodplain, which becomes public open space (Bennett & Mitchell 1983). The floodway will have to be safeguarded against pressure to redevelop it,

Table 3.1 Flood hazards and their abatement (after Cooke & Doornkamp 1974: 117, and other sources).

Critical flood characteristics	Adjustments to floods	Flood hazard abatement	
		Structural	Non-structural
Depth	Accept the loss	Levées, dykes embankments, flood walls	State laws
Duration	Public relief		Zoning ordinances
Area inundated	Abatement and control	Channel capacity alteration (width, depth, slope and roughness)	Subdivision regulations
Flow velocity			
Frequency and recurrence	Land evaluation and other structural change		Building codes
Lag time (time to peak flow)		Removing channel obstructions	Urban renewal
	Emergency action and rescheduling		Permanent evacuation
Seasonality		Small headwater dams	
			Government acquisition of land and property (creation of floodway and public open space)
Peak flow	Land-use regulation	Large mainstream dam	
Shape of rising and recession limbs	Flood insurance	Gully control, bank stabilization and terracing	
Sediment load			Fiscal methods: building financing and tax assessment
			Warning signs and notices
			Flood insurance

especially as an urban parkland location may prove very desirable to businesses or potential residents.

Lastly, flood insurance is a means of shifting the burden of losses onto the people who run the risk, namely the floodplain users and residents. Premiums should in theory be proportional to the risk, but this tends to make them unacceptably high for some subscribers and hence a compromise must be sought. If flood losses are likely to be very costly, or flood risks to involve a high degree of uncertainty from year to year, it may be necessary for government to sponsor the scheme and to back it with the resources of the state. Table 3.1 summarizes the salient characteristics of floods and the various forms of structural and non-structural hazard reduction.

Clearly, flood mitigation involves some difficult dilemmas. For instance, using dams as a structural measure of flood control is problematic. It is difficult to decide between the parallel options of a large number of headwater dams, which may exert a limited effect on a trunk river, or a major downstream dam, which may be vulnerable to failure and collapse. In general, upstream dams are best at controlling erosion and sedimentation, while downstream ones act best to restrict flooding. The latter may need to be kept half empty if they are to reduce high discharge rates, but this may limit their value for water supply, power generation or recreational uses. Perhaps for this reason, only 17 of India's 1,554 large dams were built specifically to control floods.

In highly developed regions the main constraints upon structural protection are often the availability of sites to build the levées, dams, dykes and canals, and the resulting constructional and environmental costs. Few improvements in design are possible: the basic specifications of structures have altered little over the past 40 years. Nevertheless, there is scope for adapting the design of sewers, storm drains and highway networks both to correlate with river management schemes and to reduce the impact of floods on communities downstream (by attenuating the rise of the hydrograph).

It is clear that unrestricted protection of floodplain residents and users, and the supply of relief when they suffer damage, are forms of hidden subsidy. In most cases there are few reasons why developed countries should be forced to encourage continued settlement of the major flood hazard zones, yet little attention has been given to the question of whether subsidizing floodplain users is just and valid (Mucklestone 1976). Given that there is bound to be pressure to use floodplains, a comprehensive strategy is required to manage their hazards. It may consist of the following steps.

First, the flood hazard should be recognized, assessed and mapped. This would involve compiling information on the extent of the largest recorded flood and on the cost and type of past flood damages; mapping the limits of the 100-year flood (or other "design" flood) on a topographic map; compiling profiles and cross sections of the river to show the levels of past floods; and compiling flood frequency curves and locally representative hydrographs. Secondly, flood forecasting and warning systems should be designed and set up and should be combined with emergency action plans. Thirdly, land-use controls and floodplain regulations should be enacted to ensure that occupance is restricted in the main hazard zones and buildings and other structures are adequate to the risk. Hence, existing development should be controlled and new development prevented in the principal risk areas. Fourthly, the structural approach to flood control should be applied where the cost justifies it, for example, in large cities, which have too much locational inertia to be modified substantially. Finally, the flood mitigation programme should be safeguarded against calls to reduce or abandon it and should be adjusted to take account of any change in the risk or re-evaluation of it.

Although it represents a reasonable compromise between the areas inundated by very infrequent floods and those covered by the smallest ones, the use of the 100-year floodplain is subject to criticism. For example, in the USA almost two thirds of all residential flood losses result from events that occur once every 1–10 years, which suggests that a more flexible approach to floodplain delineation may be needed at the local level.

Although most flood mitigation measures take some time to implement, short-term responses are also important. For example, after the February 1953 North Sea flood disaster, the regional authorities in East Anglia, in eastern England, instituted a flood warning system (Penning-Rowsell & Handmer 1988) consisting of the following stages:

(a) A '**flood alert**' is issued about 12 hours before high water, indicating the possibility of flooding.
(b) A '**flood danger**' notice is issued about four hours before predicted high water, indicating the probability of flooding. District authorities are informed in coastal areas likely to be affected.
(c) A '**flood arouse**' warning is given when flooding is imminent, using communications systems operated by the police. Emergency services are put on alert; boats, equipment and road transport facilities are positioned. Police loudspeaker vans are placed on standby, outlying residents are warned and manpower is called in to look after equipment and utilities that may be damaged.
(d) The '**flood alarm**' is given when the emergency phase begins, and the relief operation is put into action. Emergency management is in the hands of the county emergency office, which directs the police, road gangs, lifeboat crews, the Red Cross and voluntary organizations. The army, however, acts under its own command, independently of the civil authorities.

Coastal flooding by storm surge

Low air pressure and the friction of wind on the sea surface may cause water to mound up and flood coastal areas during a major storm, especially if this coincides with high tide and high levels of river discharge in estuaries. The **storm surge** can be defined as the increment of shoreline water level above normal seasonally adjusted average high tide levels (Jelesnianski 1978). Relative storm surge potential is a topographic variable that is a function of the central pressure (in millibars) of the storm, which is strongly correlated with storm surge height. The probable maximum elevation of surge effects must be determined from the height and location of past storm damage at particular points on the coast, although an arbitrary contour level can be selected as an interim measure while research is being conducted.

Coastal flooding caused by storms can be characterized by analogy with riverine floods. The height of the storm surge (by analogy with stage) and velocity of water movement tend to be related to the seriousness of damage. But the concept of discharge is of use only in the vicinity of river deltas, littoral currents or tidal estuaries. The range of flooding will usually be restricted to a narrow coastal zone, although in low-lying areas it is likely to be much larger: the Netherlands, Belgium and British East Anglia constitute one such area, while the coasts of Bangladesh and eastern India constitute another. Where floodplains are wide and the coast is flat, floods may be frequent as well as extensive, leading to a high and saline water table and poor drainage.

The duration of flooding may be related to the speed of drainage (which will be relatively slow, for example, from tidal marshes) or to the duration of the surge itself. Duration tends to be short in tropical storms relative to their extra-tropical counterparts, but in both cases it tends to be critically related to damage as much in an urban as in a rural coastal context. There is also an analogy with the lag time concept of riverine flooding, in that the rapid arrival of a storm and equally rapid inundation of coasts may bear some resemblance to the effect of flash floods inland. Seasonality is often a vital factor in storm surge risk. For example, 60 per cent of hurricanes that cause flooding on the east coast of the United States occur between 1 August and 15 September (Herbert & Taylor 1979).

The relationship between flood protection and the settlement of coastal lands may be direct or inverse. Structural defences may reduce the amenity value of coastal areas and hence make them less attractive to settlement, or they may lull residents into a sense of security and stimulate new developments in the protected area. Protection is seldom likely to be absolute, however, especially in estuarine areas where the effect of river floods must be added to the risk of coastal inundation. Moreover, areas that are subsiding eustatically or tectonically (such as the London Basin in England) may undergo a change in their protection requirements over time, requiring repeated investment in structural measures or renewed commitment to planning initiatives (Penning-Rowsell & Handmer 1988).

By way of an example of the seriousness of storm surge floods, consider the impact of the North Sea storm surge of 1 February 1953, which caused major damage in eastern Great Britain (Spiegal 1957). In the Netherlands 190 km of dikes were seriously damaged and 89 km were breached. Thus, 150,000 ha of polders were inundated and 400,000 buildings damaged. Although a total of 72,000 people were evacuated, nevertheless 1,835 individuals were killed, as were 47,000 cattle. The human toll and economic impact were so serious that the Netherlands embarked on a 30-year scheme for improving the national sea defences, the Delta Plan, which was completed in 1986, and was designed to close off the Ijselmeer and the Schelde estuaries. The project occupied Dutch engineers for more than 30 years and stimulated the invention of some of the world's most advanced flood control technology.

Over the period 1964–82 floods claimed 80,000 victims and affected 221 million people around the world. More detailed examples of flood hazards, and questions of their mitigation, will now be given for developed societies. The analogous case for developing nations will be considered in Chapter 8.

Flood hazards in Italy
The threshold for catastrophically intense precipitation processes will vary enormously. The sensitivity of landscape response will depend upon such factors as channel capacity, the permeability of deposits, the degree of prior saturation, slope angles and the availability of sediment capable of erosion. Disasters, when they occur, will be a function of particularly unfortunate conjunctions of circumstances and will have effects that are extremely varied in time and space.

A good illustration of such a combination is furnished by the northern Italian floods of 4–5 November 1966, in which both Venice and Florence were flooded (see Fig. 3.5 and the section of Ch. 4 on subsidence). Historical records show that the recurrence interval of major damaging floods on the River Arno is about 200 years, and there have been 62 known floods since the Romans founded the colony of Florentia nearly 2,000 years ago. However, the return period is not regular: large floods occurred in 1333, 1557, 1844 and 1966 and very nearly recurred during two successive Novembers at the end of the 1970s.

At the end of October 1966 the 5000 km^2 area of the Arno Basin became saturated with melting snow. Although only a few millimetres of rain fell on 3 November, 10–15 per cent of the average annual precipitation fell during the following night, and 15–22 per cent over the two-day period. In fact, in 48 hours about 25 per cent of the yearly average precipitation fell over northern Italy as a whole, and 44 per cent in the Ombrone Basin, to the southwest of the River Arno. Although the maximum stage on the Arno, 8.94 m, had been equalled in 1929, the flood of 4 November 1966 involved a hydrograph with an exceptionally steep rising limb, and at 5 am on that day the river burst its banks within the city of Florence.

The flood created a striking dichotomy of effects. In the centre of Florence tremendous losses were sustained to the priceless heritage of art, sculpture, books, artefacts and architecture, especially as the floodwaters contained a great deal of harmful mud, debris and heating oil. Across northern Italy, 800 communities were affected, 10,000 houses, 12,000 farms and 16,000 agricultural machines were damaged, and 50,000

Figure 3.5 About 80 per cent of Venice is located at sea level and is flooded regularly each winter. This includes the Chiesa dei Gesuiti on the island of Giudecca.

farm animals were killed or subsequently had to be slaughtered. The problem with the Ombrone catchment and parts of the Po Basin was severe erosion of bare, steep, unprotected valley slopes, which increased the sediment load of streams and rivers and hence augmented their destructive potential. Of the 112 deaths, only 32 occurred in Florence. Yet the attention of the world was riveted on the city, to the exclusion of other areas that were suffering (Alexander 1980).

Following the November 1966 floods, a computerized flood-routing model was designed for the River Arno catchment (Panattoni & Wallis 1979). Based on hypotheses about localized rainfall–runoff relationships and their translation into streamflow discharges, it was possible to simulate the effects of changes in river management techniques, such as channel armouring and reservoir construction. The model showed that, had flood-control dams existed at the time of the 1966 flood, discharge could have been restricted to 3,500 m^3/sec and overbank spillage would thus have been prevented. An on-line version of the model was proposed for immediate determination of the effect of actions such as the opening of reservoir flood-gates during peak flow or the failure of dams and levées.

Flood hazards in the USA

The American experience of floods gives an excellent illustration of how a developed society is affected by and seeks to tackle this particular natural hazard. Flooding is a major problem for the United States (Fig. 3.6). Coastal and riverine inundations affect 6–8 per cent of the nation's conterminous territories (about 300,000 km² are located in the 100-year floodplain) and more than 6.4 million structures are at risk. In 1955 the floodplains had 10 million occupants, while 30 years later in 1985 the number had doubled to 20 million, and currently about 12 per cent of the national population live in areas of periodic inundation (not all, however, within the 100-year floodplain). In fact, one sixth, or 26,400 km², of the nation's floodplains are urbanized, and they contain more than 20,000 communities susceptible to flooding, half of which have been developed since the early 1970s (Burby 1985).

The average annual loss of life in US floods has been variously estimated as 47.6, 57.2 and 89. Other sources suggest that the figure is 3.9 per 10 million of the population (or 96 per year), and is falling in response to better mitigation and preparedness (White & Haas 1975). But in 1972, 215 people died in the flood which struck Rapid City, South Dakota, when the Pactola Dam, 22 km up stream, burst, discharging

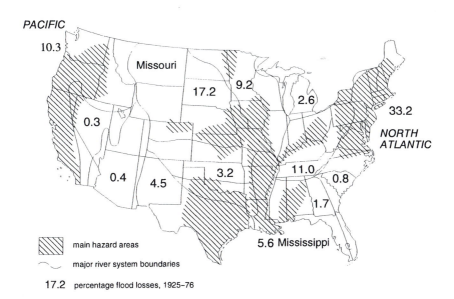

Figure 3.6 Distribution of river flood hazard in the conterminous USA (Ward 1978: 179).

water on the urban area at a peak rate of 1,400 cumecs; and in 1976, 139 people were killed by a flash flood in Big Thompson Canyon, Colorado. The monetary losses caused by floods have been rising steadily: in 1966 they were $1,000 million, in 1977 $2,200 million and in 1985 they reached $3,000–4,000 million. In the last of these years, one hurricane, Gloria, caused $1,000 million in flood damage, to which must be added insurance claims of more than $300 million. Moreover, this was not an exceptional occurrence: the 1951 floods on the Kansas River caused more than $1,500 million in damages, and, in fact, average losses have quadrupled since 1936 and continue to increase yearly. They now stand at more than $5.50 per head of the population per year (Petak & Atkisson 1982).

Floods tend to be repetitive phenomena in the United States. For example, during the eight years 1972–9, according to Federal declarations of natural disaster 1,900 communities were flooded more than once, 351 were inundated at least 3 times, 46 at least 4 times and 4 at least 5 times. This, added to the fact that the cost of adjustment is rising, means that it is hardly surprising that the Federal Government is constrained to spend $9,000 million per year on flood control and $300 million on flood forecasting.

Floodplain occupance was considered a personal or local responsibility until the Federal Government became involved in flood control projects on the lower Mississippi and Sacramento Rivers in 1917. In the case of the Mississippi, this involvement was strengthened after very serious floods in 1927, and a national policy on structural adjustment evolved in the wake of the Ohio Basin floods of 1936, which again were disastrous. Hence, over the period 1936–8, flood control programmes established dams on many large rivers, generally in conjunction with protection works in headwater areas. Such investments were made wherever it could be shown that costs were outweighed by benefits (White 1945), although long-term costs such as the gradual inundation of the Mississippi delta could not be accounted for.

By the 1960s more than 5,000 municipalities contained land liable to flood and 60 million hectares of farmland were at risk. Engineering solutions had been preferred for three decades, but the results had been equivocal: newcomers to floodplains frequently settled below vulnerable dams. Thus in 1966, while continuing its programmes of engineering protection, the Federal Government began to adopt other strategies for flood control. Floodproofing and forecasting were to be improved, Federal building on floodplains was to be curbed, and research and information dissemination were to be increased. Hence in 1969 the National Flood Insurance Act was passed, and during the next six years $8,000 of property damages were covered in 4,339 communities.

The current National Flood Insurance Programme (NFIP) was created when the 1969 Act was rewritten and passed as the 1973–4 Federal Flood

Disaster Prevention Act. The NFIP aimed to induce or coerce vulnerable communities to mitigate their flood hazards, as well as to improve the national non-structural response to floods (Platt 1976). The limits of insurance cover were doubled or tripled on those adopted in 1969. The Act contains provision for the identification and investigation of areas susceptible to flooding (about 1,000 communities produced floodplain information reports during the first year of operation of the NFIP). Federal aid is granted to states or local communities only if they participate in the NFIP and adopt ordinances to protect against flooding (otherwise aid requires a special subvention). Finally, all Federal schemes to acquire or improve property in floodplain areas are required to participate in the NFIP. In its first five years the 1973 Act prompted the writing of 1.6 million insurance policies, to a value of $60,000 million. The cost of the flood insurance scheme is subsidized by up to 90 per cent by the Federal Government.

Activities associated with the NFIP are divided between regular and emergency phases. The former is concerned to reduce property losses caused by flooding and ensure public safety, as well as to distribute costs fairly between local and central governments. By 1978 one sixth of communities with a flood hazard were participating in this scheme. Others (usually those which run a lower flood risk) participated only in the emergency phase, thus spending less on flood risk investigation and protection work.

Flood forecasting in the USA began in 1871 with the work of the US Army Signal Corps and is now the preserve of the National Weather Service. This body has set up a River and Flood Forecast and Warning System, which is linked to floodplain managers at the Federal, state and local levels, and to the news media. However, although effective action at the local level can reduce losses of life and property significantly, the gap between possible and actual performance of warning systems is large in most communities. There are several reasons for this including: ignorance of emergency procedures, of the hazard and the means of avoiding losses; preference for actions that do not encompass flood defence; lack of credibility of the local forecasting system; and inertia among the key institutions.

One of the most promising strategies is the public acquisition of land susceptible to flooding (Burby & French 1980). Since 1955, using money from the Land and Water Conservation Fund administered by the US Heritage Conservation and Recreation Service, local governments have acquired more than 400 parcels of floodplain land, representing over 15,000 ha. But many local authorities are reluctant to control floodplain use: Federal subsidies may be insufficient, tax bases may be eroded, flood hazard areas may seem too expensive to delineate and constitutional questions may be raised about the freedom of the individual. The creation of a floodway offers great incentives to abuse the programme, as property overlooking or located within the new urban park is likely to appreciate in value, thus encouraging corruption or speculation.

In conclusion, the United States has succeeded in initiating the transition from dependence on structural methods of flood prevention to a more broadly based programme of mitigation and relief. Although this approach may serve as a model for other developed countries that have a serious and widespread flood hazard, the USA has not yet solved all of its problems. The risks associated with floodplain occupance have not been reduced to minimal levels, participation in the NFIP has not been overwhelmingly high and the process of urbanization of floodplains has not been entirely halted. There is much work still to do.

Drought

Drought in the context of atmospheric processes

Drought can be defined as a condition of abnormal dry weather resulting in a serious hydrological imbalance, with consequences such as losses of standing crops and shortage of water needed by people and livestock. The human impact will depend on the extent to which a particular society relies upon the vagaries of climate to raise crops and make a living. The worst impacts tend to be felt by the simplest societies and the least organized ones, above all in cases where prolonged or persistent drought is a rare phenomenon. Where drought is recurrent, nomadism is one of the traditional responses that have been developed to combat its impact.

In the USA a **dry spell** is defined as more than 14 days with less than 1 mm of rainfall; in Great Britain the period is 15 days. In hydrological terms, a **precipitation drought** is caused by lack of rainfall, a **runoff drought** by lack of streamflow and an **aquifer drought** by lack of groundwater. If the moisture balance of an area falls, say, 30 per cent below its seasonal average value for a period of at least three weeks, then one has a possible rule-of-thumb indication of drought. This implies a failure of predicted rains and a consequent unforeseen deficit in the local moisture balance (moisture availability = precipitation − evapotranspiration − runoff ± change in surface and subsurface storage). To understand this process better it is necessary to examine its constituent parts.

Natural **evaporation** is the process by which water at the Earth's surface is converted to vapour and transferred to the atmosphere. The average rate is about 100 cm/year at the world scale, but this varies from values of 12 cm/yr in the Arctic Ocean to 138 cm/yr in the Indian Ocean (over the continents it falls in the range 36–86 cm/yr). As it is difficult to measure evaporation separately from the transpiration to the atmosphere of moisture from the leaf stomata of growing plants, on land the processes are often combined as **evapotranspiration**. **Actual evapotranspiration** (AET) depends mainly on the availability of moisture, while **potential evapotranspiration** (PET) is the maximum amount of water which could be evaporated and transpired under conditions of unlimited moisture availability (and is hence mostly a theoretical concept). In the semi-arid zone in particular, AET and PET differ substantially, as soils often lack moisture.

While evaporation can be measured independently from a pan of standard dimensions filled with water (an **evaporimeter**), transpiration must be measured as AET, using a **lysimeter**, which is effectively a weighing machine for a sample of the soil–plant system, in which changes in weight are proportional to differences in moisture content. Water, in fact, evaporates at least five times faster from an evaporimeter than from soil and vegetation. Using lysimeter data, the climatologist C. W. Thornthwaite derived a PET index for a nominal 30-day month (Thornthwaite & Mather 1955):

$$PET = 1.6 \ (10 \ T \ / \ I)^a$$

where T = mean air temperature

$$I = \sum^{12} i = \text{a heat index related to T (summed over 12 months)}$$

$i = (T \ / \ 5)^{1.514}$

a = cubic function of I

There is a close, log-linear relationship between PET and temperature, and the former is also related to humidity and wind speed and direction.

In this respect, Penman (1963) developed a formula that expresses PET (in mm/day) as a function of available radiant energy (R_N) and a term, E_a, that combines wind speed and saturation deficit, the difference between PET and AET:

$$PET = \frac{B(R_N/L) + E_a}{B + 1}$$

where $B = (\delta/\gamma) = $ Bowen's ratio

$\delta = de_s/dt = $ change of saturation vapour pressure with mean air pressure

$\gamma = $ the 'psychrometric constant', 0.486/°C

$R_N = 0.75S - L_N = $ available radiant energy

$0.75S = $ solar radiation absorbed by a surface having 25% albedo

L_N = net long-wave radiation from surface
L = latent heat of vaporization
E_a = $f(u)(e_s-e)$ [$\equiv 0.35(1 + 0.01u)$ for short grass]
u = wind speed at a height of 2 m
e_s = saturation vapour pressure at mean air temperature
e = actual vapour pressure at mean air temperature and humidity

In these equations (e_s-e) represents the "drying power' of air, independently of the availability of moisture for evaporation. Other formulae for PET can be found in Barry (1969).

Humidity is the water vapour content of the air and is measured by a **hygrometer** (a hygroscopic substance is one that is altered on contact with water vapour) or wet and dry bulb thermometers. **Absolute humidity** is the ratio of the mass of water vapour to the volume of air, the **vapour concentration**. **Relative humidity** is the ratio of atmospheric vapour pressure to saturation vapour pressure (absolute humidity divided by saturation absolute humidity) and is expressed as a percentage. It is 100 per cent in clouds and tends to decrease with altitude, especially in a cloudless sky.

Various attempts have been made to incorporate these variables into an index capable of expressing environmental moisture shortage. For example, Thornthwaite's humidity index is defined as

$$I_h = 100 \times (\text{water surplus / water need})$$

with an equilibrium value of 100. Water need is expressed by the sum of PET for the number of months when it is greater than precipitation. The aridity coefficient is the product of a latitude factor, the temperature range between the hottest and coldest months of the year, and the ratio of highest to lowest recorded yearly totals of precipitation. Its value for the middle of the Sahara Desert is 100. Finally, a radiational index of dryness has been developed, which compares the net solar radiation arriving at the ground surface with the energy required to evaporate the precipitation received.

In general terms, water surplus occurs when precipitation exceeds potential evapotranspiration locally. Water deficiency occurs when precipitation falls below PET and available stocks of soil moisture have been used up. Once rainfall begins to exceed PET again, soil moisture is recharged until a water surplus reappears. In most climates this process is seasonal, and any substantial prolongation of the period of water deficit may lead to drought conditions, represented by the lack of moisture available for growth of the normal vegetation and crops of the area. In a Northern Hemisphere climate with marked seasonality the water budget may take the form shown in Figure 3.7, in which periods of water surplus, sufficiency and deficit can be identified with respect to the amount of precipitation relative to potential evapotranspiration during each month of the year. When in the summer months precipitation falls far below PET, soil moisture is utilized for plant growth and not substantially recharged by rainfall, until a pronounced deficit occurs. During the winter, this will be remedied by rainfall in excess of PET, until soil moisture amounts to a surplus in comparison with that required to sustain plants. Drought results from a prolonged period of moisture deficiency (though, of course, it cannot be defined as a disaster unless it has a severe socio-economic impact).

One concept that is of particular use in the identification of drought conditions is **albedo**. At the global scale albedo represents the total power of the Earth's surface and its atmosphere to turn back incoming solar radiation. This quantity is expressed as the percentage of insolation that is reflected or scattered from the surface, and it takes the following characteristic values: forests 4–10 per cent, green fields 10–15 per cent, dry grass fields 15–25 per cent, dry ploughed fields 20–25 per cent, snow and ice 46–86 per cent (the wide range of values for the last of these reflects the greater reflectivity of snow when it is clean and smaller values when it is dirty). As moister conditions and denser vegetation tend to have lower albedos, drought conditions can often be detected by remote

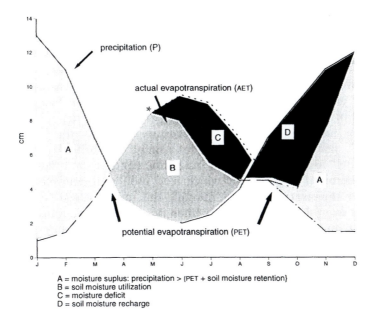

A = moisture suplus: precipitation > (PET + soil moisture retention)
B = soil moisture utilization
C = moisture deficit
D = soil moisture recharge

Figure 3.7 Typical annual variation in hydroclimatic variables in a seasonal Northern Hemisphere temperature regime.

sensing as an increase in surface reflectivity, caused by desiccation of the land surface and wilting or death of vegetation and crops (Idso et al. 1975). In practice, other indicators are needed to corroborate such findings (see below).

Characterizing drought

At Alice Springs, in central Australia, water shortage is often frequent and prolonged (the mean annual rainfall is 22.5 cm but the PET is 241.8 cm). Environmental scientists there have developed a drought index based on criteria for vegetation growth and the support of range animals. When the rainfall total of a particular month is sufficient to cause pasture growth, the index is set to zero. Evaporation is monitored by lysimeter and pans (which produce about five times as much evaporation as the lysimeters), and the daily total of AET is subtracted from each day's rainfall. If the daily water budget calculated in this way reaches zero, then the estimated AET is accumulated to provide a measure of aridity for the ensuing period. If rainfall is sufficient to cause vegetation growth, the index is returned to zero.

Arid lands usually suffer water shortage as a matter of course, whereas in humid temperate areas moisture deficits tend to be rare. Hence the impact of drought tends to be greatest in the intermediate zones, the semi-arid world, where normal growing seasons may easily be interrupted by prolonged dry spells, in which relative humidity may fall from 40 per cent to only 2–3 per cent, and which exacerbate normal water shortages. Natural scientists in the semi-arid lands have distinguished between **meteorological** and **effective** or **agricultural drought** (Nir 1974). Meteorological drought is simply the duration between two significant spells of rainfall: in the arid zone this may amount to years, whereas in the semi-arid zone it will be 5–8 months long, depending on the balance of the seasons. For unexpected water shortages to occur the dry spell must be abnormally long; but it should be borne in mind that the definition of "drought", just as that of "significant rainfall", is arbitrary.

"Effective" aridity is more of a pragmatic concept, and represents the duration between two rainfalls which feed usable groundwater supplies. The lower the average annual rainfall, the more critical to crops are deviations from that average: even when percentage deviations are small, the actual amount of water may fall short of that required for crop growth. Thus, if the saturated, or **phreatic**, zone is depleted, the **capillary fringe** may not rise up sufficiently to moisten the soil around plant roots. If the soil dries out completely to depths of more than 1 m, plants with shorter roots (including most crops) must either conserve moisture or wilt and wither. The greater the deficit of moisture in the atmosphere, the more it will absorb soil moisture by evapotranspiration, thus penetrating deeply beneath the surface.

148

According to a definition developed in Tanzania, if lack of usable moisture causes crop yields to fall by 8 per cent then a severe drought has occurred, whereas if yields are 30 per cent down then the drought is described as "major".

Wetland crops will not grow unless rainfall exceeds PET by about 50 per cent, while any area which has an annual precipitation of less than 250 mm is likely to suffer moisture shortage for dryland crop growth. In such cases, there is likely to be a dependence on groundwater for irrigation, with attendant risks of depleting the phreatic level. At deeper levels beneath the soil groundwater supplies may have accumulated under the wetter conditions of a former **pluvial** climate. These are termed **fossil groundwater** deposits, and the use of deep wells to tap them is regarded as **groundwater mining**: it represents the utilization of a non-renewable, finite resource.

In any event, it should be borne in mind that **effective water shortage** depends in part on the hydro-meteorological moisture balance and in part on shifting patterns of demand.

The indicators of drought

Drought is one of the principal natural hazards that cause food shortages in underdeveloped regions (Fig. 3.8; see also Ch. 8). Given the slowness with which food aid is usually supplied, and the logistical difficulties of mounting a major relief operation, it would be advantageous to be able to predict famine caused by drought before it becomes serious. Hence, a

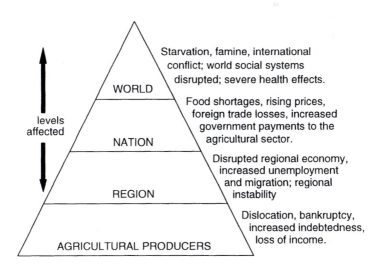

Figure 3.8 Range of possible social consequences of drought (Warrick 1983: 70).

149

series of indicators can be monitored (De Waal 1988). These include rainfall, comprising monthly totals, intensities of rain per hour and the length of "gaps" or dry spells between expected rainfalls. Also, agricultural production can be monitored in terms of the area under particular crops, the forecasts for harvests and actual harvest figures. But such information must be tempered by consideration of market prices and throughput, including the quantities sold and timing of sales of livestock (monitored by sex, age and use) and basic food grains.

Demographic data are important, as unusual population movements may indicate disruption of the rural economy, and low figures for school attendance may indicate family crisis. Lastly, health and nutrition data should be monitored; in particular, infant mortality and morbidity data, including the prevalence of measles and diarrhoeal diseases (number per thousand of the population at risk), and nutritional status (e.g. percentage of children under 5 years of age who are below 70 or 80 per cent of normal weight for height). If any of these indicators begins to appear abnormal, it can be investigated at the local level in sufficient detail to give precise information on the probability of serious famine crisis and steps can be taken to prevent this in good time.

Drought in the Sahel
One of the regions of the world most vulnerable to drought is the Sahel of Sub-Saharan Africa (Fig. 3.9a). This area extends for 2.5 million km^2 across the African continent on the southern fringes of the Sahara Desert between latitudes 13° and 17°N. Its annual mean temperature is 28–30°C and its annual precipitation averages 250–500 mm. The region is dry with a short rainy season which extends from July to September. It is also an area of relatively high population densities, widespread subsistence economies and hence extreme vulnerability to drought.

Seasonal drought in the Sahel may be defined as occurring annually from October to June, during the normal dry season. **Contingent drought**, on the other hand, occurs when the normal rainy season is delayed or abated, as took place from 1969 to 1973 and again during the mid-1980s. According to one hypothesis, global cooling in the Northern Hemisphere mid-latitudes could be responsible for this, as the increase in temperature differences between the equator and poles may reinforce the westerlies and thus suppress the winds that bring rain to the region. However, other studies show the regional distribution of wet and dry periods effectively to be random. Nevertheless, one can dispute neither the rather haphazard pattern of rainstorms during the dry season nor the failure of the intertropical convergence zone (ITCZ) to migrate sufficiently far north in early summer. The ITCZ is an area of conflicting winds at the edge of the Hadley cells of macroscopic atmospheric circulation. The atmospheric turbulence that it generates, and the moisture brought by the winds, are

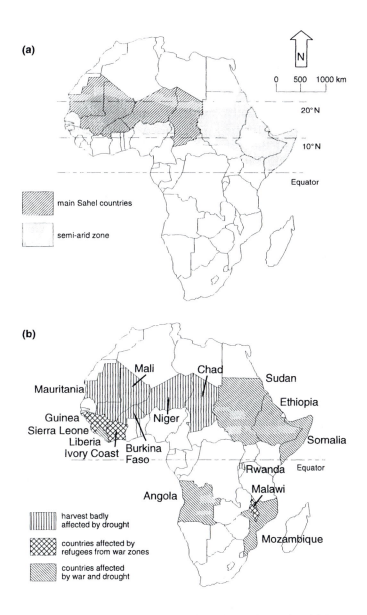

Figure 3.9 The Sahel region, drought and war in Africa.

sufficient to generate short but intense rainstorms. The absence of these during periods of contingent drought has led to great suffering among the populations of countries such as Burkina Faso, Chad, Mali, Mauritania, Niger and Senegal.

In 1973 alone, 100,000 people were killed by drought in the Sahel; 25 per cent of all cattle in the area eventually died or were slaughtered, while the herds of Mauritania were reduced by a full 80 per cent. In 1974, 200,000 people in Niger were entirely dependent on food aid, a similar number were refugees in the settlements of Mali and 250,000 Mauritanians were destitute. Throughout the region calorific intake dropped dramatically. Unfortunately, the problem of drought in Africa has not been solved and in early 1991 4.28 million people were facing starvation in the northeast of the continent.

Evidence from Sub-Saharan Africa indicates that farming households in rural areas of developing countries often evolve a survival strategy to overcome the effects of worsening drought. For example, in order to avoid starvation, an Ethiopian family might begin by selling its livestock. Later, it may turn to wage labour, though this can lead to the collapse of a precarious labour market if supply overwhelms demand and depresses wages. In times of increasing difficulty, cash or food may be borrowed from relatives, neighbours or friends. Valuables such as jewellery and firearms may be sold and capital assets may gradually be sold or lost. In the end, the repeated failure of crops and loss of all resources will lead to migration (Cutler 1984).

Several other tactics can be employed to fight and survive contingent drought in the Sahel. As far as possible, food stocks should be carefully conserved. When necessary, they should be supplemented by food aid supplied by the international relief community. Very careful seed drilling will help utilize pockets of moisture retained by the soil, while drought-resistant crops such as millet and sorghum, which have waxy surfaces, can be planted. In extreme cases, rainstorms can be followed with "instant" cultivation in order to obtain the benefit of perhaps only one or two such events in any single year.

Although it is possible to evolve a strategy to combat drought in the Sahel and other parts of the world facing similar problems, there are many obstacles. High rates of population growth in Africa mean that there are increasing numbers of mouths to feed, yet it is very difficult to increase the carrying capacity of the land. In fact, although the area of land under crops and the overall yield are both rising, yields per hectare are falling and population growth has outstripped increases in food production. Yet dependency on food aid from external sources should not be encouraged, as such supplies cannot be guaranteed indefinitely (see Ch. 8). Meanwhile, armed struggle seems to be prevalent in wide areas of the Sahel, including the subversion of food relief and the destruction of crops. Indeed, throughout the continent there is a strong correlation between war, drought and fluxes of refugees (Fig. 3.9b). This means that it is very difficult to develop the necessary socio-economic stability to consolidate a tenuous relationship with a potentially hostile natural environment (see Glantz 1976, 1987).

As the next section shows, the impact of drought tends to be very different in the developed world.

Drought in the USA
The drought hazard in America is greatest in the southwest of the country and the semi-arid parts of the Great Plains (Fig. 3.10). The latter are characterized by periodic wide shifts in precipitation, which have led to sustained droughts in the 1930s, 1950s and 1970s (Rosenberg 1978). Losses have averaged more than $700 million a year and farmers have suffered hardship, bankruptcy or geographical dislocation. Government relief and rehabilitation programmes have been required on a massive scale in order to mitigate regional economic disruption, land abandonment and rising food prices. There has also been a disturbing trend towards costly droughts in the populous urban-industrial areas of the eastern seaboard and west coast. Water shortage here can only be

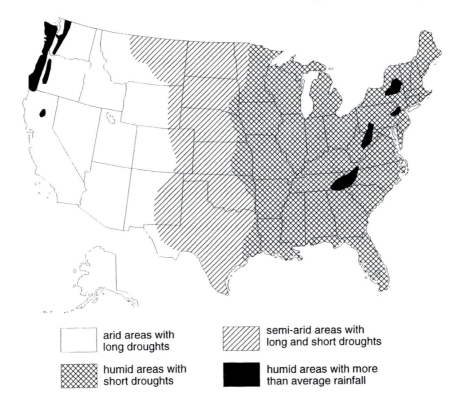

Figure 3.10 Drought potential in the USA (White & Haas 1975: 372). Copyright 1975 by MIT Press.

measured in terms of the rising demand associated with private and corporate affluence.

The agriculture of the early Midwestern settlers was more suited to humid and subhumid climates than to semi-arid ones. However, drought disasters in the 1930s stimulated improvements in irrigation and land management, diversification of farming operations and regional economies, the institution in 1938 of Federal crop insurance schemes and more liberal credit arrangements. In particular, the use of surface and ground water for irrigation has increased dramatically during the twentieth century, such that 10 per cent of all American farms now irrigate and produce 20 per cent by value of all crops. On the Prairies and Great Plains irrigation has increased six-fold. But in the western USA the evaporation loss from water surfaces is 120–240 cm per year, and seepage from canals and ditches is high. In the 17 western states, transpiration losses from phreatophytes (plants whose roots draw water from the saturated zone) may reach 25,000–30,000 million m^3 per year. Thus, irrigation efficiency on farms probably does not exceed 50 per cent.

There is also a need for better and more consistent laws governing water rights, in order to clarify the responsibilities and jurisdictions of those who draw off water. Although the area of crops insured against drought doubled from 1948 to 1974, little more than 10 per cent of all farms are currently insured. Finally, the problem of urban drought needs to be tackled more effectively by recycling water and abating demand, particularly in metropolitan areas with relatively dry climates (Wilhite & Easterling 1987).

One might add that drought also affects peripheral areas of the Commonwealth of Independent States very seriously (Fig. 3.11). This is one reason why, just as the USSR which it replaces, the commonwealth has been so dependent on imported grain and other foodstuffs, and why, paradoxically, the American economy and farming community have benefited considerably from water scarcity in Eurasia.

Hurricanes

Severe cyclonic storms that begin over the tropical seas are known as hurricanes in the north Atlantic Ocean, Caribbean Sea, Gulf of Mexico, eastern north Pacific and western coast of Mexico. They are called typhoons in the western Pacific Ocean and simply tropical cyclones in the Indian Ocean and Australasia. For clarity the term hurricane will be used in this section as the general synonym. The naming of hurricanes began in the 1940s, while in the 1980s the practice of giving them only female names was modified to include male names as well.

Hurricanes occur primarily within the tropics between latitudes 30° N

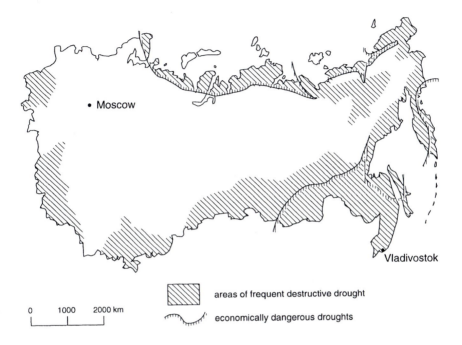

• Moscow

•Vladivostok

0   1000   2000 km

▨ areas of frequent destructive drought

〜 economically dangerous droughts

Figure 3.11 Drought hazard in the former Soviet Union (Gerasimov & Zvonkova 1974: 248).

and S, but not within the equatorial trough ±5°, where atmospheric disturbances tend to be insufficient to cause them (Fig. 3.12). For a storm to be classified as a hurricane, the sustained wind speed must exceed 65 knots (120 km/hr). In a fully developed hurricane it may exceed 200 km/hr towards the centre of the depression.

Hurricanes have a markedly different effect on communities in the developing and the industrialized worlds, as illustrated by two events which took place during 1974 (see also Table 3.2). Hurricane Fifi killed up to 8,000 people in Honduras while Tropical Cyclone Tracy killed 49 in Darwin, Australian Northern Territories. In both cases wind speeds reached 230–250 km/hr, but there was considerably more prior evacuation in Darwin than in Honduras. Also, in the latter case, deforestation aggravated the mudflows and flash floods that were responsible for many of the deaths: marginalization of the poor (see Ch. 8) was in many ways the key to the Honduran death toll.

In the United States and Caribbean the damage to property caused by hurricanes has increased substantially in recent decades, but the toll of human life has been reduced by more rigorous forecasting, warning and evacuation of coastal communities. In the USA 70 million people are

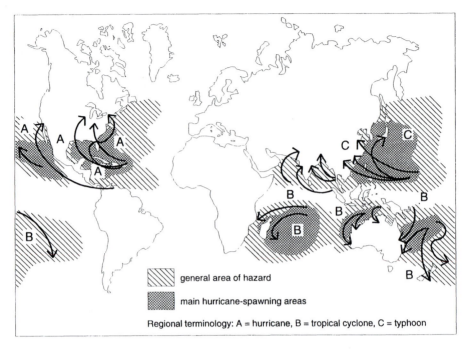

general area of hazard

main hurricane-spawning areas

Regional terminology: A = hurricane, B = tropical cyclone, C = typhoon

Figure 3.12 Areas of the world affected by hurricanes.

Table 3.2 Impact of severe storms (mainly hurricanes) over the period 1960–81 (after Wijkman & Timberlake 1984: 76)

Country	Number of storms 1960–81	Number of people killed
Countries with low-yield economies:		
Bangladesh	37	386,200
Burma	7	1,350
China	7	170
Haiti	6	5,800
India	26	24,930
Madagascar	9	930
Vietnam	6	7,480
French Caribbean	5	100
Countries with moderate-yield economies:		
Hong Kong	7	510
Mauritius	7	15
Mexico	14	1,560
Philippines	39	5,650
South Korea	10	700
Example of country with a high yielding economy:		
Italy	5	110

estimated to be at risk from hurricanes, nearly 10 per cent of them residents of Florida, the worst affected state. Death tolls average 60 per year, but are falling as a result of better preparation and warning. Property losses, however, are rising steeply and now average more than $430 million per year (White & Haas 1975). More than 30 per cent of property in America is at risk and by the year 2000 annual losses will exceed $1.90 per head of the population (they now surpass $28 per head in Florida). A study by Herbert & Taylor (1979) showed that in 1970, 80 per cent of the 28 million residents of the eastern sea coast of the United States had never experienced a hurricane, even though the area is one of major risk; many of them had moved to the coast within the previous 20 years. In the aggregate, perception and preparedness were not expected to be high. Nevertheless, when on 25–26 August 1992 Hurricane Andrew brought winds of 240 km/hr to the coasts of southern Florida and Louisiana, more than 1.5 million people were successfully evacuated and, though damages exceeded $20,000 million, only 13 deaths occurred.

The cause of hurricanes

Hurricanes develop over water and tend to lose their force over land. They form over the equatorial seas where the troposphere contains 80–85 per cent of the mass of the atmosphere and extends at altitudes of 15 km to the tropopause (Fig. 3.13e). In the hurricane-generating area surface humidity is fairly constant at 75–80 per cent, and sea surface temperatures are at least 26°C down to 60 m below the surface. This enables the right combination of heat and water vapour to sustain a thick convective layer. Moreover, a 1°C increase in sea surface temperature is sufficient to reduce the minimum pressure at which hurricanes can exist by 15–20 mb. Hence, storms are sustained by strong rates of evaporation from the sea surface and low friction of winds against it, while over land they are cut off from the main source of water vapour and slowed down by friction against a rougher surface.

One theory is that hurricanes form when latent heat release is concentrated in clusters of cumulo-nimbus clouds, resulting in a form of "organized convection". However, this is inaccurate, as the water vapour content of the lower clouds is actually increased by rainfall from the upper layers. The real cause lies in the thermal disequilibrium between atmosphere and sea surface, and between parcels of air. The stronger the contrasts are, the more severe the storm; and severe storms have warmer ascending air than moderate storms, which in turn involve warmer air than the mean tropical atmospheric temperature.

Latent heat of condensation released to the atmosphere in heavy clouds furnishes the energy for starting and maintaining hurricanes. It causes the atmosphere to warm and become less dense, so that as the air rises surface pressure falls correspondingly. This stimulates convection and

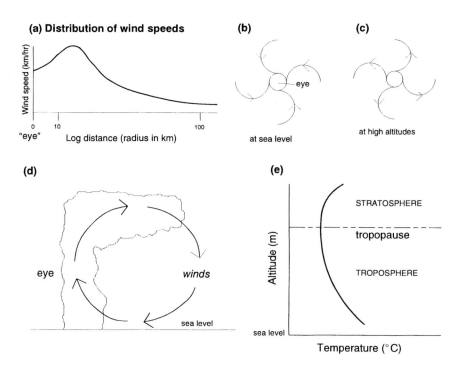

Figure 3.13 Aspects of hurricane dynamics (after Simpson 1973).

winds, which lead to further heat releases. Hence a positive feedback loop is created, as the rate of evaporation increases with rising wind speed. Only if this saturates the air throughout the inflow layer will the hurricane vortex decay rather than grow. In this way, a cold-cored tropical storm can be transformed into a warm-cored hurricane, in which energy expenditures are considerably larger. Warmed air rises and intensifies anti-cyclonic air flow in the upper troposphere (perhaps 8,000–12,000 m above sea level). Thus, high-level outflow maintains the ascent of cooling air, while low-level inflow continually generates energy from latent heat release.

Another fundamental cause of hurricanes is the presence of unstable shear winds at the convergence of the north and south Trade Winds. In the central Atlantic Ocean, this leads to the formation of "easterly waves", or deflections of air mass fronts, which eventually coagulate into rotating cyclones. However, there must be no horizontal wind shear within the area of the hurricane itself, as this tends to destroy the feedback between surface wind and sensible heat (Emanuel 1988).

A hurricane may form gradually over several days or rapidly in the space of 6–12 hours; and, in typical cases, the hurricane stage will last 2–3 days, while the storm of which it forms part may die out after 4–5 days. Yet the formation of hurricanes is much rarer than the apparent opportunities for them to occur. Only a maximum of 10 per cent of centres of falling pressure over the tropical seas become fully fledged hurricanes, and the pattern varies greatly from year to year. In a year of high incidence, perhaps 50 cyclones will develop to hurricane proportions (although not all of them will cause disasters!), while outside the tropics on an average day 20–30 cyclones exist (most of which will be considerably gentler forms of storm than the hurricane).

"Architecture" of hurricanes
The hurricane consists of a rotating atmospheric disturbance whose diameter may vary between less than 100 km and more than 1,500 km. The hurricane itself will tend to be more restricted in size than the general disturbance that surrounds it. At the centre there will be an "eye" of calm atmosphere and broken cloud cover or clear skies, which is dominated by descending parcels of air. The central calm occurs where the angular forces causing winds to converge on the centre of the storm can no longer force them inwards, and hence air rises in the "wall" of the eye. In major hurricanes the eye will be 30–50 km in diameter and in average circumstances may take between 30 minutes and 4 hours to pass a point on the surface. The eye forms during the cold-core tropical storm stage, as anvil-shaped towers of cumulo-nimbus clouds associated with local atmospheric convection redistribute the outward-flowing air and allow strong centripetal forces to develop in a sustained way around the central calm. The wall of the eye is surrounded by a band of relatively weak precipitation, which grades outwards into moderate or heavy rain. Adiabatic compression of falling air (i.e. without heat loss) renders the inner part of the eye relatively warm.

In considering the dynamics of hurricanes, it is important to distinguish between the speed of rotation of winds, which may be in the range 120–250 km/hr, and the forward motion, or tracking, of the entire cyclonic disturbance, which may occur at 40–60 km/hr (although it is not unknown for hurricanes to remain stationary for periods of time). Moving outwards from the centre of the storm, wind speeds increase rapidly from the wall of the eye to a point about 20 km away, after which they decline progressively towards the edge of the storm (Fig. 3.13a).

Internally, the hurricane is a relatively simple example of Carnot's "heat engine" (a reversible thermodynamic cycle which converts heat to mechanical energy). Air converges at sea level and rises in a spiralling motion around the eye, producing a heavy mass of cumulo-nimbus cloud of great depth (Fig. 3.13d). Outflow occurs at altitudes of 9,000–15,000 m above sea level, producing a broad canopy of cirrus cloud. In the

159

Northern Hemisphere air rotates clockwise into the depression and counter-clockwise out of it (vice versa south of the Equator; Fig. 3.13b&c), but the cyclonic circulation is much weaker at the high-level outflow than at the low-level inflow, as considerable momentum and energy are lost in creating evaporation and condensation, and in overcoming friction.

In summary, for a hurricane to form, several conditions must apply. First, there must be a mechanism which induces atmospheric pressure to fall near sea level. Condensation caused by rising moist air, which is close to saturation with evaporated sea water, must release sufficient latent heat for air density to fall. If winds are very strong at the start of the cycle, the latent heat release will not be concentrated enough to generate sufficiently intense convection. Next, a vortex must develop in the lower atmosphere, favouring the frictional transport of mass towards the centre of falling pressure. At all times, the air that rises in the centre of the cyclone must be conducted away without being impeded by other air flows, as any blocking of convection will tend to impede hurricane genesis.

Hurricane researchers have devoted much attention to the relative importance of internal and external factors in generating the storm. Regarding the former, studies of cloud microphysics are beginning to answer the question of how latent heat is released, which may shed new light on the endogenic conditions that lead to hurricane formation. During the 1950s and 1960s research concentrated on bringing radar and barometric and temperature sensing devices into hurricanes. Overflights took place at altitudes of 12,000 m and followed a clover-leaf pattern of sortie, which was an efficient means of covering the storm. This research helped determine the conditions which give rise to unimpeded lateral escape of air above the storm, thus stimulating the updraught.

The hazards associated with hurricanes
Winds of up to 88 m/s (320 km/hr) have been observed in hurricanes. They can be a severe threat to shipping, and in coastal settlements can cause structures to collapse and debris to be picked up and whirled around. Torrential rains can result in the rapid augmentation of flow in coastal stream catchments and hence in flooding from inland sources. Hurricane-driven waves can be highly erosive to coasts and shoreline structures. Moreover, the hurricane may incorporate several tornadoes (see next section). But the main hazard is caused by the **storm surge**. This results from the shearing effect of wind on water, which causes local sea level to rise and inundation to occur where the hurricane makes landfall. In fact, 90 per cent of fatalities in hurricane disasters are drownings, and the storm surge is responsible for many of these. Inundations, moreover, result in the greatest losses of property.

The level of the storm surge depends upon the following variables: the distance from the storm centre to the radius at which maximum wind speeds occur; barometric pressure in the eye (as the degree of pressure drop is proportional to the strength of winds); the rate of forward tracking of the storm; the angle at which the centre of the storm crosses the coastline; and the profile of water depths seaward along the shore platform. In addition to the storm surge, enclosed coastal sounds and bays may undergo **seiching**, which is an oscillatory movement of tsunami-like waves caused when the body of water is impacted by the hurricane (see Ch. 2). If the storm moves forward very rapidly, seiching may occur in certain open bays, and in low-lying areas it can be very destructive. For example, in Galveston Bay, Texas, seiching caused by a hurricane in 1900 drowned 6,000 people.

Various forms of damage can be caused by hurricanes (see Figs 3.18 & 8.7). Southern (1979) showed that the degree of adaptation of housing to high winds and flooding is high in areas where hurricanes have a recurrence interval of more than once a decade, low in areas with return periods of once in 25 years and highly variable in zones with recurrence intervals of 10–25 years. It is important, however, to keep damage in proportion: the Munich Reinsurance Company has argued that hurricanes (Beaufort force 12) cause an average 10 per cent loss of buildings in the areas which they pass over. Although severe extra-tropical storms (Beaufort force 8–11) cause only a 1 per cent loss, they are ten times more frequent than hurricanes and thus do equivalent amounts of damage.

Hurricane prediction

The prediction of hurricanes involves interlocking oceanographic and atmospheric research. Clusters of rain clouds must be identified and monitored and their evolution predicted if they look likely to develop into hurricanes. Most importantly, once a hurricane has appeared, the timing and location of coastal approach must be predicted, along with the probable height, extent and consequences of the storm surge. Thus, hurricane surveillance involves three stages: detection, tracking and landfall prediction.

Weather satellites are probably the most important instruments for detecting and monitoring hurricanes. These include polar orbiting and geostationary satellites, such as those of the GOES and NOAA families. For instance, a geostationary satellite positioned 35,900 km above the Equator can provide storm surveillance images once every 11 minutes. But the satellite view is that of the exhaust production of the hurricane "heat engine", rather than the convective processes which govern the future development of the storm. Nevertheless, hurricane seedlings have been classified according to the Dvorak system, which is a typology of the hook-shaped cloud clusters seen on radar and weather satellite images (Fig. 3.14).

Most hurricanes move according to well-defined paths, in which future motion is predictable on the basis of momentum derived from previous tracking (Fig. 3.15). Hence, in 70 per cent of cases hurricane paths can be forecast 24 hours in advance on the basis of directions and speeds over the previous 24–36 hours. This means, however, that 25–30 per cent of hurricanes are not amenable to simple forecasting, although the track of nine of every ten of these can be forecast using a combination of data on climatology, current meteorology and momentum (persistence), without reference to environmental influences. Overall, the World Meteorological Office has succeeded in reducing the root–mean–square error of wind forecasts from 10 to 8 m/sec, but despite this the average forecast errors for hurricane tracks are still 100, 220 and 370 nautical miles for 24-, 48- and 72-hour predictions, respectively (Sheets 1990).

In the United States, the worst hurricane hazard is experienced in the southeast of the country from the Gulf coast of Texas to Georgia and the Carolinas. The National Hurricane Center at Miami, in Florida, uses a variety of computer and other forecasting models, of which some compute dynamic processes in digital form and others make statistical calculations, based on the probability that a hurricane will follow a particular path (see Fig. 3.16 & Table 3.3). The most probable track is

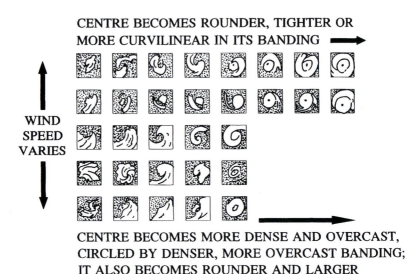

CENTRE BECOMES ROUNDER, TIGHTER OR MORE CURVILINEAR IN ITS BANDING ➡

WIND SPEED VARIES

CENTRE BECOMES MORE DENSE AND OVERCAST, CIRCLED BY DENSER, MORE OVERCAST BANDING; IT ALSO BECOMES ROUNDER AND LARGER

Figure 3.14 Dvorak classification of incipient hurricanes (Simpson 1973: 901). Copyright 1973 by the AAAS.

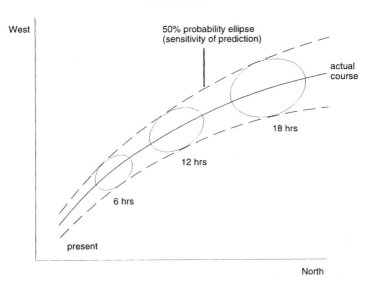

West

50% probability ellipse
(sensitivity of prediction)

actual
course

18 hrs

12 hrs

6 hrs

present

North

Figure 3.15 Prediction of the future track of a hurricane (Simpson 1973: 905). Copyright 1973 by the AAAS.

predicted up to 72 hours in advance (depending on the speed of forward movement), and the likely position of the hurricane at any point during this time interval is specified with respect to a 50 per cent probability ellipse. The width of the ellipse is indicative of the sensitivity of the forecast, and it tends to dilate as the time interval of prediction is increased. In the Caribbean Sea and western tropical Atlantic Ocean the most accurate results have been achieved with respect to hurricanes moving westward at 20 km/hr, whereas storms moving northward are less easy to forecast (Sheets 1985).

The US National Oceanographic and Atmospheric Administration has developed a numerical model designed to predict storm surges which is entitled SLOSH.
: Sea, Lake and Overland Surges from Hurricanes. Input variables include the position, size, central pressure and forward speed of the storm, the prevailing tidal factors and the configuration of offshore topography. The output consists of a mapped prediction of storm surge height, and wind speed and direction (Griffith 1986). The forecast permits an evacuation time estimate to be computed and communicated to emergency managers, which enables quantitative predictions to be made of the probability that a hurricane will strike a given coastal community. The storm surge tends to be the most predictable aspect of the hurricane: for example, a surge 7.5 m high produced in 1969 by Hurricane Camille was predicted with 98 per cent accuracy, or in other words to within 50 cm. However, there are still problems of predicting the height of flooding in estuaries (where river flows and the height of tidal influxes

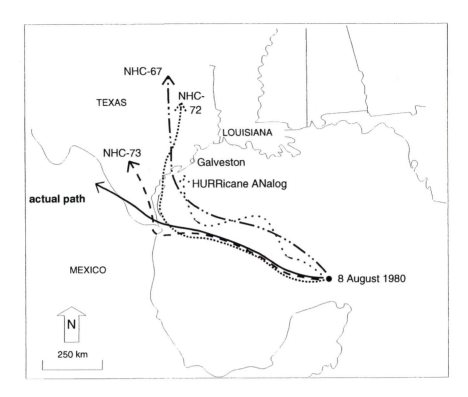

Figure 3.16 Predicted hurricane paths compared to actual tracks (Fiorino et al. 1982: 654).

must be taken into account) and along irregular coastlines.

At landfall, digital simulation models can give a clear picture of the expected maximum wind speed with distance inland, and this will probably be proportional to the damage which the storm causes. However, the exact location of the landfall itself is not easy to forecast. A 100 km error may be statistically reasonable when predicting the point at which the centre of the storm crosses the coast 24 hours later, but this will not help plan orderly evacuation and the saving of lives and property. Neither is it easy to give adequate warning to small, isolated island communities, which may be vulnerable to sudden inundation. This is a particular problem of the Pacific Basin, where many small nations and isolated inhabited islands stand in the path of hurricanes.

Logistical considerations
In the southeast USA the alert phase of severe storm monitoring is called a **hurricane watch**, while expected landfall will generate a more specific **hurricane warning**. The expected vector error for predictions of landfall is

Table 3.3 Methods of forecasting North Atlantic hurricanes (Neumann & Pelissier 1981: 523, Sheets 1990).

Method or model	Methodology	Means of prediction
Official	Subjective	Professional analysis of certain outputs
HURRAN	Climatological analogue	Tracks of other hurricanes, direction, speed, current latitude
CLIPER	Statistical	Current storm latitude and longitude, current and previous 12-hr W–E(C_x) and N–S (C_y) motion, day number, maximum sustained wind speed
Persistence	—	Previous storm motion (12–24 hr): C_x, C_y
Riehl-Haggard-Sanborn	Statistical	Actual non-horizontal wind velocity, V_{9500mb}
Miller-Moore	Statistical	Actual non-horizontal wind velocity, V_{9700mb}
NHC-64	Statistical	Sea level pressure, P_S: height of isobar surface, z_5, z_7; space between isobar surfaces, Δx, Δz; C_x, C_z; V_{9300mb}, V_{9500mb}
NHC-67	Statistical	Geopotential height, gradient and thickness, 24-hr change
NHC-72	Statistical	Same as NHC-67, but also utilizes CLIPER
NCH-73	Statistical-dynamic	CLIPER and current 1,000, 700 and 500 mb analysis
VANBAR	Barotropic-dynamic	V_{9500mb}
SANBAR	Barotropic-dynamic	100–1000 mb winds
MFM	Simple dynamic equation	P_3–three-dimensional analysis of V, T and g
Satellites	Graphic modelling	Infra-red monitoring, etc.

about 160 km. This poses a dilemma for forecasters, in that the prediction may give rise to expensive preparations and disruptive evacuations (for instance, Dade County, Florida, which contains the city of Miami, spends $2 million on preprogrammed preparation when a hurricane warning takes effect). If the hurricane nevertheless bypasses the community warned, then the effectiveness of future warnings will be jeopardized. On the other hand, applying the warning to too long a length of coastline is likely to be less effective than giving out more specific information about landfall. For these reasons, specific hurricane warnings are rarely issued more than 12–18 hours in advance of landfall, by which time coastal radar stations can assist in the tracking of the storm on a continuous basis (Simpson et al. 1985).

Rapid increases in the size of coastal populations make continual review of evacuation plans imperative. Urbanized barrier islands, for example, are vulnerable to isolation if blockages occur in the vicinity of the bridges which connect them to the mainland, especially during the evacuation of people to higher ground inland at a time when wind speeds are increasing and the coastal area is about to be inundated. Local ordinances should in any case prevent the development of a margin of land in the immediate vicinity of the shore, where the greatest damage is likely to occur during the hurricane; this is called the **setback line**.

Hurricane modification

The hurricane is the only major atmospheric disturbance powered by the condensation–precipitation process. It can be seeded with dry ice or silver iodide crystals (which replicate the form of condensation nuclei and collect supercooled water droplets) in order to decrease the liquid content of clouds and redistribute the energy of latent heat more evenly. If the temperature difference between the edge and the eye of the hurricane can be reduced, wind speed can be decreased and the energy of the storm abated. As the level of damage caused by hurricanes at landfall bears a nonlinear relationship to the strength of winds, a relatively small decrease in wind speed may cause a large reduction in property losses and potentially also in the number of fatalities. Experiments in 1969 with Hurricane Debbie showed that wind speed fell by 31 per cent after five seedings (Fig. 3.17), although it later rose again. In 1971 Hurricane Ginger, a large, diffuse storm was seeded several times and after 48 hours wind speed had fallen by 12 per cent. But again the results were inconclusive.

Hurricane seeding raises several important questions. An average

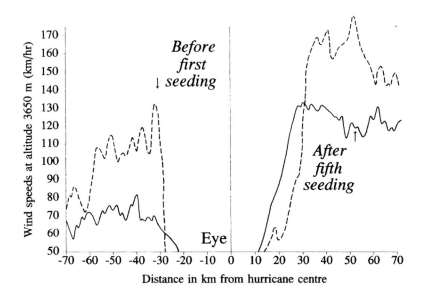

Figure 3.17 Silver iodide seeding of Hurricane Debbie, 18 August 1969 (Gentry 1970: 474). Copyright by the AAAS.

hurricane generates the equivalent of 10^{11}–10^{12} kW/hr of energy, and modifying such an energy budget may result in serious consequences if other forms of destructive storm are accidentally spawned. However, although the thermodynamic efficiency of the hurricane is about one third, its energy expenditure is lop-sided. Thus it may be possible to search for areas of the storm in which cloud seeding will have maximum impact. Furthermore, if the moisture present in clouds could be induced to evaporate instead of falling as rain, the liquid content of vortex clouds might decrease by up to 50 per cent, reducing the temperature difference between the eye and the edge of the storm and thus decreasing wind speeds.

The main problem is that hurricanes have a high degree of variability. The US Project Stormfury initiated in 1962 would only seed a hurricane if the probability of its coming within 80 km of a populated area within the subsequent 18 hours was less than 10 per cent, thus containing the risks associated with a failed experiment but reducing the practical value of the exercise. Numerical models suggest that seeding is capable of reducing maximum wind speed by 10–15 per cent, but may increase wind speed outside the maximum area by a corresponding amount or may create a second maximum hours later. In any case, the overall kinetic energy expenditure of the storm is increased, and artificially induced precipitation seems to give rise to greater natural rainfall at the margins of the storm. Other methods of stifling the hurricane, such as inhibiting evaporation by spreading the sea surface with a biodegradable monomolecular film, are of dubious practicability.

The enormous damage potential of hurricanes (Fig. 3.18) means that a small reduction in physical forces can lead to a large decrease in losses. If 10 per cent of hurricanes could be steered away from land, and the intensity of the other 90 per cent could be reduced by 15 per cent, damage costs would fall by more than a half. The benefit: cost ratios for investment in hurricane modification are 10–20:1. But a correspondingly small increase in the force of the storm could create disproportionally high increases in damage. For example, simulation models suggest that decreasing the maximum wind speed may sometimes increase, rather than decrease, the height of the storm surge. Hence, the results of experiments have been sufficiently equivocal, and the consequences sufficiently unpredictable, for a virtual moratorium to have been imposed on further attempts to modify hurricanes.

The impact of Hurricane Gilbert on Jamaica, 1988

The impact of Hurricane Gilbert on the Caribbean Island of Jamaica provides a clear illustration of how relatively simple forms of storm damage can have complex socio-economic and medical consequences. This was considered to be the worst cyclone to occur in the tropical

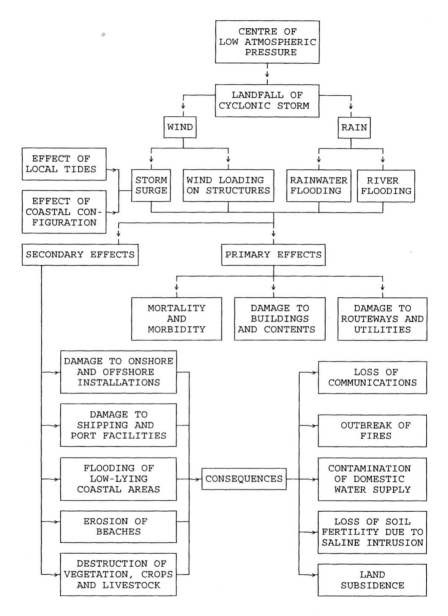

Figure 3.18 Damage potential of hurricanes (modified from White 1974: 256).

Atlantic during the twentieth century, until in 1989 Hurricane Hugo surpassed it in fury and damage.

Gilbert reached Jamaica on the morning of 12 September 1988 and made landfall at a speed of 25 km/hr. The central pressure was 963 mb and the eye was about 40–48 km in diameter. Sustained wind speeds (averaged over 10 minutes) at 10 m above ground reached 135–150 km/hr, while gusts achieved 210–225 km/hr. Jamaica has been highly active in promoting evacuation and sheltering plans, and hence loss of life nationally was only 45, including 11 people shot by the police as presumed looters. But economic losses reached $2,000–3,000 million, and may thus have exceeded the island's entire annual GDP.

About 40 per cent of the damage in Jamaica occurred to housing, with 130,000 units, or one house in four, significantly affected. Most damage resulted from loss of roofs, after which wind and water played havoc with interiors. In one instance, the roof was blown right off a church in which 400 people were sheltering. Damage was also disproportionately great with respect to certain other classes of building: for example, more than 500 of the island's 580 schools were harmed.

Hurricane Gilbert shows graphically how simple damage can have a complex impact on the functioning of essential services. Damage to ten of Jamaica's hospitals meant, not merely that expensive buildings and equipment needed to be replaced, but also that medical facilities had been lost at a time when injuries caused by the storm were augmenting the normal demand. When the country's main telephone exchange lost part of its roof, drenching rainwater put the switching apparatus out of action and cut off all international telephone connections. Similarly, internal communications were seriously compromised by the loss of microwave, radio and television towers. Many such installations were located on hill crests, where wind speeds could double at low elevations as a result of the acceleration caused when the wind is forced upwards.

Damage to industrial and commercial premises was often critical and had far-reaching consequences. Loss of income as a result of the cessation of production reduced the chances of recovery, as it then became harder to generate the capital necessary for reinvestment. Fortunately, almost 40 per cent of the risk was insured, most of it in hard currency, which allowed for prompt repayment and rapid acquisition of reconstruction materials. However, one Jamaican insurance company lost many of its records when wind and water entered through broken windows into its offices. This illustrates how, as a result of the unanticipated peculiarities of vulnerability, relatively light damage (the building itself remained structurally intact) can have consequences that are disproportionately serious.

In the agricultural sector, damage to crops (banana, citrus, coconut and coffee plantations) was compounded by destruction of storage facilities, stables and barns. The net result was a dramatic change from self-sufficiency in food to dependence on imports.

The underlying cause of the disaster was failure of structures, particularly roofs, which had been subjected to higher wind loads than they were capable of sustaining. Locally made aluminium sheeting, for example, is widely used as a building material in Jamaica, where it is relatively cheap. But in high winds it easily tears away from fasteners. On the other hand, the reinforcement of block walls and the use of hurricane straps to anchor roofs are innovations that have diffused widely across the island as a result of past hurricane damage.

Legislation has been designed that is specific both to these types of damage and to local hazard conditions. The Jamaican Building Code states that structures must be able to resist a 3-second gust of 200 km/hr, while CUBIC (the Caribbean Uniform Building Code) uses a 10-minute mean wind speed of 133.5 km/hr, which is deemed to have a 50-year recurrence interval in this area. For comparison, wind loading factors in Australian building codes involve two "design speeds", 33 and 42 m/sec (119 and 151 km/hr, respectively) which are applied selectively to houses less than 6 m high located in areas susceptible to hurricanes. In any case, damage will be reduced in future hurricane disasters only if standards are rigorously enforced and backed by appropriate experimentation and testing.

In summary, the damage occasioned by Hurricane Gilbert as well as that avoided by prior preparation both illustrate how simple measures can often help avoid complex consequences. Building codes are essential, but the parameters specified must be realistic in engineering terms and with respect to the cost and organizational difficulty of implementing them. Low death tolls and appropriate building codes indicate that Jamaica is now relatively well prepared for hurricane disasters, but on the other hand its initial vulnerability was high and much remains to be done.

Tornadoes

A tornado consists of a violently rotating column of air pendant from a cumulo-nimbus cloud and nearly always observable as an inverted cloud cone (it may, however, be obscured by rainclouds or by dust and debris picked up from the ground). Other manifestations of the same phenomenon include the whirlwind (a loose synonym for tornado), the dust-devil (a small, weak cousin of the tornado occurring in dry lands) and the waterspout (a tornado occurring over water).

Tornadoes are most numerous and devastating in the central, eastern and northeastern United States, where five are reported on an average day in May. They are also common in Australia (where 14.6/yr are reported), Great Britain, Italy, Japan and Soviet Central Asia. Most fatalities occur in the United States, where over the period 1950–78 some 689 tornadoes were killers (Galway 1981).

Mode of formation
Although tornadoes are in some respects miniature versions of hurricanes, they do not occur only in tropical ocean areas, and are probably more common over land than over sea. The urban heat island, however, tends to inhibit their formation in cities of more than 4 million inhabitants.

A series of phenomena is thought to be conducive to the formation of tornadoes. When cool, dry air occurs well above warm, humid air located at the surface, some of the former may begin to sink and thus warm up as it reaches lower altitudes. This gives rise to a temperature inversion, consisting of a warm, dry stratum immediately above the moist surface layer. Through continued **advection** (horizontal airflow), the atmosphere beneath the inversion becomes progressively warmer and moister. As the sun heats the air at the ground surface it begins to ascend and cool, forming large cumulo-nimbus clouds. But it can rise no further than the inversion, and thus results in instability of the local atmosphere.

A cold front may break into the inversion and force the surface air up through it into the cool air above. Cloud growth may be substantial and very rapid, but will not extend above the tropopause. The jet stream moves air away from the top of the cloud and convection occurs beneath it, making the storm more intense. For a tornado to develop, the position and curvature of the jet stream must be favourable and it must travel fast enough. Also, sufficient air must flow into the bottom of the storm. In addition, falling hail and descending cold air may help stimulate appropriate convection currents and energy transfers, and thus hail is often an immediate precursor of tornadoes. It is also thought that static electricity plays an important rôle in generating the vortex.

The genesis of the tornado may depend eventually on the suction vortex rotating into a position in which very high winds are created by the addition of all contributing conditions. In synthesis, tornadoes result from a combination of thermal and mechanical forces requiring layers of air with contrasting characteristics of temperature, moisture, density and wind flow (Stowe 1984). The process consists of three stages. First, a micro-cyclone forms as a result of deflection (by the Coriolis force) of airflow converging on one or more convective thunderstorms. Secondly, latent heat released in condensation creates a vigorous updraught inside the convective cell of the micro-cyclone, which aids the conservation of rotational, or angular, momentum. Thirdly, the lowering of the zone of condensation forms a wall-cloud and eventually a funnel (Fig. 3.19). As long as the updraught caused by the tangential velocity of rotation continues to increase, the radius of the vortex will decrease and the spiralling, rising winds will accelerate. This process is limited only by surface friction and the viscosity of the air.

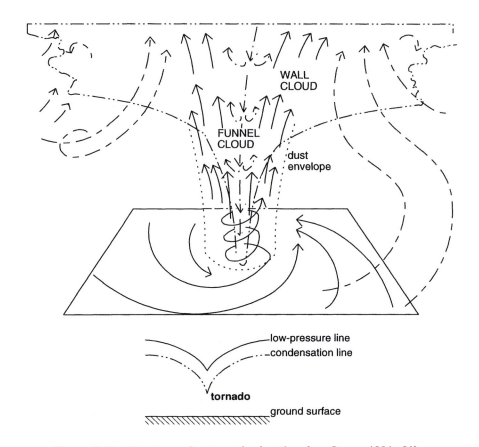

WALL
CLOUD

FUNNEL
CLOUD

dust
envelope

low-pressure line
condensation line

tornado

ground surface

Figure 3.19 Structure of a tornado (partly after Stowe 1984: 91).

Thus, like the hurricane, the tornado is the result of positive feedback in that large amounts of latent heat are released, atmospheric convection is enhanced and intense variations in pressure cause very strong winds. Rotation is a function of the direction of prevailing winds, but is usually clockwise south of the Equator and counter-clockwise north of it.

The nature of tornadoes

Tornadoes may occur either singly or in families in which several funnels are spawned simultaneously or sequentially by only one thunderstorm. Families of tornadoes were responsible for 98 per cent of the deaths which occurred in the lethal outbreak of 3–4 April 1974 in ten American states. In Texas in 1967 Hurricane Beulah gave rise to 115 tornadoes, but most tropical cyclones are capable of generating no more than ten. Tornadoes usually occur on the right side of the hurricane as it travels inland.

Tornadoes do not consist merely of a spinning funnel cloud with very low pressure at the centre. Sometimes there are many funnels, or no funnel is visible at all, especially if it is obscured by rain or clouds, or if the tornado occurs at night. No funnel was visible over the majority of the path of the large tornado which in 1925 travelled 352 km from Missouri to Indiana. It moved at 80–110 km/hr and was visible only as an approaching black cloud that appeared to roll across the ground. Certain large tornadoes have suction vortices within the walls of their main funnels. There are usually between two and five of them, and they may form, disappear and reform. Their speed of rotation may be 150 km/hr faster than that of the main funnel, which may substantially increase the instantaneous wind speed on the right side of the tornado (Fig. 3.20).

The diameter of funnels and vortices averages 300–400 m, while the path may be as narrow as 90 m, or as wide as 1,500 m, although it tends to average slightly less than 250 m. Paths are seldom longer than 75 km, and although in the most exceptional cases the path may extend for up to 750 km, mean lengths are between 15 and 65 km. The edges of the path may be very sharply defined or gradational, where damage and disturbance lessen slowly towards the periphery or end. In the latter case, damage may be ten times as widespread as the path itself, as a result of high winds and flying debris. Frequently, however, a moving tornado will devastate buildings at one point while leaving those adjacent to it untouched. Moreover, on occasion the roughness effect of the urban fabric causes the tornado to lift off the ground surface, while in other

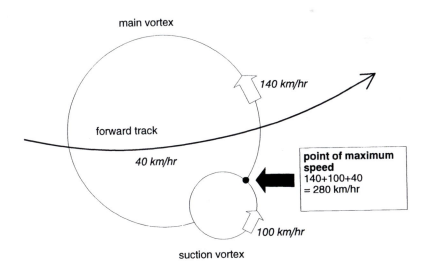

Figure 3.20 Tornado vortex rotation speed (km/hr).

cases it digs in for long distances. Hence, the path tends not to be limited by terrain or other geographical factors.

With tornadoes as well as hurricanes one can distinguish velocity of rotation from speed of forward movement. Not all tornadoes touch down, but when they do so there is usually a definable **path**, which may be evident in terms of scour marks on the land or the trail of devastation which the phenomenon causes. Many tornadoes touch down, lift and touch down again in a skipping motion.

The majority of tornadoes occuring in the USA move from southwest to northeast, although a significant number have irregular, erratic or unusual path directions. They very rarely advance westward. The speed of forward movement is usually in the range 0–120 km/hr, with a mean of 40–65 km/hr, although some tornadoes move at varying speed or remain stationary for brief periods. In the most violent tornadoes, maximum wind speeds at the rim of the funnel may reach 280–400 km/hr. The air pressure inside the funnel may be 10–20 per cent less than that of the surrounding atmosphere, and there may be a considerable frequency of static electricity discharges there.

At the simplest level, tornado path length can be described as short (<5.3 km), intermediate or long (>51.4 km).[1] The average tornado has a track length of 3.2 km, a width of 44.8 m and devastates an area of 0.155 km^2. Barely 2 per cent of all tornadoes affect an area larger than 18 km^2 or have tracks longer than 50 km. In the vicinity of Kansas City, Missouri, 0.06 per cent of the land area, or 2.36 km^2, is damaged annually by tornadoes. This means that 1 km^2 of tornado damage will be experienced every 5 months.

The more sophisticated form of tornado taxonomy uses two systems (Table 3.4). The **Fujita Scale** (F) classifies tornadoes into two categories each of weak, strong and violent force, on the basis of estimated speeds of rotation within the vortex. The **Pearson Scales** (developed by Alan Pearson at the National Severe Storms Forecast Center, at Kansas City in Missouri) classify length (P_L) and width (P_W) into six categories each. Thus, tornadoes observed in the USA are recorded with an FPP number, which can be related to the seriousness and extent of damage caused.

Tornado incidence in North America

Two circumstances combine to give the conterminous USA an atmosphere which is highly stratified at certain times of the year. Starting in January or early February, warm, moist air penetrates the continental landmass at low level northwards from the Gulf of Mexico, arriving at the Canadian border by late August or early September. Meanwhile, a jet of cold, dry air penetrates the upper atmosphere from the west, having lost its heat and moisture upon climbing the western flank of the Rocky Mountain

[1] Some sources define short paths as <0.3 km, intermediate ones as 0.3–1.5 km and long paths as >1.5 km. In any case, the distinctions are arbitrary.

Tornadoes

Table 3.4 Intensity scales for tornadoes (Fujita 1973, Howe 1974).

(a) Fujita Scale

Category	Rotation speed (km/hr)	Percentage of all tornadoes	Percentage of deaths
F_0 weak	<116	24	4
F_1 weak	117–180	38	
F_2 strong	181–253	26	30
F_3 strong	254–331	8	
F_4 violent	332–418	3	66
F_5 violent	419–512	1	
Short path ($\leqq$5.2 km)		73	11
Intermediate path (5.3–51.4 km)		25	47
Long path ($\geq$51.5 km)		2	42

(b) Pearson length and width scales for tornado paths

Path length		Path width	
Pl	<0.5 km	Pw	<5.5 m
Pl_0	0.5–1.5	Pw_0	5.5–15.5
Pl_1	1.5–5.0	Pw_1	15.5–50.0
Pl_2	5.0–16.0	Pw_2	50.0–160.0
Pl_3	16.0–50.0	Pw_3	160.0–500.0
Pl_4	50.0–160.0	Pw_4	500.0–1,500.0
Pl_5	>160.0	Pw_5	1,500.0–5,000.0

chain. Cold fronts emanating from the north and northwest cause the unstable moist air to rise, and mechanical rising is also promoted by upward-sloping terrain north of the Gulf of Mexico (Miller 1959). These conditions (Fig. 3.21) result in atmospheric instability that readily produces lines of thunderstorms which spawn tornadoes at the rate of 600 to 1,000 per year, of which 54 per cent occur in spring, 27 per cent in summer and 19 per cent each in autumn and winter (Fujita 1987).

The incidence of tornadoes is highest in the Gulf states early in the year, in the Great Plains in late spring, in the Midwest in early summer and at the Canadian border and in the northeastern states in late summer. The Gulf coastal states, especially Texas, experience a second concentration in September when hurricanes make landfall there and cause tornadoes. Nationally, the period of lowest incidence is December and January, while the peak occurs in April and May, when large spring storm systems form over the centre of the country (Fig. 3.22). At this time of year there may be one or more days on which more than 25 tornadoes occur. Although tornadoes have been recorded in all 50 states, the greatest numbers have been recorded in Texas and Oklahoma (Fig. 3.23), with a recent shift eastward, which may, however, reflect a higher incidence of reporting in states such as Louisiana, rather than any geographical shift in occurrence. The highest density of waterspouts is found off the Florida Keys.

175

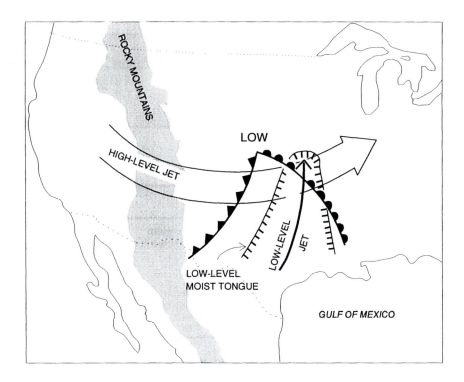

Figure 3.21 Synoptic conditions favouring severe storms and tornadoes on the Great Plains (Barry & Chorley 1987: 219).

Tornadoes in the United States can occur at any time of the day, but 82 per cent do so between noon and midnight and 23 per cent from 4 to 6 pm as a result of thunderstorm activity during the warmest part of the day (Fig. 3.24). The time of occurrence appears to be less predictable in the southeast of the country than in the Great Plains and Midwest. Otherwise, there appears to be little difference between the parameters of tornadoes in the south and north of the country: a comparison of 143 tornadoes occurring in Alabama and 219 occurring in Illinois over the period 1959–68 gave the proportion having short paths as 32 per cent and 26 per cent, respectively, and the mean path length of the rest as 20 and 21 km, respectively.

The tornado hazard
Destruction caused by tornadoes is closely related to the rotational wind speed (F number). Dust devils resulting from differential heating of the atmospheric boundary layer with the ground may last from a few seconds to about 20 minutes, but seldom reach F0 speeds and hence do not merit

classification as tornadoes. Waterspouts, on the other hand, have been known to reach F3 and can be classified as tornadoes if they come ashore. Tornadoes classified F1 are usually (but not always) harmless, but their formation may be sufficiently rapid to inhibit detection and prior warning. F2 tornadoes are capable of pushing vehicles off roads and tearing roofs off houses. Wooden housing of the kind common in North America can be severely damaged by an F3 tornado, which may also be capable of

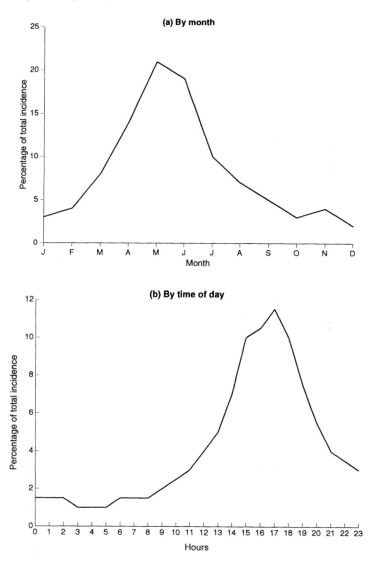

Figure 3.22 Percentage distribution of tornadoes in the USA, 1916–85 (after Fujita 1987: 46, 64).

lifting vehicles off the ground. Intensities of F4–5 may cause the destruction of wooden-framed and many masonry buildings and severe damage to steel-framed and reinforced concrete structures. Vehicles may be entrained into the wind and carried up to 1.5 km. However, the intensity of F5 is usually reached for only a few seconds, and on average the USA experiences only two such tornadoes each year. No F6 tornado has ever been observed, although one is perfectly possible.

Damage to buildings tends to be a result of three factors. The first of these is the wind speed at a point, which is a combination of rotational velocity and forward motion, which on the outside margin of movement can be additive. Secondly, debris launched into motion by the tornado may act as missiles thrown against the building. Violent tornadoes may pick up even large vehicles and hurl them through the air. Thirdly, a difference in pressure occurs between the interior of a building (ambient pressure, 1.06 kg/cm^2) and the centre of the tornado (exceptionally low pressure) as it passes overhead. If this is unable to equalize itself very rapidly it may cause the building to explode outwards. A building with a roof area of 75 m^2 may experience a pressure difference of up to 4.85 tonnes/m^2, or 52 tonnes.

Most buildings are damaged by the direct force of the wind and by missile effects, rather than by pressure differences. Hence, special

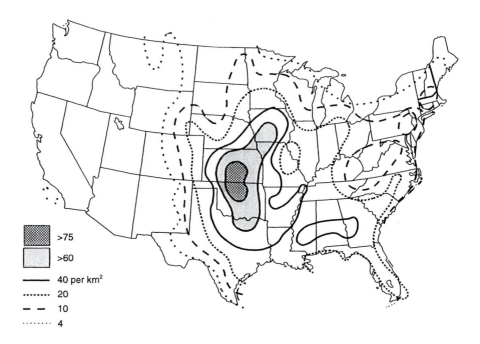

>75

>60

—— 40 per km^2

········ 20

– – 10

········ 4

Figure 3.23 Distribution of tornadoes in the USA (data from Fujita 1987).

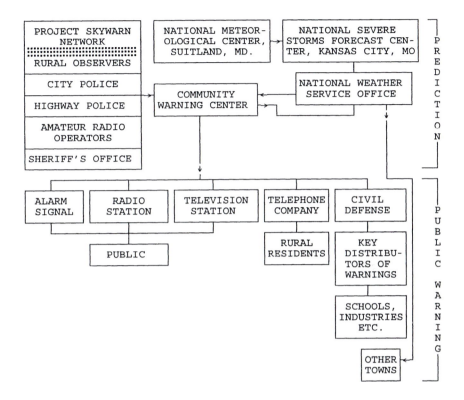

Figure 3.24 The USA's NAAA/NWS tornado warning system (Office of Emergency Preparedness 1972, Vol. 3:96).

strengthening can help them to survive weak and moderate tornadoes (the benefit:cost ratio is 20:1). Prefabricated buildings that are not tied down can be wrecked by the more common small tornadoes. Parks of mobile homes represent a significant concentration of vulnerability in tornado hazard areas and should be provided with a concrete community shelter. Eighty people survived a Kansas tornado in such a shelter while their 45 mobile homes were demolished.

There is a definite strategy for surviving a tornado. When caught in the open it is best to climb into a ditch or simply to lie on the ground if no other option presents itself. Vehicles should at all costs be evacuated. If one is indoors, an underground shelter should be used, or, if this is not possible, a centrally located room or closet on the lowest floor. In North America, the safest part of a frame house without a basement is usually the northeast section of the lowest floor, as most tornadoes move in from the southwest. If windows are open on the side of the house nearest to the approaching funnel, winds may enter the building and blow out the walls and roof. Opening windows on the other side may help to equalize the pressure, but there is seldom time to do this. In any case, flying debris may smash windows. Public buildings in tornado hazard areas should have their basements designated as shelters, which should be signposted clearly. If there is no level below ground the safest place is usually a corridor with a short roof span and strong walls. Roofs with large spans often collapse or detach from the walls as these are blown in or out. Wherever one is caught by the tornado, one should attempt to protect one's face, neck and chest, which are the most vulnerable parts of the body.

Even in the high-risk areas of the central and southern United States, the probability that a tornado will strike a given point on the land surface is less than 4 per cent, or less than once in 250 years. Nevertheless, 40 million people are at risk and nationally the death toll, although decreasing as a result of improved preparedness, still averages 125 people per year. Over the period 1950–79, 689 lethal tornadoes occurred, mostly in the late afternoon and mostly in the violent categories, F4 and F5. Since 1925, more than 9,000 Americans have lost their lives in tornado incidents, which is equal to the combined fatalities caused by floods (4,000) and hurricanes (5,000). Fatalities appear to be more numerous in the South, but neither as a result of greater incidence of tornadoes nor their impact on substandard housing (Sims & Baumann 1972). In part, southern tornadoes tend to be less visible, being more easily obscured by clouds and rain than those which occur on the Great Plains, and in part the discrepancy may result from a higher degree of fatalism and inaction in the face of a direct tornado threat.

Since 1953 the population of areas of the USA most susceptible to tornadoes has increased by more than one quarter. About $5 million per year is spent on building and renovating tornado shelters and setting up warning systems. Over the period 1961–70 damages varied from year to year from $5 million to $50 million, while the mean has since risen to $125 million. Computations made in the 1970s suggested that tornadoes cost America about $8 for every member of the population, but that by the year 2000 this would rise to $20 (White & Haas 1975).

Despite the promise once held by techniques of weather modification, nothing at present can be done to stop tornadoes from forming.

Tornado detection and monitoring

In 1953 the US National Weather Service set up its Tornado Warning System under the aegis of the National Oceanographic and Atmospheric Administration (Fig. 3.24). Detection relies first on forecasting the meteorological conditions that are tornadigenic, then on using radar stations and spotter networks to identify outbreaks of tornadoes. A "watch" is issued on days when weather conditions are right for tornado formation, and this leads to intensification of the monitoring process.

In 1977 **Doppler radar** was introduced, which is capable of measuring the velocity of raindrops within a thunderstorm system and hence can detect the rotation of a tornado (Ray et al. 1979). In addition to the capabilities of ordinary radar, Doppler instruments can measure the speed of targets moving either towards or away from them. They do so by measuring the change in frequency in the transmitted pulse caused by the motion of the target. This ability has greatly improved the process of identifying and monitoring the hook-shaped echo produced on radar screens by such storms, and hence identifying the so-called **Tornadigenic Vortex Signal** (TVS). Doppler radar has increased the probability of detecting a damaging tornado from 69 per cent to more than 94 per cent. Moreover, it has meant that the phenomenon can be detected an average of 23 minutes ahead of conventional sightings and warnings can be issued a further ten minutes ahead, which has enabled more people to take shelter. In addition, the rate of false alarms (i.e. alerting for tornadoes which do not materialize or cause a threat) has been reduced from 73 per cent with conventional methods to 24 per cent using Doppler radar. However, although many tornadoes form near the hook of storm shown as a TVS echo on a radar screen, many do not, and hence the Doppler method of detection is not infallible.

Most tornadic storms cause electromagnetic activity which interferes with a major radio frequency in the range 1–100 MHz. The impulses seem to emanate from lightning discharges within the parent cloud that produces the tornado. The number of bursts per unit time indicates the nature of the storm: light thunderstorms produce an average of 3 bursts per minute, hailstorms 10 bursts, storms with funnel clouds 13 bursts and tornadic storms at least 20 bursts per minute. However, in 1972 a test showed that only one quarter of all storms giving more than 20 bursts per minute actually produced tornadoes (Taylor 1973).

Lightning in tornadigenic storms also produces spherics which can brighten television screens in the 50–60 MHz range. Tuning a US television receiver to Channel 2 (50 MHz) may help detect tornadoes or waterspouts as far away as 15–25 km. But the peculiarities of funnel clouds, television sets and viewers' perceptions make this an unreliable method. Hence, as with other hazards, the best forms of tornado monitoring involve all possible techniques used in concert.

Case histories

Each tornado disaster is capable of contributing something to our understanding of the phenomenon, or of human responses to it. For example, seven individual tornadoes formed on 18 March 1925 and collectively scored a path that was fairly straight and uniform for over

700 km. For half of this length, they moved at speeds of 110 km/hr. The death toll was 689, mostly in Illinois (606), but also in Indiana (70) and Missouri (13). The injured numbered 1,980 and 11,000 people were made homeless. The area affected extended over more than 260 km^2, half of which was completely devastated. This outbreak is the worst on record and hence furnishes base-line data for what tornadoes can accomplish. Levels of protection and awareness were obviously much lower in 1925 than they are now, but so too were population densities and levels of urbanization in the affected area.

On 9 June 1953 suburbs of the city of Worcester, Massachusetts, were hit by a tornado which followed a northwest–southeast path for more than 10 km. There were 94 people killed and damage included a large hospital as well as many homes. The disaster was investigated by a research team from the National Academy of Sciences who set the earliest norms for shock syndromes and emergency medical organization following sudden-impact natural disaster (see Ch. 9).

Preparedness can be critical to the human toll exacted by tornadoes. On the Great Plains 115 people were killed and 500 injured by tornadoes in February 1971, largely because provisions for sheltering were inadequate. On the other hand, on Palm Sunday 1965, tornadoes killed 266 people, many of whom had been warned but had not sought shelter. In contrast, a tornado which occurred on 6 May 1975 in Omaha caused $100 million of damage, but only three deaths, largely because the population had heeded early warnings.

The degree of preparedness, and hence of human vulnerability, can change over time, especially in response to a devastating event which is capable of repeating itself. Thus, Jonestown, Arkansas, was severely affected by a tornado in 1968 when it had no preparedness plans. Thirty-four people were killed, 540 injured and damage amounted $8 million. Subsequently, spotter networks were organized, educational programmes and disaster drills instituted and an emergency operations centre and warning communications system set up. Five years later a tornado caused $50 million of property damage, rather as the 1968 event, but only two fatalities and 267 injuries, as the population had taken refuge.

Atmospheric conditions can be such that the tornado hazard is widespread rather than concentrated on a few localities. On 3 April 1974, for example, a strong, fast-moving low-pressure system formed in Kansas, while the southeast USA was invaded by warm, moist air from the Gulf of Mexico, creating atmospheric inversions at altitudes of 1,500–2,000 m, while a very rapid jet stream emanated from the southwest. As the centre of low pressure migrated into the path of these other atmospheric phenomena, three enormous lines of thunderstorms, known as **squall lines**, were formed. Tornado watches were instituted in 14 states, and 161 tornado warnings were issued, many of them on the basis of the TVS echo observed on radar screens. Of the 148 tornadoes which eventually formed on that day, most were classified at F3 or higher and six were rated at F5. The cumulative path length was 4,180 km. In all, 315 people were killed, 6,142 were injured and 33,140 buildings were damaged or destroyed. Ten states were declared disaster areas, and the total cost of damage exceeded $600 million. A study of injured survivors indicated that 73 per cent were in houses and buildings when the tornado struck, 17 per cent were in trailers or mobile homes, 7 per cent were in automobiles and 3 per cent on their way to shelters (Fujita 1979).

Lightning

The lightning hazard

The atmospheric discharge of static electricity in a lightning stroke has four principal effects: electrocution of human beings and animals; vaporization of materials along the path of the stroke; fire caused by the high temperature produced by the stroke; and sudden power surge, which can damage electrical and electronic equipment. There is also a significant hazard to aviation: the increased use of larger aircraft and solid state circuitry, which is vulnerable to power surges, has increased the potential loss of life when lightning strikes a commercial aircraft. It has been speculated, moreover, that a direct lightning strike on the wings of such an aircraft could cause the fuel tanks to explode, although there is little evidence that this has occurred.

The internal physical processes of a thunderstorm tend to concentrate a positive electrical charge in the upper part of the cumulo-nimbus cloud and a negative charge in its lower part. Two centres of charge thus develop: they may be two parts of the same cloud, two adjacent clouds or the base of one cloud and the ground beneath. When the difference in charge becomes high enough to overcome the resistance of the intervening air, a sudden and violent discharge occurs in the form of a lightning strike. Although this lasts for only millionths of a second, the temperature in the strike channel rises to 28,000°C, which causes a flash and thunderclap (Krider 1983).

Lightning hazard in the USA

In the USA thunderstorms are most common over the Florida peninsula, in the southern part of which they occur on average 80–100 days per year. In the Rocky Mountains and Wyoming their incidence is 70 days a year, but along the Pacific coast they occur on average only five days per year. On the Eastern Seaboard, thunderstorm frequency generally decreases from south to north.

The United States is one of the few countries for which any attempt has been made to quantify the hazard to human life posed by lightning strikes. Although the phenomenon has low potential for catastrophe, lightning probably kills more people than any other natural hazard in the country. The following figures are given per 10 million of the population (240 million):

	Lightning	*Tornado*	*Hurricane*
Deaths:	5.43	5.24	2.52
Injuries:	10.95	90.48	119.52

Insurance claims for the results of fire caused in buildings hit by lightning total more than $40 million per annum in the USA. Losses can be reduced by using fire-resistant materials in the construction of buildings and by installing lightning conductors. About one third of forest and grassland is subjected each year to fires caused by lightning, which number about 10,000 and cause damage valued at more than $50 million. However, forest and rangeland management to reduce the destructive effects of fire is possible (see Ch. 4), but may lead to the build-up of decaying vegetable matter that is highly flammable during periods of dry weather: hence controlled burning is often a better solution. Finally, it has been suggested that cloud seeding might be used to trigger lightning and reduce the charge differential in thunder-clouds. However, it is unlikely that seeding will be used unless there are other objectives, such as hail suppression.

Hailstorms

Although hailstorms rarely involve physical injury, their economic impact can be severe. In the USA damage to crops and property caused by hail costs at least $760 million per year. Damage appears to be a function of the intensity and duration of storms and the size of the hailstones which these produce. The damage itself is often produced not only by the impact of falling hailstones but also by the high winds and torrential rains that are part of the hailstorm.

Hail forms as a result of vigorous updraughts within convective storms. The upper part of the storm must contain an ample supply of super-cooled water droplets which, as a result of reduced atmospheric pressure at high altitudes, are still liquid at less than 0°C. Hailstones fall out of the upper part of the cloud when they have accumulated sufficient mass to defy the buoyancy forces of convective air currents. They may then be swept up again by faster-moving winds ascending from lower levels. In this way an individual hailstone will accumulate coverings of ice which resemble the layers of an onion, and may circulate vertically in the gathering storm for half an hour or more. Eventually its mass will be sufficient for it to fall out of the convection cell and reach ground level. The size of stones varies from about 2 mm to tens of centimetres. The hail tends to fall in swathes 30–250 km long and 8–50 km wide, in which the hail falls are concentrated in streaks about 1 km wide and 5–8 km long (Knight & Knight 1971, Ludlam 1961, Towery & Morgan 1977).

The principal hazard associated with hailstorms is damage to crops. Large areas of wheat, corn and soybeans represent high concentrations of vulnerability, especially if yields per hectare are normally substantial. A high percentage of value may also be lost in crops such as rye, pulses and plums. However, the risk can be reduced by careful attention to cropping patterns, which should be non-contiguous and as diverse as possible.

In the USA, where 2 per cent of the national crop production is lost annually to hail damage, hailstorms occur most frequently in southeast Wyoming, but the largest concentrations of damage are found on the Great Plains and in the Midwest. Insurance marketed by the Federal Crop Insurance Corporation is complemented by that offered by commercial mutual companies; but total national insurance liability covers only 20 per cent of the value of the nation's crops, and only 25 per cent of losses are repaid through insurance. The same situation prevails in Canada.

In Japan, 28 people are known to have died in hailstorms since 1945, but mortality is a function of high winds and sudden flooding, rather than of the hail itself. Damage to property caused by hailstones is unlikely to rise above 12 per cent of the value of damage to crops.

Avalanches

Avalanches occur when snow on slopes suffers from structural weaknesses, and these are often caused by changes within the snowpack. A large overburden of snow may remain unstable because it is anchored to a solid underlayer or occurs on a slope which is too shallow to enable it to move. On the other hand, a small layer may be unstable because it is not bonded well to snow or the ground underneath. Snow avalanche hazards are greatest on slopes in the range 25–40° (Table 3.5): lower angles do not encourage failure of the snowpack, while at higher angles there is less opportunity for the snow to build up to unstable thicknesses. For these reasons, snow avalanches are uncommon on slopes that are shallower than 15° or steeper than 60°.

Table 3.5 Relationship of slope angle to probability of avalanche release.

Small avalances		Large avalanches	
Slope	*Frequency*	*Slope*	*Frequency*
0°–15°	non-existent	0°–15°	non existent
15°–25°	rare	15°–25°	rare
25°–35°	occasional	25°–50°	common
35°–75°	common	50°–70°	less common
>75°	very rare	70°–80°	rare
		>80°	very rare

The toll of injuries and damage caused by avalanches relates to the expansion of settlement, land use and infrastructure (communication, power and transport lines, etc.) into areas which are at risk from snow-mass instability. The rising popularity of winter sports has caused the transient population of many mountain valleys to increase at times when the avalanche hazard is prominent, and this constitutes a particular problem for emergency planners and managers as to how to maintain public safety without placing unnecessary restrictions on access.

Types of avalanche
In many alpine areas there is a continuum between avalanches composed entirely of snow and ice and those that consist of nothing but rock and sediment. Rock that has already been loosened by freeze–thaw action may be carried off in considerable quantities by an avalanche, and hence many avalanches associated with failure of the snow cover may also contain debris. These movements may scour away at the surface of the land beneath, and slush avalanches are also capable of eroding efficiently, even on very gentle slopes.

Each avalanche will involve three sectors: the **rupture zone** where detachment of the snow and ice mass occurs, which is usually above the timber line; the **track** down which movement occurs, usually less steep than the rupture zone; and the **run-out zone**, in which the ground slope is shallower and movement slows, possibly spreads out, and eventually ceases. Motion often begins in hollows and shallow depressions where the snow builds up and becomes unstable under its own weight or as a result of instability factors. Frequent avalanches may excavate their tracks and free them of vegetation, whereas rare movements tend to have less easily definable tracks. Air blasts and turbulent currents may precede a rapid and substantial avalanche at up to twice the speed of the moving mass. Motion will eventually cease when slope is too low, friction too great or obstacles too difficult to surmount. In narrow valleys the avalanche may climb the opposing sideslope if it is moving fast enough.

The basic distinctions are between: (a) wet and dry snow avalanches; (b) point and areal sources; (c) movement on planar slopes from a detachment zone (**unconfined avalanches**) or down chutes and gullies, usually from a point (**channelled avalanches**); (d) movement of slabs and of powdery or slushy snow that is too incoherent to form large moving "plates"; (e) movement at ground level as debris, slabs or slurries, or through the air as powdered snow; and (f) movement of the whole snowpack, or only one of its layers (Table 3.6). Loose snow avalanches

Table 3.6 Avalanche classifications (UNESCO 1981: 29–30). Copyright 1981 by UNESCO Press.

(a) Classification scheme of the International Commission on Snow and ice

Criterion	Characteristics	
Type of rupture	*starting from a line:* slab avalanche (soft slab avalanche, hard slab avalanche)	*starting from a point:* loose snow avalanche
Position of sliding surface	*within snow cover:* surface layer avalanche (new snow fracture, old snow fracture)	*on the ground:* entire snow cover avalanche
State of humidity	*dry snow:* dry snow avalanche	*wet snow:* wet snow avalanche
Form of track	*open, flat track:* unconfined avalanche	*channelled track:* channelled avalanche
Form of movement	*whirling through air:* airborne powder avalanche	*flowing along ground:* (sliding avalanche, flowing avalanche)
Triggering action	*internal release:* spontaneous avalanche	*external trigger:* (natural, artificial)

(b) Primary and secondary avalanching

Movement	Snow type	Conditions for occurrence	Scale	Characteristics
Primary (during snowfall)	Dry	Slope steeper than friction angle of snow	Small	Surface runs of snow
Secondary (after snowfalls)	New snow on crust	External trigger	Small to intermediate	Snow block debris
	Crusted surface	External trigger	Intermediate	Large slab
	Wet surface layer	Solar radiation warm wind, rain	Small	Wet mass, slow velocity
	Depth hoar	Constructive metamorphism	Intermediate to large	Large blocks, entire layer
	Wet snow	Warm weather, weak base	Intermediate to very large	Large blocks, slow to medium velocity

mostly occur after a snowstorm and do not involve a very clear surface of sliding (Fig. 3.25). In dry snow-slab avalanches, density and packing of snow grains may be great and melting slight, making such movements particularly destructive to whatever they encounter in their path (De Quervain 1966).

In particular areas, several types of avalanche may occur as a result of variations in local conditions. For instance, during the spring thaw, slab avalanches may occur on lee slopes (where thawing simply induces weakness in some area of an otherwise coherent snowpack), while slush avalanches may dominate the windward and sunny slopes. The latter may initiate drainage channels by scouring the surface of the hillside beneath. Ground avalanches may have relatively high densities, such as 400 kg/m^3, while airborne suspensions of snow may be less dense than 10 kg/m^3.

With knowledge of snow density and velocity, it is possible to compute the approximate impact pressure of an avalanche:

$$\text{Maximum impact pressure} = \rho V^2 / g$$
$$\text{Mean impact pressure} = \rho V^2 / 2g$$

where ρ = snow density (kg/m^3)
g = acceleration due to gravity (9.81 m/sec)
V = avalanche velocity (m/sec)

Figure 3.25 Shallow powder avalanche near the Col du Grand St Bernard in Switzerland.

However, the relationship is complicated by variations in the density of the snow or ice within the avalanche (Perla & Martinell 1975).

Most avalanches are too small to be destructive and many others are too isolated to cause damage. But the impact pressure of falling or sliding snow can vary from a few kg/m^2 to more than 100 t/m^2, and the latter is perfectly sufficient to move reinforced concrete structures (see Table 3.7). Small powder avalanches may have impact pressures of 0.1–1.0 t/m^2, which is sufficient to bury, immobilize or suffocate a person, while a pressure of no more than 2.3 t/m^2 can sweep a train from the rails. Evidence suggests that an airborne powder avalanche moving at 300 km/ hr is capable of exerting an impact pressure of 100 t/m^2, if it is sufficiently large. Not merely is there a forward impetus, but there may also be vertical pressures which lift or crush structures, and the ice mass may be preceded by an air wave which is also sufficiently rapid to be highly destructive. Some of the oldest established mountain valley communities have learnt to build structures that must occupy vulnerable locations so that they have pointed, reinforced upslope elevations which both absorb and divide the impact.

An avalanche scale used in Japan relates the magnitude of the event to the logarithm (to base 10) of the mass of the snow in tonnes. Values generally range from less than 1 to a maximum of 7. In various parts of the world, a qualitative scale of hazard magnitude is sometimes used. It consists of the following categories: **sloughs** are superficial avalanches of loose, powdery snow that pose no threat to people or property; **small slides** pose a danger to life, but not property; **medium-size slides** are dangerous to people and structures; **large avalanches** are very dangerous; and **major avalanches** are exceptional in size and damage potential.

Table 3.7 Impact pressure and destruction by avalanche (Perla 1986: 398).

Size of avalance	Potential effects	Vertical descent (m)	Volume (m^3)	Impact pressure (Pa)
Sluff	Harmless	10	1–10	$<10^3$
Small	Could bury, injure or kill a human being	$10–10^2$	$10–10^2$	10^3
Medium	Could destroy a car or a wood-frame house	10^2	$10^3–10^4$	10^4
Large	Could destroy a village or forest	10^3	$10^5–10^6$	10^5
Extreme (includes ice, soil, rock, mud)	Could cause severe erosion of the land surface	$10^3–5\times10^3$	$10^7–10^8$	$10^5–10^6$

Processes that cause avalanches

The type of avalanche that develops at a particular location depends upon the process of snow metamorphism taking place in the snowpack. Such processes will differ from the very moment of snow deposition: dry, cold snowfall may lead to sloughing or flowing of surface layers, while, paradoxically, warmer conditions may allow the snow to acquire some cohesion, if the flakes adhere to one another when deposited. The density of snow may vary from less than 0.01, for loose flakes settling under calm conditions, to more than 0.25, for snow which is at temperatures close to melting point. Hence, its strength is very variable, and old, frozen snow can be up to 50,000 times more cohesive than a new deposit (Marbouty 1986).

Snow is an unstable medium sensitive to small fluctuations in temperature, heat flow and atmospheric pressure. Its viscosity increases with grain size and becomes very sensitive to temperature as the density of the medium increases. Tensile strength increases with density, decreases with rising temperature and varies with grain size, reaching a maximum in old, fine-grained deposits. Snow can behave as a viscous, elastic or rigid medium according to its characteristics. Although it has low thermal conductivity, ice has a high heat of fusion ($\approx$80 cal/g).

The way in which the physical characteristics of the snowpack change over time is fundamental to avalanche propensity, especially the processes of internal metamorphism. **Destructive metamorphism** involves the breaking down under appropriate weather conditions of the snow particles into smaller grains. The snow thus becomes denser, more compact and more cohesive, which tends to stabilize it against avalanching. Snow crystals become rounder and denser as water is transferred from their tips to their centres in the form of vapour that upon reprecipitation turns directly back to ice. This is known as **sublimation**.

On the other hand, **constructive metamorphism** involves the growth of ice and snow crystals, the enlarging of void spaces between them and the lessening of the density of the snowpack. Through vertical transfer of water vapour this may lead to the growth of a **depth hoar** at the base of the snow or at some low-density discontinuity between layers of it. This process can render a covering of compact snow particularly susceptible to slab avalanches (Fig. 3.26). If the deeper layers of the snowpack are warmer than its surface, heat may flow towards the latter, stimulating the upward movement of sublimated water vapour. This may cause small, loose crystals of ice to grow and constitute a weaker layer, above which slab avalanches can occur. Different snowfalls may create superimposed layers in the snowpack. If these events are separated by a **föhn** (a relatively warm, dry wind), the snow may form an ice crust and become

190

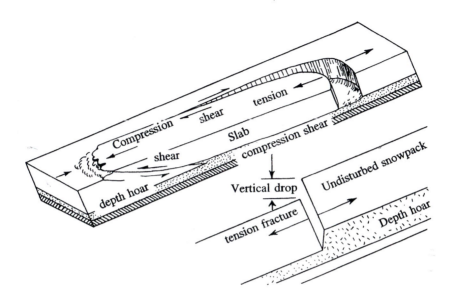

Figure 3.26 Diagram of forces at the moment of initiation of a deep slab avalanche (Bradley 1966).

stratified; hence, avalanching may occur along the layers, rather than at the base of the snowpack.

In synthesis, the creation of avalanches by destructive metamorphism is an **equi-temperature process**, in which static, mechanical stresses predominate, representing the load of the snow upon itself. Constructive metamorphism, instead, is a **temperature gradient process**, relying on thermal exchanges to cause weaknesses in the snowpack.

Loose snow may become unstable on a slope if further light, fluffy snow is added, thus increasing its weight, if metamorphism reduces its internal cohesion to a critical level, or if metamorphism lubricates its base. Slab avalanches depend on the creation of a "slab architecture" by accumulations of drifting snow or processes of metamorphism. Whether or not the slab deforms or breaks up on moving, there are usually clearly defined rupture and sliding surfaces. The shear stress that breaks the slab bond, τ, is approximately equivalent to:

$$\tau = \rho gh \sin \alpha$$

where
- ρ = density of the snow
- g = the gravitational constant
- h = thickness of the slab
- α = slope angle

If τγ exceeds the shear strength at the bond, then sliding will begin (Colbeck 1980).

The velocity of avalanches depends on slope angle, the density of the snow, its shearing strength and the length of the path over which maximum momentum can be generated. The largest, densest masses moving on relatively steep slopes may attain speeds of 200–300 km/hr, although common average velocities are about 40–60 km/hr. Experiments on a 30° slope in Japan gave average speeds of 57 km/hr for airborne avalanches and 50 km/hr for ground based ones. In general the speed of snow slides is in the range 15–25 km/hr; wet, loose snow avalanches move at less than 15 km/hr; dry movements of loose snow travel at 30–45 km/hr and dry, airborne avalanches at up to 330 km/hr. The Mount Huascarán avalanche of 1962 in Perú reached a maximum of 100 km/hr and maintained an average of 60 km/hr as it zig-zagged down-valley. However, of the 13 million m³ of material involved in this avalanche only about one quarter was snow and ice, and the rest consisted of rock and soil (Plafker & Ericksen 1978).

Search and rescue

Deaths in avalanches are a function of any of the following factors: injury caused directly by impact; fractured limbs that have been twisted or wrenched by the moving snow (especially with respect to skiers); suffocation as a result of extreme pressure of snow, inhalation of powdery snow or exhaustion of air pockets; hypothermia, especially when the subject loses consciousness; exhaustion and frostbite; or death as a result of shock.

Rescue teams have a dangerous job that often requires meticulous work in difficult conditions of weather or terrain. Searches begin with a visual survey of the avalanche area followed by rapid forays across it. If the bodies cannot be located in this way, the snow is prodded with poles up to 5 m long. Fine probing can cover 500 m² in one hour, while coarse probing is faster but less accurate. Finally, trenches or pits are dug to locate and rescue the victims. Trained dogs can be up to 40 times quicker than a human being and can make a coarse survey of 1 ha in only 30 minutes. In Switzerland they locate about half of the victims buried by avalanches, but dogs cannot scent people located more than 1 m beneath the snow surface.

Avalanche control workers should wear transmitters which send out a signal up to 90 m away, thus enabling them to be rescued rapidly if buried. Electromagnetic, mechanical and chemical methods of locating victims are all worthy of study, in order to detect anything that a buried body may radiate (such as heat, smell, heartbeat waves and vapour).

Avalanche prevention and control

Apart from avoiding the hazard by not using or inhabiting risk areas, there are two fundamental approaches to avalanche hazard mitigation: **structural** methods involve modifying terrain which may be subject to avalanching, while **non-structural** approaches involve modifying the snow to prevent or trigger an avalanche in some less harmful way.

The objective of terrain modification is to anchor the snow to the slope or direct it to areas where damage will be minimized. To this end, retention and diversion structures are capable of providing good, permanent protection, although they are expensive and they require constant maintenance. Structural methods (Fig. 3.27) comprise the following three strategies of avalanche control. First, structures can be built to prevent the formation of an avalanche. When installed in the rupture zone or track, these can help anchor the snow to the ground; when set up on the windward side of the avalanche area (**wind baffles**) they can help prevent the build-up of snow there. **Jet roofs** help prevent the accumulation of snow cornices by increasing wind velocities in lee areas. In contrast, **snow fences** prevent snow from drifting and encourage it to build up in areas where avalanches are not likely to occur, for

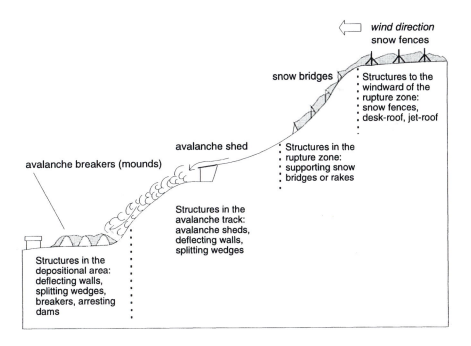

Figure 3.27 Structural methods of avalanche control (Ramsli 1974: 178).

example on very low slopes. They must be strongly built in order to withstand high winds and turbulent eddies. They also depend on the availability of bedrock, to which the structures themselves can be anchored. In this context it should be noted that winds help snow to drift if velocities are greater than 20–25 km/hr, but winds blowing at more than 65 km/hr will tend to scour the slope rather than allow snow to settle.

Installations in the track of the avalanche can help slow it down, divert it out of harm's way or divide the movement into smaller quantities. In this context, wet avalanches are more amenable to control than dry, powdery ones. Afforestation of known avalanche tracks can greatly diminish the hazard, although many movements start above the tree line. The species of pine named *pinus flexilis* is particularly well adapted to use on thin slope soils and is especially hardy. Conversely, deforestation can be a major cause of increase in avalanche activity. For example, in the Tyrol province of Austria, drastic reductions in the hectarage of forests since 1800 have caused a four-fold increase in avalanche activity. Moreover, forests swept away by large avalanches may act as battering rams for whatever they encounter downslope. Lastly, structures built in the depositional area may deflect the moving mass or divide it and break its force. Installations that are especially substantial or numerous may eventually stop small or moderate avalanches completely. Arrester dams and mounds built of earth or masonry are common defences for highways and vulnerable groups of buildings.

Terrain modification is chosen when it is necessary to protect a large area or fixed structure, such as a railway or main highway, where avalanches tend to be large and where it would otherwise be very time-consuming to clear away the deposited snow. Snow rakes, chain-link fences, sheds, bridges and tunnels are the most common installations (see Fig. 3.28). Barriers may be placed in continuous or interrupted lines, or in staggered (*en échelon*) formation. Often they must be built to withstand impact forces or weights of many tonnes.

Non-structural methods involving modification of the snowpack include the artificial triggering of avalanches before the snow builds up to a dangerous level, or before patterns of human activity (such as the tourist season) begin to create a major risk. Avalanches can be released in this way by placing dynamite on the slope or by shelling it with artillery, and cannons have been specially designed with ranges of 2,270 m and 8,180 m for this task. Often, they are trained on snow cornices and positioned in advance so that they can be fired blind during blizzards. This method of release requires evacuation of the target area and closure of routeways until the charge has been detonated. Alternatively, where weaknesses in the snowpack are the main cause of avalanches, it can be stabilized against them by mechanical compacting or chemical emendation. Although use of chemicals may help prevent the growth of a depth hoar layer, they may pollute the snow.

Figure 3.28   Reinforced concrete sheds protect the main road through the Italian Dolomites against rock, snow and ice avalanches.

Other non-structural methods to reduce or prevent avalanches include the use of statistics to predict the probability of avalanche damage to buildings. An **encounter probability** is defined in terms of recurrence intervals. Historical records must be adequate if this is to be accomplished, and hence geomorphological and vegetational evidence is often paramount. Coring of tree trunks can yield information on the damage to annual growth rings caused by avalanches in the past, which can thus be dated by dendrochronology. Biogeography can also help in the detection of historical avalanches, as vegetation scoured away by the movement of snow may be replaced by different species: for example, in Colorado, the aspen, *populus tremuloides*, often invades slopes denuded by avalanches.

Planning and forecasting

While in many mountain zones the drift to the city means that agricultural populations are stable or declining, areas affected by tourism and winter sports are usually gaining in both transient population (i.e. the seasonal influx of tourists) and permanent residents (the service population). There is thus a need to instigate avalanche zoning. In France about half of the communities that face a high risk of avalanche disaster had passed appropriate zoning laws by 1980 (De Crécy 1980). Where such laws apply, construction is banned in the high hazard (red) zone, while for the moderate hazard (blue) zone there are special norms. For example, walls facing upslope must be built of heavily reinforced concrete and must not have windows or roof overhangs. Deflection structures may be required to protect especially vulnerable constructions. Building densities and alignments may be controlled to ensure adequate spacing and orientation. Finally, forests must be inspected regularly for tree damage that could increase the risk of avalanche.

Using a similar methodology, the town of Vail, in Colorado, has designated two levels of "avalanche influence zone" (Mears 1980). In the red zone, where residential construction is banned, avalanches are predicted to have impact pressures of more than 2.9 t/m^2 or to repeat themselves in less than 25 years. In the blue zone, where structures must be built to particular standards of resistance, impact pressures are unlikely to exceed 2.9 t/m^2 and return periods are in the range 25–100 years. Before buildings can be designed for the blue zone, engineers and architects must be informed of the likely velocities, densities, flow heights and types of avalanche.

Maps at scales of 1:20,000 or smaller are useful as reconnaissance tools or for summarizing the avalanche hazard prior to drawing up zoning regulations. For example, over the period 1970–6 maps of the probable location of avalanches were made at the 1:20,000 scale for some 600,000 ha of the French Alps and Pyrénées (De Crécy 1980). More detailed work, such as the building of avalanche defence structures, requires maps that portray known avalanche tracks at scales of 1:10,000 to 1:5,000. In Switzerland the Federal statutes recommend that cantons produce and keep up to date **cadastres,** or maps at scales of 1:10,000 to 1:50,000 which record the incidence of avalanches (Frutiger 1980). For effective control of land use, such cadastres need to be integrated with land-holding registers. Avalanche zone plans can be produced by municipalities if they have enacted the necessary laws to prohibit unsafe construction; otherwise, if there is no special zoning law, avalanche hazard maps can be produced for the information of building inspectors.

In the short term, forecasting is a vital means of reducing the number of casualties caused by avalanches. Meteorological forecasts include predictions of the type and density of snowfall, the direction and speed of winds and the pattern of variations in temperature. However, forecasts of avalanche susceptibility also depend upon knowledge of the depth of old and new snow, of the conditions at its base and of the type and density of settlement of new snow. It is difficult to carry out research on snow, as its characteristics change rapidly on being handled in the laboratory or field. Nevertheless, one promising area is the use of acoustics to determine the likelihood of an impending avalanche by assessing snowmass instability, premonitory movements, slab thickness and other variables.

The many and complex conditions determining avalanche propensity (Fig. 3.29) make its assessment a difficult art, especially as there are often few records on which to base statistical forecasts (Fohn et al. 1977). In most cases there is a need to lengthen the period between the issuing of a reliable forecast and the occurrence of an avalanche. It is also vital to reduce the cost of forecasts and warnings, improve their reliability and make the information furnished to the public more easily understandable.

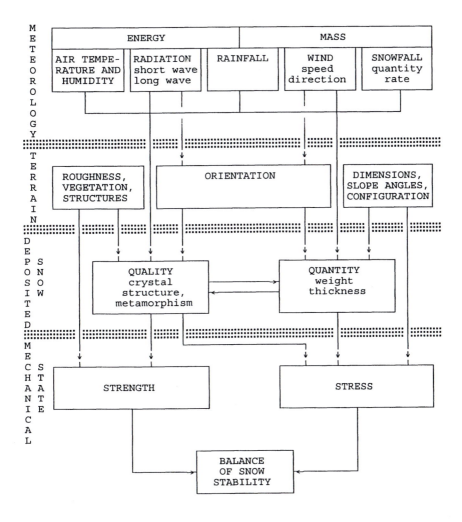

Figure 3.29 The Shōda diagram of the chain of causation in avalanche formation (after LaChapelle 1980: 79). Reproduced by courtesy of the International Glaciological Society.

Hence, it is necessary to expand the application of forecasts to newly developed areas and to carry out research on the effects of warnings. Despite these hurdles, forecasters tend to acquire a good feel for avalanche conditions and, by integrating a wide variety of observations, over time they can gain a very good idea of immediate risk (LaChapelle 1966, 1980).

A practical example is furnished by the Colorado Avalanche Warning Programme, which covers about 100,000 km² with 30 manned observation sites and several unmanned ones (Judson 1976, Bovis 1977). Local offices survey areas of 100–1,000 km² and telephone their reports daily to the central forecast office. Three types of event are forecast: **storm-induced avalanches**, which occur directly after a storm, result from the loading effect of blown or fallen snow and account for 80 per cent of all avalanches; **delayed action avalanches**, which account for about 5 per cent and are caused by increasing stresses in the snowpack (they occur in the absence of storms), **wet avalanches**, which account for the remaining 15 per cent and result from thaw and the percolation of meltwater into the snow (these are common in spring). Fortunately, storm-induced avalanches are the easiest to forecast, as they depend on specific meteorological conditions. The other two kinds tend to defy prediction because it is difficult to estimate stresses within the snowpack.

Finally, in an innovative study of value to mappers, forecasters and zoning officers, Walsh et al. (1990) used a Geographical Information System with ARC/INFO software to map snow-avalanche path location in Glacier National Park, Montana. It contained variables such as avalanche path, hydrology, structural lineaments, sedimentary lithology, sills and dikes, elevation, slope angle, slope aspect, watersheds and land cover. In addition, some 121 previous avalanche paths were identified in a 25 km by 16 km study area. The end product was a map of high, medium and low avalanche probabilities.

Avalanche problems and experience around the world
In general terms, the frequency of avalanches is highly variable from year to year, according to local climatic conditions and weather phenomena. There may be occasional periods of unusually high frequency. For example, over the winter of 1950–1 in the European Alps more than 650 people were killed in France, Switzerland and Austria, 2,500 buildings were damaged or destroyed, 6,000 ha of timber were uprooted and devastation was wreaked upon livestock and wildlife (White 1974).

In the Commonwealth of Independent States there is a close correspondence between avalanches and areas of the most abundant and intense snowfall, especially zones which are also subject to glacial activity. In the European Alps and Scandinavia the death toll caused by avalanches tends to vary from year to year. In Austria there are 2,700 avalanches each year, while in Switzerland an average of 17,480 avalanches and 24 deaths are recorded annually. In the past, however, mortality was often much higher: in 1679 in Norway, for example, between 400 and 500 people died in avalanches, while over the period 1915–18 there were 40,000 deaths in the Alps, 10,000 of them on a single day. In the latter case, however, avalanches were triggered as a tactic of warfare during troop movements along unprotected valleys.

In the USA, where about 6,800 avalanches occur each year, the average mortality is rising from a 1960s average of about 7 people per annum (Fig. 3.30), although again the figure is highly variable (118 people died when an avalanche swept away a train in the Cascades Range in 1910). The fatality rate is about 7 per cent of people caught in avalanches and 15 per cent of those partially or totally buried. Property losses average $500,000, although they too vary considerably from year to year. Unwise timber removal on upland slopes has opened up new avalanche tracks, although most hazard zones are predictable, if rather expensive to map. For prediction purposes, the west of the country is divided into three zones of broadly similar snow cover and meteorological characteristics (Table 3.8). A comprehensive avalanche forecasting network has existed

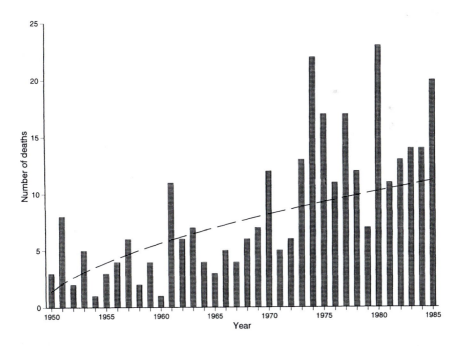

Figure 3.30 Annual death tolls in avalanches in the United States (Ferguson 1986: 3).

Table 3.8 Avalanche regions of the United States (LaChapelle 1966)

	West coast or coastal alpine	Intermontane & middle alpine	Rocky Mountains and high alpine
States	California Oregon Washington	Idaho Montana Utah Wyoming	Colorado New Mexico
Elevation	<2,500 m	2,000–3,000 m	2,500–4,000 m
Winter precipitation	15–25 m, heavy intensities and accumulations	fairly heavy	very low
Temperature	mild, mid-winter thaw	lower	low
Wind	often strong	strong and frequent	stronger, more frequent, drifting
Winter snow cover	deep, stable, often damp	dry, changeable	dry, shallow
Avalanche predominance	surface, soft slab	soft slabs, climax slabs	hard and soft slabs on lee slopes

since 1971, but there is a constant need to upgrade its capabilities: it has been estimated that during the skiing (and avalanche) season one Colorado canyon alone could contain more than 16,000 people (Ferguson 1986).

Glacier hazards

The glaciologist Lance Tufnell noted that although glaciers cover about 16.3 million km^2 of the Earth's surface, only 10,000–20,000 km of this lie near inhabited areas (Tufnell 1984). The direct hazards caused by glaciers are therefore relatively minor, although changes in climate, environment, settlement and land use could increase them.

Glaciers are capable of posing four types of hazard. First, an advancing glacier may overrun buildings or a small settlement. Villages near Chamonix in Switzerland were thus damaged over the period 1600–16 at the height of the 'Little Ice Age' (1550–1860), which was a period of increased glacier activity and expansion. **Glacial surges**, or sudden forward slip motion of glacial ice in valleys, can occur at up to 20 m a day. Although these tend to be much slower than avalanches, they are larger in scale and can overrun piedmont villages or farmland, thus causing widespread economic damage (Grove 1987).

Secondly, advancing or surging glaciers can block streams and impound lakes. Lakes may also form in the ice-free tributary valleys of a trunk glacier, behind moraines or even on the ice surface itself. Glacial dams can fail through any of several mechanisms. These include the accelerated erosion of a subglacial or englacial tunnel; hydrostatic flotation, in which the pressure of dammed-up water floats the ice off its base and seepage begins underneath it; and weakening of the ice dam by earthquakes. Rapid floods can be caused. In Switzerland spectacular examples of glacial surges in 1595 and 1818 resulted in the 200 deaths.

Thirdly, volcanic or geothermal activity in the midst of icefields can cause floods through rapid melting, and these are known by the French term **débâcles** or the Icelandic word **jökulhlaup** (glacier bursts). Discharges are often exceptionally high.

Fourthly, ice avalanches may occur from the snout of the glacier. If the snout is located at moderate altitudes, and ablation (i.e. melting and evaporation of the ice) causes it to be unstable, the avalanches can be substantial. If the ice does not separate itself completely from its parent glacier, the phenomenon is known as a **glacier thrust**.

Snow as a hazard to the urban system

The urban snow hazard
A **blizzard** is defined as a snowstorm in which air temperatures are low (generally less than $-10°C$) and winds of at least 30 knots (55.6 km/hr) blow falling snow, or that which has already fallen, such that visibility does not exceed 200 m. For a storm to be classified as a **strong blizzard**, winds must blow at a minimum of 40 knots (74.1 km/hr), temperatures must be no higher than $-12°C$ and visibility must be effectively nil. The urban and regional snowfall hazard involves disruption of normal activities and danger to travellers and outdoor workers. It results from interrelationships among snowfall, temperature, wind and patterns of human activity.

Snowfall is not measured in raingauges, but from samples collected on boards or in heated tipping-bucket gauges. These methods lack precision and the former one is unlikely to reveal fluctuations in the rate of accumulation. Wind and air turbulence can result in inaccuracies of up to ±20 per cent, and hourly snowfall rates can vary substantially: for example, in a 1947 snowstorm in New York, 48–60 cm fell over 18–19 hours at a mean rate of 2.5 cm/hr, but with a maximum of more than 6 cm/hr, which was achieved more than half way through the storm. However, in general, light snowfall involves less than 0.75 mm/hr, moderate falls are within the range 0.76–1.25 mm/hr and heavy falls involve more than 1.25 mm/hr.

The amount of snow causing the hazard is a function not merely of that which falls in any given period, but also of the rate of natural dissipation. If ground temperature is above freezing point, the snow will melt on contact. As the snow becomes dirty its albedo will fall from about 75–90 per cent towards 40 per cent, enabling it to absorb more radiant energy from the sun, some of which will be converted to heat and conducted through the snow, helping it to melt. Studies of snowpacks have shown, however, that only 15 per cent of melting results from short-wave solar radiation, while 70–85 per cent is caused by rising air temperature.

Snow may disappear through a combination of melting and evaporation. The former may be stimulated if warm rain falls, while the latter may be caused by the presence of radiant heat from the sun, heat conduction from the ground, or heat from the air which is in contact with the snow surface. Evaporation may also be stimulated if the humidity of the air is low, and especially if warm, dry winds carry away the saturated air. Melting may be increased by the high albedo of an asphalt surface, which tends to trap heat before being covered by snow, and later to release it. For this reason, cities tend to lose their snow cover more rapidly than the surrounding countryside, and also because the urban "heat island" accelerates melting, while the impurities in polluted city air tend to dirty the snow rapidly and lower its albedo. Dissipation is, however, much reduced at very low temperatures, as cold atmosphere cannot hold much water vapour and will not facilitate melting.

A classification of disruption
The disruption of the urban system caused by heavy snowfalls can be classified as **internal** when interchange is hampered within the city and **external** when interaction is restricted between a community and its surrounding area or other communities. The former may lead to the shutting down of manufacturing, construction activities and the retail trade, curtailment of transportation and communications facilities, restriction or interruption of power supplies, postponement of meetings and entertainments and temporary closure of schools. The latter may restrict or block traffic on highways and railways and from airports and may damage telephone and power lines, aqueducts and radio masts (Bendell & Paton 1981).

The level of disruption has been classified into the following (subjective) categories: I paralysing, II crippling, III inconvenience, IV nuisance, V minimal. Assessment is carried out in terms of the degree of restriction of movements, communications and activities, the closure of shops, factories, schools and institutions, danger on the roads, the proportion of activities cancelled, and the degree of media coverage given (see Table 3.9).

There is a strong and obvious relationship between the level of disruption and the physical nature of the snowfall. The frequency of higher order (I–III) disruption increases with increasing annual accumulation, but the highest order of disruption, representing complete paralysis of the urban system, is difficult to relate to annual levels of snow accumulation. Disruption, moreover, increases at a diminishing rate with

Table 3.9   Hierarchy of disruptions caused by snowfall (Rooney 1969: 114–15)

Activity	1st order: paralysing	2nd order: crippling	3rd order inconvenience	4th order: nuisance	5th order: minimal
Internal:					
Transportation	Few vehicles moving on city streets; emergency services on full alert; police and firemen available for transportation of emergency cases	Accidents at least 3 times average; traffic movement slowed	Accidents at least twice the average; traffic movement slowed	Few references in media: traffic movement slowed	No press coverage
Retail trade	Extensive closure of retail establishments	Major drop in number of shoppers in central business district; decrease in sales	Minor impact	Occasional	No press coverage
Postponements	Civic, cultural and athletic events	Major and minor events; outdoor activities forced inside	Minor events	Occasional	No press coverage
Manufacturing	Factory closures, major losses of production	Moderate worker absenteeism	Absenteeism caused by snowfall	Occasional	No press coverage
Construction	Major impact on indoor and outdoor operations	Major impact on outdoor activity; moderate indoor cutbacks	Minor effect on outdoor activity	Few mentions	No press coverage
Communication	Wire breakage	Overloads	Overloads	Few mentions	No press coverage
Power supplies	Widespread failure	Moderate difficulties	Minor difficulties	Few mentions	No press coverage
Schools	Official closure of city schools	Major lack of attendance	Attendance drops in city schools		No press coverage

Table 3.9 *continued*

Activity	1st order: paralysing	2nd order: crippling	3rd order inconvenience	4th order: nuisance	5th order: minimal
External:					
Schools	Closure of rural schools	Closure of rural schools	Major lack of attendance	Minor lack of attendance	No press coverage
Highways	Roads officially closed; vehicles stalled	Extreme driving condition warning from highway patrol; snow and ice cause accidents	Hazardous driving condition warning from highway patrol; snow and ice cause accidents	Occasional mention in on radio: e.g. 'slippery surfaces' warning	No press coverage
Railways	Departures cancelled or postponed for 12 hours or more	Trains running at least 4 hours behind schedule	Trains less than 4 hours behind schedule	Few references	No press coverage
Air Traffic	Airports closed	Commercial cancellations	Light plane cancellations; flight delayed by weather conditions	Few references	No press coverage

increasing annual snowfall. At the level of the individual storm, there is a strong relationship between the depth of snow and curtailment of social and economic activities. But municipalities that regularly experience large winter snowfalls tend to be well prepared, and therefore only exceptional snowfalls bring them to a halt (Rooney 1967).

In general, dry snow (i.e. snow at temperatures well below freezing point) is less disruptive than wet snow. The latter is stickier and denser, to the extent that snow at − 3°C weighs about twice as much as it does at − 15°C, and heavy, wet snow may break cables and fixtures. Otherwise, air temperature is difficult to correlate with the level of disruption (except that at very low temperatures snowfall is usually light). On the other hand, winds in excess of about 25 km/hr worsen the confusion by causing snow to drift back rapidly across surfaces that have been cleared (drifts up to 9 m deep have been recorded on the American Great Plains). Drifting tends to be more severe if the snow is dry and hence unlikely to stick together.

The degree of disruption is closely related to the timing and intensity of snowfall. Thick snow that falls during a weekday morning rush-hour is likely to be very much more disruptive than similar accumulations during a weekend, especially if they are forecast accurately well in advance, thus enabling most people to stay at home. The rate of accumulation of snow affects the difficulty of clearing it up and hence, in the short term, has an impact on the level of disruption (Earney et al. 1974).

Adjustment to the urban snow hazard

Although the general path of large, well-defined cyclones is easily predicted, fluctuations in the path of storms, in temperature, atmospheric moisture content and precipitation are all subject to variation that is hard to forecast. Moreover, the amount of snowfall can vary substantially, to the extent that even the best forecasts may be accurate for only about one third of the area which they are intended to cover.

Whatever the validity of warnings, the response of municipal snow-clearers will depend only in part on receipt of accurate information about an impending snowfall. If the response is to be adequate, funds must be both available and protected. If budgets are tight, there is a risk that snowfalls which occur late in the season, or protracted periods of unexpectedly heavy and frequent storms, will use up the allotment of funds before the winter is over. Sand and salt spreading, ploughing, blowing or carting away the snow all require equipment and manpower that must be used for other purposes during the warm season. This may pose intricate problems of accounting and management, and, in fact, even the simple process of removing snow may be the jurisdiction of several accounting offices in the local town hall. Additional problems occur when the likelihood of impending snowfall is uncertain, which may render difficult the decision to begin clearance, or when snowfall coincides with rush-hour, when a prompt response is required on as large a scale as possible in order to minimize disruption.

The nature of public reaction to snow hazards may be critical to the eventual seriousness of the event (Changnon 1979). Insurance may cover damage to vehicles, structural collapse or the death of livestock. Planning measures can help avoid some of the risks: thus the State of Colorado has enacted a building code which requires roofs to be able to bear the weight of a snowfall with a 30-year recurrence interval. But modifications in patterns of behaviour may help avoid these problems. The risks of crashing in light snow can often be minimized by using alternative transportation. With sufficient warning and organization, it may be possible to declare a general moratorium on activities, or **snowfest**, in the event of a major snowstorm. This appears to work best for cities with populations of at least 200,000. The moratorium must be declared at the start of the storm and hence depends upon reliable forecast information in order to avoid false alarms. Its drawbacks include the difficulty of maintaining the mobility of emergency services and the problems of shutting down continuous production processes, such as steel-making.

Snowfalls and the American urban system

In the USA mean annual snowfall increases northwards, in mountainous areas and towards the Great Lakes. But the snowfall hazard exists in all areas except Florida, Hawaii, southern California and the southwest lowlands of Arizona. In the northern parts of the country some 60 million people live in areas of high risk of winter snowfall disruption to urban areas. The period of maximum snowfalls varies with location from January–February in the northeast to March–April on the Great Plains. Disruption tends to be greatest in heavily urbanized places, although these are often the most prepared and best equipped. In the south of the country, which has 28 per cent of the national population, the risk is infrequent, but for this reason somewhat higher, as preparedness is less. In addition, more than three quarters of the population of the low snowfall risk areas live in an urban environment. In general, risks tend to be lower in rural areas, as these have small urban populations and low overall population densities. Regional systems that are heavily dependent on urban activities tend to be more easily crippled by heavy snowfalls than those where there is a fair degree of rural independence. But major snowfalls in the latter create risks of rural isolation and losses of crops and livestock.

Federal involvement in urban snow hazard management in the USA is minimal and the burden of coping devolves upon municipalities and state and local taxation. The critical areas tend to be the zones of low annual snow accumulation, where major impacts are less likely to be expected and less income and planning are devoted to prior preparation. Marginal snowfalls sometimes tend to be particularly hazardous, as they involve difficulties in perceiving the disruption until it has occurred and in deciding to initiate clear-up operations. Thus in the USA disruption tends to be higher in southern municipalities than in northern ones, although it is rarer in the former than in the latter. Hence, it tends to be related to levels of prior adjustment to the hazard of patterns of behaviour and land occupance.

Frost hazards

The frost hazard

Frost occurs when air is cooled to its dew point, at which condensation of water vapour takes place, and when this occurs below 0°C. When crops are affected, the result can be an economic disaster, which may be of the "creeping" variety if losses are cumulative. In the USA frost causes damage to crops valued at an annual average of $1,100 million, half of which occurs to orchards. Adjustments to the risk made over the past 40 years have not succeeded in mitigating losses, as the value of crops grew

206

steadily during this period. Given the increasing demand for food, frost hazards are greatest where agriculture (especially citrus fruit cultivation) has expanded into areas where climate is marginal during the growing season, or where microclimate is likely to be unfavourable (Stewart et al. 1984).

Protection measures are seldom completely effective, but can be divided into **active** and **passive** kinds (White 1974). The former are utilized before or during a freeze in order to prevent its effects, while the latter involve prior planning and preparation.

Active protection measures

Active measures against frost damage are mainly used to protect orchards where the crop value is highly concentrated and losses are potentially great. Wind machines may be used to blow warmer air into colder zones. Heaters are sometimes deployed, but the cost is best reduced by seeking out enclaves of cold air or ground to which to apply heat. Artificial fog can be generated by deploying smudge pots, thus providing a blanket of insulation, and ground-level crops can be covered by foam, plastic or glass. If the entire crop is not covered, the roots can be insulated by banking up with materials of low thermal conductivity.

Paradoxically, irrigation water can be a useful means of raising temperatures, as the process of freezing liberates latent heat when the molecules become more tightly bound in the form of ice. Hence, sprinkler irrigation is used to combat frost, although the growth of ice crystals on branches may weigh them down until they break. Ground-water can be pumped up from wells, but although this method is less costly than directly heating an orchard, it is also less effective and depends on the availability of groundwater supplies at appropriate temperatures.

Passive protection measures

Passive measures undertaken to mitigate frost hazard have been used mainly to protect field crops, rather than orchard ones. Advance planning and implementation is needed concerning the selection of crops, choice of growing season and use of cultivation techniques. The selection of a more appropriate site to cultivate involves searching for topography that is seldom subject to cold weather. In Florida, where citrus fruit is grown in abundance, there have been seven great freezes in the twentieth century. Most frosts result from the trapping of cold air near the surface during temperature inversions. Hence, the best response is to utilize hill country. However, during advective freezes plants on upper surfaces often suffer more damage than those below, as the higher sites are subject to cold winds while the valley bottoms are sheltered. Sites near lakes and coasts, on the other hand, tend to benefit from the moderating influence of surface water bodies. Having chosen the site, the most appropriate variety and rootstock can be selected in order to produce crops that mature before the orchard or field is subject to frost.

Amelioration of impacts

The pressure of demand for crops has stimulated cultivation of marginal land and the extension of the growing season, which has tended to offset protection measures. It has been calculated that mitigation of only 10 per cent of the $1,100 million US crop losses would save $913 million in 50 years (this is a plausible estimate, as there have been at least seven great freezes in Florida since 1900). The individual farmer should be encouraged to share the burden of losses with others who are threatened, and hence crop insurance should be promoted, although government relief and credit may also be needed to stimulate replanting after disaster has struck. If protection methods are unsuccessful, and impacts cannot be ameliorated at reasonable cost, the only other option may be to abandon cultivation and restart it in areas of warmer climate.

References

Alexander, D. E. 1980. The Florence floods: what the papers said. *Environmental Management* **4**, 27–34.

Barry, R. G. 1969. Evaporation and transpiration. In *Water, earth and man*, R. J. Chorley (ed.), 169–84. London: Methuen.

Barry, R. G. & R. J. Chorley 1988. *Atmosphere, weather and climate*, 3rd edn. London: Routledge.

Bendell, W. B. & D. Paton 1981. A review of the effect of ice storms on the power industry. *Journal of Applied Meteorology* **20**, 1,445–9.

Bennett, W. J. & B. Mitchell 1983. Floodplain management: land acquisition versus preservation of historic buildings in Cambridge, Quebec, Canada. *Environmental Management* **7**, 327–38.

Bovis, M. J. 1977. Statistical forecasting of snow avalanches, San Juan Mountains, southwestern Colorado, USA. *Journal of Glaciology* **18**, 87–99.

Bradley, C. C. 1966. The snow resistograph and slab avalanche investigations. Proceedings of the International Symposium on Scientific Aspects of Snow and Ice Avalanches, Davos, 1965. *International Association of Scientific Hydrology Publication* no. 69, 251–60.

Burby, R. J. 1985. *Flood plain land use management: a national assessment*. Boulder, Colorado: Westview Press.

Burby, R. J. & S. P. French 1980. The USS experience in managing floodplain land use. *Disasters* **4**, 451–8.

Changnon, S. A. 1979. How a severe winter storm impacts on individuals. *American Meteorological Society, Bulletin* **60**, 110–14.

Clarke, A. O. 1991. A boulder approach to estimating flash-flood peaks. *Association of Engineering Geology, Bulletin* **28**, 45–54.

Colbeck, S. (ed.) 1980. *Dynamics of snow and ice masses*. Orlando, Florida: Academic Press.

References

Cooke, R. U. & J. C. Doornkamp 1974. *Geomorphology in environmental management*. Oxford: Clarendon Press.

Costa, J. E. 1978. Holocene stratigraphy in flood-frequency analysis. *Water Resources Research* **14**, 626–32.

Costa, J. E. & V. R. Baker 1981. *Surficial geology: building with the Earth*. New York: John Wiley.

Cutler, P. 1984. Famine forecasting: prices and peasant behaviour in northern Ethiopia. *Disasters* **8**, 48–56.

Dalrymple, T. 1960. *Flood frequency analysis*. US Geological Survey Water Supply Paper 1543A, 1–80.

De Crécy, L. 1980. Avalanche zoning in France: regulation and technical bases. *Journal of Glaciology* **26**, 325–30.

De Quervain, M. R. 1966. On avalanche classification; a further contribution. Proceedings of the International Symposium on Scientific Aspects of Snow and Ice Avalanches. *International Association for Scientific Hydrology Publication*, no. 69, 410–17.

De Waal, A. 1988. Famine early warning systems and the use of socio-economic data. *Disasters* **12**, 81–91.

Earney, T. C., C. F. Fillmore, B. A. Knowles 1974. Urban snow hazard. In *Natural hazards: global, national, local*. G. F. White (ed.), 167–74. New York: Oxford University Press.

Emanuel, K. A. 1988. Toward a general theory of hurricanes. *American Scientist* **76**, 371–9.

Embleton, C. E. & J. B. Thornes 1979. *Process in geomorphology*. London: Edward Arnold.

Ferguson, A. 1986. [Letter] *Natural Hazards Observer* **10**: 3.

Fiorino, M., E. J. Harrison, Jr, D. G. Marks 1982. A comparison of the performance of two operational dynamic tropical cyclone models. *Monthly Weather Review* **110**, 651–6.

Fohn, P., W. Good, B. P. Bois, C. Obled 1977. Evaluation and comparison of statistical and conventional methods for forecasting avalanche hazard. *Journal of Glaciology* **19**, 375–87.

French, H. M. 1979. Permafrost and ground ice. In *Man and environmental processes*. K. J. Gregory & D. E. Walling (eds), 144–62. Boulder, Colorado: Westview.

Frutiger, H. 1980. History and actual state of legislation on avalanche zoning in Switzerland. *Journal of Glaciology* **26**, 323–4.

Fujita, T. T. 1973. Tornadoes around the world. *Weatherwise* **26**, 56–62, 78–82.

Fujita, T. T. 1979. The jumbo tornado outbreak of 3 April 1974. *Weatherwise* **27**, 116–19.

Fujita, T. T. 1987. US *tornadoes*, Part I: *70-year statistics*. Chicago: Satellite and Mesometeorology Research Project, Department of Geophysical Sciences.

Galway, J. G. 1981. Ten famous tornado outbreaks. *Weatherwise* **34**, 100–9.

Gentry, R. C. 1970. Hurricane Debbie modification experiments, August 1969. *Science* **168**, 473–5.

Gerasimov, I. P. & T. B. Zvonkova 1974. Natural hazards in the territory of the USSR: study, control and warning. In *Natural hazards: local, national, global*, G. F. White (ed.), 243–51. New York: Oxford University Press.

Glantz, M. H., 1976. *The politics of natural disaster: the case of the Sahel drought*. New York: Praeger.

Glantz, M. H. (ed.) 1987. *Drought and hunger in Africa: denying famine a future*. Cambridge: Cambridge University Press.

Griffith, D. A. 1986. Hurricane emergency management applications of the SLOSH numerical storm surge prediction model. In *Terminal disasters: computer applications in emergency management*, S. A. Marston (ed.), 83–93. Boulder, Colorado: Natural Hazards Research and Applications Information Center.

Grove, J. M. 1987. Glacier fluctuations and hazards. *Geographical Journal* **153**, 351–69.

Gruntfest, E. C. & C. Huber, 1989. Status report on flood warning systems in the United States. *Environmental Management* **13**, 279–86.

Herbert, P. J. & G. Taylor 1979. Everything you always wanted to know about hurricanes. *Weatherwise* **32**, 59–67; **32**, 100–7.

Howe, G. M. 1974. Tornado path sizes. *Journal of Applied Meteorology* **13**, 343–7.

Idso, S. B. et al. 1975. The dependence of bare soil albedo on soil water content. *Journal of Applied Meteorology* **14**, 109–13.

James, L. D. 1975. Formulation of non-structural flood control programs. *Water Resources Bulletin* **11**, 688–705.

Jarret, R. D. 1990. Palaeohydrologic techniques used to define the spatial occurrence of floods. *Geomorphology* **3**, 181–95.

Jelesnianski, C. P. 1978. Storm surges. In *Geophysical predictions*, US National Academy of Sciences (ed.), 185–192. Washington DC: National Academy of Sciences.

Judson, A. 1976. Colorado's avalanche warning problem. *Weatherwise* **29**, 268–77.

Kates, R. W. 1962. *Hazard and choice of perception in flood plain management*. Chicago: Department of Geography, University of Chicago.

Knight, C. & N. Knight 1971. Hailstones. *Scientific American* **237**, 96–103.

Krider, E. P. 1983. Lightning damage and lightning protection. In *The thunderstorm in human affairs*, E. Kessler (ed.). Norman, Oklahoma: University of Oklahoma Press.

LaChapelle, E. R. 1966. Avalanche forecasting: a modern synthesis. Symposium on the Scientific Aspects of Snow and Ice Avalanches, Davos, 1965. *International Association of Scientific Hydrology, Publication* no. 69, 350–6.

LaChapelle, E. R. 1980. The fundamental processes in conventional avalanche forecasting. *Journal of Glaciology* **26**, 75–84.

Leopold, L. B., M. G. Wolman, J. P. Miller 1964. *Fluvial processes in geomorphology*. San Francisco: W. H. Freeman.

Liebman, E. 1973. Legal problems in regulating flood hazard zones. *American Society of Civil Engineers, Proceedings; Journal of the Hydraulics Division* **99**, 2,113–23.

Lucchitta, I. & N. Suneson 1981. Flash flood in Arizona: observations and their application to the identification of flash-flood deposits in the geological record. *Geology* **9**, 414–18.

Ludlam, F. H. 1961. The hailstorm. *Weather* **16**, 152–62.

Marbouty, D. 1986. What triggers an avalanche? In *Violent forces of nature*. R. H. Maybury (ed.), 125–42. Mt Airy, Maryland: Lomond Publications.

Mears, A. I. 1980. Municipal avalanche zoning: contrasting policies in four western United States communities. *Journal of Glaciology* **26**, 355–62.

Miller, R. C. 1959. Tornado-producing synoptic patterns. *American Meteorological Society, Bulletin* **40**, 465–72.

Mogil, H. M., J. C. Monro, H. S. Groper 1978. National Weather Service's flash flood warning and disaster preparedness programmes. *American Meteorological Society, Bulletin* **59**, 690–9.

References

Muckleston, K. W. 1976. The evolution of approaches to flood damage reduction. *Journal of Soil and Water Conservation* **31**, 53–9.

NERC 1975. *Flood studies report* (5 vols). London: HMSO, for the Natural Environment Research Council.

Neumann, C. J. & J. M. Pelissier 1981. Models for the prediction of tropical cyclone motion over the North Atlantic: an operational evaluation. *Monthly Weather Review* **109**, 522–38.

Nir, D. 1974. *The semi-arid world: man on the fringe of the desert*. London: Longman.

Office of Emergency Preparedness 1972. *Disaster preparedness*, Vol. 3. Washington DC: Executive Office of the President, US Government Printing Office.

Panattoni, L. & J. R. Wallis 1979. The Arno River flood study. *EOS: American Geophysical Union, Transactions* **60**, 1–5.

Parker, D. J. & D. M. Harding 1979. Natural hazard evaluation, perception and adjustment. *Geography* **64**, 307–16.

Paulhus, J. L. H. & C. S. Gilman 1953. Evaluation of probable maximum precipitation. *American Geophysical Union, Transactions* **34**, 701–8.

Penman, H. L. 1963. *Vegetation and hydrology*. Farnham Royal, England: Commonwealth Bureau of Soils.

Penning-Rowsell, E. C. & J. W. Handmer 1988. Flood hazard management in Britain: a changing scene. *Geographical Journal* **154**, 209–20.

Perla, R. I. 1986. Avalanche release, motion and impact. In *Dynamics of snow and ice masses*, S. C. Colbeck (ed.), 397–462. New York: Academic Press.

Perla, R. & M. Martinell 1975. *The avalanche handbook*. Washington DC: US Department of Agriculture.

Petak W. J. & A. A. Atkisson 1982. *Natural hazard risk assessment and public policy: anticipating the unexpected*. New York: Springer.

Plafker, G. & G. E. Ericksen 1978. Nevados Huascarán avalanches, Perú. In *Rockslides and avalanches*, Part I, B. Voight (ed.), 278–314. Amsterdam: Elsevier.

Platt, R. H. 1976. The National Flood Insurance Program: some mid-stream perspectives. *American Institute of Planners, Journal* **42**, 303–23.

Platt, R. H. & S. A. Cahail 1987. Automated flash flood warning systems. *Applied Geography* **7**, 289–301.

Ramsli, G. 1974. Avalanche problems in Norway. In *Natural hazards: local, national, global*, G. F. White (ed.), 175–80. New York: Oxford University Press.

Ray, P. S., D. W. Burgess et al. 1979. Doppler radar: research at the National Severe Storms Laboratory. *Weatherwise* **32**, 68–75.

Rodda, J. C. 1969. The flood hydrograph. In *Water, earth and man*, R. J. Chorley (ed.), 405–18. London: Methuen.

Rooney, J. F. 1967. The urban snow hazard in the United States: an appraisal of disruption. *Geographical Review* **57**, 538–59.

Rooney, J. F. 1969. The economic implications of snow and ice. In *Water, Earth and Man*. R. J. Chorley (ed.), 389–401. London: Methuen.

Rosenberg, N. J. (ed.) 1978. *North American droughts*. Boulder, Colorado: Westview Press, for the American Association for the Advancement of Science.

Sheets, R. C. 1985. The National Weather Service Hurricane Probability Program. *Bulletin of the American Meteorological Society* **66**, 4–13.

Sheets, R. C. 1990. The National Hurricane Center: past, present and future. *Weather Forecasting* **5**, 185–232.

211

Sigafoos, R. S. & M. D. Sigafoos 1966. Flood history told by tree growth. *Natural History* **50**, 50–5.

Simpson, R. H. 1973. Hurricane prediction: progress and problem areas. *Science* **181**, 899–907.

Simpson, R. H., B. Hayden, M. Garstang, H. L. Massie 1985. Timing of hurricane emergency actions. *Environmental Management* **9**, 61–9.

Sims, J. H. & D. Baumann 1972. The tornado threat: coping styles of north and south. *Science* **176**, 1,386–92.

Southern, R. L. 1979. The global socio-economic impact of tropical cyclones. *Australian Meteorological Magazine* **27**, 175–95.

Spiegal, J. P. 1957. The English flood of 1953. *Human Organization* **16**, 3–9.

Stewart, T. R., R. W. Katz, A. H. Murphy 1984. Value of weather information: a descriptive study of the fruit-frost problem. *American Meteorological Society, Bulletin* **65**, 126–37.

Stowe, J. T. 1984. The tornado. *Scientific American* **250**, 86–96.

Taylor, W. L. 1973. An electromagnetic technique for tornado detection. *Weatherwise* **26**, 70–1.

Thampapillai, D. J. & W. F. Musgrave 1985. Flood damage and mitigation: a review of structural and non-structural measures and alternative decision frameworks. *Water Resources Research* **21**, 411–24.

Thornthwaite, C. W. & J. R. Mather 1955. The water balance. *Publications in Climatology* **8**, 1–86.

Towery, N. G. & G. M. Morgan, 1977. Hailstripes. *American Meteorological Society, Bulletin* **58**, 588–91.

Tufnell, L. 1984. *Glacier hazards*. Harlow, England: Longman.

UNESCO 1981. *Avalanche atlas: illustrated international avalanche classification.* Paris: UNESCO Press.

Vishnevskiy, P. F. & G. M. Shcherbak 1969. Characteristics of the construction of equally probable flash flood hydrographs. *Soviet Hydrology* **4**, 375–81.

Walsh, S. J., D. R. Butler, D. G. Brown, L. Bian 1990. Cartographic monitoring of snow avalanche path location within Glacier National Park, Montana. *Photogrammetric Engineering and Remote Sensing* **56**, 615–22.

Ward, R. C. 1978. *Floods: a geographical perspective*. Chichester, England: John Wiley.

Warrick, R. A. 1983. Drought in the US Great Plains: shifting social consequences? In *Interpretations of calamity*, K. Hewitt (ed.), 67–87. London: Unwin Hyman.

White, A. U. 1974. Global summary of human responses to natural hazards. In *Natural hazards: local, national, global*, G. F. White (ed.), 255–65. New York: Oxford University Press.

White, G. F. 1945. *Human adjustment to floods: a geographical approach to the flood problem in the United States*. Chicago: Department of Geography, University of Chicago.

White, G. F. 1973. Natural hazards research. In *Directions in geography*, R. J. Chorley (ed.), 193–216. London: Methuen.

White, G. F. 1974. *Natural hazards: local, national, global*. New York: Oxford University Press.

White, G. F. & J. E. Haas 1975. *Assessment of research on natural hazards*. Cambridge, Mass.: MIT Press.

Wijkman, A. & L. Timberlake 1984. *Natural disasters: acts of God or acts of man?* Washington DC: Earthscan, International Institute for Environment and Development.

Wilhite, D. A. & W. E. Easterling (eds) 1987. *Planning for drought: toward a reduction of societal vulnerability*. Boulder, Colorado: Westview Press.

Select bibliography

Allen, B., H. Brookfield, Y. Byron (eds). Frost and drought in the highlands of Papua New Guinea. *Mountain Research and Development* 9.

Arnell, N. W. 1989. Expected annual damages and uncertainties in flood frequency estimation. *Journal of Water Resource Planning and Management* 115, 94–107.

Bein, F. L. 1980. Response to drought in the Sahel. *Journal of Soil and Water Conservation* 35, 121–4.

Berke, P. & C. Ruch 1985. Application of a computer system for hurricane emergency response and land use planning. *Journal of Environmental Management* 21, 117–34.

Berling, R. L. 1978. Disaster response to flash flood. *American Society of Civil Engineers, Proceedings; Journal of the Water Resources Planning and Management Division* 104(WRI), 35–44.

Bernhard, T. 1983. *Frost*. Frankfurt: Insel.

Brun, E., E. Martin, V. Simon, C. Gendre, C. Coleou 1989. An energy and mass model of snow cover suitable for operational avalanche forecasting. *Journal of Glaciology* 35, 333–42.

Burton, I. & R. W. Kates 1964. The floodplain and the sea-shore: a comparative analysis of hazard zone occupance. *Geographical Review* 54, 366–85.

Buser, O., P. Fohn, W. Good, H. Gubler, B. Salm 1985. Different methods for the assessment of avalanche danger. *Cold Regions Science and Technology* 10, 199–218.

Butler, D. R. & S. J. Walsh 1990. Lithological, structural and topographic influences on snow-avalanche path location, eastern Glacier National Park, Montana. *Annals of the Association of American Geographers* 80, 362–78.

Changnon, S. A. 1971. Hailfall characteristics related to crop damage. *Journal of Applied Meteorology* 10, 270–4.

Changnon, S. A., B. C. Farhar, E. R. Swanson 1978. Hail suppression and society. *Science* 200, 387–94.

Chatterton, J. B., J. Pirt, T. R. Wood 1979. The benefits of flood forecasting. *Institution of Water Engineers and Scientists, Journal* 33, 237–52.

Conway, H. & J. Abrahamson 1988. Snow-slope stability: a probablistic approach. *Journal of Glaciology* 34, 170–7.

Costa, J. E. 1978. The dilemma of flood control in the United States. *Environmental Management* 2, 313–22.

Cuny, F. C. 1991. Living with floods: alternatives for riverine flood mitigation. In *Managing natural disasters and the environment*, A. Kreime & M. Munasinghe (eds), 62–73. Washington, DC: Environment Department, World Bank.

De Freitas, C. R. 1975. Estimation of the disruptive effect of snowfalls in urban areas. *Journal of Applied Meteorology* 14, 1,166–73.

Dingman, S. L. & R. H. Platt 1977. Floodplain zoning: implications of hydrologic and legal uncertainty. *Water Resources Research* 13, 519–23.

Dunn, G. E. & B. I. Miller 1964. *Atlantic hurricanes*. Baton Rouge, Louisiana: Louisiana State University Press.

Eagleson, P. S. 1972. Dynamics of flood frequency. *Water Resources Research* 8, 878–94.

Fairbridge, R. W. (ed.) 1967. Tornadoes and hurricanes. *The encyclopaedia of atmospheric sciences*, 1003–6. Stroudsburg, Pennsylvania: Dowden, Hutchinson & Ross.

Fendell, F. E. 1974. Tropical cyclones. *Advances in Geophysics* **17**, 2–100.

Ferguson, S. A., M. B. Moore, R. T. Marriott, P. Speers-Hayes 1990. Avalanche weather forecasting at the Northwest Avalanche Center, Seattle, Washington, USA. *Journal of Glaciology* **36**, 57–66.

Flora, S. D. 1953. *Tornadoes of the United States*. Norman, Oklahoma: University of Oklahoma Press.

Ford, D. T. & A. Oto 1989. Floodplain management plan enumeration. *Journal of Water Resource Planning and Management* **115**, 472–85.

Glantz, M. H. 1984. Floods, fires and famine: is El Niño to blame? *Oceanus* **27**, 14–20.

Godschalk, D., D. Brower, T. Beatley 1989. *Catastrophic coastal storms: hazard mitigation and development management*. Durham, North Carolina: Duke University Press.

Golde, R. H. (ed.) 1977. *Lightning*. Vol. 2, *Lightning protection*. New York: Academic Press.

Hager, W. H. & R. Sinniger 1985. Flood storage in reservoirs. *Journal of Irrigation and Drainage Engineering* **111**, 76–85.

Hall, A. L. 1978. *Drought and irrigation in northeast Brazil*. Cambridge: Cambridge University Press.

Handmer, J. (ed.) 1987. *Flood hazard management: British and international perspectives*. Norwich, England: Geobooks.

Heathcote, R. L. 1969. Drought in Australia: a problem of perception. *Geographical Review* **59**, 175–94.

Howard, R. A., J. E. Matheson, D. W. North 1972. The decision to seed hurricanes. *Science* **176**, 1191–202.

Jiusto, J. E. & H. K. Weickmann 1973. Types of snowfall. *American Meteorological Society Bulletin*, **54**, 148–62.

Johns, R. J. 1989. The influence of drought on tropical rainforest vegetation in Papua New Guinea. *Mountain Research and Development* **9**, 248–51.

Kennedy, E. 1992. The impact of drought on production, consumption and nutrition in southwestern Kenya. *Disasters* **16**, 9–18.

Krzystofowicz, R. & D. R. Davis 1983. A methodology for evaluation of flood forecast-response systems: Part 1: Analyses and concepts, Part 2: Theory, Part 3: Case studies. *Water Resources Research* **19**, 1,423–54.

Lansford, H. 1979. Tree rings: predictors of drought? *Weatherwise* **32**, 194–9.

Linsley, R. K. 1982. Social and political aspects of drought. *American Meteorological Society, Bulletin* **63**, 586–91.

Liu Fengshu & Wang Xinian 1990. A review of storm surge research in China. *Natural Hazards* **2**, 17–30.

Liverman, D. M. 1990. Drought impacts in Mexico: climate, agriculture, technology and land tenure in Sonora and Puebla. *Annals of the Association of American Geographers* **80**, 49–72.

Mathur, K. & N. G. Jayal 1992. Drought management in India: the long term perspective. *Disasters* **16**, 60–5.

Mogil, H. M., M. Rush, M. Kutka 1977. Lightning: a preliminary assessment. *Weatherwise* **30**, 192–201.

Moore, H. E. 1958. *Tornadoes over Texas: a study of Waco and San Angelo in disaster*. Austin, Texas: University of Texas Press.

Select bibliography

Morgan, R. 1985. The development and application of a drought early warning system in Botswana. *Disasters* **9**, 44–50.

Mortimore, M. 1989. *Adapting to drought: farmers, famines and desertification in West Africa*. Cambridge: Cambridge University Press.

Nalivkin, D. 1983. *Hurricanes, storms and tornadoes: geographic characteristics and geological activity*, B. B. Bhattacharaya (trans.). Rotterdam: A. A. Balkema.

National Severe Storms Forecast Centre 1980. Tornadoes: when, where and how often? *Weatherwise* **33**, 52–9.

Newark, M. J. & D. McCulloch 1992. Using tornado climatology to help plan a Doppler radar network. *Natural Hazards* **5**, 211–19.

Oaks, S. D. & L. Dexter 1987. Avalanche zoning in Vail, Colorado: the use of scientific information in the implementation of hazard reduction strategies. *Mountain Research and Development* **7**, 157–68.

Parker, D. J. & E. C. Penning-Rowsell 1983. Flood hazard research in Britain. *Progress in Human Geography* **7**, 182–202.

Penning-Rowsell, E. C., D. J. Parker, D. M. Harding 1986. *Floods and drainage: British policies for hazard reduction, agricultural improvement and wetland preservation*. London: Unwin Hyman.

Perry, A. H. & L. Symons 1980. The economic and social disruption arising from snowfall hazard in Scotland: the example of January 1988. *Scottish Geographical Magazine* **96**, 20–5.

Peterson, R. E. 1979. Life cycle of a tornado. *Weather* **34**, 316–19.

Pielke, R. A. 1990. *The hurricane*. London: Routledge.

Platt, R. H. 1986. Floods and man: a geographer's agenda. In *Geography, resources and environment*. Vol. 2, R. W. Kates & I. Burton (eds), 28–68. Chicago: University of Chicago Press.

Ray, P. S. & D. W. Burgess et al. 1979. Doppler radar: research at the National Severe Storms Laboratory. *Weatherwise* **32**, 68–75.

Riebsame, W. E., S. A. Changnon Jr, T. R. Karl 1991. *Drought and natural resources management in the United States: impacts and implications of the 1987-89 drought*. Boulder, Colorado: Westview Press.

Riehl, H. 1963. On the origin and possible modification of hurricanes. *Science* **141**, 1,001–10.

Rodier, J. A. & M. Roche 1984. *World catalogue of maximum observable floods*. Washington, DC: International Association of Scientific Hydrology.

Rooney, J. F. 1969. The economic implications of snow and ice. In *Water, earth and man*, R. J. Chorley (ed.) 389–401. London: Methuen.

Rydant, A. L. 1979. Adjustments to natural hazards: factors affecting the adoption of crop-hail insurance. *Professional Geographer* **31**, 312–20.

Sadowski, A. F. 1966. Tornadoes with hurricanes. *Weatherwise* **19**, 71–5.

Salkey, A. 1966. *Drought*. Oxford: Oxford University Press.

Salm, B. & H. Gubler (eds) 1987. *Avalanche formation, movement and effects*. Washington, DC: International Association of Hydrological Sciences.

Schaerer, P. A. 1981. Avalanches. In *Handbook of snow*, D. M. Gray & D. H. Male (eds), 475–518. Oxford: Pergamon.

Sewell, W. R. D. 1969. Human response to floods. In *Water, earth and man*, R. J. Chorley (ed.), 431–51. London: Methuen.

Simpson, R. H. & J. S. Malkus 1964. Experiments in hurricane modification. *Scientific American* **211**, 27–37.

Simpson, R. H. & H. Riehl 1981. *The hurricane and its impact*. Baton Rouge, Louisiana: Louisiana State University Press.

Smith, D. I. & J. W. Handmer 1984. Urban flooding in Australia: policy development and implementation. *Disasters* **8**, 105–17.

Somerville, C. M. 1986. *Drought and aid in the Sahel: a decade of development co-operation*. Boulder, Colorado: Westview Press.

Swanson, E. R., S. T. Sonka, C. D. Taylor, P. J. Van Blokland 1978. An economic analysis of hail supression. *Journal of Applied Meteorology* **17**, 1,432–40.

Uman, M. A. 1984. *Lightning*. New York: Dover.

US Forest Service 1968. *Snow avalanches: a handbook of forecasting and control measures*. Washington, DC: US Department of Agriculture.

US National Research Council 1988. *Estimating probabilities of extreme floods: methods and recommended research*. Washington, DC: Water Science and Technology Board, National Research Council, National Academy Press.

US National Research Council 1990. *Snow avalanche hazards and mitigation in the United States*. Washington, DC: National Academy Press.

Viemeister, P. E. 1961. *The lightning book*. Cambridge, Mass.: MIT Press.

Wall, G. & J. Webster 1980. Consequences of and adjustments to tornadoes: a case study. *International Journal of Environmental Studies* **16**, 7–15.

Ward, R. C. 1978. *Floods: a geographical perspective*. New York: John Wiley.

Wood, D. W., T. C. Gooch, P. M. Pronovost, D. C. Noonan, 1985. Development of a flood management plan. *Journal of Water Resources Planning and Management* **111**, 417–33.

Zegel, F. H. 1967. Lightning deaths in the USA: a seven-year survey from 1959 to 1965. *Weatherwise* **20**, 169–73.

CHAPTER FOUR

Disasters and the land surface

This chapter deals with a range of hazards that varies considerably with regard to speed of impact. Accelerated soil erosion and desertification are slow-onset disasters, if indeed they ever reach disastrous proportions. Subsidence, soil heave and coastal erosion are usually gradual processes, though sometimes rapid enough to accomplish much destruction in a single episode. Seismic landslides and wildfires are generally sudden and swift, but non-seismic landslides can vary in speed and duration from interminably slow and long drawn-out to exceptionally rapid and brief (i.e. from years to seconds). In this chapter the fate of the biosphere can be seen to influence that of the lithosphere: the health and composition of vegetation cover is often a guide to the strength of erosional processes and, especially, to the progress of desertification. Conversely, eroded or desertified land is by definition infertile. Essentially, the linkages are subtle enough, and the range of impacts broad enough, for the following sections to deal with an exceptionally wide variety of impacts, adjustments, monitoring technologies and mitigation techniques.

Soil erosion

Problems of erosion are underestimated in the United Kingdom and northern Europe, taken for granted in much of the Mediterranean Basin, persistently serious in the USA and verging on the catastrophic in China. **Accelerated soil erosion** involves layered organic and mineral complexes that are fertile in terms of plant growth and are removed by running water, flowing wind or moving ice at rates faster than their averages over geological timespans of, say, 10^5–10^6 years.

The hazard of soil erosion is strongly linked to that of landsliding. Shallow landsliding often represents the loss of soil material by sliding or

217

flowing along boundaries within the soil profile. Other hazards can also provoke intense erosion. For example, hurricanes may destroy forest cover at the same time as they give rise to high-intensity rainstorms which cause vigorous surface runoff with high ability to scour (see Fig. 8.7). Hailstorms and high-intensity thunderstorms involve great expenditure of kinetic energy, which can lead to substantial overland flow or rainsplash erosion. Turbulent floods – especially flash floods – can result in high bed shear and high rates of entrainment in streams. In drainage basins with soil mantled slopes, such events may produce thick, viscous mudflows of high destructive potential.

The erosional system

Erosion can be defined as the movement of rock or sediment impelled by the energy of relief, gravity and climatic factors such as precipitation or wind flow. The **erosional system** can be divided into a series of components, beginning with **detachment**, the dislodging of particles of sediment by splash impact or bed shear as a result of the movement of wind, water or ice. A critical threshold of energy is passed at which the material enters into movement by a process of **entrainment** into the flow. It then undergoes **transportation** in suspension, as bed load or, in the case of stream erosion, also as dissolved load.

Eroded sediment can also undergo periods of temporary **storage** or more permanent **deposition**, in which the transported material is consolidated into sedimentary deposits or rocks. These are characterized by bedding and other sedimentary structures related to their mode of deposition. For example, the valley fill deposited by streams tends to be sorted and stratified and is termed **alluvium**, while the more chaotic, unsorted deposits that accumulate at the base of slopes are called **colluvium**. Fine silt deposited uniformly by abating wind flow is termed **loess**.

In the case of fluvial erosion, the system boundary is the **watershed** or **drainage divide**. Aeolian erosional systems usually have much less distinct boundaries. System inputs can be divided into those which are static, or passive (such as energy of relief, soil deposits and instability zones), and those (such as meteorological processes) which are dynamic (see Fig. 4.1).

Several subsystems can be identified. One of these is the soil, which consists of differing layers, or **horizons**, of which the uppermost ones, the O and A horizons, are dominated by organic material, the intermediate layer (B horizon) is composed of mineral grains and mineralized organic matter and the lower or C horizon consists of weathered particles of the underlying parent rock. Under the US classification scheme, a **soil series** is formed of groups of soils with similar horizons, while a **soil association** consists of groups of series.

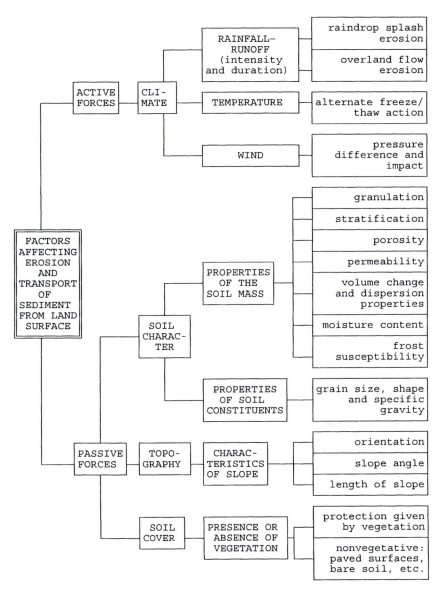

Figure 4.1 Factors affecting the erosion and transportation of sediment (Guy 1964: E4).

The dominant part of each soil consists of the A and B horizons and is known as the **solum**: it has a definite crumb structure. To be fertile, the solum must hold plant nutrients and moisture and allow them to circulate throughout the root zone. Soils must therefore combine porosity and permeability with stability and strength. As a result of the slowness with which organic material is combined with weathered mineral particles, soils take decades, centuries or millennia to form, but they can be eroded by the most aggressive forms of scour in a matter of minutes. Erosion not only removes nutrients contained in the organic material of the A horizon, but it also tends to expose the B horizon, which, in soils of low fertility, may contain excessive amounts of clay (**argillic**), be resistant to root penetration (**fragic** or **duric**), be acidic or salty (**salic**) or saturated with aluminium (**bauxitic**).

Besides soils, perhaps the most ubiquitous geomorphological elements at the Earth's surface are slopes. In fluvial systems they form a distinct subsystem which interacts with basal channels (another subsystem) in a continuing process of undercutting during valley deepening, which leads to slope collapse and shallowing, which supplies sediment for the channel to transport away. Although oscillatory, this process tends to create stable and characteristic morphological relationships between the two connected subsystems.

Many erosional systems are dominated by **negative feedback**, which results in self-maintenance and preservation of stable parameters. For example, eroding valleys which experience some violently disruptive event such as a major earthquake or flood may be subject to a series of phases of trenching and filling, as fluvial erosion and deposition compensate for the shock to the system. **Positive feedback** is also present in erosional systems and involves the crossing of **thresholds**, for example from sheetwash on planar slopes to intensified gullying and the formation of arroyos or wadis. In the American Southwest such a transition was initiated in the latter part of the nineteenth century as a result of the intensification of winter rains, plus the effects of overgrazing by cattle and the increases in runoff caused where they compacted the land surface.

Forms of erosion

Soil erosion on hillslopes may involve rainsplash, sheetwash, rilled and channelled erosion or aeolian transport.

The kinetic energy of a rainstorm is a function if its intensity:

$$E = 1.213 + 0.890 \log_{10} I$$

where E = kinetic energy in kg m/m^2 mm
I = rainfall intensity in mm/hr

The total kinetic energy expended can be computed by accumulating values of E over the duration of the storm or other time unit. As they expend very large amounts of kinetic energy, extremely intense rainstorms can cause large amounts of erosion, as happens in West Africa. But in the aggregate (to coin a pun!) most soil appears to be shifted by storms that yield 30–60 mm. Threshold intensities vary between 6–10 mm/hr for Western Europe to 25 mm/hr for the tropics.

The kinetic energy of raindrops is proportional to their fall velocity, and thus **rainsplash** is a significant cause of erosion (Selby 1976, Stocking & Elwell 1976). It will loosen and displace soil particles in random directions, but with a net downslope trend. It can also compact bare soil surfaces, thus making them resistant to rainwater infiltration and more susceptible to overland flow. Raindrop diameter may vary between 0.5 and 6.0 mm, while fall velocity may be in the range 2–9 m/sec. Splash erosion can be expressed as a function of soil characteristics:

$$E = K (D / APS)$$

where   K = constant of proportionality
                 D = soil dispersion by rain (or its cohesiveness)
                 A = infiltration capacity (e.g. 0.25–0.5 mm/hr for a clay loam, 25–50 mm/hr for a sand loam)
                 P = soil permeability (hydraulic conductivity)
                 S = sediment size

Overland flow of water on slopes occurs primarily when rainfall intensity exceeds the infiltration capacity of the soil. Infiltration capacity tends to be high in a dry soil, but will settle down to a residual value when the soil becomes saturated. This may be in the range 0.25–0.5 mm/hr for a clay loam, to 25–50 mm/hr for a sandy loam. Rainfall intensity may vary between 5 mm/hr for a gentle storm and 100 mm/hr for one of extremely high energy expenditure. Thus, the more intense storms tend to be those which cause surface flow on slopes. When this occurs by the laminar movement of the water, as at the start of overland flow, **sheetwash erosion** is absent. But when a critical depth and velocity are reached the flow becomes turbulent and erosion begins. Overland flow velocity may be 200–300 m/hr and the erosion which it creates can be measured using pins driven into the surface of the slope or troughs for collecting eroded sediment.

Flow in channels may be **perennial**, if the stream is always present, or **ephemeral**, if the channel is sometimes dry (in which case the flow may be seasonal in character). Channel erosion may occur by the cutting of rills and gullies in erodible soils and sediments (Fig. 4.2), the incision of streams and collapse of their banks or the extension of headcuts into erodible hillsides. The rate of headcut advance is a function of drainage area, channel slope and the total rainfall of significant storms. The rate of channel erosion can often be assessed visually or quantitatively by examining **drainage density**, which is expressed as the ratio of length of channels (in kilometers) to the area that they drain (in km^2). The average density of channels in humid, temperate mid-latitude areas is in the range 1.4–4.5km/km^2, but **badland** gullies and rills may exceed 1,000km/km^2.

Figure 4.2 A rilled slope in eroded clays at Hallett Cove, South Australia, illustrates the fine texture of accelerated soil erosion resulting from slopewash processes.

The figure may be nearer 10,000 km/km^2 if micro-rills are created seasonally as a result of concentration of sheetwash erosion on slopes.

Channel entrenchment may occur when a stream cuts through sediments to form an arroyo, wadi or canyon. This has a variety of causes and consequences. Overgrazing may destroy plant cover, concentrate runoff on compacted tracks and reduce surface infiltration capacity at such points. Increases in rainfall intensity associated with slight shifts in climate may lead to higher rates of overland flow. Droughts may reduce water tables and dry the soil out more frequently so that in desiccating it loses some of its resistance to erosion. Concentration of surface water flow leads to increased flood runoff and higher rates of shear on channel beds and banks. In terms of consequences, the transition to channel trenching exposes a greater surface area to erosion, with less surface crusting, thus increasing the throughput of sediment in the stream channel system. In this way, crop and grazing lands may be destroyed (Cooke & Reeves 1975).

Wind erosion is a function of a soil's erosion potential, land surface roughness, length of the erodible surface along the prevailing wind direction, average wind strength, general climatic conditions and degree of vegetation cover (Chepil & Woodruffe 1963). Of these variables, only the climatic and wind factors cannot be controlled to prevent erosion, although trees can be used to provide a windbreak. Soil stabilization in order to prevent wind erosion is particularly expensive. At the global scale, aeolian erosion, just as ice scour, is of only minor importance, even though locally it may be severe. Under current world climates, most sediment is carried as the solid load of rivers and streams (Table 4.1).

Conceptualizing and measuring erosion

The cohesiveness and coherence of soils and sediments, in other words their resistance to detachment and transportation, is termed their erodibility (Bryan 1968). Consolidated clays (with particles of diameter less than 0.002 mm) are cohesive, in that electrochemical bonds hold individual particles together and compaction may give the sediment an almost rock-like character. But they may also resist infiltration, hence encouraging overland flow and slopewash erosion. Whereas loose sand is neither cohesive nor coherent, indurated sands have a degree of coherence, although not to the extent of more massive hard rocks, such as unweathered granite. Many sediments that are rich in calcium carbonate produce a weak natural cement which increases their resistance to erosion, while in an analogous manner volcanic tephra produce welded and fused tuffs, which are known as ignimbrite.

Soils containing 30–35 per cent clay are generally cohesive and form stable aggregates that resist raindrop impact and splash erosion. Open-textured soils tend to be the most erodible, as they dry out quickly and have low cohesion or cementation. Thus, the most erodible soils are those with silt contents of 40–60 per cent and a clay content of no more than 30 per cent. These are light enough to be entrained but not cohesive enough to stick together and resist the shearing force of water or wind.

Erosivity is represented as an index combining all factors which define the susceptibility of an area to erosion. Hence, it consists of erodibility plus atmospheric processes. Rainfall, for example, can be characterized in terms of annual, monthly or daily amounts, duration of storms in hours,

Table 4.1 Estimates of world erosion rates (Judson 1968)

Source of sediment	10^9 tonnes/yr
River-borne:	
Dissolved load	2.7–3.9
Solid load (suspended and bed load)	8.3–58.0
Total dissolved and solid load	3.8–17.5
Mean	9.3
Wind-borne:	0.06–0.36
By glacial ice:	0.1

It is assumed that all eroded material reaches the oceans.

and intensity in mm/hr. Although the correspondence is not perfect, high erosivity is often associated with very high values of drainage density developed in areas of weak sediments; in other words, with the evolution of badland landscapes (Stocking & Elwell 1976).

Erosion can be measured as volume or mass per unit area per unit time, or as rate of surface lowering. The units must be selected to give a level of sensitivity that is appropriate to the scale of measurement and rate of erosion; for example, $m^3/m^2/yr$, $kg/km^2/yr$, tonnes/km^2/1,000 years, mm/yr, m/1,000 years, etc. Units are broadly convertible: for example, a soil loss of 1.3 t/ha/yr is equivalent to 1 cm of surface lowering per decade. In streams, bedload is measured as kg/m or kg/Q (where Q is discharge in m^3/sec), suspended load is given as mg/litre, and dissolved solids are expressed for samples taken from the stream as parts per thousand or per million in relation to water. Yield of sediment can be expressed as a function of stream discharge, Q, using the **sediment rating curve**:

$$y = aQ^b$$

where a and b are empirical parameters.

In practical terms it is difficult to measure erosion without interfering with the process and hence biasing the data obtained. Streamlined bottle samplers are used to collect suspended sediment in streams; slotted traps or cages are used for bedload; nails, troughs and tanks measure slopewash erosion; and radioactive tags, paint or dye can help trace sediment movement. At a more theoretical level, nomographs are available that enable erosion rates to be computed using rainfall, discharge, drainage basin size, sediment type, mean particle diameter and other variables (Goldman 1986).

Net erosion is known as **sediment yield**, and is measured at a particular point in the drainage basin, usually where the stream flows out of it. The movement of sediment on slopes tends to be sporadic and uneven, while storage occurs in floodplains and valley fills, and hence it is usually difficult to obtain an accurate measure of the overall rate at which soils or sediments are being stripped off all slopes in the catchment. Thus sediment yield is a measure of the erosion taking place in the basin, minus storage, and quantifies the rate of sediment exported out of the local erosional system. Where individual slopes can be monitored sufficiently to obtain an idea of the rate of actual erosion occurring there, sediment yield can be divided by the total production of sediment to give the **sediment delivery ratio**. This decreases nonlinearly as drainage area increases according to the following approximate relationship:

$$S_d = 1/A^{0.2}$$

At the same time, the areal average denudation rate decreases with increasing drainage basin size (Table 4.2).

Investigation of a 159 km^2 drainage basin in the State of Maryland, USA, showed that 52 per cent of eroded sediment was stored on slopes, 14 per cent was stored in floodplains and 34 per cent of total upland erosion was delivered to the outflow point as sediment yield. Sediment yields in the southeast USA have been measured variously as 20 tonnes/km^2/yr from forested catchments, 300 t/km^2/yr from rangeland and 150–10,000 t/km^2/yr from construction sites where land is temporarily exposed in a highly disturbed and erodible form (Wolman 1967, Wolman & Schick 1967; see Ch. 5).

Soil loss in farmed upland area (A, in tonnes/km^2/yr) can be expressed by the **Universal Soil Loss Equation** (USLE – Wischmeier & Smith 1978):

$$A = R.K.L.S.C.P$$

where
R = precipitation (or the erosion potential of rainfall)
K = soil erodibility
L,S = slope length and angle (topographic factors)
C = cropping management factor (fallow, etc.)
P = conservation practice factor (strip cropping, fallow, etc.).

The accuracy of this versatile descriptive measure can be increased by considering the sediment delivery ratio, in other words, the volume of sediment eroded from cropland which is stored on slopes and in channels and floodplains.

According to Wischmeier (1976), the USLE can be put to various uses. For example, it can provide soil loss estimates on which to base conservation programmes for either cultivated or non-agricultural land. Regarding the former, it is well able to predict the average annual soil loss from a field slope with given land-use conditions. It can also indicate the cropping and management practices (including conservation measures) that help conserve particular soils or reduce erosion rates on

Table 4.2 Estimated rates of denudation for drainage basins of different sizes (various sources).

Drainage basin area (km^2)	Denudation rate (m/1000 years)	Sediment delivery ratio
2,5000,000–15,600,000	0.02–0.047	<0.02
37.0–3280.0	0.03–0.06	0.13–0.03
3.9	0.03–0.1	0.27
0.08	0.06–0.22	0.56
0.003	2.55	0.85
0.0003	12.6	0.95

particular slopes, or it can predict the results of changing these methods. Finally, the USLE can help determine what alterations should be made in cropping or land management practices, without undue loss of soil, in order to intensify cultivation.

The counterpart of the USLE for aeolian conditions is the **Wind Erosion Equation** (WEE), which is based on the following relationship:

$$E = f (I, K, C, L, V)$$

where
I = soil erodibility
K = surface roughness and shape
C = wind speed and duration
L = field length in the direction of wind blow
V = vegetation cover

Like the USLE, the WEE was designed to predict erosion rates over 10 years or more.

The Erodibility Index (EI) is based on the Universal Soil Loss Equation (USLE) or the Wind Erosion Equation (WEE) and **soil loss tolerance levels (T)**. The WEE is much less accurate than the USLE, tending to overestimate wind erosion and to result in higher values of T. The T value is defined by the US Soil Conservation Service as the maximum level of soil erosion that will permit a high level of crop productivity to be sustained economically and indefinitely (Alexander 1988). T varies with climate and with soil characteristics, such as depth and fertility. Values fall within the range 2.5–11.2 t/ha/yr. The method has certain drawbacks: first, it does not take off-site damage (such as water pollution) into account; and, secondly, T is usually less than the rate of soil formation, and so a soil that is eroding at the value judged to be tolerable may slowly be depleted.

In order to determine whether a soil is highly erodible, the T value for each map unit is substituted in the USLE, and the land-use and management factors (C and P) are removed. The equation is rewritten as $EI = RKLS/T$, where R, K, L and S are fixed physical factors. A land unit is considered highly erodible if $EI > 8$.

The effect of eroded sediment

If used wisely, eroded sediment can enrich lowland soils, but it also has a series of negative impacts (Duda 1985). For instance, it depletes water storage by silting up reservoirs. In this respect, the trap efficiency of water dams may reach 1–5 per cent per year, and hence the life of a reservoir tends to be finite (see Ch. 5). The solutions, moreover, are expensive: annual dredging may cost $2–5 per cubic metre of soil removed, and similar costs apply to the removal of silt from harbours and navigable waterways. Navigability is not the only problem in channels, where silt may cause aggradation, thus increasing the flood risk. At the same time, the aquatic ecosystem may be impaired, water supplies may be polluted and the costs of water purification greatly increased. Lastly, water erosion strips off valuable topsoil and removes land from productive agricultural use, unless the nutrients and structure of the soil can be restored by artificial means. This is not usually possible and is invariably expensive.

Erosion control

Various methods of erosion control have been invented and the strategies are now well developed (Morgan 1986). One can distinguish between mechanical or engineering methods and chemical or agronomic ones (agronomy signifies the use of agricultural technology).

The principal scope of mechanical methods is to reduce the shearing effect of wind or water on the soil surface, and hence stop the entrainment of soil particles or cause their deposition. One of the most widely used of such methods is contour ploughing, which involves following the level of the land, rather than ploughing up and down slopes in a manner that tends to channel runoff. The method can be supplemented by building **contour bunds**, or earth banks up to 2m wide, which are planted with grass or trees in order to stop the downslope movement of soil from the rows between them. Terraces (Fig. 4.3) are particularly useful where slopes are too steep to be ploughed with the contours, and often the soil behind their stone or earthen walls is imported from flatter areas. They tend to increase infiltration rates, reduce runoff and stop the migration of soil downslope. But terraces are very labour-intensive, as they must be well built and constantly maintained. In the absence of maintenance, concentrated infiltration can increase pore-water pressures behind the terrace front until it bursts, and in a flight of terraces the damage can be progressive downslope. Larger terraces consist of subvertical risers and horizontal shelves (which are cultivated), while miniature versions can be cut out of a soil-mantled slope with a hoe.

Among engineering solutions, gully check-dams reduce the depth and longitudinal gradient of eroding channels. If sufficiently well anchored, they can prevent upstream gullies from oversteepening, diminish the erosive energy of streamflow and reduce sedimentation and trenching down stream. But their value as sediment traps is severely limited as they tend to fill up rapidly; and if the weight of sediment behind them is greater than their strength, they may burst and release a highly erosive slug of debris into the channel. Moreover, their effect on erosional processes up stream is probably limited to a few tens of metres from the lip of the structure.

Mechanical measures designed to reduce wind erosion usually function to reduce wind velocity at the surface, entrap moving soil, increase the

Figure 4.3 Agricultural terraces constructed by people of the Inca culture at Pisac in central Perú. The terraces carefully follow the natural contours of the land and, if the stone facings are properly maintained, they are an effective means of soil conservation. They also served a defensive purpose.

resistance to entrainment of soils and maintain the vegetation cover. Windbreaks, or shelter belts, are usually placed at a right angle to prevailing winds and should be permeable, to avoid creating erosive vortices. The spacing, height, width, length and shape of belts all influence their effectiveness. As wind velocity tends to increase immediately behind the belt, it should have a tapering end. But windbreaks may cause problems of shade and weed growth which affect crop yields in their vicinity.

Agronomic measures of soil conservation are based on the rôle of plant cover in reducing erosion rates. Thus, controlled grazing allows the vegetation cover of an area to regenerate after it has been browsed or grazed and reduces the effect of soil compaction by trampling. On arable lands, crop rotation and careful ploughing can both be used to minimize erosion. There are also techniques such as **minimum tillage** (in which ploughing is avoided altogether and seeds are drilled directly into crop residues) and barley row intercropping, which stabilizes the soil and stops it eroding from between crop rows. Strip cropping requires small-scale machinery, rather than highly mechanized farming.

Through plant selection species can be planted that resist erosion, for example by intercepting falling raindrops efficiently, resisting fluid flow at ground level, trapping transported sediment, adding organic matter that binds the soil and increasing the rate of infiltration along their roots. In general, erosion is greatest under row crops (especially grape vines, but also wheat), is less on close-seeded crops and least under grass and forage crops. The difference can be reduced by **mulching**, which involves covering the soil with crop residues such as straw, maize stalks or wheat stubble, and it simulates the effect of plant cover in binding the soil together. Later in the annual production cycle the mulch is replaced with a real crop.

Organic or inorganic additives can be mixed into soils to increase their strength or reduce their erodibility. The aim is to create a resistant soil structure where one did not originally exist. When this is done by chemical means it is termed **soil emendation**. Thus, highly sodic soils must be neutralized or they will disperse easily in overland flow (i.e. they tend to go spontaneously into suspension in water). Lastly, gullies can be stabilized against further erosion using hardy plants capable of growing vigorously in conditions of shade and infertile soils.

Experiments conducted in the 1930s showed that, in respect of slopes ploughed up and down, soil loss is reduced 43 per cent by contour ploughing, 75 per cent by rotation strip ploughing and 97 per cent by careful terracing (Mitchell & Bubenzer 1980). In another set of trials, a 26 ha catchment in Tennessee, denuded of forest in the 1930s and 1940s, was subjected to gully control and contour-ploughing measures in the 1950s. Water yield was halved, peak flows reduced by 90 per cent and sediment yield reduced by 96 per cent. Moreover, the US Department of Agriculture has estimated that extension of minimum tillage from 10 per cent of American croplands (in 1974) to 80 per cent would reduce soil erosion by one half (Carter 1977).

Soil erosion in a world context
Since agriculture began in earnest about 3,500 years ago, world erosion rates in 480,000 km of main channels have increased from an estimated 9,300 to 24,000 million tonnes per year (Judson 1968). The vast majority (96 per cent) is fluvially eroded and only 2 per cent of eroded soil comes from aeolian sources. Rates of surface lowering that were 2–3 cm/1,000 years prior to human influences have now doubled or, in highly localized instances, increased by several orders of magnitude.

The total land area of the world is 14,477 million hectares, of which 13,241 million ha are free of ice, but only 11 per cent (1,500 million ha) are cultivated (El-Hinnawi & Hashmi 1987). Potentially cultivable land occupies another 1,700 million ha, but its global distribution is very uneven. Hence, 92 per cent of cultivable land is being tilled in Southeast Asia, but only 15 per cent of that in South America. One of the vital reasons for this is difference in land tenure. Pressure on land is as uneven as its use and is the main cause of erosional crises. For instance, according to the United Nations Environment Programme (UNEP), the Ethiopian highlands are losing 1,000 million tonnes of topsoil a year as a result of overuse (Hurni 1988).

Soil degradation causes 5–7 million ha (0.3–5 per cent) of land to be taken out of cultivation each year, but again the global distribution of impacts is very uneven. Thus, although soil erosion has been identified in virtually every country of the world, it tends to concentrate in some of the poorest, such as El Salvador, where 77 per cent of the land area is seriously eroded and Colombia, three quarters of which is affected by erosion, one third severely so.

In general terms, erosion rates reach a maximum, and environments are most fragile, in seasonally arid, or semi-arid, lands (Fig. 4.4). With annual precipitation totals of 250–300 mm, rainfall is insufficient to maintain a continuous vegetation cover (Fig. 4.5), but in wet periods instantaneous intensities can be sufficient to cause fierce erosion (I recall being submerged up to my knees in runoff thick with eroded *terra rossa* during a particularly violent thunderstorm in the Province of Almeria, in the semi-arid part of Spain, during the winter of 1973–4).

Soil erosion in the USA

The US Department of Agriculture has estimated that 70 per cent of the nation's soil erosion is anthropogenic in origin. There are some very clear examples of the relationship between human activity and soil loss: for instance, the introduction of cattle into Colorado in the late 1800s increased average erosion rates on prairie from 0.2–0.5 mm/yr to 1.8 mm/yr. Moreover, in broad terms, sediment yield in America doubles for every 20 per cent reduction in forest cover.

In the USA sediment sources have been quantified as follows: 40 per cent from agricultural lands, 26 per cent from streambank erosion, 12 per cent from pasture and rangeland, and 22 per cent from miscellaneous sources. Suspended sediment in streams is the greatest environmental pollutant by weight and volume in the country. It causes $1,000–2,000 million per year in damage and treatment costs, and arrives in the nation's waterways at a rate of 4,000 million tonnes per year.

Water runoff in the United States erodes about four times as much soil as does wind, yet there have been some spectacular episodes of aeolian erosion. "Dust Bowl" conditions of the 1930s were primarily caused by a series of hot, dry years that depleted the vegetation cover, dried out soils and hence decreased their weight. Overgrazing and lack of soil conservation accounted for the rest of the problem, especially the rapid

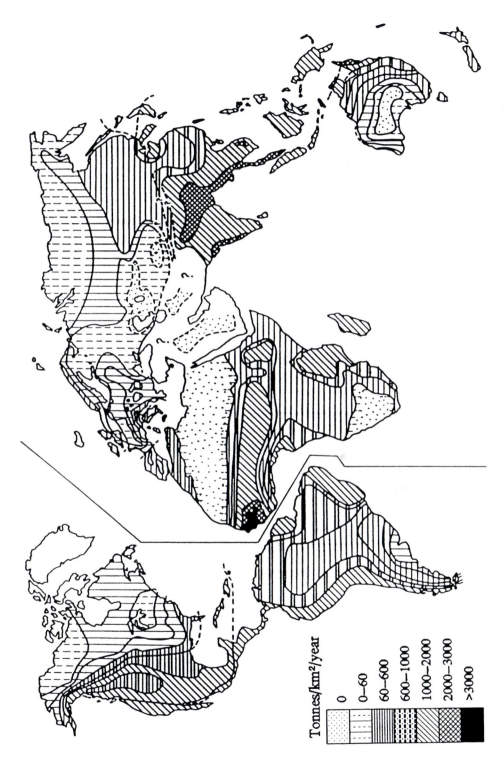

Figure 4.4 World distribution of erosion rates (Cooke & Doornkamp 1974: 23, after Fournier 1960).

Tonnes/km²/year

0
0–60
60–600
600–1000
1000–2000
2000–3000
>3000

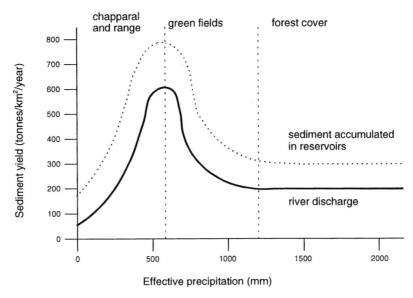

Figure 4.5 Relationship between precipitation and sediment yield (Langbein & Schumm 1958: 1077–8). Copyright 1958 by the American Geophysical Union.

expansion in hectarages of wheat, which had increased from 2 to 5 million ha over the period 1910–19 in Kansas alone.

In 1932 high temperatures and scarce rains led to drought on the Great Plains. Next year, 179 dust storms were reported by western weather stations, and they gained steadily in intensity the following year. By May 1934 a cloud 5,000 m high stretched from Alberta to Texas, and Montana to Ohio. It transported soil more than 2400 km, suffocated birds in flight and coated the decks of ships 500 km off the coast with dirt. In 1935 a single storm destroyed 1.9 million ha of wheat in Kansas, Nebraska and Oklahoma, while another storm deposited 12 million tonnes of dust on Chicago. Silica in the air caused death through aggravated respiratory diseases. About 375,000 km^2 of Colorado, Kansas, Nebraska and Oklahoma were devastated over the period 1933–40. For example, at Amarillo, Texas, an average of 36 storms occurred from January to April in each of three consecutive years. Drifts of eroded soil up to 7.5 m high accumulated, burying barns and houses. In the worst affected areas farms and ranches failed; unemployment, destitution and migration were common.

A committee set up in 1937 recommended planting 5.8 million ha to grass instead of cultivating them. But ownership patterns impeded land conversion. The founding of the US Soil Conservation Service helped the diffusion of soil conservation measures, but with varying levels of success (Steiner 1990). Soil conservation schemes promoted by USDA have unfortunately not been sufficiently well linked to efforts to give farmers the financial stability and thus the economic freedom they require in order to implement erosion controls. Although the US Environmental Protection Agency now requires each state to plan to reduce water pollution, including suspended sediment loads, state laws often exempt the causes of soil erosion from control.

Although America has not witnessed a full-scale repetition of the Dust Bowl conditions of the 1930s, there is ample scope for these to recur (Borchert 1971). Drought, which has a 20–25 year cycle on the Great Plains, returned in the 1950s, when 3.8–6.1 ha of land again underwent severe erosion. And again, in 1977 wind erosion scoured more than 642,000 km^2 in New Mexico and lifted 25 million tonnes of soil from only 2,000 km^2 of the San Joaquin Valley, California.

An archetypal soil, if there is such a thing, forms at a rate of about 3.7 t/ha/yr. If erosion exceeds the rate of formation, the structure of the soil will become progressively impoverished. Even in the short term, deep soils cannot afford to lose more than 11.2 t/ha/yr, which is the maximum soil loss tolerance value set for any American soil by the US Department of Agriculture (values may be lower on poor quality soils). A random sample of Corn Belt farms showed that 84 per cent were losing more than 3.7 t/ha/yr, while in 1977 the maximum T values were exceeded on 454,000 km^2 (27 per cent) of American croplands. Large areas are losing 22–30 t/ha/yr, and the maximum values recorded have reached 150 t/ha/ yr, a rate which can be sustained only for a very short period until the soil is entirely gone (Smith & Stanley 1965).

Soil erosion is the main conservation problem on about half of America's croplands (see Fig. 4.6), but only one third of the Corn Belt's 11.7 million km^2 has been subjected to long-term soil conservation. As a rule, the desire to maximize crop yields in the short term has taken preference over the advantages of soil conservation. Essentially, high grain prices have promoted the abandonment of crop rotation patterns in favour of continuous wheat cultivation, ploughing methods have left the soil bare of residues and weeds that might give it cohesion, fields have been levelled to facilitate spray irrigation, rangeland has been converted to grain production and cropland has been ploughed before the erosive winter rains. All of these factors, which are dedicated to achieving higher short-term productivity of the land, have increased erosion rates. The fundamental causes are as much economic as they are agronomic. Overseas demand has inflated wheat prices, while farmers have become more indebted as a result of greater reliance on economics and technology in the cultivation cycle.

The essence of the problem is that fertility can be maintained in the short term artificially, despite the loss of soil structure. Nitrogen, phosphorus and potassium nutrients can be replaced with fertilizers for

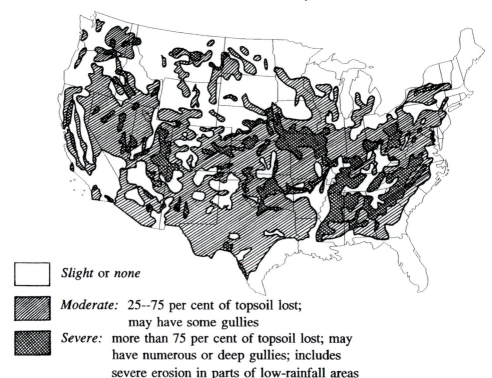

☐ *Slight* or *none*

▨ *Moderate:* 25--75 per cent of topsoil lost;
  may have some gullies

▩ *Severe:* more than 75 per cent of topsoil lost; may
  have numerous or deep gullies; includes
  severe erosion in parts of low-rainfall areas

Figure 4.6 Distribution of soil erosion in the United States (Cooke & Doornkamp 1974: 26).

only about $60/ha/yr (at 1970s prices), whereas soil conservation measures tend to be more expensive, even though they may be more effective in the long term. However, nationally, losses of these nutrients cost more than $1,000 million per annum, and the leached chemicals pollute surface and phreatic water supplies. Indeed, in 1983 the US Government spent more than $20,000 million on supporting farmers, a sum which for the first time exceeded the nation's total net farm income.

With adequate protection measures, the total area of US cropland could be increased from 168 million to 269 million hectares, but half of the reserve land is susceptible to erosion, which could subtract more than 8 per cent from production in a century. In fact, it is estimated that 22 million hectares of US cropland could be lost to erosion by AD 2083, which, at the very least, would cause large areas of marginal land to be pressed into cultivation. This would pose a major challenge to the use and organization of erosion control methods.

Soil erosion is part and parcel of a wider environmental problem – desertification – which is restricted neither to deserts nor to the world's poorer nations.

Desertification

Desertification is a process of ecological degradation that occurs primarily in arid, semi-arid and subhumid lands and causes the biological productivity of the land to be lost or substantially diminished: grazing lands cease to produce adequate pasture, dryland agriculture fails and irrigated fields are abandoned as a result of salinization, waterlogging or other forms of soil deterioration. The phenomenon can take the form of extension of desert margins or intensification of desert conditions within dry regions. Whereas drought should be considered an essentially reversible phenomenon, desertification tends to be permanent, or at least requires substantial inputs of capital and resources to remedy the damage. However, when prolonged periods of drought are coupled with ecological mismanagement, they may end in permanent degradation of the land. Visible scars (Fig. 4.7 and Table 4.3) are not essential marks of the process, as loss of nutrients and consequent decline in crop yields may be sufficient to betoken desertification. The phenomenon can therefore be regarded as a "creeping disaster", which may go unrecognized until profound and serious changes have occurred in the fertility of the land (Biswas & Biswas 1980).

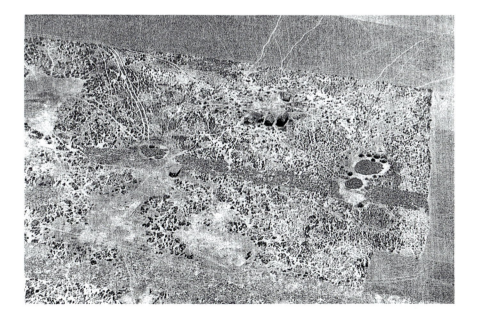

Figure 4.7 Land degradation amounting to desertification in rural Bangladesh.

Table 4.3 Processes of erosion in dry-land regions (Warren & Maizels 1977: 233).

Land region	Water erosion	Wind erosion
Mountains	landslides, coarse washout deposits	little, except on summits
Hills	gullying, sheetwash	little
Piedmonts	severe gullying, sheetwash, coarse washout deposits	soil loss, pavement formation, wind mounds
Plateaux	gullying and landslides on scarps, some sheetwash	soil loss, pavement formation
Plains	piping, some gullying	soil loss, pavement formation, wind mounds, scalding
Alluvial valleys	some gullying along terrace edges, sheetwash, silt accumulation	pavement formation, wind mounds, scalding
Enclosed depressions	silt acumulation	soil loss, wind mounding, scalding
Aeolian sands	rarely eroded – only on subhumid margins	dune formation, wind mounds

The scope of desertification

World surveys have classified desertification as follows: 18 per cent slight, 53.6 per cent moderate, 28.3 per cent severe and 0.1 per cent very severe (Fig. 4.8). However, the phenomenon is not easily identified with precision: estimates of the rate at which ecological degradation is spreading may be little more than guesses. There are several reasons for this. Land deterioration can take many forms (Table 4.4), and it may be "spotty", rather than advancing by a clearly identifiable front of land transformation. Thus, although the average rate of advancement of the Sahara Desert in the Sudan is about 5 km per year, the process is more one of patchy degradation within a broad area of 100–200 km^2. Moreover, soil degradation damages areas of greater size than those which can be regarded as fully "desertified". In addition, desertification tends to increase the levels of sediment transport by streams and wind, and hence to be closely related to erosion and sedimentation.

Amounts of desertification are not easily estimated, as at least a decade may be required before it can adequately be distinguished from drought. However, the annual loss of land is thought to be about 60,000 km^2, distributed among about 100 countries (Fig. 4.9). Of these countries, 27 are in Africa, and since 1925 on the southern fringe of the Sahara more than 650,000 km^2 of land have ceased to be productive. It is estimated than 14 per cent of the world's population (comprising 600–700 million people) live in areas of threatened drylands, and more than 60 million people are already affected by desertification. To a greater or lesser extent, one third of the land surface and one seventh of the population of the world are affected directly, but the rest of the global population must

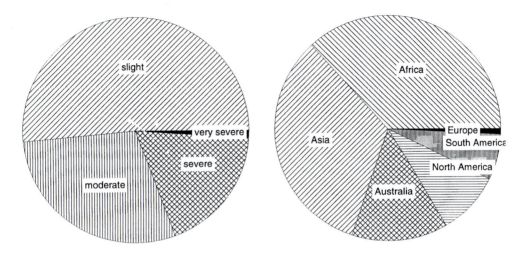

Figure 4.8 Land area in four desertification classes, by continent (data from El-Hinnawi & Hashmi 1987: 39).

Table 4.4 Forms and severities of desertification (Warren & Maizels 1977: 216)

Form	Severity: Slight	Moderate	Severe	Very severe
Water erosion	rills, shallow runnels	soil hummocks, silt accumulations	piping, coarse washout deposits, gullying	rapid reservoir siltation, landslides, extensive gullying
Wind erosion	rippled surfaces, fluting and small-scale erosion	wind mounts, wind sheeting	pavements	extensive active dunes
Water and wind erosion			scalding	extensive scalding
Irrigated land	crop yield reduced less than 10 per cent	minor white saline patches, crop yield reduced 10–50 per cent	extensive white saline patches, crop yield reduced more than 50 per cent	land unusable through excessive salinization, soils nearly impermeable, encrusted with salt
Plant cover	excellent – good range condition	fair range condition	poor range condition	virtually no vegetation

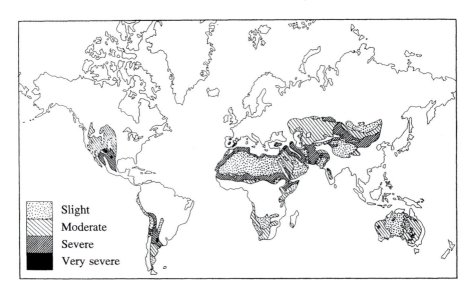

Figure 4.9 World distribution of desertification (Dregne 1983: 16).

bear the indirect effects of diminished food production and increased suffering which must be alleviated by international aid.

Hence desertification depletes world food resources by reducing the productive capacity of land. Increased transportation of dust in the atmosphere may adversely affect the global radiation balance and deplete food production even further. In the marginal and degraded areas there will be a tendency for genetic resources to be diminished, as the number of surviving plant and animal species will decline. The loss of nutrients may be sufficient to cause significant decreases in the productivity of the land. For human beings, the worst possible consequence is unrequited suffering and eventual starvation, or unplanned mass migrations with highly negative social and economic repercussions.

Causes of desertification
Climatologists have given much thought to the rôle of climatic change in increasing the aridity and decreasing the biological productivity of large land areas. Quaternary climatic oscillations have resulted in continuing adjustments in the drainage pattern of arid areas, such that river capture may deprive certain subcatchments of moisture. Like the climate that controls them, deserts are extraordinarily dynamic phenomena when viewed in the long term. In Africa over the past 5,000 years there has been a general increase in aridity, but with many oscillations on time-scales of different lengths.

238

Hence, the desert may expand or contract as precipitation varies over time. This process can be charted by various means. Dendrochronology of the annual growth rings in trees can identify seasons when moisture availability was restricted and growth lessened. Limnology can indicate the expansion and contraction of lakes over time, while geoarchaeology can indicate changes in early settlement patterns in relation to water availability (thus Roman irrigation systems in Tripolitania, North Africa, have silted up as the climate is now too dry for them to be used). Moreover, cycles of aridity have been related to the 11- or 22-year incidence of sunspots. Yet despite these efforts, no single natural agency seems to be responsible for climatic perturbations over 10,000–1,000 years and a natural cause cannot be proven for most desertification.

Instead, it appears certain that unwise use of the land by mankind is the main reason for its degradation (Cloudsley-Thompson 1978). Through evolution, xerophytic ("drought-loving") plants in drylands have evolved strategies for conserving moisture and overcoming the exceptional dry years, but the extra stress placed on them by humanity's demands upon the land may break this tenuous symbiotic relationship. The degradation of soils may take the form of a slow process of decline. For example, the land between the Jordan River and the Mediterranean Sea has been farmed for thousands of years, but there has been an almost imperceptible, yet inexorable, change to the current state in which soils are severely impoverished.

The loss of vegetation at the margins of deserts causes albedo to increase and produces biophysical feedback which diminishes rainfall, increases the dust content of the atmosphere and accelerates the rate of desert spread. According to the **Otterman hypothesis**, overgrazing in areas of marginal rainfall leads to removal of vegetation, which increases the albedo of the land surface and results in desertification. When soils are denuded, the resulting increase in albedo reduces surface temperatures (because there is increased reflection and diminished absorption of solar radiation into the ground). As surfaces radiate less heat to the lower atmosphere, which becomes cooler, the lapse rate of temperature with altitude (i.e. the decrease in temperature with height above the ground) is diminished and this reduces the convective activity which creates clouds and rainfall and is therefore indirectly responsible for plant growth. On the other hand, Jackson & Idso (1975) found that denuding soils increases, rather than decreases, their average temperature. Otterman countered that infra-red remote sensing, one of the principal means by which land surface temperature is monitored, tends to over-estimate heat emissions. Evaporation leads to the storage of latent heat and thus to reductions in temperature at very low altitudes. All things considered, it appears that much depends on the type of plants and their biogeographical community structure (e.g. whether they congregate in clumps) and the extent to which they are degraded.

The human impact

Both the causes and the consequences of desertification are linked to social and political factors. Wise land use may be at least partly a function of political stability and management decisions, while land degradation may result in unplanned mass migrations which have serious social and political consequences. Conditions may also be worsened by the loss of genetic resources, seedstock and livestock breeding inherent in the decline in productivity of the land. In respect of these considerations, desertification may be viewed as the interaction between socio-economic pressures and ecosystem fragility (Nnoli 1990).

As a general rule, dry-farmed land where there is a lack of erosion control tends to suffer water erosion in areas dominated by winter rains and wind erosion in zones that are subject to summer rains. In many desert margin areas the twentieth century has been a period of dramatic increases in population. In Algeria, for example, nomadic peoples have become dependent on borehole irrigation, which has reduced their mobility and made them dependent on locationally fixed resources. Excessive population pressure on these can lead to environmental degradation and loss of productivity (Dregne 1978).

In Egypt the construction of Lake Nasser and the Aswan High Dam was intended to control the annual flooding on the River Nile and provide a constant supply of irrigation water. Before this scheme came into effect, drifting sand that encroached on the fringes of the fertile corridor of the Nile Valley during the dry season was stabilized by the river-borne silts deposited during the annual flood. Now that the flood has been eliminated, sand encroachment is very much more difficult to control as it overwhelms the irrigated lands at the western edge of the valley.

Conversely, irrigation schemes in Western Nubaiya have caused waterlogging and salinization. The water table rose alarmingly over a period of years, bringing soluble salts with it, which precipitated at the surface under the influence of evaporation. Similarly, in Pakistan in the early 1960s more than 20 per cent of irrigated land underwent profound salinization, while almost 4 per cent of cultivated land in India has suffered from contamination with salts or alkalis. Desertification by salinization appears, in fact, to be a world-wide problem, as similar instances have been reported from the Commonwealth of Independent States, Iraq, Mexico, Pakistan, Syria and the southwest USA.

One other frequent cause of desertification is overgrazing, which results from too many animals feeding on a limited supply of forage. The animals may be accompanied by too many people and land degradation is often consequent upon total lack of range management. The problem is not merely one that occurs in Third World countries: in the western USA, overgrazing has led 70 per cent of rangelands to produce no more than 50 per cent of their original forage potential. Drought may lead to the decimation of herds and outmigration of populations, which may lead to a reversion to better plant communities, but only if range management is adequate during the period of recovery. In this context, it is necessary to define the **carrying capacity** of the range, which is the number of foraging animals it can support (or the amount of fodder it will produce) without degradation of the vegetation.

Recent opinion suggests that the significance of overgrazing has been exaggerated (Mace 1991). Where rangeland is degraded to the point that woody species replace grasses, the result is not necessarily a herder's nightmare. In fact, the people of Sub-Saharan Africa are used to having flocks and herds of different kinds of animals and, as the population of cattle declines, so that of the browsing animals, camels and goats, may increase.

Identifying, monitoring and tackling desertification
As the phenomenon of desertification is one of gradual deterioration, it is vital that it be monitored effectively in order to detect serious changes as they occur. Sequential satellite images can help to do this, in that recorded increases in albedo or changes in the "green index" of the land surface may be related to loss of vegetation (see Ch. 6). Agricultural production figures are also useful for determining the location and severity of desertification. However, censuses of livestock can be misleading, as the market value of beasts (rather than the fertility of the land) may set the stocking level, productivity per beast may change with new techniques of husbandry, the range may not necessarily be stocked at its carrying capacity or livestock may be difficult to count over vast areas of prairie or bushland. Thus the threshold of decline in carrying capacity at which desertification occurs may in the end be set quite arbitrarily.

It is estimated that desertification costs the world economy at least $26,000 million each year, most of which is the result of rangeland deterioration, and that it could cost $90,000 million to halt current trends towards ever greater losses of land productivity. The costs of reversing moderate desertification were estimated in the 1970s as $20 per hectare for pasture, $100/ha for dryland farms and $675/ha for irrigated land (Dregne 1978). Thus, it is clearly much less laborious and expensive to prevent desertification than it is to rectify it.

Each incidence of desertification needs an individual strategy to ameliorate it (Ahmad & Kassas 1987). Degrading oases may benefit from the establishment of a "green belt" around them, while careful planning of water movement and percolation on irrigated land are needed in order to prevent the accumulation of salts. In very degraded drylands, retrenchment may be the only solution and recovery may be out of the question in the short term. In less damaged ecosystems, rotational cropping, the culling of livestock, redistribution of ownership rights, fallowing, careful ploughing or the establishment of irrigation may be appropriate strategies. The critical factor will be the length of time over which agricultural investment must compensate itself and loans be repaid. If land is so damaged that it cannot be put to productive use, then proportionately more investment will be required, or at least some other source of support will be needed in the meantime. Unfortunately, unless there is considerable outmigration, the pressure of population on the land may remain the same during this period.

In synthesis, the difference between drought and desertification is clearly a matter of the permanence of ecological degradation, but it is not easy to establish in the short term (in fact at least a decade of monitoring is required). Recovery from drought may be slow and the damage semi-permanent. Alternatively, a spell of drought may irreversibly damage an ecosystem that was already undergoing serious degradation. In any case, change is often so slow and subtle that the seriousness of the threat is not recognized until the damage has been done. Measures used to combat land degradation need to be integrated with broad programmes of social and economic development, which build conservation of the environment and resilience into society. Unfortunately, the countries which through poverty and underdevelopment are least able to combat desertification are often those which are most vulnerable to it.

Landslides

The landslide hazard consists of loss of life, injury, property damage, disruption of communications, supplies and socio-economic activities, and loss of productive soil and land. Greater scientific and public awareness, coupled with increasing ability to predict landslides, is reducing the hazard worldwide, but at the same time increased construction in the vicinity of unstable slopes is increasing it. In many parts of the world population pressures are high at the base of slopes (for example, in coastal cities flanked by mountains), in canyons (among mountain ranges) and at the unstable edges of plateaux. There is in general a tendency towards greater concentration of physical capital in areas of high landslide activity as more roads, homes and other structures are built there. Landslides are ubiquitous phenomena of which a high proportion are induced by human activity. However, only the largest slides involve government intervention to ameliorate the damage.

Basic definitions
There are two types of hillslope (Carson & Kirkby 1972). On **weathering-limited** slopes debris is removed by a process of *in situ* rock disintegration as quickly as it can be generated. In contrast, the development of

242

transport-limited slopes depends more on the ability of erosional processes, including landsliding, to remove weathered debris. These slopes often have a thick cover of soil or disintegrated rock material, called **regolith**. By virtue of their regolith cover, transport-limited slopes tend to be more susceptible to landslides than weathering-limited ones.

Although legitimized by common usage, the word "landslide" is not the most precise term, as the phenomenon covers falling, toppling, sliding, flowing and subsidence of soil, rock or sediments under the influence of gravity and other factors (see below). The correct term is **mass movement**, all the more so as very often a combination of types of motion is involved, such as when internal deformation occurs in a sliding mass so that it undergoes slip-flowage. The lowering or wasting of the land surface by mass movements and other forms of erosion is termed **mass wasting**. It results in reduction of the vertical relief of the land surface, although this can be counteracted by tectonic and other forces promoting uplift of the land or fall in sea level.

Mass movement will occur wherever a slope is steepened beyond its **threshold angle of stability**, which is the steepest angle at which it can maintain itself. At higher angles the profile of the slope will alter itself to restore stability by undergoing slope failure.

The causes of mass movements

The forces that promote mass movement can be divided into external and internal categories (Table 4.5). The **exogenic** causes of slope instability include steepening or heightening the profile, removing the lateral or underlying support (especially through the effect of stream or road cuts), and loading the upper edge as a result of construction, landfill dumping, or landsliding. They also include changes in either relative relief (local differences in altitude) or slope gradient as a result of faulting, tectonic uplift or the creation of artificial slopes by grading with construction machinery.

These factors are complemented by the **endogenic**, or internal, causes of landslides. For example, weathering involves disintegration that weakens soil and decreases its resistance to shearing. Deforestation or other kinds of devegetation can also weaken a slope (see below), as the roots of plants tend to hold soils together, accounting for up to 90 per cent of stability on certain slopes. The stabilizing effect persists even for a certain period after the plants have died, until the roots decay. But most important among endogenic causes is increased infiltration of water, which can lead to soil saturation. It may result from ploughing or from poorly organized drainage on a slope that has been modified by deforestation or urbanization. Saturation increases **pore water pressure** which exerts a positive force that may cause the slope to fail.

In addition to the distinction between exogenic and endogenic

Table 4.5 Causes and parameters of landslides and slope instability (after Cooke & Doornkamp 1974, and other sources).

Factors that contribute to increased shear stress:

- Removal of lateral support by erosion of the lower part of the slope, by artificial cuts, etc.
- Surcharge: loading of the slope crest with an external load
- Internal increase in the weight of the slope material (as by water saturation)
- Ground vibrations and the earthquake mechanism of landslide generation
- Undermining of the slope
- Lateral pressure in cracks (for example, by water freezing)
- Tectonics of regional tilting

Factors that contribute to reduced shear strength

- Properties of clay, such as shrinking and swelling
- Gross structure of rock (faults, joints, bedding, etc.)
- Pore pressure effects
- Freeze-thaw effects
- Drying and desiccation
- Loss of capillary tension
- Breakdown of soil structure (weathering reduces effective cohesion c')
- Deterioration of intergranular cement (reduction of cohesive strength)

Parameters of landsliding:

- Total vertical fall
- Slide mass
- Velocity of movement
- Horizontal distance of movement
- Average slope over which the slide moves
- Cohesion of sliding mass
- Topography and geometry of bed surface
- Pore pressures within the sliding mass
- Shear strength and shear stresses
- Type of failure plane under the sliding mass, and its shape
- Mechanism of sliding (q.v.)

processes, it is useful to distinguish between immediate and long-term causes of slope failure. Immediate causes include vibrations, ground shaking (for example, during earthquakes), heavy rainfall, and freezing and thawing. Long-term causes involve the slow, progressive weakening or steepening of the slope: for example, it may, be simultaneously deprived of some of its strength by weathering and undercut by streams.

The rôle of human activities

There are several ways in which urban development can cause or augment instability in hillslopes. Sliding along pre-existing slippage planes may be aided by loading or the addition of moisture through poor drainage of the site. Alternatively, movement may occur along weak layers in sediments or discontinuities in rocks whose presence has not been noted by the developer or counteracted in the building process. Slumping may be caused by the settlement of fill that has not been compacted or engineered properly, often because it is too thick. And finally, excessive soil moisture may accumulate as a result of failure to remove debris, or the slope may become disaggregated through failure to remove either vegetation or compressible soils.

In forests timber harvesting can have a substantial impact on slope stability, but one that depends on the nature of the bedrock and soils, the size of the harvested area, the methods used to remove tree trunks and the level of disturbance caused when access roads are built (Furbish & Rice 1983). Generally, tractors do more damage than hauling logs by cable, as runoff is concentrated along the wheelings. Slope failure occurs 5–8 years after clear-cutting, when tree roots cease to bind the soil together, and at this point the high input of water and sediment into basal channels can cause slopes to be undercut and fail as streams shift their course. On the other hand, trees can help destabilize soils by transmitting wind forces to them or by the wedging action of roots as they grow.

Fundamental forces in slopes
The study of the behaviour of slope materials is part of the science of rheology and involves the concepts of mechanical force and resistance. The process of movement along planes is known as **shear**, and it usually involves slippage against which friction is mobilized in order to retard it. Applied forces are known as **stresses**, and slopes usually fail as a result of **shear stresses** acting along straight or curved **shear planes** (Fig. 4.10). The deformation caused by movement is called **strain**, and, logically, if it results from shear stresses it is known as **shear strain**. In general terms, the amount of resistance to movement that the slope can mobilize is a measure of its **strength**, and the component of this which is directed against shear stresses is termed the **shear strength**.

Shear strength is a function of three main factors. The **angle of internal friction**, ϕ, is a gross measure of how a granular mass deforms under shear stresses and is measured with respect to the principal direction of compression. For a set of forces and resistances with defined directions and magnitudes, particles will slide over one another at a given angle to the main direction of force. Secondly, **effective normal stress** involves a component of the weight of the slope material that acts to mobilize friction and hold the slope together. Thirdly, the packing, cementation or electrochemical bonds that hold a granular material together represent its **cohesion**. The equivalent for solid units of rock is their **coherence**, which is a function of their mass and homogeneity, and is likely to be high if there are large thicknesses of uninterrupted solid rock.

Solid materials are brittle, that is, they resist the forces that try to deform them (usually by a combination of rigidity and elasticity, or fully recoverable deformation) until they snap by a process of crack propagation. Plastic materials, on the other hand, are ductile. This means that when sufficient force is applied to them they will deform to a certain

Figure 4.10 Sliding in variegated clays at Scalo Brindisi di Montagna in southern Italy. The striped pole is resting on an exposed part of the underlying shear plane.

extent before cracking and will maintain the shape imparted to them by the principal force as far as other stresses (such as gravity) permit. Unless they are confined, fluids such as water, which exhibits a linear relationship between rates of stress and strain, deform immediately force is applied to them, such that shearing occurs everywhere within the material.

Many substances, and especially sediments that undergo landsliding, show a rheology that varies with their water content. For clays and similar materials, this can be characterized by **Atterberg's limits**. Here, the transition from solid to plastic is known as the plastic limit, M_{PL}, and that from plastic to liquid as the liquid limit, M_{LL}. Both are expressed as percentage content of water in the test sample. The interval between the two is known as the plasticity index

$$PI = M_{LL} - M_{PL}$$

while the natural groundwater content of a drained sample, M, can be related to M_{PL} and PI by the liquidity index

$$LI = (M - M_{PL})/PI$$

which is the ratio of the actual moisture content to the liquid limit. Essentially mudflows have an LI > 1, while LI < 1 for slumps.

The factor of safety and its calculation

Geotechnical engineers define the **Factor of Safety**, **F**, as the *ratio of forces which resist slope failure to those which drive it*. Hence, the value F = 1.0 represents equilibrium, or the critical threshold at which elements contributing to stability equal those leading to instability. It can never be maintained more than momentarily. Values of F < 1.0 lead to failure of the slope; F ≈ 1.35 is fairly stable and F ≥ 1.5 is definitely stable. As a rule of thumb values in the range 1.0 ≤ F ≤ 1.25 represent a conditionally stable slope, whose strength must be increased or shear stress decreased in order to make it permanently stable.

The driving forces of slope stability include gravity (the result of a steep slope or heavy sliding mass), shear stresses and high pore water pressure. The last of these factors is the result of the incompressibility of water. When the pores, or **voids**, between solid particles become saturated, the water takes up some of the stress imposed by the weight of the soil or sediment, which is transferred from grain-to-grain contacts between solid particles to the fluid itself. This reduces the stabilizing effect of weight, and in the end the soil mass can only adjust to such forces by moving downslope to a more stable position. The resisting forces include granular interlocking, cohesion of a soil mass, cementation of sediment grains and other characteristics which impede mobilization of a rock, soil or sediment mass. The engineer Karl Terzaghi suggested that increases in the driving forces are cyclical (as a result of rainfall events which raise pore-water pressure, or of earthquakes), while loss of soil or rock strength is progressive and takes place over a longer timespan. Eventually both factors will combine to reduce the slope material below the F = 1.0 threshold, at which point failure will occur and raise the F value anew (Terzaghi 1950).

In practice, the factor of safety is determined for a slope under differing conditions of weight distribution, angle, shearing resistance and pore water pressure (Fig. 4.11). The length, location and form of the shear plane or planes are among the first parameters to be postulated in slope stability analysis, and they will govern the nature of the models chosen to represent slope processes – for example, whether circular or non-circular shear planes are assumed. The models are approximations to actual slope conditions: generally, short, steep slopes are most likely to undergo rotational failure upon a roughly circular plane of slip, and long slopes will undergo a more translational (elongated) and hence non-circular movement (Chowdhury 1978).

The **method of moments** for strictly rotational failure of a single mass (Fig. 4.11a) postulates a resisting moment

$$\tau_c L R$$

where τ_c = shear strength per unit of sliding surface
             L = length of circular arc
             R = radius of arc
and an acting moment
        Wx

(a) Method of moments

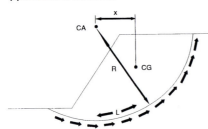

(b) Fellenius' method of slices

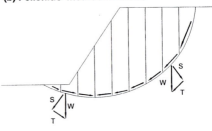

(d) Non-circular slip plane analysis

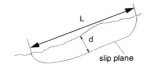

(c) Terzaghi and Peck method

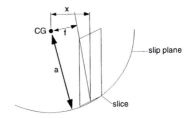

(e) Terzaghi's diagram of cyclical progression towards failure

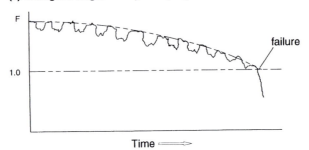

Figure 4.11 Analysis of rotational slope failure.

where
W = total weight of potential sliding mass
x = horizontal distance from centre of arc (CA) to centre of gravity (CG) of the sliding mass

Thus,

$$F = \frac{\text{resisting moment}}{\text{acting moment}} = \frac{\tau_c L R}{W x}$$

The evident crudeness and inaccuracy of this approach led W. Fellenius (1936) to propose the **method of slices** (Fig. 4.11b&c), in which

$$F = \frac{\text{resisting forces}}{\text{driving forces}} = \frac{\Sigma S \ \tan\phi \ + \ \Sigma C}{\Sigma T}$$

where
- S = normal component of slice weight, W
- ϕ = friction angle for each slice (triaxial or shearbox test)
- C = cohesion along arc
- T = tangential component of slice weight (vector resolution of S and W).

The choice of slice width is arbitrary and may be governed by the position of boreholes or other sample points, or by peculiarities of the slope in question.

This approach, as stated here, ignores the refinements of effective stress and the calculation of inter-slice force. With regard to effective stress, the **Mohr–Coulomb failure criterion** states that

$$\tau_f = \text{shear strength of the mass at failure}$$

and
$$\sigma_n = \text{normal stress (perpendicular to the slope).}$$

As pore water pressure, u, reduces normal stress, the **effective normal stress** is:

$$\sigma_n' = \sigma_n - u$$

and the Mohr–Coulomb criterion is

$$\tau_f' = C + (\sigma_n - u) \ \tan\phi$$

So far we have dealt only with **total stress analysis**. We may use the above observations to broaden the approach to **effective stress analysis**.

Using the Mohr–Coulomb failure criterion, but ignoring inter- and intra-slice forces, we can state the simplified method of Alan W. Bishop (1955) as follows:

$$F = \frac{\Sigma\{[C'b + (W-ub) \ \tan\phi'] \ (1/m_\alpha)\}}{(\Sigma W \ \sin\alpha)}$$

where
- $m_\alpha = \cos\alpha \ [1 + (\tan\alpha \ \tan\phi'/F)]$
- C' = cohesion intercept in terms of effective stress (i.e. taking pore water pressure, u, into account)
- ϕ' = angle of internal friction in terms of effective stress
- b = width of slice
- α = basal angle of slice (along slip plane).

F is obviously present on both sides of the equation. Hence, it is solved iteratively by substituting an estimated value of F in the calculation of mα, calculating F in the main equation, substituting the result in the m_α equation, recalculating F, and so on until successive F values converge to a satisfactorily low level of change between iterations. A method invented by Karl Terzaghi and Ralph B. Peck (Terzaghi & Peck 1967) calculates the resolution of forces for each slice in a manner similar to that of the simplified Bishop method, but allowing for inter-slice forces as well as effective stresses.

Such methods are concerned exclusively with circular slip planes. Janbu (1977), however, invented a "method of slices" for non-circular planes that is probably much more realistic when long slopes are involved. One of his methods is as follows:

$$F = f_o \sum \{[C'B + (W - ub)\tan\phi']/\cos\alpha\ m_\alpha\}/(\sum W \tan\alpha)$$

where f_o = correction factor based on the orthogonal depth:length ratio for the landslide, d/L (Fig. 4.11d)

C' = cohesion intercept in terms of effective stress (bars)

b = width of slice (metres)

W = weight of each slice $(W = \gamma bd_s)$(kiloNewtons)

γ = unit weight of soil [kg m^{-2} to kN m^3] ($\times$ 0.009807)

u = pore water pressure (kg cm^{-2})

ϕ' = angle of internal friction in terms of effective stress

α = inclination of base of sliding to horizontal (degrees)

m_α = cosα [1 + (tanα tanφ'/F)]

d_s = depth of slice (metres)

Once again, the equation is solved iteratively, using an initially postulated value of F.

But to what extent are these methods capable of predicting slope failure in real situations? All models are necessarily simplifications of reality and are therefore approximations. The researcher must decide how much approximation is legitimate. First, more accurate models require fewer assumptions, but more data and longer calculations. There is an enhanced risk of equifinality (similar outcomes produced by different processes) or other forms of indeterminacy in the model results. Secondly, simpler models lack sensitivity and may wildly misestimate F values in cases where the sliding mass (and slip plane) are very inhomogeneous or where stability is delicately balanced around F values of unity (Nash 1987).

Thirdly, circular shear planes are assumed in most rotational models. In an isotropic situation, momentum would tend to keep the moving mass on a circular path. The lower end would push upwards, such that the final slices act as a resisting force against the tangential component of weight of all the other slices. This is satisfactory for many short artificial slopes at angles in excess of, say, 20–25°, but it is seldom as reasonable on longer, shallower natural slopes. At the very least, a non-circular slip plane may be needed, and the assumption that slices are not deformable may be invalidated if shearing coincides exactly with mudflow or if there is a shear *zone* (zone of deformation) instead of a shear *plane*. Limit equilibrium analysis by slices cannot cope with many such complications: a finite element solution may be needed, involving differential calculus.

Finally, these methods refer only to the moment at which sliding begins. As mentioned above, wetting and drying cycles and, in some areas, periodic seismic stresses are among the factors that lead to a progressive reduction in the factor of safety as time passes (Fig. 4.11e).

Types of mass movement
Whatever the level of simplification chosen for engineering analysis, landslides tend to be complex and varied phenomena. They differ according to the degree of sliding, flowing, creeping, falling or toppling involved, according to their wetness or dryness, and according to their speed of movement (Table 4.6). As actual landslides often show a combination of these diagnostic phenomena (especially if the lithology of the slope is complex), it is very difficult to combine all descriptions into a standard taxonomy. Classifications have been published by, among others, Blong (1973), Nemcok et al. (1972), Nilsen et al. (1979), Skempton & Hutchinson (1964) and Varnes (1978). Partial classifications also appear in the literature, such as that of rock topples by Goodman and Bray (see Evans 1981) and that of mudflows by Kurdin (1973). As it is based on the logical ordering of field examples of general relevance, the scheme proposed by Varnes (Fig. 4.12) appears to have gained the greatest acceptance, but in no sense has it resolved all problems of classification.

Essentially the following ten groups of phenomena fall under the heading of mass movements:

(a) As discussed, the classic form of landslide is the rotational slide,

Table 4.6 Rate of movement scale for landslides (Varnes 1978)

		extremely rapid
3.0	metres per second	-----------------------------------
		very rapid
0.3	metres per minute	-----------------------------------
		rapid
1.5	metres per day	-----------------------------------
		moderate
1.5	metres per month	-----------------------------------
		slow
1.5	metres per year	-----------------------------------
		very slow
0.06	metres per year	-----------------------------------
		extremely slow

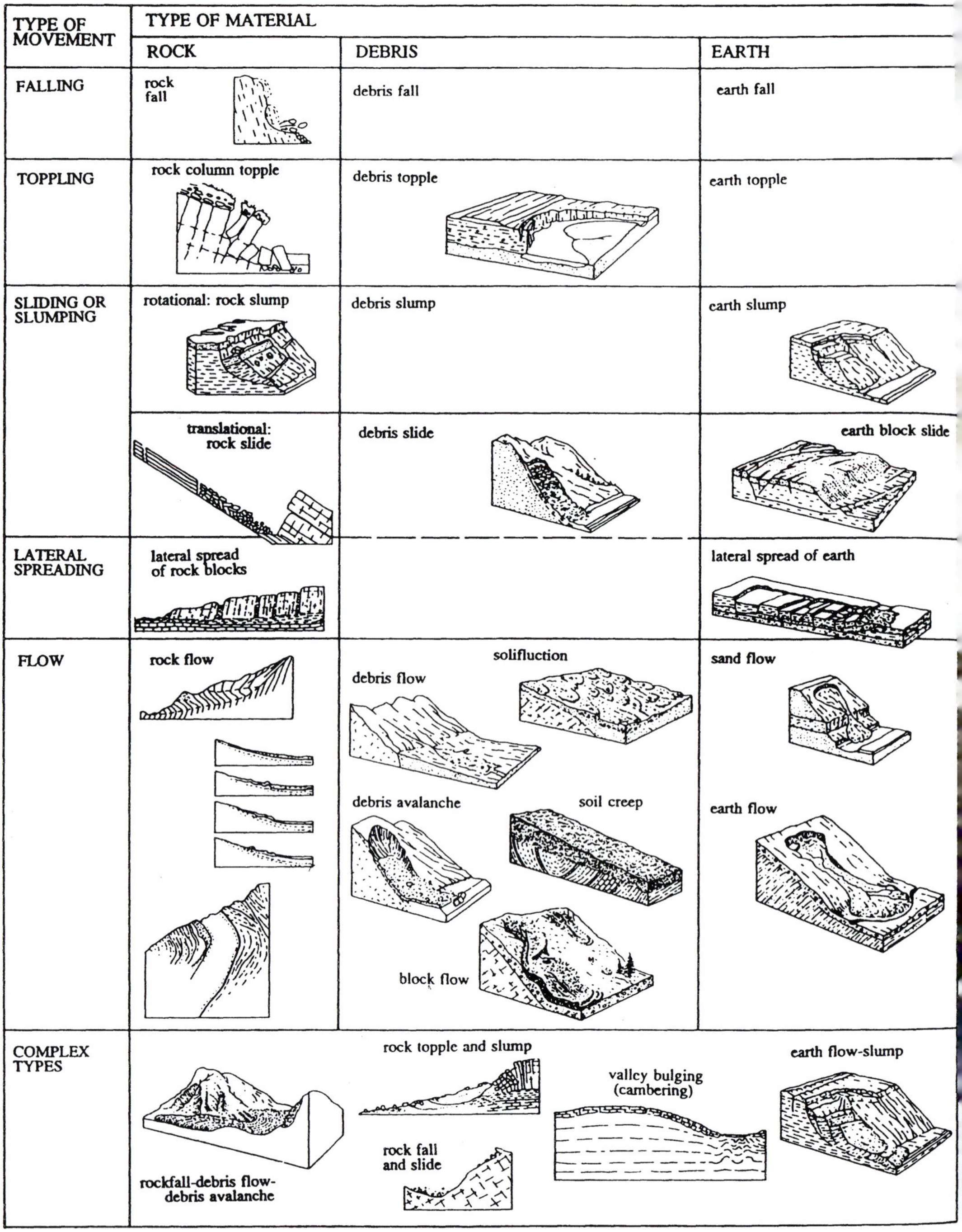

TYPE OF MOVEMENT	TYPE OF MATERIAL		
	ROCK	DEBRIS	EARTH
FALLING	rock fall	debris fall	earth fall
TOPPLING	rock column topple	debris topple	earth topple
SLIDING OR SLUMPING	rotational: rock slump	debris slump	earth slump
	translational: rock slide	debris slide	earth block slide
LATERAL SPREADING	lateral spread of rock blocks		lateral spread of earth
FLOW	rock flow	debris flow	solifluction	sand flow
		debris avalanche	soil creep	earth flow
		block flow		
COMPLEX TYPES	rockfall-debris flow-debris avalanche	rock topple and slump	rock fall and slide	valley bulging (cambering)	earth flow-slump

Figure 4.12 Varnes's landslide classification (Varnes 1978).

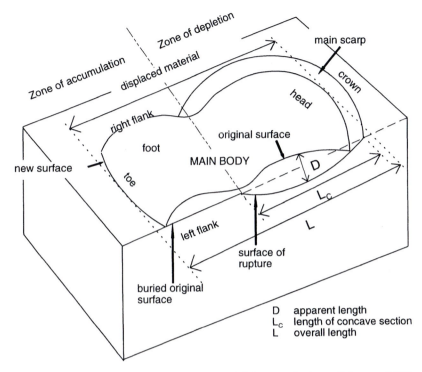

Figure 4.13 Terms used to describe landslide morphology (Varnes 1978).

which has a curved surface of rupture and produces **slumps** by backward slippage. The terminology associated with such slides (headscarp, trunk, toe, etc.) is given in Figure 4.13. Some rotational slides are **multiple regressive** phenomena, in that continued instability causes new headscarps to develop progressively further up the slope.

(b) Relatively flat, planar movements along surfaces are called **transla-tional slides**. These are common where the bedding planes of sedimentary or metamorphic rocks dip in the direction of slope and blocks detach and slide along them. A special case in steep rock slopes is the **wedge failure**, in which the sliding planes are constituted by orthogonal joints that intersect the slope at an angle.

(c) More complex landslides involve a combination of slip along a circular arc and a flat plane, and these are termed **roto-translational**.

(d) In some instances of **soil slab failure**, resistance to downslope movement changes from sliding friction to viscous flow as a slab of saturated regolith changes to a thick liquid and accelerates to avalanche speed. Thus it may start moving at, say, 30 cm/sec and accelerate to 10 m/sec.

(e) **Debris slides** or **avalanches** occur in surficial deposits of granular

253

material (Fig. 4.14). The rupture surface is roughly parallel to the inclination of the bedrock surface. Slumping occurs at the head of the deposit and flow-sliding at its foot, which creates a toe of debris that may become fan-shaped if it spreads out as it comes to rest.

(f) **Debris flows** occur after sliding is initiated by the saturation of debris with water. For example, the eruption of Mount St Helens in Washington State in May 1980 created a debris flow 2.83 km^2 in area on the north fork of the Toutle River. Debris flows are characterized by their internal deformation, which amounts to uniform viscosity. But if part of the sliding mass behaves as a rigid solid (and perhaps flow only occurs at its base) the phenomenon is one of **plug flow**, which is a combination of sliding and flowing.

Figure 4.14 This debris slide occurred in 1986 on a 45° slope at Cuyocuyo in the Cordillera Oriental of southern Perú. Several houses were destroyed and their occupants narrowly escaped with their lives.

(g) **Falls** occur through the air; for example, from jointed fractured rock exposed in vertical cliffs.

(h) **Topples** involve the outward rotation (or inward buckling and basal collapse) of angular blocks or rock columns that become detached from cliffs. These are likely to be defined by the intersection of joints or other fractures, and to have their basal stability disturbed by erosion.

(i) **Mudflows** have been variously defined, but must contain between 20 per cent and 80 per cent of fine sediments (which are usually an admixture of sand, silt and clay) and be saturated with water. Viscous movement produces friction and turbulence which generate dispersive stresses capable of lifting and transporting very large boulders. Hence these are often highly destructive phenomena. In 1941, at Wrightwood in California, mudflows containing blocky debris and 25–30 per cent water moved at 11–18 km/hr over slopes of less than 10° (Sharp & Noble 1953).

(j) One of the least destructive mass movement phenomena is **soil creep**, which tends to be slow and superficial. However, many other forms of landslide can undergo creep and may gradually do serious damage to buildings and structures.

At the other extreme, catastrophic mass movements are extremely difficult to explain. If more than 100,000 m³ of debris is involved, such movements can occur with extreme rapidity and very low coefficients of friction. According to Shreve (1968) "Blackhawk-type" landslides (or **sturzstroms** in the terminology of Hsü 1975; see Table 4.7) acquire so much momentum in the initial fall of debris that they leave the ground, overriding and trapping a cushion of compressed air upon which they ride at speeds of up to 400 km/hr with very little friction. Leakage of air into the debris is less than 10 per cent and more than one minute is eventually needed to disperse the air cushion. This model is supported by both the frequency with which such movements occur after rock falls into glacial cirques (which have rims that favour the air launch) and the morphology of the deposited debris tongue or fan.

Kent (1966) put forward an alternative hypothesis based on fluidization within the rock mass. Stress transmitted among agitated rock particles helps to compress air in the pores between them and transform the debris into a fast-moving stream. Heim (1932) argued for elasto-mechanical collisions rather than fluidization, emphasizing the importance of stresses exchanged between solids rather than fluid effects. This appeared to be corroborated when Howard (1973) identified such movements on the Moon and Lucchitta (1978) found them on Mars, which rather put paid to the fluidization hypothesis. Other models (Erismann 1979, Hewitt 1988) tend to emphasize the tremendous amount of energy created by these landslides and its capacity to melt and fuse rock and thus reduce basal friction. Nevertheless, as fewer than a hundred such events have been described in the literature, and very few observed first-hand, the problem of correctly identifying the mechanism remains an open one.

Landslide amelioration and hazard reduction
Various engineering measures can be utilized to prevent landslides (see Table 4.8), and generally speaking the importance of pore-water pressure in generating landslides makes slope drainage one of the most significant. It can be achieved by constructing underground wells or tunnels and surface channels or ditches, by pumping out groundwater, or even in some cases by electro-osmosis. But given the ubiquitous nature of slope failure and the high cost of stabilizing the ground, it is often more

Table 4.7 Data on three major debris avalanches in the USA.

Blackhawk landslide:

Location:	Mojave Desert, California
Age or date:	prehistoric
Size:	282,000,000 m^3
Lithology:	crystalline limestone
Structure of source:	undercut thrust block
Deposit area:	14 km^2
Coefficient of friction:	0.13
Height of slope climbed:	60 m+
Maximum speed:	120–235 km/hr
Observations:	material dropped 13,000 m and ran out 9,700 m

Gros Ventre landslide:

Location:	Wyoming
Age or date:	23 June 1925
Size:	40,000,000 m^3
Lithology:	weathered clay, sand and sandstone
Structure of source:	undercut dip slope
Deposit area:	2 km^2
Coefficient of friction:	0.19
Height of slope climbed:	110 m
Maximum speed:	165 km/hr
Observations:	largest landslide in US history; 640 m drop, 3,000 m runout

Madison Canyon landslide:

Location:	Montana
Age or date:	17 August 1959
Size:	28,000,000 m^3
Lithology:	Dolomite and weathered schist
Structure of source:	buttressed dip slope
Deposit area:	1 km^2
Coefficient of friction:	0.27
Height of slope climbed:	130 m
Maximum speed:	180 km/hr
Observations:	set in motion by a magnitude 7.1 earthquake; 400 m vertical drop, 1,500 m runout

Table 4.8 Some methods of preventing and controlling slope failure (after Cooke 1984, Kockelman 1986).

Major solution	Specific methods
Avoidance	*Control of location, timing and nature of development:* • Remove, bridge or bypass unstable land • Land-use restrictions and subdivision controls • Hazard mapping and land-use zoning • Engineering geology or soil surveys before, during and after development • Redevelopment or moratoria on unsafe land uses • Use of sanitary codes to restrict development • Seasonal limitations to slope development • Grading and hillside development regulations • Acquisition, restructuring or removal of property • Warning and public education measures • Disclosure of hazard to property buyers • Establishing legal liabilities of property owners • Insurance against hazard • Financial assistance to promote hazard reduction (loans, tax credits, lower tax assessments)
Reduced shear stress	*Control of cut and fill:* • Limit or reduce slope angles, cut and fill • Limit or reduce slope unit lengths • Remove unstable material • Provisions against expansive soil hazards
Reduced shear stress and increased shear resistance	*Improved drainage:* • Surface drainage: terrace drains, other drains and methods of drainage • Subsurface: drains, drain wells • Control irrigation
Increased shear resistance	*Retaining structures:* • Buttress or counterweight at slope foot • Cribs or retaining walls • Piling, tie-rods, anchors or other foundation engineering methods
Mainly increased shear resistance	*Protect surface:* • Control vegetation cover • Harden surface (e.g. concrete cover) • Chemical treatment or electro-osmosis *Compaction:* • Control fill compaction

advantageous to concentrate on preventing the consequences of mass movements. The "non-structural" measures are as follows.

Prior knowledge of the landslide hazard may allow evacuation (especially as many mass movements are not exceptionally rapid), but vigilance, forecasting and monitoring are all necessary preliminaries. It may be possible to correlate the duration and intensity of storms with critical slope conditions, such as the saturation of regolith with rainwater, or the growth of a perched water table above a bedrock surface. Given the lag time between peak precipitation and the attainment of maximum pore pressures, this may permit evasive action such as evacuation of vulnerable property.

Landslides and the hazards which they create can be mapped, especially as many are recurrent phenomena. In fact, evidence of past instability is often the best guide to future landsliding. Not only can actual or potential mass movements be mapped, but so can vulnerable rock formations, soils and slope angles (Varnes 1984a). Hence, the cartographic representation of landslide risk involves a series of components or overlays. First, a structural and surface geology map should be drawn to show the different geological units likely to suffer from slope failure. Next, slope inclination and aspect should be mapped. In the Northern Hemisphere, for example, southwesterly-facing slopes may receive more solar insolation and undergo enhanced wetting and drying cycles, which may contribute to the production of an unstable mantle of weathered debris. North-facing slopes may accumulate more snow in winter and become saturated with water during the spring thaw.

Then, actual and potential landslides should be mapped, using different symbols to represent different instability phenomena – e.g. mudflows, shallow soil slips, deep-seated rotational slides, and so on. Landslides may be classified on the basis of type of material, type, speed, cause and location of movement, volume or area occupied, geometry and morphology, character of sliding surface and water content. Finally, on the basis of the strength of geological deposits, slope angle and orientation and the presence of landslide scars, a map of slope stability categories can be assembled. It will consist of "severity zones", such as nil, low, moderate and high. By drawing up such maps with respect to various mass movement phenomena, it is possible to depict spatial variations in the likelihood of different types of landslide.

Crozier (1986) devised a probability classification for mass movements, based on the visible evidence of potential sliding and whether the slope in question had, in the recent past, been active or stable (Table 4.9). Stevenson (1977) defined landslide risk, R, in mappable terms as

$$R = (P + 2W)\ (S + 2C)\ (U)$$

where P = a clay factor based on the plasticity index
              W = a water factor based on the height of the piezometric surface
              S = slope angle
              C = slope complexity
              U = land use (woodland, built-up, etc.).

The term $(P + 2W)$ is a measure of hazard, while (U) is the indicator of vulnerability (compare with equations in Ch. 1).

Briggs (1974) mapped landslide susceptibility according to an overdip

Table 4.9 Landslide probability classification (after Crozier 1986: 212).

Class	Description
I	Slopes which show no evidence of previous landslide activity and which by stress analysis, analogy with other slopes or by analysis of stability factors are considered highly unlikely to develop landslides in the foreseeable future.
II	Slopes which show no evidence of previous landslide activity but which are considered likely to develop landslides in the future. Landslide potential indicated by stress analysis, analogy with other slopes or by analysis of stability factors; several subclasses may be defined.
III	Slopes with evidence of previous landslide activity but which have not undergone movement in the previous 100 years.
IV	Slopes infrequently subject to new or renewed landslide activity. Triggering of landslides results from events with recurrence intervals greater than five years.
V	Slopes frequently subject to new or renewed landslide activity. Triggering of landslides results from events with recurrence intervals of up to five years.
VI	Slopes with active landslides. Material is continually moving, and landslide forms are fresh and well defined. Movement may be continuous or seasonal.

criterion in which the main hazard areas occur where slopes exceed 30 m in height and 8.5° (15 per cent) in angle and the rock layers dip at more than 7.6 m/km with an azimuth of no more than 45° from the direction in which slopes dip. Thus slides are most likely where steep slopes are paralleled by steeply dipping strata.

Once landslide areas and unstable slopes have been identified in an area, and stability ratings have been assigned for planning purposes, the geotechnical and geomorphological information can be integrated with planning decisions in order to choose which zones to develop and in which to prohibit development or to concentrate slope stabilization works. The bottoms and mouths of steep ravines, and the bases of steep slopes, should not be developed if possible. Disturbed ground and hummocky slopes often give a visible sign of instability which must be investigated thoroughly.

It has been estimated that in the most promising cases regional, community and site investigations could obtain enough technical information to reduce landslide costs by 95–99 per cent. Hence, in San Mateo County, California, landslide zoning regulations were adopted in 1975 on the basis of a hazard map developed by the US Geological Survey (Baldwin et al. 1987). On 1,055 properties developed in the subsequent ten years (370 of which were on hillsides) no landslide damage has occurred.

A real-time landslide warning system

The real-time landslide warning system developed for the San Francisco Bay region of California (Keefer et al. 1987) is based on the assumption that debris flows are the most hazardous form of mass movement experienced in the area. In fact, 25 people died in such events during the storms of January 1982 (see below), and one died in the February 1986 storms. The hazards appear to be greatest where houses containing sleeping occupants are destroyed at night by fast moving streams of debris (Cotton & Cochrane 1982).

Worldwide, the lowest threshold of rainfall intensity, I_r in mm/hr, and duration, D in hours, that will initiate widespread debris flows is

$$I_r = 14.82 \ D^{-0.39}$$

For debris flows to be common, antecedent moisture must be high and a significant proportion (perhaps 30 per cent) of average annual rainfall must fall during the storm that triggers them (Cannon & Ellen 1985).

The contributing causes of shear strength in a slope, as given above, can be formulated into an equation

$$s = c' + (p - u_w) \ \tan \phi'$$

in which c' is the effective cohesion of the material, ϕ' is the effective friction angle, p is the total stress normal to a potential slip surface and u_w is pore-water pressure. The forces that resist movement are thus c' and ϕ', the strength variables, while those that promote slippage are the height, angle and length of the slope and the distribution of pore pressures. The equation shows that pore pressures augmented by saturation of the voids between sediment grains reduce the effective overburden stress which mobilizes friction in the slope $(p - u_w)$. Antecedent moisture helps neutralize the suction present in dry soils and raise soil moisture to field capacity, the maximum water content which can be sustained in conditions of free drainage (dependent on the texture of the soil).

Under the simplest conditions, the pore-water pressure necessary to cause movement is

$$u_{wc} = Z\gamma_t \left(1 - \frac{\tan\theta}{\tan\phi'}\right)$$

where Z is the depth of the shear plane, γ_t is the total unit weight of the slope material, and θ is slope angle. The critical volume of soil moisture per unit area of slope is

$$Q_c = \frac{u_{wc}}{\gamma_w} \ n_{ef}$$

where γ_w is the unit weight of water and n_{ef} is the effective soil porosity. Under the simplifying assumption that all rainfall infiltrates, drainage I_o is directly related to the critical soil moisture Q_c:

$$(I_r - I_o)D = Q_c$$

The values of the threshold parameters I_o and Q_c are determined for local conditions (steepness and form of slope, resistance of rock materials, etc.).

Rainfall forecasts are obtained by real-time monitoring using an ALERT (Automated Local Evaluation in Real Time) network of telemetered raingauges. If telemetered piezometers are also installed in sample slopes, the relationships given above can be calibrated with reasonable accuracy ($I_o = 6.86$ mm/hr and $Q_c = 38.1$ mm/hr in the most reliable estimates for the San Francisco Bay area). Warnings can be issued as conditions become critical for periods of 6–60 hours. Although forewarning of the precise location of individual mass movements is not usually possible, debris flow activity does correspond with periods in which the rainfall and infiltration thresholds are exceeded.

Landslide hazards around the world
Worldwide, the average death toll in landslides may be about 600 per year. Destructive impacts seem to be concentrated in the tropics, especially in areas susceptible to hurricanes and earthquakes and in high mountain zones. If any such characteristics coincide with the presence of dense communities of poor urban settlers the result is frequently catastrophic, as in Caracas (Venezuela), Cusco (Perú), Ponce (Puerto Rico) and Rio de Janeiro (Brazil).

In Japan landslides are primarily related to periods of annual snow melt and the occurrence of typhoons, but earthquakes, volcanic eruptions and tsunamis also play a part in causing slope instability. In the Japanese archipelago the geological formations most susceptible to mass movement are fractured crystalline rocks, mudstones and shales, the last of which develop mudflows. Hot-spa areas are particularly susceptible to mud- and debris flows.

Landslides in the Commonwealth of Independent States predominate in areas of the most recent mountain building. Snowmelt, heavy rainfall and earthquakes are all important causes of mudflows and other mass movements in the Caucasus and in Soviet Central Asia.

In the Mediterranean Basin, as in the former Soviet Union, landsliding has been intensified by deforestation and a legacy of poor environmental management. The deep clay and sand deposits around the rim of the basin are easily dissected, and many landslides are an indirect effect of gully entrenchment as a response to Quaternary sea level falls. Land-sliding has had particularly catastrophic consequences in Italy, which has large areas of highly tectonized, relatively young sedimentary deposits on which urbanization has taken place (Fig. 4.15).

In the tropical areas of the Third World overgrazing and population pressure on land have in many places caused the threshold to be crossed between shallow drainage and deep gullying of thick weathered mantles. Steeper slopes and the sides and headcuts of gullies (such as the dongas of Africa) are often unstable. Population pressure is particularly acute in Hong Kong, where steep, unstable slopes in deeply weathered granite are often highly urbanized. Of mass movements caused by a storm in June 1966, 70 per cent were associated with road sections, buildings and land disturbed by cultivation.

Especially if it is made compulsory, insurance forces awareness of the hazard and itemization of the costs associated with the risks entailed. It appears that New Zealand is the only country to have developed a comprehensive programme of landslide insurance, which is integrated with its flood, earthquake and fire insurance schemes (Falck 1991). France, Germany, Italy and Japan have all initiated landslide mapping and hazard reduction programmes (the maps are usually made at the 1:25,000 scale). Indeed, it is probable that, together with the growing

Figure 4.15 Landslide damage amounting to total or partial collapse. Nine buildings were destroyed and eight people lost their lives in the landslide of 19 July 1986 at Senise in Basilicata Region, southern Italy.

internationalization of the scientific study of landslides, such initiatives will lead to more countries adopting an active policy of systematic landslide mitigation.

Landslide hazards in the USA

Virtually every American state suffers some landslide risk, but certain areas are particularly affected (see Fig. 4.16). These include mountainous regions such as Appalachia (Pennsylvania, Virginia, etc.) and the Cascades ranges of California, Oregon and Washington, which contain active volcanoes with unstable flanks. Like the Californian Sierras, the Rocky Mountains encompass some partially urbanized or heavily used areas (such as the Colorado Front Range) where both landslides and avalanches are especially dangerous and damaging. Mass movements are also a problem in lowlands, such as the Mississippi Valley, where they are caused by stream bank erosion, and on many coastlines, including those of the Great Lakes, where wave action causes rockfalls. Finally, parts of Montana, South Dakota and Texas are at risk through particular combinations of tectonic and erosional conditions.

Although ground failure is the most expensive hazard to afflict the USA (Hays 1981), loss of life in landslide disasters has been relatively low, averaging no more than 25 deaths per year for most of this century. Estimates by the National Transportation Research Board put the total cost of US landslides at $1,000–1,500 million per year (Schuster & Krizek 1978). This figure breaks down into the following proportions:

    Damage: Property and activity costs, 40%
               Reconstruction, 20%
    Adjustment: Maintenance of protection works, 38%
               Engineering protection, 1%
               Acquisition of property, 1%

Three counties bear a disproportionate amount of the burden: landslides cost $4 million per year in Allegheny County, Pennsylvania (the Greater Pittsburgh area), $5.2 million per year in Hamilton County, Ohio (the Greater Cincinnati area), and $5.9 million per year in the San Francisco Bay region of California. In the early 1970s it was estimated that mass movements cost an average of $1.60 per person per year with regard to private property damages and $3 per person per year with regard to all property damages (Fleming & Taylor 1980).

In California, it is estimated that landslide damage over the period 1970–2000 will cost $9,900 million, which will be 18 per cent of the total cost of all natural hazards in that state. It has further been estimated that an expenditure of $1,000 million, if concentrated on the appropriate type of mitigation work, could reduce the annual total of damages by 90 per cent, which represents a benefit to cost ratio of 9:1 and makes landslide

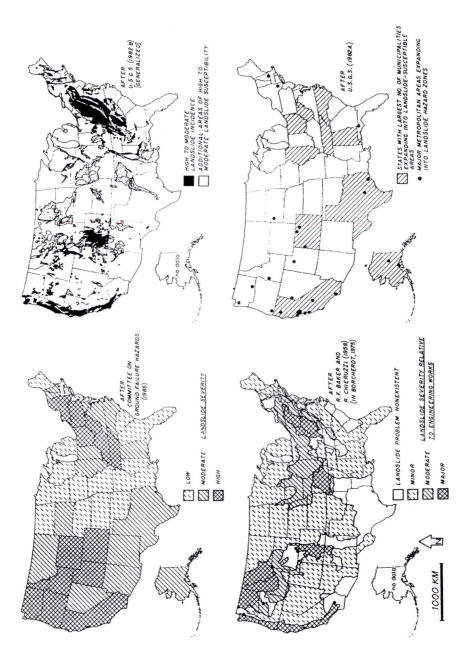

Figure 4.16 Four contrasting views of slope stability in the United States (Alexander 1989: 168–9, and various sources).

amelioration and avoidance highly desirable in economic terms (Leighton 1976).

Landslides are a particular problem in southern and central California, especially where they coincide with urbanized and suburban communities, such as those built in canyons and on alluvial fans at mountain fronts (the former are subject to debris slides and flows, while the latter suffer from mudflows and floods). The rainy season occurs from November to April in southern California and from October to May in the northern part of the state. Most storms that produce mass movements have a 3–4 day duration and follow an earlier wetter period. In the San Francisco Bay area it has been estimated that 125–150 mm of rain must fall in a period of 12–24 hours if debris flows are to occur, as they have done at least 13 times since 1900. Widespread mass movements occurred in southern California following storms in 1938, 1961–2, 1969 and 1982, but the mean recurrence interval is probably much shorter, and has been estimated at 6.4 years (Ellen & Wieczorek 1988).

In the Santa Monica Mountains of southern California soil slips occur widely once or twice a year after rainfall at intensities of perhaps no more than 6 mm/hr spread over 2–3 days. Major slides occurred in January 1969, when saturation of regolith and colluvium occurred after 25 cm of rain had fallen. Vegetation had been burned off the slopes by wildfire and this reduced surface stability prior to the storms. Slides occurred on volcanic rocks, shale, slate and sandstone, and 12 fatalities were caused in eight incidents comprising individual mudflows, debris flows or debris avalanches (Campbell 1975).

Over the period 3–5 January 1982, 500–600 mm of rain fell on the San Francisco Bay region. Debris flow-slide avalanches killed 26 people in Santa Cruz, San Mateo and Marin Counties, 10 of whom died in a block glide incident in Santa Cruz County (Cotton & Cochrane 1982). In many cases, the mass movements followed stream courses, having developed in unstable areas of ephemeral stream headcuts, which initially consisted of no more than slight depressions on the hillsides. They began as flow-slides and turned into viscous flows that were able to reach velocities of 10 m/sec (36 km/hr). As in the 1969 disaster, most of the victims died during the night while asleep in their homes, which underlines the need not merely for proper planning controls concerning where urbanization can take place, but also for a system of forecasting and warning for debris flow-slides.

The US National Flood Insurance Act provides insurance against mudflow damage associated with riverine flooding, but not against most other types of landslide. Communities that participate in the National Flood Insurance Programme must initiate hazard mitigation measures, but landslide mitigation is largely the preserve of individual counties or metropolitan areas. For example, in Fairfax County, Virginia, developers must obtain engineering geological advice before building on slopes. The City of Los Angeles developed its grading regulations over the period 1952–63 and subsequently refined them to take account of improvements in grading technology, thus achieving a 90 per cent reduction in damages. In fact, the California Environmental Quality Act of 1970 requires that landslide hazards be considered in all zoning changes, general plans and major public and private projects.

Example of a landslide disaster: Aberfan, South Wales, 1966
The Aberfan disaster of 21 October 1966 involved a mudflow that resulted from the collapse of part of a gigantic tip heap of rock waste derived from coal mining activities (Aberfan Tribunal 1967). Early on that Friday morning subsidence of about 3–6 m occurred on the upper flank of the tip. At 9.15 am more than 150,000 m^3 of debris broke away and flowed downhill at high speed. Under the duress of liquefaction at its head, the supersaturated rock waste moved as a series of viscous surges, which eye witnesses described as "dark, glistening waves". Some 120,000 m^3 of debris were deposited on the lower slopes of the mountain and approximately 42,000 m^3 came to rest in the urban area in a slurry 7–9 m high.

The movement overwhelmed two cottages, killing their occupants. Before coming to rest it engulfed and destroyed a school (which was occupied at the time) and 18 houses and damaged another school and various other dwellings (Fig. 4.17). There were no survivors. The dead numbered 144, 116 of them school children, most of whom were between the ages of 7 and 10. A further 29 children and 6 adults were injured, some of them seriously. Sixty houses had to be evacuated, 16 of them because they were damaged by mud from the flow.

The tip had never been surveyed, but right up to the time of the mudflow it was continuously being added to in a chaotic and unplanned manner. A commission of enquiry determined that the root cause of the disaster was negligence on the part of British National Coal Board officials, as proper supervision of waste tipping and adequate monitoring of the stability of the tip could have prevented the disaster. The event underlines the need for the formulation and application of regulations governing the construction and maintenance of large, man-made landforms.

Seismically generated landslides

It is often impossible to keep property developers or the urban poor off past or potential seismic mass movement zones, especially in areas of

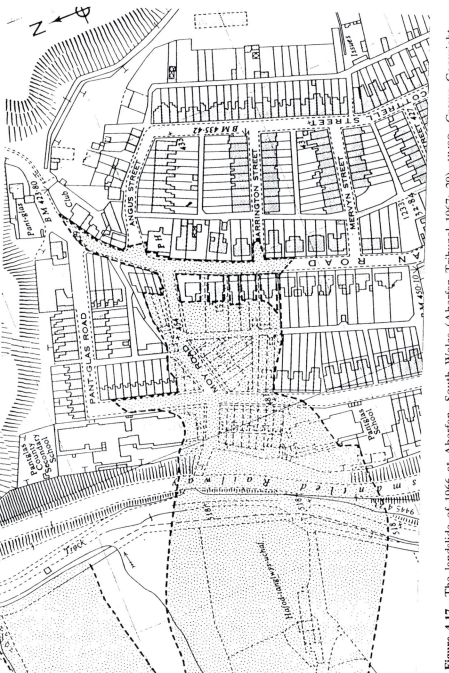

Figure 4.17 The landslide of 1966 at Aberfan, South Wales (Aberfan Tribunal 1967: 29). HMSO, Crown Copyright.

rapid urbanization such as coastal California and the canyons at the periphery of Guatemala City. Landslides generated or set off by earthquakes have been recorded since time immemorial: Livy's account of the Battle of Trasimeno during the Second Punic War makes reference to one. But it is only since the 1960s that geologists and geomorphologists have really understood the power of seismic tremors to destabilize the ground. Full appreciation of the problem has resulted at least in part from careful investigations of ground failure in some Latin American earthquakes, including Chile, 1960, northern Perú, 1970, and Guatemala, 1976. Large amounts of data on ground failure were also obtained after the 1964 Alaskan earthquake (US National Academy of Sciences 1973), while the geomorphological traces left by the enormously powerful earthquakes of 1811–12 at New Madrid, Missouri, have been investigated belatedly.

In order to promote adequate planning it is vital to know which terrains are at risk from ground failure during earthquakes, and what the geomorphic effects are likely to be.

The geomorphological effects of earthquakes
Earthquakes are capable of provoking a wide variety of the effects on the ground surface. Surface fault rupture may or may not be directly related to the fault whose movement initiated the main earthquake. The majority of seismogenic faults do not have a surface trace, being mantled by sediments or other rock units, but many are sufficiently linked to surface fault systems to cause these to move in adjustment to the main fault displacement. Normal, reverse or oblique faulting at the surface creates **fault scarps**, which are generally less than 5m in height, unless they are the artefact of repeated earthquakes, in which they can be tens or hundreds of metres high. In either case they can be tens of kilometers long.

Strike-slip faulting creates **offset channels**, trough-like valleys, linear ridges and **sag ponds**. The subsidence features may be either a direct result of differential block movements on each side of the fault or of selective erosion of bands of rock crushed and brecciated by the stresses that created the fault. The severity of alteration of the ground increases with the length of the rupture zone, which may vary from less than 1 km to longer than 300 km (the surface trace of the San Andreas Fault in California extends over about 1,000 km). The zone of geomorphological effects may be only 200–350 m wide, but it is not uncommon to find subsidiary fault ruptures perhaps 3–5 km from the main fault. Even where surface faulting is not manifest, surface cracks may develop, which are shallow (only centimetres thick) but may reach 1 km in length. These often result from compaction of loose sediments or from permanent horizontal displacement of the ground as it shakes.

The intense stresses that accumulate in the locked part of a fault (a 'seismic gap') may cause ground upwarping. For instance, in response to seismic stresses the so-called Palmdale Bulge in the Los Angeles area of southern California has risen 15–25 cm since 1960. Conversely, subsidence may result from the downward movement of structural units after faulting (settling), from the compression or reduction of pore spaces (compaction) or from the loss of groundwater (drainage).

Important as they are, seismically generated landslides usually do not differ in their morphology and internal processes from those generated under non-seismic conditions. However, they tend to be more widespread and sudden and, on occasion, also more catastrophic. In particular, rockfalls may occur in incoherent rock masses dislodged by the tremors. One form of mass movement usually limited to earthquakes is liquefaction failure (see below), which can cause fissuring or subsidence of the ground. It can also cause the emergence of **sand boils**, when water laden with coarse sediments is vented from subsurface layers of sand or silt in which artesian pore-water pressures develop during the liquefaction process.

The nature of seismically generated landslides
Earthquakes produce ground motions which in turn induce large inertial forces of an oscillatory nature in slopes and embankments (Fig. 4.18). These forces are of short duration and alternate in direction many times. Even though the load is repetitive, it is brief, and hence permanent deformation does not necessarily occur. As the earthquake waves pass through and along the ground, loading with symmetrical stresses may occur in one direction or two. The actual stresses undergone by the ground consist of two components. Initial stresses are present in the material of the slope before the earthquake and are a function of the gravitational load and the stress history of the slope. In the case of a man-made embankment, the latter can include constructional factors. Earthquake stresses are superimposed on the initial stresses.

Eventual failure of the slope may be caused by liquefaction of sands or silt deposits (see below) or by decreased strength. More important than losses of strength are increases in shear stresses applied to the slope during the earthquake, which induce cyclic mobility (Castro & Poulos 1977). In general, the likelihood of slope failure as a result of a given level of shear stress increases with the number of cycles applied. Larger cycles reduce the number required to cause failure, although long duration of ground motions may be needed to induce catastrophic landslides. Moreover, in many cases the velocity of ground motion is as important as the number and amplitude of cycles.

269

(a) Unidirectional stress pulse loading

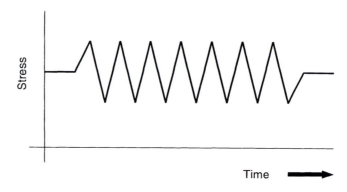

(b) Bidirectional stress pulse loading

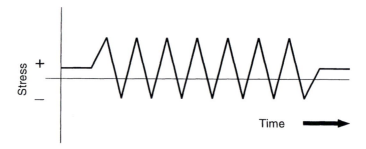

Figure 4.18 Earthquake stress-pulse loading in slopes.

Often sliding occurs hours or even days after strong motion of the ground has ceased. This is especially true when rainfall increases pore-water pressure in slopes (particularly in those composed of clays) which have not been able to liberate all effects of the seismic stresses and have undergone surface fissuring which increases their infiltration capacity. Relatively small aftershocks (e.g. $M_L \leq 4.0$) may be sufficient to set off the mass movements.

The cyclic mobility of clays
Clays are able to bear much stronger strains (deflections of 1–2 per cent) than sands (≤ 0.4 per cent) without permanent deformation. They therefore manifest a higher degree of elasticity, and their cohesive strength allows them to adjust slowly to increased pore-water pressure. Catastrophic failure of homogeneous clay deposits during or after an earthquake is rare, although certain kinds of clay tend to liquefy when disturbed (see below). But sudden increase in the load on a clay may

cause creep and later failure, especially as the cohesion inherent in clay means that it cannot dissipate pore-water pressures as quickly as can sands (Ansal & Erken 1989). The creep tends to generate a broad failure zone rather than a narrow shear plane.

The greatest stress provoked by an earthquake in a clay slope is called the peak or **yield acceleration**. Its value is a function of the geometry of the slope and maximum strength of the clay. The resistance that the clay brings to bear is called its **transient strength**, and the values of this depend upon the rate of strain and the amplitude of ground motions, rather than directly upon the static undrained strength. If yield acceleration exceeds transient strength, then failure of the slope results.

The sensitivity of clays
Clay sensitivity is the effect of remoulding on the strength and consistency of the material. A sensitive clay is one whose shear strength is decreased to a very small fraction of its former value on remoulding at a constant moisture content. Clays that show a marked difference in rheological properties between their undisturbed and remoulded states are known as **quick clays** (Kerr 1963). Upon remoulding, they tend to lose nearly all their shearing resistance and to show no appreciable gain in strength. Thus, when disturbed they go through a metamorphosis from a relatively brittle material to a liquid mass. This change is a particular characteristic of glacial and glacio-lacustrine clays initially deposited in exceedingly tranquil conditions. On drying out, such clays acquired a "house of cards" structure in which, seen under an electron microscope, their platy particles are randomly oriented and often supported on the thinner edges. Remoulding "collapses" this structure and it cannot be regained.

Quick or **sensitive clays** are found among the glacial deposits of Alaska and other parts of northern North America, such as the Leda Clays of Ontario and Quebec. At Saint Jean-Viannay in Quebec in 1971, 18 people were killed in a spontaneous liquefaction slide, albeit not one caused by earthquake (Tavernas et al. 1971). In glacial terrains, quick clays are often found in combination with sand and gravel strata and lenses, and the differential stresses which earthquakes set up between these deposits can create very destructive mass movements. These frequently dam rivers and lakes and tend to have a high transporting capacity capable of rafting large boulders and structures such as houses.

Liquefaction of sand deposits
Liquefaction involves the temporary loss of strength of sands and silts which behave as viscous fluids rather than as soils. It occurs when seismic waves pass through a saturated granular layer of uncohesive sediments, distorting the granular structure and causing some of the void spaces to collapse. The ground shaking is then partially transferred from the grains

Table 4.10 Classes of structural instability induced by liquefaction (after US National Research Council 1985: 194). Reprinted by permission of the publishers, National Academy Press.

Types of structural instability	Structures most often affected
Loss of foundation bearing capacity	Buried and surface structures
Slope instability slides	Structures built on or at the base of the slope Dam embankments and foundations
Movement of liquefied soil adjacent to topographic depressions	Bridge piers Railway lines Highways Utility lines
Lateral spreading on horizontal ground	Many types of structure, especially those with slabs on grade Utility lines Highways Railways
Excess structural buoyancy caused by high subsurface pore water pressure	Buried tanks Utility poles
Formation of sink holes as a result of sand boils	Structures built on grade
Increase of lateral stress in liquefied soil	Retaining walls Port structures

of sediment to the pore water. High pore pressures are not dissipated as quickly as they are generated, and the increased pore-water pressure results either in drainage or in failure. When pore-water pressures exceed the normal stress imposed by the weight of the sediment column, the material becomes fluid and deformations can occur easily (Seed 1968).

Liquefaction is generally restricted to areas of sands and silts deposited less than 10,000 years ago, in which the water table is less than 10 m below the surface. It has several possible consequences (see Table 4.10). For instance, the loss of bearing strength may result in the collapse of foundations or the settling of structures. Thus, as a result of the earthquake of 1964 at Niigata in Japan, four-storey buildings rotated 60–80° out of the vertical, while remaining structurally intact, and underground tanks rose up above the ground surface.

Liquefaction may cause **lateral spreads**, which commonly occur over distances of 3–5 m on slopes of 0.3–3°, though earthquakes of long duration can cause spreading of 30–50 m on slopes of similar angle. Damage is seldom catastrophic but is usually destructive: for example, in Alaska the 1964 earthquake caused lateral spreads which damaged 200 bridges. These structures spread with the underlying soil. Lateral spreads also caused water mains to rupture during the 1906 San Francisco earthquake, and hence they were indirectly responsible for the devastation caused in that city by fire. Where slopes exceed 3°, flow failure may occur in liquefied soil, and on occasion blocks of intact soil or rock may ride upon the liquefied material. Such mass movements can be rapid and far-reaching. Flow failures under the sea have destroyed port foundations and created destructive tsunamis. They have also been devastating on land: for example, flow failures occurred over 1.5 km after the 1920 Kansu earthquake in China, and about 200,000 people were killed by them.

Alternatively, sand boils may occur (see above). During the earthquakes that affected the northeastern part of the Indian subcontinent in 1757, 1897 and 1950, liquefaction subsidence and sand boils were extremely widespread. Some of the latter involved sizeable fountains of water and sand which were forcibly ejected from the ground. The diameter of the phenomenon, however, seldom exceeds 1–3 m. Sand boils are often indicative of clay interbedded with sands, a combination highly vulnerable to catastrophic failure during earthquakes. Sand or silt seams or layers in clay may liquefy and cause very dense composite mass movements. Slides develop progressively as the top layers are stripped off, and the reduction in pressure on the lower layers helps them fail as they liquefy. Lastly, during earthquakes, sand lenses in cohesive sediments (clays) generate high stresses at their margins, leading to exceptionally rapid liquefaction there, which spreads towards the centre of the lens. As a result of the curved, differential stresses, liquefaction tends to be much faster in lenses than in strata of sand.

Limiting factors
The earthquake which occurred in Guatemala City in 1976 had a Richter magnitude of 7.5 and caused 10,000 mass movements of sizes greater than 15,000 m^3 and 11 landslides whose volume exceeded 100,000 m^3 (Harp et al. 1981). Many shallow rock falls and debris slides also occurred in the canyons that surround the city. It was found that the distribution of landslides before the earthquake could not be used to explain the pattern of seismically generated slides. Moreover, the seismic intensity data were not consistent with the distribution of mass movements at the smallest scale. Instead, the governing factors were slope steepness and topography.

At Guatemala City in 1976, the regional distribution of seismic landslides could be explained with respect to various seismic, geological and geomorphic factors. For example, landslides occurred where seismic intensity exceeded VI on the Modified Mercalli scale. However, no other correlations with intensity could be verified. Predictably, 90 per cent of mass movements occurred in Pleistocene pumices, which were the most unstable lithology in the area. Regional fracture trends influenced local slope instability, while, in general, debris slides were predominant in the range 30–50°, and rock slides and falls were most common above 50°.

273

Lastly, topographic amplification of ground motion occurred as a result of the presence of canyon morphology.

In an analysis of 40 major earthquakes and many less important ones that generated landslides Keefer (1984) observed that events with magnitudes greater than 6.5 generated proportionately more rock falls and slides and soil falls. Keefer used the taxonomy of Varnes (1978; see Fig. 4.12) to classify the landslides and relate them by type to frequency, areal extent and the minimum magnitude capable of inducing them. Generally, landsliding is distributed over irregularly spaced areas, but is strongly related to magnitude (Table 4.11).

Precautions against seismically induced mass movements
The risk of seismic mass movement can be reduced by various means, including planning measures and foundation and structural engineering techniques. It is preferable that zoning be employed in order to avoid susceptible terrains. Prior field experience and hazard mapping are necessary, such that a microzoning map is produced on which to base land-use decisions. In many cases, including that of Anchorage (Alaska) in 1964, zoning and microzoning have been conducted in response to, rather than before, a major earthquake. However, if potentially unstable land must be built upon, it can be consolidated by compaction, grouting or drainage, though this is sufficiently expensive to be feasible only for small areas. Alternatively, or additionally, displacement-resistant foundations can be constructed, possibly allowing a maximum permanent shift of

Table 4.11 Threshold conditions of various types of seismically generated mass movement and relative abundance (after Keefer 1984).

Type of mass movement (Varnes 1978, 1984)	Threshold earthquake magnitude (M_L or $M_s{}^*$)	Common threshold mm scale intensity	Minimum threshold mm scale intensity	Abundance in 40 documented earthquakes
Rock falls	4.0	VI	IV	$> 10^5$
Rock slides	4.0	VII	V	$> 10^5$
Disrupted soil slides	4.0	VI	IV	$> 10^5$
Soil falls	4.0	VI	IV	10^3–10^4
Soil block slides	4.5	VII	V	10^4–10^5
Soil slumps	4.5	VII	V	10^4–10^5
Soil lateral spreads	5.0	VII	V	10^4–10^5
Rock slumps	5.0	VII	V	10^3–10^4
Rapid soil flows	5.0	VII	V	10^3–10^4
Rock block slides	5.0	VII	V	10^2–10^3
Slow earth flows	5.0	VII	V	10^2–10^3
Subaqueous landslides	5.0	—	—	10^2–10^3
Rock avalanches	6.0*	VI	IV	10^2–10^3
Soil avalanches	6.5*	VI	IV	10^4–10^5

up to 30 cm, or foundations can be laid very deep in order to transfer the load to more stable layers not subject to liquefaction. (Pilings can be used to do this.)

In the earthquakes studied by Keefer (1984), at least 90 per cent of landslide-related deaths were caused by rapid soil flows, rock avalanches and rock falls, i.e. the most sudden and destructive forms of mass movement. This implies that such landslides deserve special attention, especially in mountainous seismic zones. Most of the deaths were caused by the burial of houses (or entire settlements) in the runout zone, which suggests that hazard zoning should be employed as a defence.

Some examples of seismic landslides

The New Madrid earthquakes, which occurred in December 1811 and January 1812 in the Mississippi Valley in Missouri, were observed by very few settlers, as the area was sparsely populated at the time. However, it is known that they were among the most powerful seismic events to occur in North America during historical times, and that the area in question does not coincide with a plate margin, which is the usual form of seismogenic structure involved. Examination of contemporary sources and relict landforms has shown that major landsliding occurred on slopes and along river banks, and liquefaction of sand deposits was widespread (Obermeier 1988).

The Yakutat Bay, Alaska, earthquake of 1899 caused 1,200 km^2 of land to be uplifted and warped to a maximum of 14 m, which represents the greatest recorded differential shift in land level. The western margin of the Americas appears to be highly susceptible to abrupt changes in land level that are induced by earthquakes, and the 1964 Prince William Sound earthquake in Alaska caused 100,000 km^2 to subside. Concomitantly, 160,000 km^2 of land were uplifted an average of 1.8 m and to a maximum of 11.6 m. In this event 60 per cent of the estimated $500 million of damage was caused by ground failure, five landslides causing a total of $50 million of damage in the city of Anchorage. One of these was the notorious Turnagain Heights landslide, which involved failure on a very large scale of layered glacial sands and clays in a residential suburb built on a steep slope undercut at its base by the sea (Seed & Wilson 1967).

In 1949 in Tadzhikistan, in the former USSR, a seismic landslide moving at up to 360 km/hr overran the town of Khait and killed 12,000 people. Similarly, the earthquake of 31 May 1970 at Ancash in Perú set off an avalanche of rock, mud and ice, which fell 3.66 km vertically and moved 11 km horizontally from the peak of Mount Huascarán, burying the towns of Ranrahirca and Yungay with the loss of 18,000 lives (Browning 1973). A snow and ice avalanche containing some rock debris had already killed 4,000 people in these villages in 1962 (see Ch. 3).

Conclusion
Insufficient data exist worldwide to determine once and for all whether the probability of slope failure during earthquakes is effectively random or highly predictable in terms of a variety of parameters. In this respect, the significance of reactivated landslides is especially acute. There is, moreover, a danger of interpreting all landslides observed in an area after an earthquake as caused by that event. Given the difficulty of assessing the exact age of even some of the freshest slides, it is vital to gain an idea of the pre-earthquake distribution of mass movements if the impact of the tremors on slope stability is to be assessed.

Subsidence

Subsidence is a type of mass movement that involves the downward displacement of surface material caused by natural or artificial removal of underlying support (collapse) or by compression of soils (consolidation). Subsidence becomes a disaster if it occurs either very rapidly and causes injury or destruction, or if it overstresses large buildings (or bridges, etc.) situated on the subsiding ground to the point that these structures fail. It has a series of possible anthropogenic and natural causes (Fig. 4.19).

Consolidation
The natural compression of saturated sediment or rock under the influence of static load is called consolidation. It occurs in two or three phases. During **primary consolidation**, pore water is stressed in the sediment (which is usually a clay of sufficient elasticity to absorb the stress without failing) and pressure increases to expel water from the voids. The overburden load is therefore partially transferred from the soil water to the solid matter, via grain-to-grain contacts. **Secondary compression** involves reduction in the volume of the soil mass caused by adjustment of its internal structure after most primary consolidation has occurred. This takes place as a reduction in the ratio of voids to solids or, in other words, as an increase in the soil's **bulk density**. Thirdly, as water is an incompressible medium, the removal of the overburden load may result in expansion and shattering leading to the absorption of water by the soil or sediment, which is then known as **overconsolidated**.

The engineer Karl Terzaghi introduced the principle of **effective stress** (Terzaghi 1936), which we have already encountered in connection with landslides. The total load (comprising lithostatic pressure, or total stress), σ_c, of a vertical column of soil of unit cross section is sustained in part by the hydrostatic pressure, p, of pore water (if the material is porous) and in part by intergranular pressure, the effective stress, σ_z. The forces transmitted between particles in mutual contact have the following relationship:

$$\sigma_c = p + \sigma_z$$

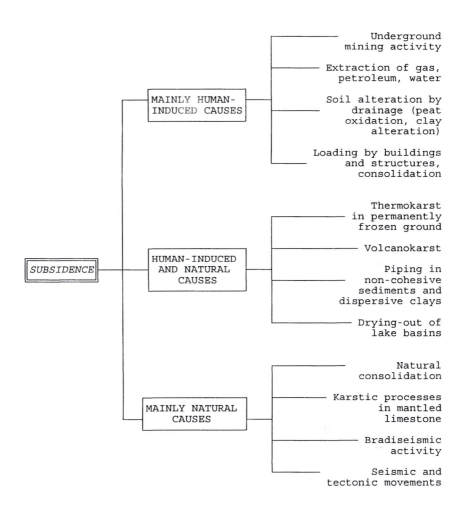

Figure 4.19 Causes of subsidence (Costa & Baker 1981: 285). Copyright © 1981 by John Wiley.

The total load per unit area, σ_c, is constant for any given depth. Thus, when water migrates through the column of soil or sediment, the hydrostatic pressure, p, will vary and therefore σ_z will also change. Hence, if pore-water pressure diminishes in a particular stratum, intergranular forces must increase by way of compensation. While variations in p have little effect on the state of the sediment, increases in σ_z tend to result in the compaction of strata.

Variations in interstitial pressure, U, are correlated with the coefficient of consolidation of the soil, C_v, with the thickness of the deposit, z, and with time, t, according to Terzaghi's equation:

$$\frac{\delta U}{\delta t} = \frac{C_v . \delta^2 U}{\delta z^2}$$

In other words, the variation over time in interstitial pressure is equal to the coefficient of consolidation multiplied by the rate of variation of the rate of variation of U with depth. The equation for the time, t, necessary to achieve consolidation is:

$$t = \frac{H^2}{C_v} = \frac{m_v . \gamma_w . H^2}{k}$$

where m_v = volumetric compressibility of the sediment
         γ_w = specific weight of water in the pores
         H = half the thickness of the strata undergoing consolidation
         k = permeability of the sediment.

Consolidation is at its worst more of a creeping disaster: it is a slow process that to be complete requires the dissipation of about 90 per cent of pore-water pressure, which may take years or centuries (see Fig. 4.20). Undrained loading, however, may render the sediment incompressible and lead to mass movement (as a result of high pore-water pressures) rather than subsidence.

The withdrawal of fluids

The withdrawal of fluids from the ground is a very common cause of subsidence (Poland & Davis 1969). The amount occuring in a sediment depends upon its mineral composition, sorting, level of prior consolidation and cementation, degree of confinement, porosity and permeability (Fig. 4.21). The greatest hazard comes from compressible beds, such as confined Tertiary and Quaternary sediment that is well sorted and has montmorillonite as the dominant clay mineral. Silt and clay undergo non-elastic compaction: they do not rise up when fluid is restored to them (the degree of rebound in sands is, however, limited).

Permeable strata that contain water and are bounded underneath by impermeable layers are known as aquifers. Where they occur high on a hillside (for instance, where the lithology of the hill consists of alternating strata of sand and clay) they are known as **perched aquifers**, and water may seep out of them in springs. In fact, wherever the water table intersects the ground surface springs are likely, and if the theoretical (hydrostatic or potential) level of water is higher than the level actually reached, flow occurs under pressure and wells drilled into the aquifer undergo **artesian flow**, without the need for pumping. Aquifers that have no impermeable capping are termed **unconfined**, and conversely **confined aquifers** are those in which the migration of groundwater is impeded by the presence of impermeable strata and the water is thus likely to be held under pressure.

278

Figure 4.20 Pre-revolutionary houses in the Tatar city of Kazan', 800 km east of Moscow. The buildings are situated at the confluence of the Rivers Kazana and Volga and are built on soft fluvial sediments which have lost their bearing capacity through consolidation. This has led to subsidence of up to one metre.

Subsidence of the ground above aquifers may result from water extraction at a rate which exceeds that of recharge by percolating or migrating groundwater. Pore-water pressures are lowered and effective stresses are increased in the sediment as groundwater no longer takes up a proportion of them. The level of water reached in a pipe tapping the aquifer is referred to as **hydraulic head**. Pumping causes drawdown, and a cone of depression in the water level spreads around the point of extraction. The lag time involved in falling hydraulic head, and the intensity with which this occurs, depend upon both the geometry and geological configuration of the basin and the hydromechanical properties of the sediments, including their permeability and compressibility. The amount of compaction which follows drawdown depends on the compressibility of the compacting unit and the weight of the overburden. The compaction of many granular materials is more or less irreversible, depending on the percentage of fine-grained sediments. At best, halting fluid withdrawal can stop subsidence and cause a slight rebound, given that the load effect of water on the underlying rocks is diminished (as the capacity of the aquifer will have been reduced by compaction of the sediments).

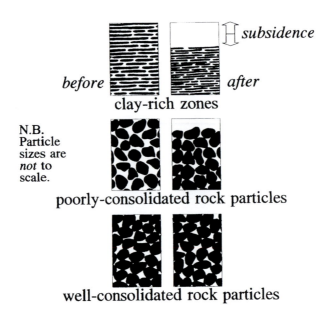

before   ⊞ *subsidence*

after

clay-rich zones

N.B.
Particle
sizes are
not to
scale.

poorly-consolidated rock particles

well-consolidated rock particles

Figure 4.21 Effect of fluid withdrawal on consolidation of sediments (Marsden & Davis 1967: 95).

Subsidence as a result of over-exploitation of aquifers is particularly common in the vicinity of large or rapidly expanding metropolitan centres and areas of extensive irrigation (as a proportion of the irrigation water is lost to evaporation). Impacts tend to be more costly in the former case, and examples include London (England), Tucson (Arizona) and Mexico City. Whereas in the chalk and clay basin in which London is situated there has been relatively little damage (the main problem being one of water supply), in Mexico City subsidence has reached a maximum of 7 m. Buildings can survive such effects if they are not extensive or tall and if subsidence under and around them is not differential such that critical stresses are set up and they require an impossible amount of flexibility. Hence in Mexico City the buildings at the edge of the lake bed are those that underwent the most differential subsidence and hence suffered the worst damage.

Fossil fuel withdrawal is also a notable cause of subsidence. At Niigata in Japan, for example, natural gas extraction has caused coastal land to sink up to 150 cm (Fig. 4.22). In Texas, oil extraction has caused sinking of the coast around Galveston Bay, which is now more vulnerable to flooding by hurricane or tidal surges and seiches. Losses in the Houston–Baytown area were valued at $109 million over the period 1943–73. At Long Beach in California, subsidence resulting from oil and

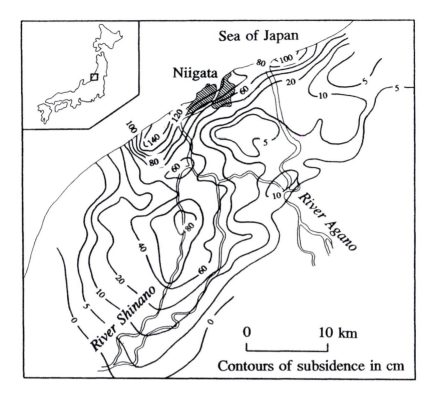

Figure 4.22 Subsidence in the vicinity of Niigata, Japan, caused by natural gas extraction (Whittow 1980: 191).

gas extraction reached a maximum of 70 cm/yr. Serious damage was done to port facilities, and the ensuing differential stresses in the ground were indirectly responsible for the Baldwin Hills dam disaster (see Ch. 5).

The case of Venice

Subsidence of between 9 and 15 cm has occurred at Venice, largely because of groundwater and methane extraction in the Marghera-Mestre area of the Venetian lagoon (Fontes & Bortolami 1973). As a result the city is now more vulnerable to surge tides and to corrosion by rising damp and salt incrustation, for which the bow waves of motor boats and ships are partly responsible.

Although pumping has now stopped and wells have been capped (water supplies now come from inland by a recently constructed aqueduct), over 85 per cent of the subsidence at Venice is unrecoverable, and rebound will only be of the order of 1–2 cm by the year 2000 (Gambolati et al. 1974). Nearly 70 per cent of the historic city lies at less than 1.25 m above mean sea level (see Fig. 3.5). Eustatic sea-level rise adds 15 cm/100 years to the rate of inundation, while in previous centuries subsidence had already reached 10 cm/100 years. Inundations of the city, the *acque alte*, occurred with a frequency of once in 5 years over the period 1875–1940, once per year over 1940–65 and three times per year over 1965–75. Normal water level fluctuations are about 0.61 m in the Venetian lagoon, to which must be added 5–15 cm where currents are constrained at the point of outflow. Periods of unusually low atmospheric pressure cause the water level to rise proportionately by 10–30 cm; moreover, a wind of 10 km/hr raises the water level 10 cm, while one of 55 km/hr raises it by up to 50 cm. Seiches in the Adriatic Sea Basin can raise the water level at Venice by 37–65 cm, with a possible maximum of 90 cm. Thus, the worst flood problems are likely to occur when some of these phenomena coincide.

Over the period 3–5 November 1966 winds of more than 100 km/hr occurred in the Adriatic Basin and the resulting storm surge at Venice reached 1.9 m above datum and flooded 80 per cent of the historical city. The waters remained at least 1.1 m higher than the seasonal norm for 24 hours and $70 million of damage was caused. Storm surges are regular events at Venice in winter, but their impact has been worsened considerably by subsidence. Moreover, the 1966 floods have a calculated return period of less than 250 years. Thus it is to be hoped that, despite the enormous cost involved, the Italian Government will implement plans to build a flood barrier (which can be raised or lowered) across the mouths to the lagoon, especially as the problem is unlikely to clear up of its own accord.

Hydrocompaction subsidence

Hydrocompaction is the reduction in the dry strength of sediments by adding water to them. The process it mostly affects loose alluvium, alluvial fan or mudflow deposits, and moisture deficient aeolian sediments such as loess (thus it can be particularly important in arid environments). The dry strength of sediments may be a result of cohesive bonding in clay, interlocking in granular and, especially angular, deposits, or partial cementation with compounds that are produced naturally. Subsidence occurs when the bonds are broken during wetting of the sediment and consolidation results under the weight of the overburden. Irrigation in dry climates is an especially important cause, and thus 500 km^2 of irrigated land in the San Joaquin Valley, California, has subsided.

Materials susceptible to hydrocompaction tend to have a high void ratio (bulk density <1.3 g/cm^3) and a clay content of about 12 per cent. If the percentage is lower, void spaces will be insufficient, while if it is much greater, the sediment will tend to swell. Hydrocompaction is also common where montmorillonite is present in the sediment, as this clay mineral solidifies considerably when it dries out, and becomes viscous when it hydrates. Lastly, to be susceptible, the deposit must be dry, and must generate its maximum strength at moisture contents of less than 10 per cent.

Piping

Piping is most common in materials that are permeable, but that contain cracks in which permeating water can concentrate. Sediment is transported and erosion occurs wherever interflow is augmented in this way, and the result is a vertical or inclined pipe that may eventually collapse. Loess, tuff, volcanic ash, fine-grained alluvium and colluvium are all susceptible. In addition to a suitable material, the process requires sufficient water to saturate material above base level (which locally is represented by the level of streams), sufficient hydraulic head to cause subsurface flow as a response to the weight of water at unequal levels and an outlet capable of concentrating flow.

A special case of subsurface erosion is given by dispersive piping. Dispersive clay contains a sufficiently high proportion of exchangeable sodium cations to enter readily into suspension when in contact with water. Consolidated clays, and especially those that are overconsolidated, tend to contain tension cracks, and these may act as centres for dispersive pipe development. Particularly aggressive forms of piping such as this may undermine roads and the foundations of structures until these collapse. It often goes undetected until it gives a full surface manifestation, by which time the damage may already have occurred.

Karst and pseudokarst development

Karst is the name given to areas of massive limestone deposits that erode mainly by a combination of surface and subsurface solutioning. The surface of such outcrops tends to be distinguished by dry valleys and incomplete drainage networks (as streams easily dive underground) and enclosed depressions as a result of concentrated solutioning activity and the collapse of underground cavities. The most hazardous form of cavern collapse in limestone is that which occurs in **mantled karst**, where a layer of soil and sediments overlies the limestone and masks the development of depressions and unstable cavities. The abrupt formation of **dolines** (also known as **sinkholes**) in mantled karst caused substantial damage in a residential and commercial district of Winter Park, Florida, in May 1981, when the initial collapse of the doline was followed by its gradual enlargement as the sides fell in and saturation and drainage of the base occurred sporadically. A similar incident in South Africa caused several deaths, as the collapse occurred at night and swallowed up some houses.

In general sinkholes have both natural and anthropogenic causes. The former include heavy rainfall, solutioning, compaction of overburden material and cavern collapse. The latter comprise artificial drainage, overloading with water and explosion shocks. For the phenomenon to present a hazard, soluble bedrock such as massive limestone or dolomite must be mantled by a thick layer of unconsolidated surface deposits. Subsurface drainage must occur via pipes and caverns that have an arched structure capable of collapsing.

In areas of permanently frozen ground, subsidence can result from what is (by analogy with limestone areas) called **thermokarst**. Differential melting in frozen terrains can lead to differential drainage, the emergence of frost-heave polygons and settling of drained ground. The causes can be either natural or anthropogenic (Fig. 4.23). Even if they are not heated internally, buildings and structures located on permafrost can alter the thermal regime of the ground underneath them such that thawing and settling gradually occur (Ferrians et al. 1969). Heated buildings need to be well isolated from the ground surface, for example, by placing them on

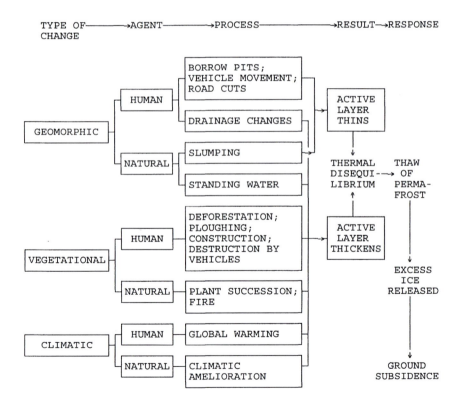

Figure 4.23 How subsidence can be caused by permafrost degradation (French 1979: 149).

pilings. Piping in volcanic terrain produces **volcanokarst**, which together with thermokarst is a form of **pseudokarst**.

Other causes of subsidence

Geological faulting is a cause of subsidence in some areas. Normal faults which are activated seismically or by creeping movements may be responsible for the sinking of fault-bounded rock blocks where the stress regime is extensional rather than compressional. Subsidence tends to be very limited in area along strike-slip faults, but can be extensive in graben. Houston, Texas, has been damaged in this way. Fault movement can also cause earthquake-induced subsidence. For instance, in the 1964 Alaskan earthquake 160 km^2 of land subsided by 0.75–2.25 m to the north and west of the seismogenic fault line, while other areas of land settled as a result of the compacting effects of the tremors. However, earthquakes are also capable of lifting up those blocks of crust which are affected by compressional tectonics.

Economic activities associated with geological deposits are well known as a cause of subsidence. Coal or other mines, where the workings are not backfilled with spoil material or supports are inadequate, may eventually subside as rock-wall bursts occur and roof props collapse. Unless the mines are very well mapped, it may be difficult or impossible to predict where and to what extent surface effects are likely. This is a particular problem in places with a long history of chaotic subsurface extraction of minerals, especially those areas that were associated with the nineteenth-century Industrial Revolution. Similar but less well-known activities that can result in subsidence include leaching of intergranular elements (such as uranium and copper) from subsurface rock formations, coal gasification and oil shale retorting. In the United States, subsidence induced by mining results in losses of about $30 million per year (Allen 1978).

Mitigation of subsidence

Monitoring of subsidence rates can be carried out in restricted areas using large-scale strain gauges. Precise geodetic levelling can also be employed, but it needs a stable base level that is not exposed to subsidence. Laser surveying is particularly promising in this respect as it can be carried out to high levels of accuracy over long distances. But few areas have long-term records of subsidence derived by repeated survey or measurement of the phenomenon.

Fluid extraction wells can be repressurized with water to halt subsidence, and coal mines can be backpacked with shale and spoil material. However, from a legal perspective it is often difficult to establish a direct link in terms of cause and effect between human activities and subsidence of the ground. Nevertheless, some governments have enacted laws requiring fluid and mineral extraction concerns to safeguard against the problem and, in some cases, to compensate victims. The problem derives ultimately from the fact that subsurface mineral or water rights and surface property rights may be separate from each other and in the hands of radically different groups.

In a very few cases subsidence has had a beneficial, even a planned, effect. Controlled mining under Duisburg Harbour in West Germany caused it to subside by an average of 2.4m and saved much money in dredging costs. It has been suggested that reservoirs could be built in this way, although the idea has not been tested.

Soil hazards

Expansive soils are those that vary substantially in volume in relation to their water content (Dudley 1970, Gromko 1974). Most clays are susceptible to this process (see Table 4.12). Thus, at the microscopic level, montmorrillonite clays adsorb water molecules onto and into their lattice structures, causing them to swell substantially upon hydration. Conversely, on desiccation they will shrink proportionately and may crack up to 1.5 m deep. The swelling pressure in montmorrillonite and bentonite clays may reach 0.16–0.6 MN/m^2 (where MN = megaNewton), which is sufficient to damage small buildings substantially. Expansive soils are therefore subject to heave and in pedology are termed **vertisols**. Where undisturbed by human activity they may form polygonal crack-and-heave structures known as **gilgai**.

The propensity of an expansive soil to undergo changes in volume may depend on the proportion of clay-sized sediments ($\leq$2 mm diameter) that it contains, the range of particle sizes and the degree of activity of sodium cations (Na^{++}). Dense or consolidated soils swell most, and dry soils swell more than damp ones; however, remoulded soils swell more than undisturbed ones, even though they may be less dry and compact. In any

Table 4.12 Free swelling potential of clay minerals (Mielenz & King 1955).

	Free swelling (per cent)	
	2000	
	1000	sodium montmorillonite
	500	
illite	100	
kaolinite		calcium montmorillonite
	50	
	20	— kaolinite
		— illite
	10	
	5	— kaolinite

case, for a given mixture of chemical and pedological characteristics, lightly loaded expansive soils pose more of a threat to structures on top of them than those held down by substantial weights.

The pressure of expansive soils can be classified in terms of the hazard to buildings that it poses:

< 0.15	MN/m^2	non-critical
0.15–0.17	"	marginal
0.18–0.25	"	critical
>0.25	"	very critical

The hazard is not merely one of lifting and disturbing structures built on such soils (Mitchell 1986). Concentration of runoff and infiltration may lead to differential swelling, which will crack rigid structures. The most flexible buildings, such as those with wooden frames, will survive best. Emending soils with hydrated lime can reduce the problem by substituting highly active sodium ions with more inert calcium ones. Alternatively, the site can be lined with impervious clays of the rarer inert varieties.

In the USA expansive soils caused damage valued at $5,600 million per annum. Building owners whose property has been affected have the right to sue architects and engineers, if it can be established that the building designers should have been aware of the hazard.

Coastal erosion

Coastal erosion is not necessarily a disastrous phenomenon, but problems occur when erosional processes and human activity come into conflict. An average shoreline recession rate of more than 6 m/yr may pose no threat in an unpopulated area, while a rate of less than 0.5 m/yr may cause severe damage to a densely settled area. Both erosion and accretion of coasts are influenced by human-induced changes as well as by natural processes (Fig. 4.24), while erosion can have many different negative impacts on both the physical and the human environments (see Table 4.13). World-wide increases in the scale and density of human occupation of shorelines have increased the vulnerability of coastal populations and made the need for better management policies imperative.

Causes of coastal erosion
There are five main influences on erosion potential at the coast: exposure of rocks and sediments to the action of waves and currents; the supply of sediments (beach **starvation** or **nourishment**); the topography of the coast and neighbouring continental shelf; the tidal range and intensity of currents; and the climate of the coastal area. In addition, human activities

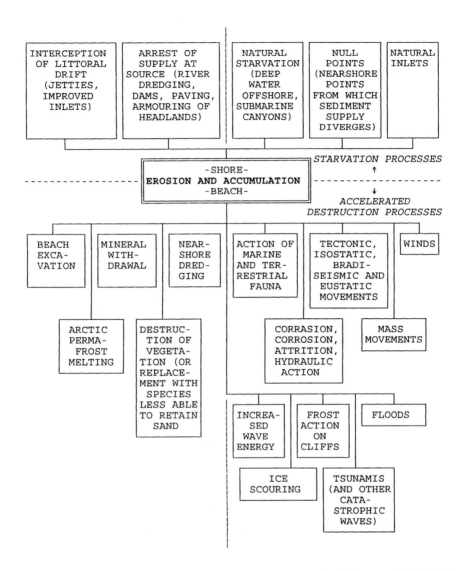

Figure 4.24 Natural and human factors in coastal erosion (Sorensen & Mitchell 1975: 3).

can increase the impact of wave erosion in several ways. These include the destruction of the coastal vegetation that helps to stabilize unconsolidated coastal sediments. Natural defences are weakened when beach deposits are removed for uses such as building or road construction, in which they serve as aggregate for cement mixes. In restricted areas the lowering of coastal water tables can desiccate sand

Table 4.13 Summary of coastal erosion damage potential (Sorensen & Mitchell 1975: 27).

Immediate effects (frequently or potentially damaging):

- Removal of small island
- Breaching of barrier islands: formation of washover channels and new inlets
- Migration of existing inlets
- Loss of beach sediments
- Recession of dunes, bluffs and cliffs
- Destruction of existing habitats of shore flora and fauna
- Weakening and destruction of sea walls, revetments, bulkheads and other coastal defences
- Loss of agricultural land, lawns and gardens
- Deposition of sediment on streets and productive lands
- Isolation of jetties and groynes from mainland
- Undermining and destruction of boat-launching facilities, septic tanks, swimming pools, utility lines and building foundations
- Disruption of communications causes by collapse of seafront, bridge, road and railway foundations
- Shoaling of offshore waters, silting of navigation channels, harbours, shellfish beds and coral reefs

Secondary effects:

- Loss of income from beach recreation
- Reduction of protection against future storms
- Creation of unstable cliff slopes
- Pollution of beaches caused by broken sewer lines or eroding spoil tips
- Alteration of lagoon ecosystems as a result of silting and invasion of saline water behind breached defences
- Flooding of land areas behind breached defences

dune systems, thus making them vulnerable to aeolian erosion. Elsewhere, the withdrawal of underground water, oil or gas deposits, or mining coal beneath the shore, can cause land subsidence and coastal inundation. Lastly, to a minor extent, dredging steepens the offshore slope profile and thus causes it to absorb more of the energy of erosive waves.

Beach starvation is one of the commonest causes of erosion of sand and shingle coasts. It can be promoted by natural features (submarine canyons, capes, headlands, etc.), which act as barriers to longshore drift or as a means of channelling sediment away into the open ocean basins. But it can also be an effect of artificial structures (such as groynes, jetties and navigation channels), which impede longshore sediment supply. Moreover, soil conservation and damming of rivers both tend to impede the movement of river-borne sediment to the coast and thus eventually to cause beach starvation. Hence, where beaches exist downdrift of a major estuary, it may be difficult to strike a balance between the need for reservoirs, which tend to trap sediment, the importance of soil conservation in the upstream catchment, and the need for constant supplies of sediment in the littoral drift zone.

Both the losses resulting from coastal erosion and the hazard itself can be the subject of mitigation (Table 4.14). Adjustment can be classified into four categories: structural control and protection works, land-use management practices, warning systems and initiatives for public relief and rehabilitation.

Structural control and protection works
In many circumstances, if the shoreline is left undisturbed by mankind, it will develop its own defences against marine erosion. A broad berm stabilizes a coastline by absorbing wave energy. Sand dunes serve as a source of sediment for beach replenishment and as a barrier, preventing overwash from being so frequent that the shoreward vegetation has no time to recover between storms. Barrier islands and beaches are likely to be breached by the occasional very large storm, and hence will develop brackish lagoons and salt marshes on their shoreward sides. These will be colonized by salt-tolerant vegetation, and serve useful functions as habitat for many aquatic and avian species and as natural sediment traps. They

Table 4.14 Adjustments to the hazard of coastal erosion (White & Haas 1975: 364). Copyright 1975 by MIT Press.

Adjustments to loss	Modifications of loss potential	Modifications of erosion hazard	Adjustments affecting cause of hazard
Major:	*Major:*	*Major:*	*Major:*
• loss bearing • insurance	• coastal zoning • building codes • public purchase of eroding land • land fill	• dune stabilization • groynes • bulkheads • sea walls • revetments • beach nourishment and "perched" beaches • breakwaters	• prohibition of beach excavation and harbour dredging • sand bypassing
Minor:	*Minor:*	*Minor:*	*Minor:*
• emergency public assistance	• moving endangered structures • installing deep pilings • storm warning and forecasting systems	• regulations against destruction of dune vegetation • phreatophyte removal • artificial seaweed • bubble breakwaters • emergency filling and grading • grading slopes	• removal of river dams • biological control of river fauna • reduction in soil conservation activities • storm track modification

will be protected by the shoreward line of barrier islands, beaches and offshore bars.

The process of littoral drift is one in which sediment migrates laterally across the shore. Its movement results from the fact that waves may break at an angle to the shore, but will retreat orthogonally to it, thus moving sediment in a push–pull zig-zag motion. The process operates mainly in a narrow zone from the breaker line to the beach, and sand transport here tends to be greatest where waves break in conjunction with topographic highs. The direction of sediment movement may vary between individual storms, but over time there is likely to be a definite trajectory of drift. Hence, at a given point, material moving downdrift is replaced by that which moves in from updrift. On a stable shoreline, the sediment budget is essentially balanced and the processes of supply and removal will evolve towards dynamic equilibrium over a period of years. Under such conditions, man-made structures upset the balance and alter the distribution of material (Rosenbaum 1981). Thus, severely eroding shorelines are often found downdrift of artificial structures: field, laboratory and wave-tank models have all determined that the effects are quite predictable.

Artificial means are widely used to stabilize eroding coastlines. For instance, sea walls are intended to reflect wave energy in order to prevent retreat of the coastline during eroding storms (Kenney 1978). They tend, however, to concentrate the reflected force of the waves in the nearshore zone and hence to be vulnerable to undermining leading to collapse of the wall (Fig. 4.25). Permanent revetments, such as a rubble fill, cause less erosion of the nearshore profile, as the structures are partially permeable to sea water and, through their rough profile, are capable of trapping sediment. However, material may be trapped at the expense of downdrift zones.

Detached offshore breakwaters are designed to protect the shore against the destructive action of waves, while at the same time preserving longshore drift. The lack of turbulence between the structure and the shore will cause deposition of sand and possibly even a **tombolo**, which is a sand bar that extends from the shore to the structure. In order to function effectively, the structure must be located at a distance from the beach that is 3–6 times its length; hence, for effective protection breakwaters tend to be very large and very distant.

A **groyne** is a solid (although occasionally permeable), narrow barrier projecting seaward perpendicular to the coast, which is designed to trap sediment and ensure that the beach has a certain fixed width. Initially, a very wide sand accumulation is produced updrift of the groyne. If they have ever existed, offshore sand bars disappear and sand starvation occurs downdrift for 3–5 times the length seaward of the structure. Eventually, having moved in an arc to the end of the groyne, sediment

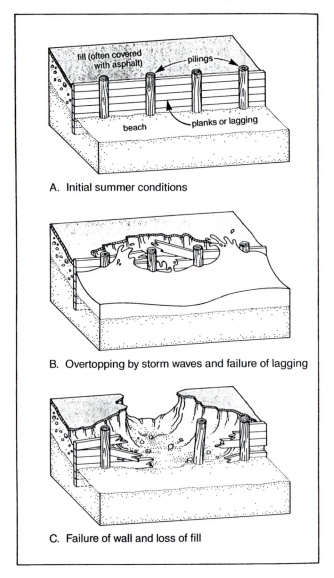

A. Initial summer conditions

B. Overtopping by storm waves and failure of lagging

C. Failure of wall and loss of fill

Figure 4.25 Progressive lateral failure of a wooden sea wall (Fulton-Bennett & Griggs 1987: 30).

moves past it and recreates the offshore bar, but the shore itself goes on eroding indefinitely downdrift. Second- and third-generation structures may be required to stabilize the coast, but these simply propagate erosion further and further downdrift.

Artificial filling of the accretion zone updrift of the groyne may be partially effective in reducing downdrift erosion: it allows the uninterrupted flow of sand past the groyne. The artificial fill must periodically be renourished in order to avoid starvation on the downdrift beach. However, it is often difficult to find a source of artificial beach sediments, as the material must be hydrodynamically stable. Removing sediments from other parts of the beach system may render these unstable as well. The process of **sediment bypassing** involves mechanically or hydraulically moving sediment across a shoreline structure (such as a harbour entrance) towards the downdrift zone that is vulnerable to starvation (Tornberg 1968). This is intended to stop the structure from detaining too much sediment or channelling it into deeper water where it cannot be utilized by the beach system. Bypassing systems are expensive to construct and maintain: the relatively deep, quiet waters at a harbour entrance are a good settling ground for the sediment carried in littoral drift. Although widely practised, dredging of harbour or waterway entrance channels and dumping of the sediment out at sea tend to exacerbate the problem, and in the end harbour mouths may cause net erosion of sand beaches up to 12 km downdrift.

Beach nourishment is the corollary of sand bypassing, and may present the most effective way of strengthening depleted beaches against erosion. It will maintain the recreational capacity of the beach while minimizing the effect on littoral drift and the aesthetic impact of human intervention. However, dredging for a source of sand nourishment may not only create sand deficits elsewhere, but also disrupt local ecology. Moreover, beach nourishment may turn out to be more expensive than the erosion itself: in the United States beach erosion seldom costs more than $50 per metre of shore in direct costs, whereas sand nourishment may cost $165–$1,000 per metre. Hence, it is a practicable solution only where the amenity value of the beach is very high. Artificially nourished beaches may also be susceptible to heavy storm damage.

Natural vegetation is resistant to salt spray and flooding, will grow in salt marshes and on windswept sand flats and acts to dissipate wave energy and trap sand. It can help to reverse the direction of feedback in the coastal erosion system, causing shorelines and barrier islands to build rather than erode. Where sand dunes have been removed to accommodate high-density building construction, it is generally not feasible to contemplate restoring them. Dune stabilization can be accomplished using artificial objects that trap sand, and then planting the dunes with grasses (these must continuously be replenished).

Although a good idea in theory, not all dune stabilization programmes have been beneficial. On the coast of North Carolina, for example, high dunes have been created with a straight, steep-sided profile. These have increased wave turbulence and therefore the stresses caused by storms, and the ensuing erosion has undercut many dunes. Where they have not been eroded away, these dunes have resulted in a drastic increase in the width of the beach behind the steep seaward face. This has changed the vegetation of the backshore area to species that are less tolerant of floods and burial by beach sand. When there are no floods, the vegetation grows to tall, dense thickets that require intensive management, including cutting and burning; but when the dunes are finally overtopped such growth may easily be killed off. Dune stabilization of this kind has destroyed the brackish habitat that many local plant and animal species, especially birds, require. Instead, colonization been carried out by coastal residents, whose domain is now at risk under the increased erosional regime that dune stabilization has created. Stable dunes do, however, minimize the erosive effects of high water, although they can be overtopped, breached or levelled by hurricane surges or the waves of major storms (Gares 1990).

Faced with the permanent disequilibrium of coastal erosion systems, the best course of action would often be to leave nature to restore the balance and not to intervene. This should involve leaving a wide buffer zone between the shoreline and urban development of the littoral zone. But given the intensity of human use of coastal zones, this is often not practicable. The most commonly used alternative is to build new and different structures, and although they may be successful in halting coastal retreat in the short term, in the long run they tend to complicate the problem. Private owners of beachfront properties tend not to have sufficient resources to invest in very substantial bulkheads, landfills, groynes, pilings or dune stabilization programmes. Inadequately designed structures break up easily under the duress of wave action and exacerbate the problems of coastal erosion even further. The best solution may be to adapt to shoreline erosion, rather than try to modify its consequences.

Non-structural methods

Erosion warnings can sometimes be issued in conjunction with storm-surge and high-tide advisories. But predicting water levels does not necessarily lead to accurate identification of the quantity of erosion to be expected during a given period. Warnings may be given usefully 6–12 hours before the onset of inundation, but potential victims of coastal erosion tend not to act on such advice, and indeed there is little that they could do in the short term except evacuate beachfront property. Moreover, sustained periods of onshore swell may be caused by the passage of a distant storm at sea, and erosion risks may be high although other forms of coastal hazard are absent and hence no warnings are issued. In general, although it is episodic, coastal erosion is too slow and inexorable a process for transient warnings to be particularly useful.

Relief and rehabilitation are seldom a prominent counter-measure against coastal erosion, and are likely to be given principally where the problem is closely linked to storm damage and flooding. In the United

States, for example, erosion *per se* is not considered worthy of a Presidential Disaster Declaration and hence does not usually stimulate relief disbursements. Coastal erosion is beyond the scope of many natural hazard insurance schemes, such as the US National Flood Insurance Act of 1968. But the succeeding Flood Protection Act of 1973–4 offers insurance protection against the damages caused when the undermining of structures by waves and currents exceeds expected seasonal levels. Hazard reduction measures must be applied at the community level in order to maintain eligibility.

Land-use management
Given the heavy cost of effective engineering works, and the expense of rehabilitation, the best strategy is to institute a coastal management policy that prohibits or removes from the shoreline development that would be threatened by erosion. Planning laws should establish conservation zones and management policies for eroding coasts. The type, location and intensity of development should be controlled. Severely eroding zones can be the subject of building moratoria, zoning laws and the establishment of setback lines, shoreward of which building is not permitted (Gares & Sherman 1985). Conservation districts and special tax zones can be declared, dune conservation and public open space ordinances enacted, sanitary codes enforced and sea-wall regulations imposed. Generally, however, the high value of waterfront property will prohibit on economic grounds any extensive programme of public acquisition of threatened property: seafront lots tend to be expensive, small in size and densely occupied. Moreover, private owners may be able effectively to resist legislation concerning acquisition, management or public access to their properties.

Local, regional or national programmes to safeguard eroding shorelines need to be securely based on accurate studies of the risk. They must also be integrated with complexes of other hazards, including coastal flooding, and high intensities of land usage. Finally, plans must be implemented and policed effectively in the face of likely demands for more liberal development of the coastline being managed.

Coastal hazards in the USA
About 25 per cent of the American coastline (32,800 km) is significantly affected by erosion. Critical effects and damage have occurred over 4,320 km, chiefly along the heavily populated Atlantic and Great Lakes coasts (US National Research Council 1990). The current average annual costs are $300 million, consisting of losses of both property and protection structures. Deaths caused by coastal erosion are extremely rare, but losses are rising with the increasing use and urbanization of vulnerable

coasts. The main threat is to communities of no more than 10,000 people and to rural and sparsely settled areas; but these may depend heavily on tourism, which may be threatened if erosional damage is heavy.

Excluding Alaska, private owners are responsible for about 69.5 per cent of the American shorefront, and they seem to prefer structural defences to non-structural methods (although they often lack the financial resources and technical expertise to put such methods into effect). The US Army Corps of Engineers has concluded that the 7 per cent of coastline which is presently undergoing erosion in the conterminous USA would benefit from more structural protection. But under current legislation the two thirds of coastline privately owned is not eligible for financial aid from the Federal Government. In line with their limited resources and jurisdictions, local communities tend to prefer cheaper methods, such as dune stabilization, rather than more expensive structural engineering programmes. However, even these require constant maintenance and protection against incursion by unscrupulous developers.

Land management schemes probably have a brighter future than structural measures at the American coastline (Ricketts 1986). The problem is perhaps most easily tackled in the case of the 11 per cent of critically eroding shorelines (465 km) that are Federally owned and the 18.5 per cent (770 km) owned by states and local authorities. A variety of planning laws has been enacted for the almost 70 per cent that is in private hands, and most of the measures listed above have been applied at some locality. For example, in 1973 Florida enacted a Beach and Shoreline Preservation Act, which established setback lines that prohibit construction or excavation within 15 m of high water marks (however, there has been some debate as to whether the setback should not be extended further). Connecticut, Massachusetts and North Carolina have regulations governing the construction of buildings and protection works at the shore, as well as beach and sand supply alteration, the preservation of coastal wetlands, and so on. Notwithstanding these measures, 11 per cent of US coastal communities have enacted no legislation whatsoever for erosion control.

Forest and range fires

The uses and significance of fire

The earliest archaeological evidence of the use of fire by *homo sapiens* comes from the Palaeolithic of Africa, some 60,000 years ago. Since then, mankind has regarded fire variously as a source of fascination, a tool, a weapon, a cultural device and a threat. It has acquired a diverse symbolism, including elements of chastisement, punishment, purification, keenness and inner reflection. Fire gave early man security at night against predators, while communication, social living and intellectual development probably began at the fireside.

From prehistory to modern times, fire has played a leading rôle in environmental change (Goudie 1983). It has been used to clear forest for agriculture or improve the quality of grazing land for domestic animals. It has served to attract game, deprive it of cover or drive it from cover during hunting, and also to kill or drive away predatory animals and parasites, such as ticks and mosquitoes. In times of conflict it has served as a weapon to repel attacks by enemies, or to drive them out of their refuges. In times of peace it has provided warmth and enabled food to be cooked, and has played a fundamental rôle in industry, including the making of tools, pottery and charcoal, and the smelting of ores. In areas of dense vegetation it has served to expedite travel; and, simplest of all, it has satisfied the pyromane's love of spectacle.

Fire has been widely utilized as an integral part of land cultivation systems, such as the **ladang** of Malaysia and Indonesia, and the **milpa** of the Latin American Maya. In these, the land was prepared by felling or deadening the forest. The debris was then left to dry during the hot season and was burnt before the first rain, after which holes for planting seeds and roots were dibbled among the ashes of the soft earth. Land that had recently been burnt in this way was rich in nutrients, but it tended to lose these after a long spell of cultivation, while land that was burnt too frequently became overgrown by perennial grasses which inhibited agriculture. Hence, although still practised widely in the tropics, the slash-and-burn technique is only adequate where population densities are low, and tends to cause environmental damage where the carrying capacity of the land is exceeded.

Causes of forest and range fires

Natural fires have been a relatively frequent and regular phenomenon in range, bush and forest lands. Dendrochronology has been applied to trees of great longevity which were not killed by repetitive fire, and its findings indicate that in the USA, prior to settlement forest fires occurred with a return period that varied from 7 to 80 years.

In certain ecosystems heavy accumulations of dead vegetation may become compact, rotted and fermented, thus generating heat and resulting occasionally in spontaneous combustion. But the two main causes of environmental fires are lightning and human activity (either deliberate or unintentional). Lightning strikes the surface of the Earth on average 100,000 times per day and is capable of igniting dry vegetative matter. But the importance of lightning as a cause of fire varies considerably. In the forest lands of the western USA it has been regarded as the cause of between 10 per cent and 50 per cent of fires; in the pine savannah of Belize over half the fires result from lightning strikes, but it causes only 8 per cent of bushfires in Australia, while nearly all fires in the maquis of southern France have anthropogenic causes.

The nature of environmental fires

The size, duration, intensity, temperature and frequency of environmental fires all vary. In 1963 in the Brazilian Paraná 20,000 km² of forest were consumed in only three weeks, but this may be an extreme case. Some fires are relatively quick and cool, only affecting ground-level vegetation. These are known as **surface fires**, and they burn in grass, low shrubs and plant debris such as fallen bark and leaves (Figs 4.26 & 27). They may travel at high speed but are relatively easy to control. **Crown fires** affect the whole forest up to crown level and generate very high temperatures. **Dependent crown fires** occur when the heat and flames

(a) Ground fire   (c) Dependent crown fire

(b) Surface fire   (d) Running crown fire

Figure 4.26 Types of environmental fire (modified after Butler 1976: 86–7).

Figure 4.27   Range fire in the Province of Sassari, Sardinia, in August 1988. Four people were encircled by the flames and died before they could be rescued by firefighters.

from surface fires ignite the crowns of trees. The fire in the crowns travels at the same speed as the surface fire. These tend to occur where trees are well-spaced (as in savannahs) and winds are low. **Running crown fires** occur when winds are hot and strong and the vegetation is very dry. They can be rapid, unpredictable and devastating, as the crown fire travels ahead of its accompanying surface fire. Strong convection currents may transport burning material ahead of the advancing flames, causing new outbreaks known as **spot fires**. Crown fires tend to develop where conditions allow the fire to be particularly hot when it starts. Lastly, **ground fires** occur in subsurface organic materials, such as peat in bogs and humus in forests. They spread slowly and unobtrusively and can kill the roots of plants and trees. The progress of a fire and its repetition by spreading are shown schematically in Figure 4.28.

   The direction of propagation of a fire, and its shape as it progresses, are influenced by the strength, direction and variability of the wind. The type of fuel also governs the speed of diffusion of the fire, as the inflammability of species varies. Topographical factors are important as well: for example, a fire that burns uphill may be stimulated by an increase in convection and radiation. Thus, grassland fires often spread in relation to wind direction and speed. But draughts and irregularities of

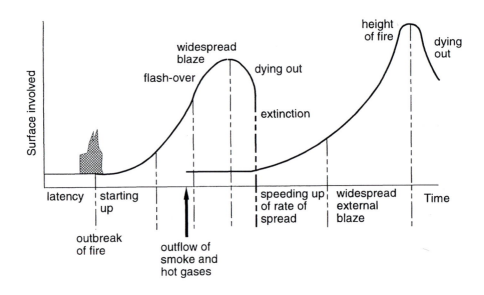

Figure 4.28 Schematic diagram of the progress of a fire (ICDO).

topography and fuel load tend to cause the spreading front to become uneven. Along with wind speed, the stature of grasses is the main determinant of the height of flames.

The maximum temperatures reached in the fire depend on the type of vegetation that is burning. Maquis (chaparral), consisting of small trees, woody grasses and oleose, xerophytic shrubs, tends to reach temperatures between 540°C and 1,100°C, depending on local circumstances. Savannah reaches a more restricted range of temperatures (between 540°C and 850°C); grassland may burn at up to 900°C and pine forests at 800°C. However, in comparison with forest conflagrations, temperatures tend to be low in grassland fires, especially if the flames move rapidly. But the so-called "backburns" generate greater heat and are thus more destructive than forward-moving grass fires.

The ecological effects of fire

It has been estimated that in the USA forest fires release 35 million tonnes of particulates into the atmosphere each year, about 15 per cent of which are less than 5 μm in diameter. Major fires can destroy nutrients and the structure of the soil, although in most cases burning forest and scrub vegetation releases as ash large amounts of potassium, magnesium, calcium and phosphorus, which are then taken up by the soil (although some nutrients are inevitably lost by volatilization). Toxic substances,

300

litter and competitive organisms may be removed by volatilization, while the decomposition of dead organic material is hastened. Fire also causes the pH of the soil to rise, such that upon burning acidic soils become alkaline. However, the loss of vegetation on slopes that have burnt may lead to landslides, especially as soil erosion is likely to be greater there. The water content or yield of such slopes may increase by up to ten times, given that moisture is no longer taken up by vegetation. Thus after a 1959 chaparral fire in Arizona runoff increased sixfold and sediment yield 270-fold. And the burning of 265 km^2 of forest in the Australian Alps in the early 1970s produced a fourfold increase in runoff and up to 100 times more suspended sediment. Months after the fire sediment loads during peak flows had increased to 115,000 tonnes per day from a fraction of that amount.

Grasslands are mainly composed of deciduous microphyte plants that grow rapidly and decompose slowly, readily producing the fuel for fires. They are mainly found on broad, level plains which enable surface fires to spread rapidly as humidity is likely to be low, winds relatively strong, insolation intense and precipitation slight. The fine leaves and erect stems of grasses ignite and burn with ease; as a result, fire is the main agent of decomposition and nutrient recycling in many grasslands. In fact, the ash from burning, and the bacteria and fungi that colonize the burnt area, provide mineral nutrients, while the dark surface may have a higher temperature, fewer frosts and thus better and more prolonged growing conditions. Therefore the soil is improved during the first few years after burning, which is when plants grow more vigorously and increase in size. This positive outcome is possible because grasses lose only one year's growth when burnt in a rangeland fire, because the growing point of dormant grasses, from which the following year's growth comes, is protected from the fire by being underground and because seeds are produced only one or two years after germination.

Some of the world's savannahs, grasslands and maquis have come into existence because of the impacts of fire on other types of vegetation, such as forest. But the effect on trees is not always negative: in species like the Jack pine, fire causes seeds to be released and may also assist seed germination. It alters seedbeds by removing humus and litter, and by supplying ash, which is beneficial to species like the Douglas fir and the sequoia. In other ways it stimulates vegetative reproduction in some species of woody and herbaceous plants, controls the population of forest insects, parasites and fungi and stimulates the flowering of many shrubs and herbs. But although many shrubs sprout vigorously after burning, repeated loss of their top layer (which is exposed to fire) keeps them small. Moreover, fires that are frequent enough to inhibit seed production in woody shrubs restrict their area of colonization.

Birds and mammals usually do not panic or show fear in the presence of fire; indeed, they may even be attracted to burning landscapes. Landscapes that are dependent on fire, or initiated or maintained by it, tend to show the greatest diversity of fauna. The picture is different, however, in managed landscapes. Fire in the National Parks of South Africa has stopped bush from encroaching on rangeland and has thus reduced its carrying capacity for browsing animals.

Mankind's efforts deliberately to suppress fire may lead to the accumulation of enormous quantities of inflammable materials, which will allow conflagrations of the crown type to develop. Moreover, fire suppression in forests has allowed them to be invaded by species that are intolerant of fire, and to develop a denser vegetation, which means that they burn more readily when fire eventually breaks out (McLean 1991). There are signs also that mosquito invasion may be related to the suppression of fire, if these insects can breed easily among the denser vegetation that results.

Countermeasures

The damage caused by fires tends to vary from year to year, but can be sufficient to make substantial investment in fire control measures worthwhile (Egging & Barney 1979). In Canada over the period 1968–77, for example, the amount of commercial timber burnt in forest fires equalled one quarter of the annual harvest. And in 1981 the burnt area actually exceeded the hectarage logged.

The extinction of an environmental fire depends upon removing one of the three conditions necessary for its existence: heat, oxygen and fuel. Heat can be lowered by pumping water onto the fuel, the supply of oxygen can be cut off by beating the flames (where this is practicable) or covering the fuel with earth, and fire-breaks can starve the fire of its fuel. Fire-fighting involves a combination of these methods. In small fires the advancing front can be attacked and halted, whereas in large fires it is usually more effective to work on the flanks and to create a fire-break to stop the advance of the front.

Lookout towers and aeroplane patrols can help detection, and aerial photography can assist with monitoring fires and areas that are susceptible to them. Strips of land devoid of combustible material can be bulldozed through forests. When fire eventually breaks out, water or chemicals can be sprayed on it to retard the flames, or water can be dropped from helicopters, which have skimmed it from lakes. Helicopters can also help lay hoses, and aeroplanes can drop parachutists in appropriate places. Indeed, a major fire may require more than 2,000 people to work at fighting it, and their safety must be ensured if there is any chance that the direction of travel of the fire may alter, for example if the wind direction changes. Spot fires pose a particular hazard to fire fighters, as wind and convective air currents may lob burning brands ahead of them and eventually trap them between two converging sources of fire. Vigilance and a good chain of command are probably the only remedies for this.

One way of ensuring that the beneficial effects of grassfires (such as the maintenance of nutrient supplies, biomass and grass species diversity) are not outweighed by the negative effects is to use controlled burning. This is best carried out in conditions of adequate soil and surface moisture, in order to ensure that the fire does the minimum amount of damage and does not get out of hand. Fire can be used in conjunction with rangeland management to ensure an adequate quality of pasture, but only if planned carrying capacities are based on minimum fodder availability and allowance is made for the temporary loss of forage.

Environmental fire in the south of France

In the south of France from June to August rainfall is slight or absent and maximum temperatures frequently exceed 30°C. Soil moisture reserves decline five-fold and the dry *mistral* wind that blows strongly and turbulently from the north and northwest is channelled down the Mediterranean valleys at very high speeds. Each year 2,000–3,000 fires affect 25,000–35,000 ha of forest (Wrathall 1985, Malanson & Trabaud 1988). About 1.25 per cent of woodland is lost overall, rising to 4 per cent of the most susceptible forests, which are inland from the Côte d'Azur and Provençal coast. Fires are most common in the three summer months and during the hottest part of the afternoon. Major burns occur on average once every six years, but individual stands of forest seem to regenerate the fuel that renders them susceptible to burning within 12–20 years. Despite the cyclical nature of events, there are marked variations from year to year in the impact of forest fires. In part this is because major fires, which account for 80 per cent of the surface burned, constitute only 2 per cent of events. The other 98 per cent are fires that consume only a few hectares.

The vegetation in Provence is largely secondary forest, maquis (chaparral) and garrigue scrub. The rise of tourism and decline of agriculture have resulted in a combination of abandonment, overexploitation and mistreatment of the land, which accounts at least in part for the current precariousness of the ecosystem. Human carelessness or wanton behaviour, which is responsible for starting many fires, is somewhat counterbalanced by the precautionary measures taken by forestry authorities. In one experimental fire control zone, firebreaks 200m wide divide the forests into 3,000–5,000 ha lots, which are in turn partitioned by tracks into 1,000 ha stands. Reservoirs, water tanks and helicopter landing pads are positioned at strategic intervals. Finally, wide-ranging proposals are being debated to restructure the forests with fire-resistant species, rehabilitate agriculture, control urbanization and create buffer zones of unforested land. But it remains to be seen whether the desire to conserve will prevail over the pressure to develop.

Australian bushfires

The causes of environmental fire in Australia include lightning, the carelessness of travellers and holiday-makers, and the effect of deliberate small-scale burning of the vegetation, which runs out of control. The necessary fuel is abundant in 40 per cent of the country, and seasonal differences in vegetation growth mean that fire risk varies from place to place during the year. The Australian southeast has one of the world's worst environmental fire risks, which results from its particular combination of climate and vegetation (Cheney 1979). Long, hot summers occur with periods in which hot, dry northerly winds blow from the interior of the continent. Winter rainfall is low and desiccation is complete during the summer. In the hills around Adelaide there are many eucalyptus trees, which are extremely fire prone as they have combustible barks and produce volatile oils and large quantities of litter. The use of eucalyptus groves as a screen for residential property, and the planting of stands of pine (which are also highly combustible, but lack the innate fire resistance of eucalyptus), have both augmented the damage.

Urban fires

Although not strictly a part of natural hazards studies, urban fires are considered here briefly because they relate to other sections of this book, for example, earthquakes, high-rise buildings and medical emergencies.

The phases in the development of an urban fire are **outbreak**, **development**, **spreading** and **extinction**. Besides cataclysmic natural events, urban fires are caused by electrical faults (in an estimated 35–40 per cent of cases), human carelessness or malevolence, the lack of supervision of a naked flame, or the uncontrolled product of a chemical or physical reaction. Once the flame is ignited it will either go out by itself or develop in its initial surroundings at a variable speed. The materials encountered are likely to vary in terms of their abundance, flammability and calorific potential. As it develops, the fire will produce heat, and hot gases which are potentially toxic or chemically aggressive.

The spread of fire in urban areas depends on the structure and materials of buildings, their volumes and partitions or spacing. Doors, windows, façades, ducts and holes may provide shafts that act as vectors for the spread of the fire. The risk to people, rather than property, depends on their location and concentration, their perception of the hazard, and the potential for rapid evacuation via safe routes. Particular problems are likely to result from the rapid destruction early in the fire of "nerve centres", such as communications headquarters and the nodes of warning systems.

The risk can be reduced greatly by structural and non-structural measures. Extinguishers, sprinklers, hoses, hydrants, evacuation routes and fire sensors are all well-known structural approaches. Evacuation drills designed for cases of fire hazard can be combined with those created for natural hazards such as earthquakes. Fire hazards can be investigated in terms of all phases, producing information on where and when fires are likely to break out, how they are likely to develop and spread and with what degree of rapidity this is likely to occur. Fire-fighting plans must be based on this sort of knowledge and prediction.

References

Aberfan Tribunal 1967. *Report of the tribunal appointed to inquire into the disaster at Aberfan on October 21st, 1966.* London: HMSO.

Ahmad, Y. J. & M. Kassas 1987. *Desertification: financial support for the biosphere.* West Hartford, Connecticut: Kumarian Press.

Alexander, D. E. 1989. Urban landslides. *Progress in Physical Geography* **13**, 157–91.

Alexander, E. B. 1988. Strategies for determining soil-loss tolerance. *Environmental Management* **12**, 791–6.

Allen, A. S. 1978. Basic questions concerning coal mine subsidence in the United States. *Bulletin of the Association of Engineering Geologists* **15**, 147–62.

Ansal, A. M. & A. Erken 1989. Undrained behaviour of clay under cyclic shear stresses. *Journal of Geotechnical Engineering* **115**, 968–83.

Baldwin, J. E., II, H. F. Donley, T. R. Howard 1987. On debris flow/avalanche mitigation and control, San Francisco Bay area, California. In *Debris flows/avalanches: process, recognition and mitigation*, J. E. Costa & G. F. Wieczorek (eds), 223–36. Boulder, Colorado: Geological Society of America.

Bishop, A. W. 1955. The use of the slip circle in the stability analysis of slopes. *Geotechnique* **5**, 7–17.

Biswas, M. R. & A. K. Biswas (eds) 1980. *Proceedings of the United Nations Conference on Desertification.* New York: Pergamon.

Blong, R. J. 1973. A numerical classification of selected landslides of the debris slide-avalanche-flow type. *Engineering Geology* **7**, 99–114.

Borchert, J. R. 1971. The Dust Bowl in the 1970s. *Annals of the Association of American Geographers* **61**, 1–22.

Briggs, R. P. 1974. Map of overdip slopes that can affect landsliding in Allegheny County, Pennsylvania. *US Geological Survey Miscellaneous Field Studies Map* MF543.

Browning, J. M. 1973. Catastrophic rock slide, Mount Huascarán, north-central Perú, May 31, 1970. *Bulletin of the American Association of Petroleum Geologists* **57**, 1,335–41.

Bryan, R. B. 1968. The development, use and efficiency of indices of soil erodibility. *Geoderma* **2**, 5–25.

Butler, J. E. 1976. *Natural Disasters.* Richmond, Australia: Heinemann Educational.

Campbell, R. H. 1975. Soil slips, debris flows, and rainstorms in the Santa Monica Mountains and vicinity, southern California. *US Geological Survey Professional Paper* **851**, 1–51.

Cannon, S. H. & S. Ellen 1985. Rainfall conditions for abundant debris avalanches, San Francisco Bay region, California. *California Geology* **38**, 267–72.

Carson, M. A. & M. J. Kirkby 1972. *Hillslope form and process*. Cambridge: Cambridge University Press, Chs 6 & 7.

Carter, L. J. 1977. Soil erosion: the problem persists, despite the billions spent on it. *Science* **196**, 409–11.

Castro, G. & S. J. Poulos 1977. Liquefaction and cyclic mobility factors. *Proceedings of the American Society of Civil Engineers, Journal of the Geotechnical Engineering Division* **103**(GT6), 501–16.

Cheney, N. P. 1979. Bushfire disasters in Australia, 1945–1975. In *Natural hazards in Australia*, R. L. Heathcote & B. G. Thom (eds), 88–90. Canberra, ACT: Australian Academy of Sciences.

Chepil, W. S. & N. P. Woodruffe 1963. The physics of wind erosion and its control. *Advances in Agronomy* **15**, 211–302.

Chowdhury, R. N. 1978. *Slope analysis*. Amsterdam: Elsevier.

Cloudsley-Thompson, J. L. 1978. Human activities and desert expansion (the Sahel Symposium). *Geographical Journal* **144**, 416–23.

Cooke, R. U. 1984. *Geomorphological hazards in Los Angeles*. London: Allen & Unwin.

Cooke, R. U. & J. C. Doornkamp 1974. *Geomorphology in environmental management*. Oxford: Clarendon Press.

Cooke, R. U. & R. W. Reeves 1975. *Arroyos and environmental change in the American southwest*. Oxford: Oxford University Press.

Costa, J. E. & V. R. Baker 1981. *Surficial geology: building with the earth*. New York: John Wiley.

Cotton, W. R. & D. A. Cochrane 1982. Love Creek landslide disaster, January 5, 1982, Santa Cruz County. *California Geology* **35**, 153–7.

Crozier, M. J. 1986. *Landslides: causes, consequences and environment*. London: Croom Helm.

Dregne, H. E. 1983. *Desertification of arid lands*. New York: Harwood.

Duda, A. M. 1985. Environmental and economic damage caused by sediment from agricultural non-point sources. *Water Resources Bulletin* **21**, 225–34.

Dudley, J. H. 1970. Review of collapsing soils. *Proceedings of the American Society of Civil Engineers, Journal of the Soil Mechanics and Foundations Division* **96**(SM3), 925–47.

Egging, L. T. & R. J. Barney 1979. Fire management: a component of land management planning. *Environmental Management* **3**, 15–20.

El Hinnawi, E. & M. H. Hashmi 1987. *The state of the environment*. Oxford: Pergamon.

Ellen, S. D. & G. F. Wieczorek (eds) 1988. *Landslides, floods and marine effects of the storm of January 30, 1982, in the San Francisco Bay region, California*. US Geological Survey Professional Paper 1434, 1–310.

Erismann, T. H. 1979. Mechanisms of large landslides. *Rock Mechanics* **12**, 15–46.

Evans, R. S. 1981. An analysis of secondary toppling rock failures: the stress redistribution method. *Quarterly Journal of Engineering Geology* **14**, 77–86.

Falck, L. R. 1991. Disaster insurance in New Zealand. In *Managing natural disasters and the environment*. A. Kreimer & M. Munasinghe (eds), 120–5. Washington, DC: Environment Department, World Bank.

Fellenius, W. 1936. Calculation of stability of earth dams. *Transactions of the Second Congress on Large Dams* **4**, 445.

References

Ferrians, O. J., R. Kachadoorian, G. W. Greene 1969. *Permafrost and related engineering problems in Alaska.* US Geological Survey Professional Paper 678, 1–37.

Fleming, R. W. & F. A. Taylor 1980. Estimating the cost of landslide damage in the United States. *US Geological Survey Circular* 832, 1–21.

Fontes, J. C. & G. Bortolami 1973. Subsidence of Venice during the last 40,000 years. *Nature* **244**, 339–1.

Fournier, F. 1960. *Climat et erosion.* Paris: Presses Univèrsitaires de France.

French, H. M. 1979. Permafrost and ground ice. In *Man and environmental processes*, K. J. Gregory & D. E. Walling (eds), 144–62. Boulder, Colorado: Westview Press.

Fulton-Bennett, K. & G. B. Griggs 1987. *Coastal protection structures and their effectiveness.* Sacramento, California: California Department of Boating and Waterways, and Marine Science Institute, University of California.

Furbish, D. J. & R. M. Rice 1983. Predicting landslides related to clearcut logging, northwestern California, USA. *Mountain Research and Development* **3**, 253–9.

Gambolati, G., P. Gatto, R. A. Freeze 1974. Predictive simulation of the subsidence of Venice. *Science* **183**, 849–51.

Gares, P. A. 1990. Predicting flooding probability for beach/dunes systems. *Environmental Management* **14**, 115–23.

Gares, P. A. & D. J. Sherman 1985. Protecting an eroding shoreline: the evolution of management response. *Applied Geography* **5**, 55–69.

Goldman, S. J., 1986. *Erosion and sediment control handbook.* New York: McGraw-Hill.

Goudie, A. S. 1983. *The human impact: man's rôle in environmental change.* Oxford: Basil Blackwell.

Gromko, G. J. 1974. Review of expansive soils. *Proceedings of the American Society of Civil Engineers, Journal of the Hydraulics Division* **100**(GT6), 667–87.

Guy, H. P. 1964. *An analysis of some storm-period variables affecting stream sediment transport.* US Geological Survey Professional Paper 462E, 1–46.

Harp, E. L., R. C. Wilson, G. F. Wieczorek 1981. Landslides from the February 4, 1976, Guatemala earthquake. *US Geological Survey Professional Paper* 1204A, 1–35.

Hays, W. W. (ed.) 1981. Facing geologic and hydrologic hazards: earth science perspectives. *US Geological Survey Professional Paper* 1240B, 1–108.

Heim, A. 1932. *Bergsturz und Menschenleben.* Zürich: Vierteljahrschreibe und Naturforschung Geselschaft.

Hewitt, K. 1988. Catastrophic landslide deposits in the Karakoram Himalaya. *Science* **242**, 64–7.

Howard, K. A. 1973. Avalanche mode of motion: implications from lunar examples. *Science* **180**, 1,052–5.

Hsü, K. J. 1975. Catastrophic debris streams (sturzstroms) generated by rockfalls. *Bulletin of the Geological Society of America* **86**, 129–40.

Hurni, H. 1988. Degradation and conservation of soil resources in the Ethiopian highlands. *Mountain Research and Development* **8**, 123–30.

ICDO 1986. *International Civil Defence* **374/5** (newsletter). Geneva: International Civil Defence Organization.

Janbu, N. 1977. State-of-the-art report on slopes and excavations in normally and lightly overconsolidated clays. *Proceedings of the Ninth International Conference on Soil Mechanics and Foundation Engineering, Tokyo*, **2**, 549–66.

Judson, S. 1968. Erosion of the land, or what's happening to our continents? *American Scientist* **56**, 356–74.

Keefer, D. K. 1984. Landslides caused by earthquakes. *Bulletin of the Geological Society of America* **95**, 406–21.

Keefer, D. K., R. C. Wilson, R. K. Mark et al. 1987. Real-time landslide warning during heavy rainfall. *Science* **238**, 921–5.

Kent, P. E. 1966. The transport mechanism in catastrophic rock falls. *Journal of Geology* **74**, 79–83.

Kenny, M. 1978. Sea wall attack. *Oceans* **11**, 30–5.

Kerr, P. F. 1963. Quick clay. *Scientific American* **209**, 132–42.

Kochelman, W. J. 1986. Some techniques for reducing landslide hazards. *Bulletin of the Association of Engineering Geologists* **23**, 29–52.

Kurdin, R. D. 1973. Classification of mudflows. *Soviet Hydrology* **12**, 310–16.

Langbein, W. B. & S. A. Schumm 1958. Yield of sediment in relation to mean annual precipitation. *Transactions of the American Geophysical Union* **39**, 1,076–84.

Leighton, F. B. 1976. Urban landslides: targets for land-use planning in California. In *Urban geomorphology*. D. R. Coates (ed.), 37–60. Boulder, Colorado: Geological Society of America.

Lucchitta, B. K. 1978. A large landslide on Mars. *Bulletin of the Geological Society of America* **89**, 1,601–9.

Mace, R. 1991. Overgrazing overstated. *Nature* **349**, 280–1.

Malanson, G. P. & L. Trabaud 1988. Computer simulations of fire behaviour in garrigue in southern France. *Applied Geography* **8**, 53–64.

Marsden, S. S. & S. N. Davis 1967. Geological subsidence. *Scientific American* **216**, 93–100.

Mielenz, R. C. & M. E. King 1955. Physical-chemical properties and engineering performance of clays. In *Clays and clay technology*, J. A. Pasli & M. D. Turner (eds), 196–254. Sacramento, California: Division of Mines and Geology.

Mitchell, J. K. 1974. Natural hazards research. In *Perspectives on environment*, I. Manners & M. Mikesell (eds), 311–41. Washington, DC: Association of American Geographers.

Mitchell, J. K. 1986. Practical problems from surprising soil behaviour. *Journal of Geotechnical Engineering* **112**, 255–89.

Mitchell, J. K. & G. D. Bubenzer 1980. Soil loss estimation. In *Soil erosion*, M. J. Kirkby & R. P. C. Morgan (eds), 17–62. Chichester, England: John Wiley.

Morgan, R. P. C. 1986. *Soil erosion and conservation*. Harlow, England: Longman.

Nash, D. 1987. A comparative review of limit equilibrium models of stability analysis. In *Slope stability*, M. G. Anderson & K. S. Richards (eds), 11–75. New York: John Wiley.

Nemcok, A., J. Pasek, J. Rybár 1972. Classification of landslides and other mass movements. *Rock Mechanics* **4**, 71–8.

Nilsen, T. H. et al. 1979. *Relative slope stability and land-use planning in the San Francisco Bay region, California*. US Geological Survey Professional Paper 944.

Nnoli, O. 1990. Desertification, refugees and regional conflict in West Africa. *Disasters* **14**, 132–9.

Obermeier, S. F. 1988. Liquefaction potential in the central Mississippi Valley. *US Geological Survey Bulletin* **1832**, 1–21.

Poland, J. F. & G. H. Davis 1969. Land subsidence due to withdrawl of fluids. In *Reviews in engineering geology*, Vol. 2, D. J. Varnes & G. Kiersch (eds), 187–269. Boulder, Colorado: Geological Society of America.

References

Ricketts, P. J. 1986. National policy and management responses to the hazard of coastal erosion in Britain and the United States. *Applied Geography* **6**, 197–221.

Schuster, R. L. & R. J. Krizek (eds) 1978. *Landslides: analysis and control.* Washington, DC: Transportation Research Board.

Seed, H. B. 1968. Landslides during earthquakes due to soil liquefaction. *Proceedings of the American Society of Civil Engineers, Journal of the Soil Mechanics and Foundations Division* **94**(SM5), 1014–122.

Seed, H. B. & S. D. Wilson 1967. The Turnagain Heights landslide, Anchorage. *Proceedings of the American Society of Civil Engineers, Journal of ths Soil Mechanics and Foundations Division* **93**(SM4), 325–53.

Selby, M. J. 1976. Slope erosion due to extreme rainfall: a case study from New Zealand. *Geografiska Annaler* **58A**, 131–8.

Sharp, R. P. & L. H. Noble 1953. Mudflow of 1941 at Wrightwood, California. *Bulletin of the Geological Society of America* **64**, 547–60.

Shreve, R. L. 1968. *The Blackhawk landslide.* Boulder, Colorado: Geological Society of America.

Skempton, A. W. & J. N. Hutchinson 1964. Stability of natural slopes and embankment foundations: state-of-the-art report. *Seventh International Conference on Soil Mechanics and Foundation Engineering, Mexico,* 291–335.

Smith, R. M. & W. L. Stanley 1965. Determining the range of tolerable erosion. *Science* **100**, 414–24.

Sorensen, J. H. & J. K. Mitchell 1975. *Coastal erosion hazard in the United States: a research assessment.* Boulder, Colorado: Institute of Behavioural Sciences, University of Colorado.

Steiner, F. R. 1990. *Soil conservation in the United States: policy and planning.* Baltimore: The Johns Hopkins University Press.

Stevenson, P. C. 1977. An empirical method for the evaluation of relative landslip risk. *Bulletin of the International Association for Engineering Geology* **16**, 69–72.

Stocking, M. A. & H. A. Elwell 1976. Rainfall erosivity over Rhodesia. *Transactions of the Institute of British Geographers, New Series* **1**, 231–45.

Tavernas, F., J.-Y. Chagnon, P. LaRochelle 1971. The Saint Jean-Viannay landslide: observation and eye-witness accounts. *Canadian Geotechnical Journal* **8**, 463–78.

Terzaghi, K. 1936. The shearing resistance of saturated soils. *Proceedings of the First International Conference on Soil Mechanics and Foundation Engineering* **1**, 54–6.

Terzaghi, K. 1950. Mechanism of landslides. *Bulletin of the Geological Society of America, Berkeley Volume,* 83–122.

Terzaghi, K. & R. B. Peck 1967. *Soil mechanics in engineering practice.* New York: John Wiley.

Tornberg, G. F. 1968. Sand bypassing systems. *Shore and Beach* **36**, 27–33.

US National Academy of Sciences 1973. *The Great Alaska earthquake of 1964* (8 vols). Washington, DC: National Academy Press.

US National Research Council 1985. *Liquefaction of soils during earthquakes.* Washington, DC: National Academy Press.

US National Research Council 1990. *Managing coastal erosion.* Washington, DC: National Academy Press.

Varnes, D. J. 1978. Slope movement types and processes. In *Landslides: analysis and control,* R. L. Schuster & R. J. Krizek (eds), 11–33. Washington, DC: Transportation Research Board.

Varnes, D. J. 1984. *Landslide hazard zonation: review of principles and practice.* Paris: UNESCO Press.

Warren, A. & J. Maizels 1977. Ecological change and desertification. In *Desertification: its causes and consequences*, UN Conference on Desertification (ed.), 171-260. Oxford: Pergamon.

White, G. F. & J. E. Haas 1975. *Assessment of research on natural hazards.* Cambridge, Mass.: MIT Press.

Whittow, J. 1980. *Disasters: the anatomy of environmental hazards.* Harmondsworth, England: Penguin.

Wischmeier, W. H. 1976. The use and misuse of the Universal Soil Loss Equation. *Journal of Soil and Water Conservation* **31**, 5–9.

Wischmeier, W. H. & D. Smith 1978. *Predicting rainfall erosion losses: a guide to conservation planning.* Washington, DC: US Department of Agriculture.

Wolman, M. G. 1967. A cycle of sedimentation and erosion in urban river channels. *Geografiska Annaler* **49A**, 385–95.

Wolman, M. G. & A. P. Schick 1967. Effects of construction on fluvial sediment, urban and suburban areas of Maryland. *Water Resources Research* **3**, 451–64.

Wrathall, J. E. 1985. The hazard of forest fires in southern France. *Disasters* **9**, 104–14.

Select bibliography

Albini, F. A. 1984. Wildland fires. *American Scientist* **72**, 590–7.

Ambraseys, N. N. & S. Sarma 1969. Liquefaction of soils induced by earthquakes. *Bulletin of the Seismological Society of America* **59**, 651–64.

American Society of Agronomy 1982. *Soil erosion and conservation in the tropics.* Madison, Wisconsin: American Society of Agronomy.

Anderson, M. G. & K. S. Richards (eds) 1987. *Slope stability: geotechnical engineering and geomorphology.* New York: John Wiley.

Anderson, M. G., K. S. Richards & P. E. Kneale 1980. The rôle of stability analysis in the interpretation of the evolution of threshold slopes. *Transactions of the Institute of British Geographers, New Series* **5**, 100–12.

Batie, S. S. 1983. *Soil erosion.* Washington, DC: The Conservation Foundation.

Beasley, R. P., J. M. Gregory, T. R. McCarty 1984. *Erosion and sediment pollution control.* Ames, Iowa: Iowa State University Press.

Beck, B. F. (ed.) 1984. *Sinkholes: their geology, engineering and environmental impact.* Rotterdam: A. A. Balkema.

Benbrook, C. M. 1988. First principles: the definition of highly erodible land and tolerable soil loss. *Journal of Soil and Water Conservation* **43**, 35–8.

Bishop, A.W. 1973. The stability of tips and spoil heaps. *Quarterly Journal of Engineering Geology* **6**, 335–77.

Biswas, M. R. & A. K. Biswas 1978. Loss of productive soil. *International Journal of Development Studies* **12**, 189–97.

Blaikie, P. 1985. *The political economy of soil erosion in developing countries.* Harlow, England: Longman.

Blair, M. et al. 1984. *When the ground fails: planning and engineering response to debris flows.* Boulder, Colorado: Natural Hazards Research and Applications Information Center.

Boardman, J., I. D. L. Foster, J. A. Dearing (eds) 1990. *Soil erosion on agricultural land.* New York: John Wiley.

Borunov, A. K., A. V. Koshariov, V.V. Kandelaki 1991. Geoecological consequences of the 1988 Spitak earthquake (Armenia). *Mountain Research and Development* **11**, 19–35.

Brabb, E. E. 1991. The world landslide problem. *Episodes* **14**, 52–61.

Bremer, H. & K. M. Clayton (eds) 1989. Coasts: erosion and sedimentation. *Zeitschrift für Geomorphologie, Supplementband* **73**.

Britton, N. R. 1986. An appraisal of Australia's disaster management system following the Ash Wednesday bushfires in Victoria, 1983. *Australian Journal of Public Administration* **45**, 112–27.

Bromhead, E. N. (ed.) 1986. *The stability of slopes*. London: Chapman & Hall.

Brown, A. A. and K. P. Davis 1973. *Forest fire control and use*. New York: McGraw-Hill.

Brunsden, D. & D. B. Prior (eds) 1984. *Slope instability*. Chichester, England: John Wiley.

Bruun, P. 1968. Beach erosion and coastal protection. In *Encyclopaedia of geomorphology*, R. W. Fairbridge (ed.), 68–70. New York: Van Nostrand Reinhold.

Bull, W. B. 1974. Geologic factors affecting compaction of deposits in a land subsidence area. *Bulletin of the Geological Society of America* **84**, 3,783–802.

Carrara, A. 1991. GIS techniques and statistical models in evaluating landslide hazard. *Earth Surface Processes and Landforms* **16**, 427–45.

Christensen, N. L. 1989. Interpreting the Yellowstone fires of 1988. *BioScience* **39**, 678–85.

Costa, J. E., G. F. Wieczorek (eds) 1987. *Debris flows/avalanches: process, recognition and mitigation* (Reviews in Engineering Geology, Vol. 7). Boulder, Colorado: Geological Society of America.

Dolan, R., B. Hayden, J. Heywood 1979. Analysis of coastal erosion and storm surge hazard. *Coastal Engineering and Administration* **2**, 41–53.

Dregne, H.E. 1978. Desertification: man's abuse of the land. *Journal of Soil and Water Conservation* **33**, 11–14.

Dudal, R. 1982. Land degradation in a world perspective. *Journal of Soil and Water Conservation* **37**, 245–6.

Dudley, J. H. 1970. Review of collapsing soils. *Proceedings of the American Society of Civil Engineers, Journal of the Soil Mechanics and Foundations Division* **96**(SM3), 925–47.

El-Ashry, M. T. 1971. Causes of increased erosion along United States shorelines. *Bulletin of the Geological Society of America* **82**, 2,033–8.

El-Swaify, S. A., W. C. Maldenhauer, A. Lo (eds) 1985. *Soil erosion and conservation*. Ankeny, Iowa: Soil Conservation Society of America.

Fowler, L. C. 1981. Economic consequences of land surface subsidence. *Proceedings of the American Society of Civil Engineers, Journal of the Irrigation and Drainage Division* **107**(IR2), 151–9.

Gambolati, G., G. Ricceri, W. Bertoni, G. Brighenti, E. Vuillermin 1991. Mathematical simulation of the subsidence of Ravenna. *Water Resources Research* **27**, 2,899–918.

Garwood, N. C., D. P. Janos, N. Brokhaw 1979. Earthquake-induced landslides: a major disturbance to tropical forests. *Science* **205**, 997–9.

Geli, L., P.-Y. Bard, B. Jullien 1988. The effect of topography on earthquake ground motion: a review and new results. *Bulletin of the Seismological Society of America* **78**, 42–63.

Glantz, M. H. (ed.) 1977. *Desertification: environmental degradation in and around arid lands*. Boulder, Colorado: Westview Press.

Goudie, A. S. (ed.) 1990. *Techniques for desert reclamation.* Chichester, England: John Wiley.

Grainger, A. 1990. *The threatening desert: controlling desertification.* London & Washington, DC: Earthscan.

Hallsworth, E. G. 1987. *The anatomy, physiology and psychology of erosion.* New York: John Wiley.

Healey, D., F. Jarrett, J. M. McKay (eds) 1985. *The economics of bushfires: the South Australian experience.* Melbourne: Oxford University Press.

Helms, D. & S. L. Flader (eds) 1985. The history of soil and water conservation: a symposium. *Agricultural History* **59**, 103–341.

Holzer, T. L. (ed.) 1984. *Man-induced land subsidence.* Boulder, Colorado: Geological Society of America.

Holzer, T. L. 1986. Land subsidence. *Earthquakes and Volcanoes* **18**, 131–7.

Holzer, T. L. 1989. State and local response to damaging land subsidence in United States urban areas. *Engineering Geology* **27**, 449–66.

Innes, J. L. 1983. Debris flows. *Progress in Physical Geography* **7**, 469–501.

Jackson, R. D. & S. B. Idso 1975. Surface albedo and desertification. *Science* **189**, 1012–13.

Johnson, A. I., L. Carbognin, L. Ubertini (eds) 1986. *Land subsidence.* Washington, DC: International Association of Hydrological Sciences.

Johnson, L. C. 1987. Soil loss tolerance: fact or myth? *Journal of Soil and Water Conservation* **42**, 155–60.

Keefer, D. K. 1984. Rock avalanches caused by earthquakes: source characteristics. *Science* **223**, 1288–9.

Keefer, D. K. & A. M. Johnson 1983. Earthflows: morphology, mobilization and movement. *US Geological Survey Professional Paper* 1,264.

Kemmerly, P. 1981. The need for recognition and implementation of a sinkhole-floodplain hazard designation in urban karst terrains. *Environmental Geology* **3**, 281–92.

Kerr, P. F. 1979. Quick clays and other slide-forming clays. *Engineering Geology* **14**, 173–82.

Kirkby, M. J. & R. P. C. Morgan (eds) 1980. *Soil erosion.* New York: John Wiley.

Knight, D. H. & L. L. Wallace 1989. The Yellowtone fires: issues in landscape ecology. *BioScience* **39**, 700–6.

Komar, P. D. (ed.) 1983. *Handbook of coastal processes and erosion.* Boca Raton, Florida: CRC Press.

Lal, R. (ed.) 1988. *Soil erosion research methods.* Ankeny, Iowa: Soil and Water Conservation Society.

Larson, W. E., F. J. Pierce, R. H. Dowdy 1983. The threat of soil erosion to long-term crop production. *Science* **219**, 458–65.

Leschchinsky, D. 1990. Slope stability analysis: generalized approach. *Journal of Geotechnical Engineering* **116**, 851–67.

Lyles, I. 1975. Possible effects of wind erosion on soil productivity. *Journal of Soil and Water Conservation* **30**, 279–83.

Mabbutt, J. A. (ed.) 1980. *Case studies on desertification.* Paris: UNEP, UNESCO Press.

Mather, J. R., R. T. Field, G. Yoshioka 1967. Storm damage hazard along the eastern coast of the United States. *Journal of Applied Meteorology* **6**, 20–30.

McLean, H. E. 1991. The changing threat of wildfire. *American Forests* **97**, 50–4.

Middleton, N. J. 1992. *World atlas of desertification*, D. S. G. Thomas (ed.). Sevenoaks, England: Edward Arnold.

Mitchell, J. K. 1975. *Fundamentals of soil behavior*. New York: John Wiley.

Nilsen, T. H. & E. E. Brabb 1975. Landslides. In *Studies for seismic zonation of the San Francisco Bay region*, R. D. Borcherdt (ed.), 75–87. US Geological Survey Professional Paper 941A.

Olshansky, R. B. 1990. *Landslide hazard in the United States: case studies in planning and policy development*. New York: Garland.

Orme, A. R. et al. 1980. Coasts under stress. *Zeitschrift für Geomorphologie, Supplementband* **34**.

Otterman, J. 1974. Baring high-albedo soils by overgrazing: a hypothesized descrtification mechanism. *Science* **186**, 531–3 [discussion, 189, 1,013–15].

Paylore, P. & R. Haney (eds) 1976. *Desertification: process, problems, perspectives*. Tucson, Arizona: University of Arizona Press.

Peacock, W. H. & H. B. Seed 1968. Sand liquefaction under cyclic loading simple shear conditions. *Proceedings of the American Society of Civil Engineers, Journal of the Soil Mechanics and Foundations Division* **94**(SM3), 689–708.

Peck, R. B. 1967. Stability of natural slopes. *Proceedings of the American Society of Civil Engineers, Journal of the Soil Mechanics and Foundations Division* **93**(SM4), 403–17.

Phillips, J. D. 1989. Predicting minimum achievable soil loss in developing countries. *Applied Geography* **9**, 219–36.

Pimental, D., J. Allen, A. Beers et al. 1987. World agriculture and soil erosion. *BioScience* **37**, 277–83.

Poland, J. F. 1986. *Guidebook to studies of land subsidence due to groundwater withdrawal*. Paris: UNESCO Press.

Popescu, M. E. 1986. A comparison between the behaviour of swelling and of collapsing soils. *Engineering Geology* **23**, 145–64.

Porter, S. C. & G. Orombelli 1981. Alpine rockfall hazards. *American Scientist* **69**, 67–75.

Reining, P. (ed.) 1978. *Desertification papers*. Washington, DC: American Association for the Advancement of Science.

Robbins, J. R. 1990. Burning issue: the politicization of a bushfire. *International Journal of Mass Emergencies and Disasters* **8**, 325–40.

Rollins, K. M. & H. B. Seed 1990. Influence of buildings on potential liquefaction damage. *Journal of Geotechnical Engineering* **116**, 165–85.

Rosenbaum, J. G. 1981. Shoreline structures as a cause of shoreline erosion. In *Environmental geology*, 3rd edn, R. W. Tank (ed.), 198–210. New York: Oxford University Press.

Royster, D. L. 1979. Landslide remedial measures. *Bulletin of the International Association of Engineering Geologists* **16**, 301–52.

Saxena, S. K. (ed.) 1979. *Evaluation and prediction of subsidence*. New York: American Society of Civil Engineers.

Schechter, J. 1977. Desertification processes and the search for solutions. *Interdisciplinary Science Reviews* **2**, 36–54.

Schuster, R. L. & R. W. Fleming 1986. Economic losses and fatalities due to landslides. *Bulletin of the Association of Engineering Geologists* **23**, 11–28.

Seed, H. B. 1967. Slope stability during earthquakes. *Proceedings of the American Society of Civil Engineers, Journal of the Soil Mechanics and Foundations Division* **93**(SM5), 1,055–122.

Spooner, B. & H. S. Mann (eds) 1982. *Desertification and development*. London: Academic Press.

United Nations 1977. *Desertification: its causes and consequences*. Oxford: Pergamon.

Van Nao, T. (ed.) 1982. *Forest fire prevention and control*. The Hague: Martinus Nijhoff.

Veder, C. 1981. *Landslides and their stabilization*. New York: Springer.

Vogl, R. J. 1979. Some basic principles of grassland fire management. *Environmental Management* **3**, 51–7.

Vogt, G., 1990. *Forests on Fire*. New York: Impaci.

Voight, B. (ed.) 1978. *Rockslides and avalanches* [2 vols]. Amsterdam: Elsevier.

Walls, J. 1980. *Land, man and sand: desertification and its solution*. New York: Macmillan.

Wood, W. L. 1991. Managing coastal erosion. *American Scientist* **79**.

Yool, S. R., D. W. Eckhardt, J. E. Estes, M. J. Cosentino 1985. Describing the brushfire hazard in southern California. *Annals of the Association of American Geographers* **75**, 417–30.

Youd, T. L. 1973. *Liquefaction, flow and associated ground failure*. US Geological Survey Circular 688.

Záruba, Q. & V. Mencl 1982. *Landslides and their control*, 2nd edn. Amsterdam: Elsevier.

The human impact and response

CHAPTER FIVE

Damage and
the built environment

The damage caused by natural disasters can be considered in various ways. In terms of how a building performs under stress, the hazard that has received the most attention is that posed by earthquakes, which are therefore considered in this chapter once again to provide a sort of prototype for the study of other disasters. Standard means of assessing the costs and damages caused by disaster are described, as are questions of how public safety can be assured in conditions of prolonged risk. Finally, three particular sources of vulnerability to natural catastrophe are assessed. As the greatest concentrations of buildings, infrastructure and population occur in major urban areas, cities usually manifest distinctive impacts when affected by, for instance, floods, landslides or earthquakes. With respect to individual structure, tall buildings and dams provide examples of how advances in engineering design technology and construction methods have created very distinctive kinds of vulnerability: to hurricanes and earthquakes in the case of high-rise buildings, and to floods, earthquakes, landslides and sedimentation in the case of dams.

Earthquake damage

The nature of earthquake damage to the built environment
Construction failure is the principal threat to life and limb during an earthquake (Page et al. 1975). Seismic damage to buildings often seems inexplicable: the visual impression may be one of total destruction with no apparent pattern. However, damage is not random, although it is likely to be complex and perhaps poorly understood.

Seismic damage to buildings is subject to four main influences, the first of which is the seismic energy of the earthquake, as represented by the magnitude, duration and acceleration of strong motions of the ground. The second factor is distance from the epicentre, including the response of surface rock or sediment along the path of seismic waves that reach any particular point. Thirdly, surface geology is important in terms of how variations in soil, rock, topography, faults and folds affect the transmission of seismic waves. Finally, damage varies in relation to construction type, with special reference to the regularity of form and state of maintenance of a building. With these factors in mind, it would be helpful to know the answers to the following questions:

(a) How do buildings of particular designs, sizes and materials perform during seismic shaking?
(b) How should damaged buildings be buttressed to ensure their temporary safety pending reconstruction?
(c) How can damaged buildings be repaired permanently and anti-seismically?
(d) How can new buildings be designed in order to minimize structural collapse in earthquakes?

The nature of earthquake damage has been known in principle since 1860 with respect to masonry buildings (Tobriner 1984), and since the early 1900s with respect to steel-frame and reinforced-concrete construction. **Anti-seismic design** is a mixture of adapted engineering and architectural practice incorporating accumulated experience about the seismic performance of buildings, hypotheses about their response to seismic forces and design techniques to combat weaknesses induced in them during earthquakes (Green 1987, Key 1988).

The damage potential of ground shaking
Each earthquake yields unique combinations of intensity, duration and type of vibration. The ground waves comprising this mixture vary from those with periods of seconds to those so fast that they set up an audible hum. The damage potential of ground shaking is governed by the amplitude, duration and frequency of waves, and each earthquake will yield a unique combination of these (Thiel & Sutty 1987). Damage tends to increase with amplitude and duration of ground motion, but in a complex way that is hard to predict. Peak accelerations of up to 1.25 g (where $g = 9.81$ m/sec^2) have been recorded, but few buildings are capable of resisting the speed and displacement caused by an acceleration of more than about 0.6–0.7 g. In fact, accelerations of 0.1–0.3 g are more common in earthquakes of moderate power (magnitudes 5–7).

Surface waves travelling along the Earth's surface manifest the strongest perceptible vibrations and cause most damage. Loose sediments tend to transmit seismic waves with much greater amplitude than they

317

would have on travelling through hard rock, which tends to produce more regular and resonant vibrations that are up to four times smaller. Thus, in California in the 1971 San Fernando Valley earthquake, and also the 1989 Loma Prieta disaster, the worst damage occurred in buildings situated on older, poorly engineered fill deposits, semi-consolidated fill and loose alluvial material, while the least damage was sustained by buildings located on bedrock and carefully engineered fill (McNutt & Sydnor 1990). Moreover, convex outcrops of rock, scarps and ridges tend to radiate seismic waves with a corresponding destructive effect on buildings, especially if these are arranged down the side of the hill, such that collapse may be progressive.

Building collapse may result from the cumulative impact of earthquake motions, which progressively weaken the structure. In this respect, the duration of the earthquake may be critical, and it is not uncommon for buildings to remain standing in earthquakes that last, say, 30 seconds, whereas strong motions lasting 100 seconds would cause their eventual collapse. In tall buildings inertial sway occurs, which concentrates the force of the shaking at ground level in the form of a highly destructive basal shear (Fig. 5.1). This often causes settling and compression of lower floors before the rest of the building has been weakened to the point of collapse (Osteraas & Krawinkler 1989). Other forms of inertia may be present, depending on the distribution of weight and strength in the building: for instance, vertical compression may interact with inertia at the top of the building to crush the middle floors (Meli & Avila 1989). Or the greater amount of sway (and hence distortion) that they experience may cause the top floors to collapse first. Unless anti-seismic building codes are in force, it is relatively rare for buildings to be designed to resist horizontal shear forces.

The fundamental period and effects of shaking

Both buildings and soil or rock formations have a **fundamental period** (Dobry et al. 1976), consisting of a wavelength at which they tend to oscillate more strongly (just as a bell oscillates with a constant tone when struck by its clapper). Damage is increased when the amplitude of vibrations is accentuated (as waves are additive) at the fundamental period, or by conflict between frequencies, for example between the fundamental period of the building and that of its foundations. The impact on the building of the tremors sets it vibrating at its fundamental period, T, which broadly conforms to a simple function of its height and width:

$$T = 0.5H\sqrt{b}$$

where H is the height of the building and b is its width.

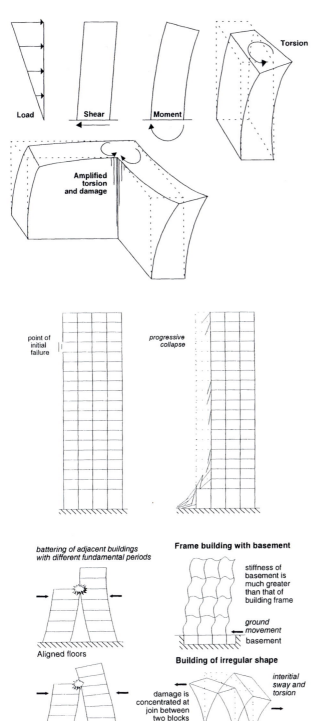

Load Shear Moment Torsion

Amplified torsion and damage

point of initial failure

progressive collapse

battering of adjacent buildings with different fundamental periods

Aligned floors

Floors not aligned

Frame building with basement

stiffness of basement is much greater than that of building frame

ground movement
basement

Building of irregular shape

interitial sway and torsion

damage is concentrated at join between two blocks

ground movement

Figure 5.1 Effects of earthquake shaking on high buildings (after Bolt 1988: 212, and Hammond 1989).

These observations give rise to six possible circumstances:

(a) If the building or other structure vibrates with the same frequency as the seismic waves that strike it, the vibrations may double in amplitude, or at least be reinforced, with damaging consequences.

(b) Adjoining buildings may have different fundamental periods and hence may vibrate out of phase with each other. If the buildings are less than 1 m apart, the resulting interference can cause mutual destruction. Progressive collapse may occur in this way, for example if buildings are closely spaced down a hill.

(c) Taller buildings tend to have longer fundamental periods, and to be more seriously damaged when located on soft ground that also has a long fundamental period. Low buildings with shorter fundamental periods are most vulnerable on firmer ground that has higher frequencies, for example on bedrock outcrops.

(d) During earthquakes, horizontal accelerations usually exceed vertical ones, but most buildings are designed to resist little more than the vertical component of their own weight. When by design or default a building is constructed with only static criteria in mind, it can be described as **aseismic**, rather than anti-seismic.

(e) Waves that travel along the Earth's surface manifest the strongest perceptible vibrations and cause most damage.

(f) Buildings constructed on loose or unconsolidated fill or sediment are likely to suffer more damage than those situated on compacted, well-engineered fill or deposits.

The seismic performance of structures

A building's seismic performance is affected by the following variables: its mass and stiffness; its ability to absorb energy (its **damping capacity**); its margins of stability (under what duress will it eventually fall down?); and its structural geometry and continuity (is it regular or irregular in plan, poorly integrated, tall and thin, etc.?). The building's earthquake resistance depends upon it having an optimum combination of strength and flexibility, as it must both resist and absorb the impact of the earthquake waves.

The urban fabrics that actually exist in towns and villages pose problems in terms of their seismic resistance. Historic buildings tend to be less flexible and more irregular than modern ones (although this is not always so, for the more idiosyncratic structures may have been destroyed by earthquakes long ago). Historic buildings may be in a poorer state of repair than modern ones, thus having many points of weakness which an earthquake can exploit. Nevertheless, no historic building or structure need be demolished simply because it is old: although the cost is sometimes very high, techniques exist for rendering most structures anti-seismic, and for upgrading their states of maintenance to acceptable levels (Fielden 1980).

A less tractable problem is posed by mixed structures, where each part may act independently with different resonance, flexibility and stiffness, according to the construction type and materials. Buildings of this kind are often the first to collapse in earthquakes. Before this happens, architectural mouldings may fall off façades and exterior elevations, and hence building regulations should prohibit them or ensure that they are well anchored.

The oscillatory forces produced by earthquakes cause a series of characteristic failure patterns in buildings (Tiedemann 1984). Tension results in the thinning or open cracking of structural members, which is a particular problem in materials that lack ductility, such as concrete. Compression can cause wide members to be crushed and slender ones to buckle. Buckling is the more dangerous of these two processes as it tends to occur suddenly. As they deform, beams suffer compression on the concave side and tension on the convex one. Walls can shear through and overturn, while the connection between columns and floor plates can result in a "punch-through" and collapse of the horizontal member. Vertical members suffer concentration of shear forces at their tops and bases, which may cause them to hinge at these points. Finally, bolts may shear through their cross sections.

When shaking starts, the loose structural members in a building may be the first to fail. If walls are not securely bonded to each other they may simply burst; or if floor and roof joists are not well tied into them, the beams may come out and wedge the walls apart or batter them down (see Fig. 5.2). A good solution to this problem is to construct a steel or concrete ring beam at roof level, that will prevent the structure from collapsing outwards. Horizontal members should be tied to vertical ones by supports of ample width and sufficient flexibility to prevent them from shearing right through during the shaking.

Field experience shows that the majority of problems encountered with vernacular housing relate to roofing material which is either too heavy (thus creating inertial forces that tend to destroy the walls and floors beneath) or too flimsy (thus capable of being carried or blown away). Experience throughout the Third World suggests that, especially in earthquakes and windstorms (Minor 1984), solving the problem of roof stability is often the key to the safety of the entire building (Park et al. 1987).

In the last resort, damaged buildings can be buttressed and their condition monitored until repair can be effected (Fig. 5.3). Shoring systems for collapsed or precarious structures must be light, portable, flexible, reliable and tolerant of changes in load. The basic weights to be supported may be known in advance: for instance, a wood floor should weigh about 50–100 kg/m^2, simple masonry weighs 500–750 kg/m^2 and a normal concrete floor weighs 600–750 kg/m^2. Shoring systems include

(a) Deformation mechanisms which cause total collapse

structure prior to earthquake

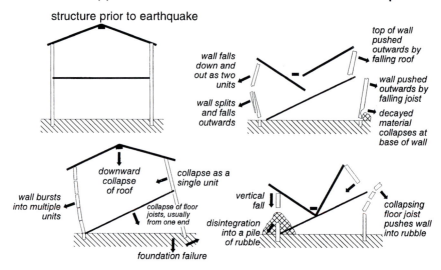

(b) Mechanisms which cause partial collapse

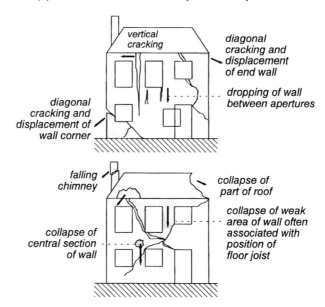

Figure 5.2 How small masonry buildings collapse in earthquakes (Hughes 1981: 412–13).

Figure 5.3 This building at Barrea in the Abruzzi of central Italy was damaged by earthquake in May 1984. Pending repair, it has been braced with steel trusses running through it and wooden buttresses wedged against the sides.

wooden posts, adjustable steel pipes, screw or ratchet jacks, hydraulic or compressed air jacks, cribbing (i.e. stacks of wooden baulks) and scaffolding. At the same time, two kinds of indicator can be used to monitor cracks that have appeared in buildings. The simplest method is to cement a piece of glass across the crack; this will break if the fissure enlarges. Alternatively, two pieces of glass, each with appropriate markings, can be attached to opposing sides of the crack and superimposed, thus indicating how much movement has taken place.

Finally, little is known about the spatial patterns of earthquake damage. These are likely to be complex, as, although seismic energy may decay fairly uniformly away from the epicentre, the distribution of buildings will be clustered rather than uniform, and local geology is likely to vary from place to place. In the 1948 Fukui earthquake in Japan ($M_L = 7.3$), in which 5,390 people were killed, the distribution of damage was assessed for certain structures. Wooden houses were damaged or destroyed to a radius of 30 km from the epicentre, with a sigmoidal decline in the number of structures affected with distance. Railway bridges were also affected up to 30 km away, but the incidence of damage was much more sporadic (Nakabayashi 1984).

Types of building construction

The seismic performance of buildings can be related to the materials and methods of their construction (Lagorio 1990; Fig. 5.4). Masonry buildings are of three types: random rubble, where the stones are held together by lime or cement mortar; dressed stone, with or without interstitial mortar; and brick. The latter is a high risk category, because it performs with great rigidity, low strength and little ductility (the ability to undergo change in form without brittle fracture). Rubble masonry is the weakest type (Fig. 5.5); it must be thick and heavy in order to resist even static forces, and thus it has much inertia. However, its strength can be improved by inserting a ring beam, injecting cement into cavities in the wall and cementing soldered steel grids to the walls. Often, floors and roofs are poorly tied to rubble masonry walls, and hence this is another factor that demands attention. Patches and reinforcements in such buildings should have the same strength and ductility characteristics as the original masonry, as "hard spots" can be as dangerous in earthquakes as can zones of weakness, given that stresses tend to concentrate at such points (Hughes 1981).

The position of apertures in masonry walls is often critical to their seismic resistance. For instance, if windows or doors are too near roof level, or too near each other, this is where cracking will begin. Physical condition is another critical determinant of stability: if damp or frost penetrate the structure they will weaken it. Older buildings can often withstand brief, intense earthquakes, but not prolonged shaking. Certain architectural elements, such as arched vaults, may be too rigid to survive

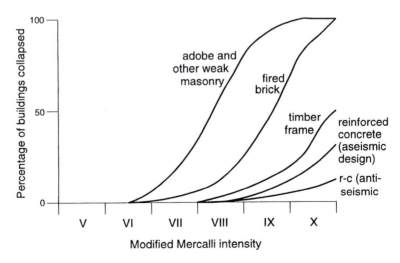

Figure 5.4 Impact of shallow-focus, near-field earthquakes on different building types (Coburn et al. 1989: 113).

Figure 5.5 Partial collapse of a house at Salvitelle in the epicentral area of the 23 November 1980 southern Italian earthquake. The age of this building is uncertain, but it was clearly constructed of unreinforced random rubble masonry using traditional aseismic building techniques.

for long during strong motions. Subsidence or heave is capable of seriously compromising masonry buildings if foundations are inadequate or absent and the subsoil is soft.

With the exception of wooden buildings, frame structures are of two main kinds: steel beam and joist, and reinforced concrete. Both types, if in good condition, can resist the large inertial forces produced by powerful earthquakes, although tall, steel-framed buildings tend to sway too much. On the other hand, the ductility of steel means that it can be overstressed and severely deformed without losing all its strength and failing completely.

The critical points are the joints between horizontal and vertical members of the load-bearing structure (Fig. 5.6). If it is not strong and flexible, concrete will turn to powder when it is hammered by the

Figure 5.6 During earthquakes reinforced concrete buildings tend to show up their weakest points, namely the joints between columns and beams or floors, which is where distortion will be concentrated.

tremors. Reinforcing bars will burst out of columns if they are not tied in well, and stairs may collapse if they are suspended from top and bottom only. As the rectangles of the frame distort into a parallelogram, the infilling (which may be brick or concrete panels, for example) will crack and fall out. The oscillatory motion of earthquakes produces a characteristic X-shaped cracking pattern in such panels.

Several construction systems are used to provide frame buildings with earthquake resistance. Reinforced concrete structures must have particular patterns and sizes of beams and columns, and the horizontal and vertical members must be tied into each other in particular ways. Structural partitions that help stabilize the frame, known as **shear walls**, can provide rigid surfaces that counteract the forces which cause deformation. **Moment-resistant frames** involve a large, dense vertical network of beams and columns made of steel or strongly reinforced concrete, which is designed to impede horizontal displacement. Most diagonally braced frames are made of steel and include X-shaped bracing within the rectangular frame (Borg 1983).

Wooden buildings have reasonably good seismic resistance, as they are quite flexible during shaking. But the structure must be well tied together, not top-heavy, and must be properly anchored to the foundation plinth (which must be stable in its own right). Adobe buildings consist of mudbrick or rammed earth in which the only form of bonding is the weak natural cement produced by dried mud. They are a cheap, traditional form of housing in many parts of the Third World, and afford reasonable insulation against extremes of heat and cold, especially if they have a heavy thatched or wooden roof. But they are extremely aseismic and have an innate propensity to collapse during earthquakes, with catastrophic results. To some extent they can be secured, for example using steel reinforcing rods, but they pose a continuing problem of high vulnerability.

Concluding remarks

Most forms of building can be safeguarded against all but the most severe earthquake shocks by proper planning and regulation of new construction and by the seismic retrofitting of existing buildings. It is vital to establish and seek to enforce adequate anti-seismic building codes, although the cost of implementing them may be an insurmountable obstacle. The wrecking of anti-seismic building laws and plans through corruption, bribery and speculation is something that occurs in all types of culture and economy, regardless of the presence or absence of public wealth. However, this should not deter us from struggling to ensure the safety of the built environment, for the technical obstacles can be surmounted, and so eventually can the political ones.

In the aggregate, the strength of buildings is not the only factor that determines the seismic death tolls per thousand of the local population (see Ch. 7). Although deaths, in particular, tend to be concentrated in buildings that collapse totally, even close to the epicentre most earthquakes will not cause every building to fall down. Hence, low rates of occupance (small family sizes) can result in relatively low death tolls whatever the overall size of the population at risk.

Having given some consideration to how damage is created, we will now examine the ways in which it is appraised, and some of the issues associated with conceptualizing it.

Costing and surveying damage

The concept of "Act of God" now has little relevance to natural disasters: regardless of who caused them, in the final analysis they are recurrent, broadly foreseeable events that are a function of human vulnerability patterns. However, it is not easy to determine when a victim should be held responsible for damage patterns and when, conversely, these are the responsibility of society in general (Huffman 1986). Unless this question

Figure 4.3 Agricultural terraces constructed by people of the Inca culture at Pisac in central Perú. The terraces carefully follow the natural contours of the land and, if the stone facings are properly maintained, they are an effective means of soil conservation. They also served a defensive purpose.

resistance to entrainment of soils and maintain the vegetation cover. Windbreaks, or shelter belts, are usually placed at a right angle to prevailing winds and should be permeable, to avoid creating erosive vortices. The spacing, height, width, length and shape of belts all influence their effectiveness. As wind velocity tends to increase immediately behind the belt, it should have a tapering end. But windbreaks may cause problems of shade and weed growth which affect crop yields in their vicinity.

Agronomic measures of soil conservation are based on the rôle of plant cover in reducing erosion rates. Thus, controlled grazing allows the vegetation cover of an area to regenerate after it has been browsed or grazed and reduces the effect of soil compaction by trampling. On arable lands, crop rotation and careful ploughing can both be used to minimize erosion. There are also techniques such as **minimum tillage** (in which ploughing is avoided altogether and seeds are drilled directly into crop residues) and barley row intercropping, which stabilizes the soil and stops it eroding from between crop rows. Strip cropping requires small-scale machinery, rather than highly mechanized farming.

Through plant selection species can be planted that resist erosion, for example by intercepting falling raindrops efficiently, resisting fluid flow at ground level, trapping transported sediment, adding organic matter that binds the soil and increasing the rate of infiltration along their roots. In general, erosion is greatest under row crops (especially grape vines, but also wheat), is less on close-seeded crops and least under grass and forage crops. The difference can be reduced by **mulching**, which involves covering the soil with crop residues such as straw, maize stalks or wheat stubble, and it simulates the effect of plant cover in binding the soil together. Later in the annual production cycle the mulch is replaced with a real crop.

Organic or inorganic additives can be mixed into soils to increase their strength or reduce their erodibility. The aim is to create a resistant soil structure where one did not originally exist. When this is done by chemical means it is termed **soil emendation**. Thus, highly sodic soils must be neutralized or they will disperse easily in overland flow (i.e. they tend to go spontaneously into suspension in water). Lastly, gullies can be stabilized against further erosion using hardy plants capable of growing vigorously in conditions of shade and infertile soils.

Experiments conducted in the 1930s showed that, in respect of slopes ploughed up and down, soil loss is reduced 43 per cent by contour ploughing, 75 per cent by rotation strip ploughing and 97 per cent by careful terracing (Mitchell & Bubenzer 1980). In another set of trials, a 26 ha catchment in Tennessee, denuded of forest in the 1930s and 1940s, was subjected to gully control and contour-ploughing measures in the 1950s. Water yield was halved, peak flows reduced by 90 per cent and sediment yield reduced by 96 per cent. Moreover, the US Department of Agriculture has estimated that extension of minimum tillage from 10 per cent of American croplands (in 1974) to 80 per cent would reduce soil erosion by one half (Carter 1977).

Soil erosion in a world context

Since agriculture began in earnest about 3,500 years ago, world erosion rates in 480,000 km of main channels have increased from an estimated 9,300 to 24,000 million tonnes per year (Judson 1968). The vast majority (96 per cent) is fluvially eroded and only 2 per cent of eroded soil comes from aeolian sources. Rates of surface lowering that were 2–3 cm/1,000 years prior to human influences have now doubled or, in highly localized instances, increased by several orders of magnitude.

The total land area of the world is 14,477 million hectares, of which 13,241 million ha are free of ice, but only 11 per cent (1,500 million ha) are cultivated (El-Hinnawi & Hashmi 1987). Potentially cultivable land occupies another 1,700 million ha, but its global distribution is very should be regarded as 100 per cent effective, and the risk will thus have been increased, although impacts may be rarer, as a result of the increase in vulnerability. Moreover, it seems that when a community adopts one particular set of adjustments it is then less likely to seek other forms of protection. In most cases, however, it is best to employ a range of different mitigation strategies.

The question of costs and losses is examined from a more formal economic point of view in Chapter 9.

Damage survey

The costing of damage is a necessary prelude to the dispensing of relief and aid after disasters. But damage cannot be assigned a monetary value until it has been assessed by field survey (Reitherman 1985). During the aftermath of a major impact, there will be a strong need for interaction between those professionals who deal with the built environment: planners, architects, engineers, insurers and local government and public works officials. It will be necessary to adopt a unified approach to damage, first by looking at a particular hazard from all its angles (architectural, insurance, structural engineering, etc.), and secondly by looking at how several hazards interact (e.g. earthquakes, landslides and seiches). One way of convincing governments and professional associations to interest their members in concerted action against disasters may be to persuade them that such events are expensive in terms of their impacts on society.

But there are several reasons why estimates of damage costs are difficult to obtain. First, in order to know what has been destroyed, it is necessary to have some form of prior inventory of damageable items, with an assessment of their current value. Urban environments are very complex and municipalities have usually been discouraged from keeping full records by the magnitude of the task: land registry records may not necessarily furnish the appropriate information, even where they are accessible. Secondly, estimation of the costs of disaster is likely to be as imprecise as it is incomplete. For example, it is difficult to assess the cost of cleaning up very widespread or dispersed destruction, or damage to complex machinery or equipment (water and sewerage mains, for instance, may require months of adjustment to ensure their functionality). The costs of lost production or missing tourist revenues are hard to assess, especially if the normal values are subject to fluctuation.

Thirdly, the long-term costs of disaster depend on some assessment of long-term behaviour (Geipel 1990); for example, to what extent reconstruction will incorporate new development, the financing of which is not strictly part of the cost of the disaster. Shifting patterns of employment and migration, as well as fluctuations in the world economy, mean that many assessments of the costs of disaster are wildly inaccurate.

330

The residual costs of the disaster are difficult to separate from costs related to pre-existing, long term socio-economic ills (such as bad housing and unemployment) and post-disaster factors (such as the desire to settle or expel a transient population). Fourthly and finally, in countries such as the United States, many assessments of the costs of disaster are based in part on the cost of litigation in the courts (for example, over whether upstream changes in a drainage basin contributed to flood damage downstream). The costs of legal claims and services may in the end be greater than the true cost of the disaster.

Post-disaster inspection and structural survey
One of the principal needs after disaster is for substantial levels of information. As part of this, it will be necessary to inspect damage in systematic detail (Hays 1986, Anagnostopoulos et al. 1989). An example of how this is done is given in Table 5.1, which lists the principal factors sampled in post-earthquake surveys of housing and the categories of damage assigned. House-to-house damage survey can be used to certify dwellings as safe for occupance, or in need of demolition or of particular kinds and levels of structural repair. It thus serves the immediate purposes first of ensuring public safety when the urban fabric is in a potentially dangerous state and, secondly, of helping to calculate the need for evacuation and alternative shelter.

In some countries, damage survey is legally required of local authorities before the central government will disburse relief moneys or provide alternative housing for homeless survivors. But it involves a series of particular difficulties. For example, appropriate personnel may not be trained to do this sort of work in the time, in the way and on the scale demanded by the exigencies of the post-disaster situation. Professional architects or engineers, among others, may rightly be unwilling to risk their reputations on inadequate structural surveys. But whatever the level of training and willingness, there may in fact be no register or inventory of available personnel; the post-disaster situation is always urgent, but it takes time to locate and co-opt surveyors and engineers. Moreover rapid survey runs a high risk of superficiality. Thus, a large public building that has been damaged will take a small team of engineers and architects several hours to survey with the minimum of rigour needed to ensure its stability. If housing stock is variable in age, construction type and quality, the imposition of uniform standards and safety criteria during structural survey may pose considerable problems.

Another problem is that the methodology of damage surveys may not exist or be known in the disaster area. It may be necessary to develop and disseminate among the inspectors of structures a series of standard methods appropriate to the type of damage found there. On more than one occasion in the past a vital factor has been left out and only

331

Table 5.1 Typical post-earthquake damage census for use in areas of small, vernacular buildings (not for use on large public buildings).

Property type (100% census):

- Dwelling
- Commercial property
- Artisan professional use
- Store-room or cellar
- Other
- Σ Total (number and percentage in each category)

Damage level:

0 No damage
1 Irrelevant damage: building is habitable and usable; repair is not urgent
2 Light damage: habitable or usable; to repair
3 Moderate damage: to evacuate partially; repairable pending reoccupation
4 Severe damage: to evacuate fully; repairable at high cost
5 Very severe damage: to evacuate and demolish
6 Partially collapsed: to repair and demolish
7 Totally collapsed: site must be cleared of rubble
Σ Total number of buildings by category, overall total

Type of horizontal structure (walls):

- Masonry: ashlar (dressed stone blocks)
  random rubble with mortar
  brick (note *type* of brick)
- Reinforced concrete (is it a well-integrated structure?)
- Steel frame (is it an integrated structure?)
- Wood frame and laths
- Adobe, cobb (stones and mud), pisé (rammed earth) or mudbrick (is it rendered – i.e. covered with cement, stucco or other surface?)

Date of construction:

- Before 1900 (often too difficult to assess more precisely)
- 1901–45 (early twentieth century)
- 1946–65 (early post-war period)
- After 1965 (late post-war period)

Miscellaneous:

- Urban or rural building
- Number of floors, rooms and residences
- Does the building have stalls, garage or a communal frontage with the buildings next to it on the street?
- Total number of residential and non-residential properties
- Total number of rooms, occupants and evacuees
- Survey of emergency accommodation, requisitioning of buildings

recognized as time progressed. But adjusting a structural survey to take account of afterthoughts or new conditions may stop it from being uniform and therefore reduce the comparability between its parts.

Surveys take place against a background of flux. A surveyor's home and family may be at risk, and therefore absenteeism is often high among people who have been co-opted for this sort of work. This being said, Dynes & Quarantelli (1976) found that such rôle conflict is a much less serious problem than has often been assumed. Nevertheless, extraneous social questions may penetrate into the work of damage assessment. In situations where the results of a survey are necessary preliminaries for obtaining grants or loans for repair and reconstruction, bribery and intimidation may be used to influence the outcome. Alternatively, while damage survey may be a public service, corrupt surveyors may offer their expertise privately for high sums of money to families who wish to expedite the process. This is essentially a problem of supply and demand, in which there are too few surveyors and there is too much demand for their services.

Many countries send teams of logistical, scientific, technical and engineering experts to major international disasters in order to assess their effect. This may be done as a form of aid, if the foreigners lend their professional services to the relief effort, or it may be a means of deriving general lessons from the experience which can be taken into account in the team's home country. But there may be a language barrier, or a lack of knowledge of local customs, geography and geology; and above all there will be limitations of time. Thus, the picture obtained by international survey teams is invariably partial and may run the risk of superficiality. There is little time to collect the large quantities of data needed to obtain an objective picture and no time to analyze them.

Information gained abroad may be fed back to a team's country of origin in a very partial or inefficient way. Moreover, its visit may use up local resources in the disaster area and among the host personnel it may divert attention from vital tasks of survey, relief administration and analysis of information on damage. There is a need for foreign personnel to develop a longer-term interest in disasters abroad – especially those in the Third World – in order really to get to grips with the problems which these pose and to bring a fresh perspective to bear on them, in addition to the local or indigenous one.

In general, one can conclude that damage assessment during the aftermath of a natural disaster depends first upon the level of prior preparedness and organization, so that surveys will be well organized, rigorous and consistent, and, secondly ,on the level of inter-disciplinary communication, which will govern the extent to which different sorts of expertise are brought to bear on the problem (Borg 1983). Such assessment is, however, a necessary ingredient in the long and complex process of ensuring public safety.

Buildings, structures and public safety

The concept of structural integrity
The **structural integrity** of a building is its physical adequacy for its intended function. The opposite is failure, the partial or total collapse or other form of destruction of the building. Although structural design can

rarely ensure absolute safety, it can always be improved in order to safeguard human lives. Ideally, design and construction should seek to meet safety levels that have been established on the basis of known intensities of hazard, and to do so at the lowest practicable cost.

The earthquake of July 1976 at Tangshan in China furnishes a good example of what can happen if structural integrity is not maintained: of 352 multi-storey buildings in that city of about 1 million inhabitants, 177 were completely destroyed, 85 collapsed partially, 86 were very seriously damaged and only 4 maintained their structural integrity in full. Highways, bridges, dams and vernacular housing were also seriously affected. However, it should be remembered that this was a very high-magnitude seismic event, whose acceleration levels probably exceeded the resistance that most earthquake-proof buildings could have mustered (Cheng Yong et al. 1988).

The literature on past earthquakes is, however, replete with examples of structural failure in which buildings were not designed to maintain their structural integrity under conditions of abnormal loading. During the 1964 Alaska earthquake, façade plates made of reinforced concrete panels detached from the structure of a department store and collapsed into the street beneath (Steinbrugge 1982). At the same time, the residential suburb of Turnagain Heights was torn apart by slumping and liquefaction flow failures (see Ch. 4). In the former case the construction type was inappropriate, while in the latter the siting and drainage of the subdivision were ill conceived (Seed & Wilson 1967).

Methods of ensuring structural integrity
There are four principal methods by which structural integrity can be ensured (Foster 1980). They each have advantages and limitations, but can to some extent be combined in a comprehensive approach towards hazard reduction in the urban environment.

First, if sufficient data can be obtained, it may be possible to calculate the probability that the maximum forces likely to be generated by a particular type of disaster in a given area exceed the resistance of a structure. Hence, one obtains the probability that load will exceed the strength of the structure during its lifetime, leading to failure. This approach has several possible drawbacks. It assumes that the physical forces are known – in other words, that the magnitude and frequency of impacts have been established with certitude. Moreover, imperfect knowledge of the physical forces may mean that errors in calculating the required strength of the building are incorporated into its construction. In addition, changes in building materials, techniques and regulations may affect the resistance of a building that is repaired, altered or added to over time. New means of improving the resistance of the structure may come to light after construction has taken place. Generally, however, it is not economically feasible to design a building to be perfectly safe against hazards, and hence some risk of failure is assumed, although it should be kept small and made explicit to architects, engineers and building users. Regardless of the merits of design, errors of construction may render a building unsafe in a manner that is difficult to appreciate until failure occurs. Such errors can be safeguarded against by rigorous quality control during construction.

Secondly, safety can be improved by identifying the specific causes of building failure in past disasters and investigating them in an objective, interdisciplinary way. Learning by experience is very common in the architectural and engineering professions, and building codes and regulations are often revised on the basis of misfortunes that graphically illustrate the inadequacy of current practice. Corruption, ignorance and speculation are three evils of the building trade common to all forms of society. Materials may be wrongly used and construction techniques wrongly applied, especially if these are new and unfamiliar; builders or building users may ignore safety criteria, or clerks of works may be susceptible to bribery by unscrupulous builders who are keen to save money on safety measures. Such faults can only be counteracted by exposing them, whatever the risks are.

Thirdly, structures can be designed to **fail safe**, so that the loss of certain critical components does not cause a major disaster. In some instances the main structural members should be reinforced so that overall collapse is not the building's principal response to a geophysical impact. It is especially important to design large or tall buildings to resist the tendency to collapse progressively. This can be done by first identifying **load pathways**, or the trajectory that components of the structure will take as they detach and fall, topple or settle. If such damage is inevitable and cannot be prevented by reinforcement or better design, the load should be channelled away from where it can do further harm, and building users should be protected. After the 1985 Mexico earthquake, for example, it was suggested that the utility wells of tall buildings in Mexico City be strengthened so that they function as strong points during major seismic shaking. This would render them less likely to collapse than the rest of the building and they therefore could be used as refuge areas by the occupants.

A forgiving environment for failures is created when planners and architects assume a relatively high incidence of destruction and damage and attempt to mitigate the consequences, rather than the causes. For example, if it is likely that an earthquake will rupture fuel storage facilities or cause explosions in petrochemical works, their siting near housing areas or transportation lines should be prohibited. Furthermore, rolled earth dams that run some risk of leakage from the reservoir behind should be made with clays that swell up and self-heal when penetrated by

335

water (although it is exceedingly difficult to apply this approach to other forms of dam hazard). Many catastrophic failures begin as small problems: slight breaching of a river levée leads to the inundation of many hundreds of square kilometers. Fail-safe devices can either prevent the spread of damage likely to result in compound failure, or they can draw attention to themselves so that they can be rectified before the failure becomes more serious.

An analogous approach is to design structures to fail or collapse harmlessly during the physical impact, in such a way as to reduce its physical force. Dams and barriers have been designed to collapse when hit by avalanches, mudflows or landslides in order to break the momentum of the movement. Similarly, warehouses placed parallel to the waterfront represent uninhabited buildings whose destruction by a tsunami is less catastrophic than that of housing, and which break some of the force of the wave. The aim here is to reduce the force of the impact by inducing it to expend part of its energy on destroying something harmless.

The fourth approach is to analyze the human consequences of structural failure, and thus to reconcile the high cost of safe design with relatively low probabilities of failure. Risk analysis demands that the greater the probability of loss of life, and the larger the predicted death toll, the more stringent should be the measures taken to prevent it (Cornell 1968). This is one of the rationales behind building codes.

Building codes and public safety
In the tropical cyclone of 1977 in Andhra Pradesh (India) and southern Italian and El Asnam (Algeria) earthquakes of 1980 the failure of reinforced concrete construction was widespread (Wood 1981), indicating that both design norms and material standards are necessary. These are provided by the **building codes**, which are laws that represent the minimum requirements of urban and structural design essential to the protection of public safety. They are intended to establish minimum safeguards during the construction of buildings, prohibit unhealthy and insanitary conditions and protect building users against fire, structural collapse or other hazards. Hence, in areas subject to natural disasters, the local building codes must establish criteria to safeguard structures against the appropriate impacts. Codes should be applied equally to existing, new and planned construction, and they should ensure that social changes, such as increases in population density or changes in the uses of buildings, do not increase the risk.

One of the principal problems of building codes as a means of ensuring public safety is that they usually exhibit a fair degree of inertia and inflexibility, when they should instead adapt rapidly to changing social and technical circumstances (Hageman 1983). Most building codes have evolved gradually in response to new materials, as these have come into use, and to past experience of structural failure or insanitary conditions. Having perceived the need, it is relatively easy to update the relevant norms, but much less so to alter the condition of existing buildings. In Los Angeles, for example, between 20,000 and 50,000 older buildings do not meet modern standards of anti-seismic construction (although this is not a large proportion of the city's buildings stock), as appropriate norms were first introduced after the 1933 Long Beach earthquake (Alesch & Petak 1986).

Nevertheless, at the world scale there is an insurmountable problem of retrofitting buildings to conform to modern standards of earthquake resistance. Given that the cost of doing so is prohibitive, the public is constrained to live with a large risk of failure. This will tend to increase as structures weaken with age, although it can be mitigated by committing resources to upgrading the resistance and state of maintenance of buildings and ensuring that standards are enforced as new structures are erected. The problem is most easily tackled with respect to relatively minor modifications which can produce a large increase in safety: for example, prohibiting the application of decorative cornices to buildings in seismic zones may prevent lethal collapses into the street beneath and perhaps not be excessively expensive in terms of modifying the façades of existing buildings.

Prior microzoning can be used to define areas of the urban fabric where vulnerability is particularly high: for example, neighbourhoods full of older buildings which have not been upgraded or retrofitted. Not only should such areas receive attention and funds to improve the quality of buildings, but they can be targeted for special evacuation plans or early emergency attention when disaster strikes.

Particular risks are posed by high-rise buildings, which require protected refuge areas and evacuation routes. They should also have helicopter landing pads on their roofs, which should be strengthened appropriately, painted with identification marks and fitted with emergency illumination. Construction should restrict the use of flammable materials, such as plastics which give off noxious gases when they catch fire, and the potential effect of falling glass if windows should splinter (fragments may reach a terminal velocity of up to 125 km/hr). Stairwells should be insulated and filled with pressurized air to stop the entry of smoke. Designers should also take account of the behaviour patterns of building users and, when designing evacuation routes, should not assume that this will be orderly and rational.

The effectiveness of building codes is greatly reduced if they are haphazard, difficult to understand and enforce, and slow to change in response to new conditions. In the USA there are no less than 5,000 local building codes. Although local circumstances will vary from place to place, there is a distinct need for homogeneity in order that information on hazard reduction can be shared between all levels of government and the professions.

Natural and technological hazards

It should be recognized that natural and man-made hazards may coincide (Harriss et al. 1978). For example, tsunamis may cause fires or slicks if they capsize oil tankers, and earthquakes may start urban fires or cause radiation leaks from nuclear power stations. When an activity such as burning, processing or manufacturing a volatile material involves a threat to public safety the risks of interference by natural disaster impact should be evaluated carefully and mitigated. The locations of power stations, chemical plants and pipelines should be chosen with reference to maps of the distribution of natural hazard impacts. In this context, fixed activities should be distinguished from mobile ones. Approximately 1,700 hazardous substances are regularly transported and hence moving them may coincide with avalanche, landslide and flood risks, or earthquake damage to bridges. Road tanker drivers and the captains of ships can be trained to recognize and cope with the risks, and, often, routes can be planned to avoid the locations of risk such as avalanche zones, and centres of vulnerability, such as population concentrations. The movement of hazardous materials should be carefully monitored as it takes place, so that the exact location of problems can be pinpointed as soon as they occur.

Prior mitigation as the route to safety

Clearly, building construction should seek to accommodate both the commonplace and the exceptional (with regard to the latter, for instance, drought may desiccate houses until they crack). Forgiving environments are most necessary in public places, where there are high rates of usage and large numbers of occupants (Foster 1980). But the current situation of high vulnerability and the sorry catalogue of failures both suggest that, in general, architects and engineers are little preoccupied with creating a forgiving environment for failures, and that they need more education and propaganda concerning structural mitigation (Lagorio 1990).

Building designers should also participate in the wider process of hazard mitigation. Thus, vulnerability to natural catastrophe can be reduced if, prior to disasters, qualified professionals examine the hazardous geological conditions of particular building sites (e.g. vulnerability to liquefaction or subsidence). They should also consider the historical vulnerability of local construction practices to the types of disaster which threaten the area in question, and the quality of local materials, workmanship and building maintenance. Finally, they should

338

investigate the suitability of housing to its environment (when particular techniques have been adopted in areas where they are unsafe).

Where it is feared that damage to housing may be widespread, four types of information should be acquired before the event. First, the distribution of geophysical hazard should be mapped, as should the distribution of buildings, routeways and services and their vulnerability to damage (see the first section of this chapter). Secondly, housing demand should be estimated in terms of local demography and population dynamics, building occupancy levels, existing homelessness, economic conditions and employment levels. Thirdly, the sociological make-up of the local population should be appraised, including the way in which it uses its building stock and its likely ability to cope with a future disaster. Fourthly, the size, resources and skills of the local building industry should be assessed with special regard to its probable ability to participate in a future reconstruction programme.

One of the main concerns of hazard managers should be to limit new or reconstructed development to safe land. Where private property and democratic institutions are the rule, this cannot be done without due process, but a variety of methods exists (see Godschalk & Brower 1985). **Land-use planning** can specify the areas in which particular residential, commercial, industrial and infrastructure functions are allowed and those where they are banned. Thus, **zoning ordinances** (or planning by-laws) can specify the dimensions of buildings, often in relation to land area and density of habitation. They can also stipulate the remedies for unacceptable land-uses. It can be argued (Driscoll 1984) that, unlike compulsory purchase, zoning regulations deprive people of compensation for having to forego those uses of their land which proper hazard management prohibits. On the other hand, the restrictions do save landowners from potential losses during disaster impacts. In conjunction with zoning, **subdivision regulations** (or specific land development plans) set the standards for urbanization and the provision of sanitary and other services. For example, public amenities such as schools and parks may be required of developments of a given size. Furthermore, it may be necessary to build hazard mitigation into new development projects.

Measures to control what can or should be built must be combined with legislation that determines how structures should be built. Building codes represent the safety standards for the materials and design of buildings and structures in force to protect the public against fire, disease and, potentially, recurrent natural hazards. Performance codes stipulate how the design of buildings must respond to various exigencies, again potentially including natural hazards, while specification codes define the necessary qualities of construction materials. **Public health regulations** should be integrated with building codes as a means of ensuring factors such as water supply and waste disposal. They should also include provisions to avoid serious disruption of these services during disaster impacts.

Finally, development itself offers a window of opportunity for hazard mitigation. **Capital improvement programmes** and public facility construction can be used by government to ensure that state-owned buildings are not excessively vulnerable to natural hazards. **Land acquisition programmes** and compulsory purchase may serve to acquire vulnerable sites and redevelop them so as to reduce the risk. Generally, the original owners must by law be compensated. Where full-scale public purchase is not contemplated, it may be necessary to compensate owners for banning hazardous land uses or for the risks caused by inadequate protection. This can be done by reducing their local tax assessments.

It is clear that most of the observations made so far in this chapter have special relevance to urban areas, as these are where most buildings are concentrated. This prompts an examination of the degree to which the vulnerability inherent in cities is responsible for particular forms and intensities of natural catastrophe.

Hazards and the urban environment

Almost all natural hazards – geological, biological, hydrological and atmospheric – may affect urban areas. From a human point of view cities constitute an enormous source of risk simply because a great many human beings and a great deal of capital are concentrated into a space that is small enough to be particularly vulnerable to a large-magnitude or high-intensity natural event, but spread out enough to put itself at risk in many different places at once (and often to occupy dangerous terrains which are unsuitable for urbanization). The particular risk constituted by the world's largest cities is summarized in Table 5.2, which shows that many of them are situated in the Third World and are subject to multiple types of natural hazards, moreover repetitively (Havelick 1986).

Seismic hazards and landslides are highly significant when they occur in an urban context, but only three hazards create problems that are distinctly urban. Flooding is exceptional because the urban fabric is generally the most impervious category of terrain. Sedimentation is distinctive because urban construction involves the heaviest known rates. Pollution is a special urban problem, as cities involve some of the greatest concentrations of sewage, detergents, airborne particulates, gases and domestic refuse, most of which emanate from point sources. In addition, several natural and man-made hazards have particular manifestations in cities. Groundwater contamination as a result of the migration of toxic leachate, subsidence caused by fluid (especially water) withdrawal and landslides caused by urban construction are three examples (Tank 1983).

Table 5.2 Natural disasters and large cities (after Mitchell 1990).

(a) Sudden-impact natural disasters in large cities, 1946–88

Country	Number of disasters	Types of disaster	Examples of cities affected
Brazil	15	floods, landslides	Rio de Janeiro, Recife
Philippines	8	hurricanes	Manila
Mexico	7	earthquakes, floods	Mexico City
Japan	6	hurricanes, earthquakes	Tokyo, Nagoya
India	6	hurricanes, floods	Madras
Indonesia	6	floods	Jakarta, Bandung
Rep. of Korea	5	floods	Seoul
Argentina	5	floods	Buenas Aires
Pakistan	3	floods, storms	Karachi
Perú	3	floods	Lima
Portugal	3	floods	Lisbon
Others	25	earthquakes	Bucharest, Tangshan, Tashkent
Total	92		

(b) "Super-cities" affected by natural hazard risk

City and country	Population (millions)	Natural hazard risk
Mexico City, Mexico	19	Earthquake, subsidence
Chongqing, China	12	Earthquake, tsunami, flood, hurricane
Calcutta, India	11	Flood, hurricane
Beijing, China	11	Earthquake
Jakarta, Indonesia	8	Earthquake, flood, hurricane
Canton, China	8	Hurricane
Lima, Perú	8	Earthquake, flood
Delhi, India	7	Flood
Hong Kong, UK/China	7	Hurricane, landslide

Natural hazards and urbanization: the American case

As American examples show, unwise land-use is one of the roots of the public policy problem posed by natural hazards and disasters, and urban land tends to be the most vulnerable to all types of damage. About 3.4 per cent of the land area of the conterminous USA is now urbanized, and the total is increasing at about 0.04 per cent (3,900 km²) per year in a very haphazard way. Land is a limited and irreplaceable resource, for which competition is fierce, especially near expanding urban centres. Over the period 1950–80 agricultural land use decreased by 9 per cent, and hence abandonment must be added to urbanization as a cause of changing geographical patterns in America.

The basic criterion for land use is an economic one, and is not a function of the physical character of the land or the hazards that it bears. The regulation of land use in the USA is the responsibility of the individual states. Nearly every one of these has expanded its list of regulations, although this conflicts with the traditional idea of private property and an owner's complete right to determine the use of his land.

The most heavily applied land-use control is **zoning**, which constitutes a community decision that a particular tract of land will consist of residential units, commercial units, public open space or an industrial site. The objective is to control the rate of growth and the location of development. Zoning ordinances stem from powers delegated by state to local government. The first attempts at this were made in 1632 in Cambridge, Massachusetts, but the method was not really implemented until the 1920s. Since then, the courts have acted both to protect and to challenge public and environmental interests against the complaints of developers and unwise land users. Overriding the rights of private landowners in the interests of public welfare is not an easy decision. It must be backed by careful planning; and it cannot be done in response to a vague threat of natural disaster, but must instead respond to a definite, verified risk.

Historically, common law has defined natural hazards as "Acts of God", thus protecting governments from culpable liability. According to a 1916 court decision, "Acts of God" ". . . are those events, accidents, or manifestations of nature which proceed from natural causes, and which are unusual and unprecedented in character and cannot be reasonably anticipated or guarded against by the exercise of ordinary care". This viewpoint is outmoded, and nowadays local government can be held responsible for the impact of natural hazards if it fails to inform the public of the problem and take steps to mitigate it (Coates 1985). The issue has been forced by the growth of scientific knowledge, which has increased our knowledge of the location and timing of natural disaster impacts. Thus, because they are responsible for drawing up, adopting and enforcing building codes, zoning ordinances and emergency plans (for the protection of public safety), public bodies such as cities and counties are inherently vulnerable to liability.

Flooding

As a general model, older residential areas tend to suffer less from flooding than newer ones, despite often having higher densities of population and dwellings (Cooke 1984). This is because problems have been corrected after previous floods: flood control structures have been built and non-structural mitigation measures instigated slowly over time. In latter years urban sprawl has often preceded the necessary environmental protection measures, perhaps in part because of speculation and in part because planning authorities are ignorant of the risks involved. In other cases the pressure on land and the rates of expansion of particular cities have simply been too great to resist, or alternatively the growth itself has been unplanned and unfettered by zoning regulations.

In the Third World high rates of demographic expansion and migration to urban areas have caused many primate cities to double in size, and double again, in the space of a very few years. The poorer inhabitants may live on floodable canyon floors, or on or at the base of unstable slopes in very precarious positions. But even in southern California, the archetypal affluent environment, thousands of hectares of alluvial fans, arroyo beds and active alluvial plains, where floodflows once ran relatively unconfined, have been covered with houses, orchards, roads, factories and commercial premises. Forest and brushfires have denuded the hills of the vegetation that once stabilized them against debris production, such that floods are inevitably accompanied by immensely destructive debris flows.

The problem is particularly acute where development takes the form of urban sprawl. This can be defined as the rapid expansion of suburban development without complete planning for the optimum control and development of associated land and water resources. Damage to the urban sprawl usually results from both inundation by floodwaters and the impact of landslides or mudflows. Mass movements can not only crush or carry away a building but also undermine its foundations, and they often occur when rain saturates upland or canyon soils and develops high pore-water pressures there.

The lack of infiltration into paved ground, and the pre-eminence of pipe and channel flow, mean that urban areas create a hazard to whatever lies downstream. A very high proportion of runoff (perhaps 90 per cent) is fed to channels. Hence, given the low roughness of urban culverts and channels, the rising limb of the hydrograph is steep and the lag time between rainfall and peak surface flow is short. This can cause severe erosion down stream or can feed a flood in another urban area. In this respect the urban drainage network is usually based upon the pre-existing natural one. It normally feeds water to a single outflow channel, which is generally one that nature has designed to accommodate smaller discharges (as the water yield of the upstream basin would have been less before urbanization decreased infiltration levels). Engineers have not always remembered to enlarge such channels, or to find ways of reducing the input to them.

Flooding and drainage are two associated problems in an urban context, as human land use has increased the rate of flooding in urban zones. For example, in the metropolitan areas of Washington, DC, Rock Creek, a tributary of the Potomac River, had 103 km of channels in 1913 but has only 43 km above ground now that it has become urbanized. As a rule of thumb, fully sewered areas have a total discharge that is 2½ times greater than that of corresponding natural areas, and peak discharge is increased by 70 per cent (it may be up to eight times greater than in areas with absolutely no impervious cover). Sewerage modifies the infiltration rate, reduces rates of groundwater recharge and changes the cross-sectional morphology of streams, usually by expanding it and making it potentially unstable.

Flooding in urban areas of the USA
The current rate of urban expansion into floodplains of the USA means that more than 6.4 million homes are at risk. In 1973, 16 per cent of urban areas were located in the 100 year floodplain (and in only 7 per cent of the nation's total land area). Throughout the country the majority of flood damage costs pertain to urban areas: in New York State, for example, 79 per cent of communities with populations of more than 2,500 have reported flood and drainage problems.

In southern California, antecedent moisture conditions play a vitally important rôle in the flood hazard (Rantz 1970). Sustained precipitation leads to saturation of upland basins and eventually to floods accompanied by large-scale movements of debris. Flood control basins have been created in order to trap water and sediment and reduce the risk to urban zones downstream (mainly by lowering the peak rate of water flow or sediment delivery; see Fig. 5.7), but the problem is not so easily solved. In storms during the year 1969 the hydrograph of the Santa Ana River was extremely peaked at the inflow to control basins, where the peak discharge was 2,100 m³/sec. The basins reduced this to 165 m³/sec, spread over a longer period, and they also trapped 1,530,000 m³ of sediment. But despite this, 10,000 people were driven from their homes

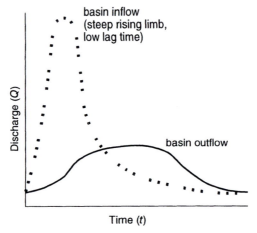

Figure 5.7 Use of flood control basins to modify shape of urban hydrograph.

and 92 were killed. Damage amounted to $62 million, between 65 per cent and 80 per cent of which occurred in urban areas.

Erosion and sedimentation

The preparations for urbanization include clearing land, grading or levelling it and digging foundations. This high level of disturbance of the land surface can lead to considerable increases in the sedimentation rate (see Fig. 5.8 and examples listed in Table 5.3). Afterwards, however, sedimentation rates usually decline to a minimum as a result of the land surface being rendered impermeable. Thus 80 ha of land in Virginia, used for highway construction over a period of three years, yielded an estimated 37,000 tonnes of sediment (Vice et al. 1969). Road building took place on only 11 per cent of the basin but contributed 94 per cent of its total sediment yield. In extreme cases, the tonnage of sediment produced during construction is up to 40,000 times that derived from analogous farmland (Wolman 1967). The effect may be noticeable in terms of increased turbidity of rivers downstream of the construction site.

Increased erosion and sedimentation provoked in urban areas give rise to a whole catalogue of problems. For example, raindrop impact tends to disperse soil particles and seal the land surface against infiltration. It is easily enhanced on slopes that have been denuded and compacted by

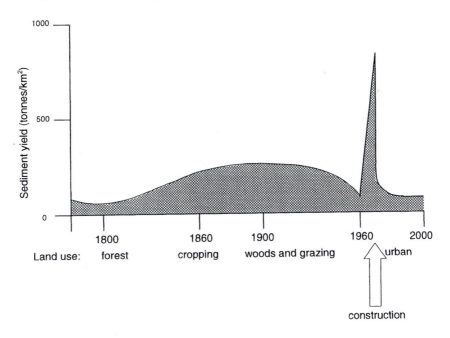

Figure 5.8 Rate of erosion during change in land use (Wolman 1967: 386). Copyright 1967 by Scandinavian University Press.

Table 5.3 Differences in sediment yield (tonnes/km^2/year) in relation to construction and urbanization in the USA (after Wolman 1967, Wolman & Schick 1967, and other sources).

State	Forest	Agriculture and rural	Non-urban activity	Construction	Urbanized	Construction/urbanized rate
DC				575		
Kentucky	5		22	411		
Maryland	7	83	412	380	19–39	× 10
	13	114		663	37	× 19–27
	29	116–309		830		× 990–2,030
	45	117		2,010		
	74	150–960		4,050		
		168		8.600		
		179–327		25,800		
		185		28,700		
		289–649		50,200		
Michigan			642	11,650	741	× 23
				17,000		
Virginia	9	1,876				
Wisconsin		<1		19		

construction activities. The results include increased surface water yield and lower rates of groundwater recharge. Moreover, where cut-and-fill grading is being practised, sheet, rill and gully erosion can exceed 75,000 m^3/km^2 over short periods on steep slopes cut in unconsolidated material (Fig. 5.9). At the same time, aeolian erosion of bare soil surfaces may be a problem at construction sites. Deposition can also occur: for example, in Metropolitan Chicago dust precipitates reach 100–280 tonnes/km^2/year and constitute 70 per cent by weight of road sweepings.

High sediment loads greatly increase the damaging effect of floodwater. Floods that occur in the vicinity of building sites can rapidly flush out large volumes of loose sediment disturbed by the construction process. Eroded sediment may then block natural water courses, man-made channels or drains, the dredging or protection of which tends to be costly. Hence, the streams and lakes that constitute an amenity in urban areas can be damaged and rendered unsightly by siltation. This can also make them liable to flooding if bed levels are raised by the deposition of sediment. In addition, reservoir siltation and its downstream counterpart, clear-water erosion, are often associated with urban areas (see below). Erosion and deposition in channels can cause bridges to collapse or culverts to disintegrate, for example through increased scour. The water may become anaerobic (lacking in free oxygen) and ecological changes may take place; for example, reduction in the population density of organisms, or changes of species composition.

346

Figure 5.9 The grading of unprotected slopes during urban construction can lead to rill development and vast increases in sediment yield.

Increased sediment transportation can cause public health problems (Guy & Jones 1972). High rates of sediment deposition can create marshy ground where mosquitoes may breed. Harmful bacteria, toxic chemicals and radionucleides may be adsorbed onto sediment particles (having been relatively harmless in their previous residence). For example, urban atmospheres contain relatively high concentrations of heavy metals, including arsenic, cadmium, lead, mercury and nickel. Sediment carrying the harmful substances may enter the water supply, or may cause eutrophication of waters. The rate of transport of adsorbed pollutants may be several orders of magnitude slower than that of dissolved pollutants, leading to longer residence times and therefore lower dispersion and higher environmental levels of these substances. It is thus hardly surprising that water treatment costs are greatly increased when rivers carry a high sediment concentration, and it may be considerably cheaper to reduce the solid load at source than to trap the sediment and purify the water.

In synthesis, urban erosion and sedimentation can easily be ignored until they are chronic problems which are very expensive to put right. They are part of more general and complex questions of environmental quality and management in urban areas. The solution necessitates full recognition and proper evaluation of the problem. Unfortunately, the costs of urban sedimentation are rarely calculated and are seldom considered when estimating the cost of flooding. Urban streams and conduits are not generally considered as analogous to natural channels in the way that they carry water and sediment. Moreover, sediment loads may be hard to predict, especially during phases of urbanization, when they are likely to fluctuate substantially.

The strategy for controlling erosion and sedimentation in urban areas must first involve collecting data and using them to educate the public and local officials on the dangers involved. A proper public policy must be instituted, especially regarding land-use control as an instrument for ameliorating the hazard. Suitable erosion control measures during the critical phases of urban construction include temporary and permanent revegetation of bare slopes, stream diversions and other forms of site rehabilitation. Slopes can be terraced while they are being graded, and structures installed for the abatement of erosion or stabilization of slopes. Measures can be taken to store excess rainfall before it is fed into vulnerable urban channels, and storm drainage systems in the urban area should be carefully maintained. In the drainage basin upstream of the urban zone floodwater retarding structures should be constructed and vegetation used to stabilize stream banks in areas where runoff is critical to the capacity for eroding.

Urban landslides: the Californian example
As described in Chapter 4, in January 1982 landslides in the Santa Cruz Mountains of central California killed 26 people and caused $282 million of property damage. The contributory factors included steep slopes, weak sedimentary rock, active faults, unstable soil covers, inadequate maintenance of the vegetation cover, and intense precipitation. The example in question shows that landslides can be as widespread as urban development, which through destabilizing the fragile equilibrium of slope systems is often one of their main causes. The intense rainfall of early January 1982 produced slope failures of almost all types, from rockfalls to mudflows, in areas where hundreds of houses had been constructed without prior geological site investigations. Logging, burning of the vegetation, grading, loading with buildings and the digging of road cuttings all exerted a vital influence upon slope stability in the region (Furbish & Rice 1983).

As basin saturation is closely related to high pore-water pressures, flooding and landsliding both result from high intensity prolonged rainfall in California. Since the beginning of the twentieth century, destructive events have had an irregular 6–8 year recurrence interval. The main problem, however, is inadequate land-use planning, prohibiting ribbon development in valleys at the base of steep slopes, and neighbourhoods on coastal cliffs or unstable inclines. This deficiency is highlighted by the fact that in many cases the hazard has been mapped, seismic intensity has been related to patterns of slope instability, and weak Cenozoic and Mesozoic sedimentary rocks and unstable geological structures have all been identified.

In the California Master Plan some attempt has been made to calculate the economic significance of each geological hazard, by calculating the value of property, lives and infrastructures at risk in each urban unit. Risks have been assessed for several intensities of disaster impact. They have been related to current and projected population densities, in order to derive land use recommendations. When high levels of urbanization are coupled with high risk of natural disaster priority is given to hazard mitigation.

Yet for any given event it is difficult to predict the exact location of abrupt slope failure from spurs of land, ridges, uncompacted fills, unstabilized road cuts and poorly drained slopes in the vicinity of urban areas. Fatalities usually occur at night, when sleeping residents are unaware of the risk, and the slope failures which do the damage are often small and highly localized, especially those that emanate from the heads of first-order tributary gullies (Baldwin et al. 1987). Yet small, abrupt mudflows and debris slides can be lethal, and their cumulative impact on urban drainage and water supply is substantial. The number of fatalities could be reduced by the use of warning systems of the kind recently established in California, which detect the threshold conditions for slope failure, and when rainfall events are sufficient to permit these to be surpassed; and the damage can be abated by intensive slope management and more effective zoning laws (Nilsen et al. 1979).

Post-earthquake urban fire

Fires are not an inevitable consequence of earthquakes that devastate large urban areas. However, there have been some notable instances in which they have played a major part in the disaster. In the 1906 San Francisco earthquake 80 per cent of the damage was caused by fire: the city's water mains ruptured as a result of soil spreads and liquefaction failures, while the methods used to control the fire (especially the dynamiting of buildings to create firebreaks) actually helped spread it. In the earthquake of 1923 in Japan, 38.2 km^2 of the Tokyo area went up in flames, while the 1972 earthquake in Nicaragua involved a large fire in the centre of Managua.

Japanese researchers have developed models of post-earthquake urban fire, which relate seismic intensity to building type and density, wind velocity and deterioration of the fire-fighting response (Scawthorn et al. 1981). Their aim has been to develop a ratio of the costs of destruction caused directly by ground shaking to those that result from fire. The following predictions were obtained for the city of Osaka (losses have been converted into millions of US dollars):

Local magnitude of earthquake	5.5	6.5	8.0
Distance of epicentre from Osaka (km)	20	40	160
Losses caused by fire (US$ million)	54	188	13,000
Losses caused by seismic shaking	70	158	17,000
Losses caused by ground failure	0	3.4	0

The post-earthquake fire hazard has increased as a result of the rising scale and density of development and variety of architectural styles and building materials, as well as the increasing forms and usages of energy in the urban environment. But modern fire regulations and building codes have improved the situation somewhat, and hence fire is by no means an inevitable consequence of earthquake damage (Nakano & Matsuda 1984).

The performance of urban areas during natural disaster can also be viewed in terms of individual buildings, of which high-rise structures are potentially the most vulnerable.

High-rise buildings in disaster

High-rise buildings are particularly vulnerable to disaster because of their size and height, the number of occupants they may contain and the problems of evacuating them safely. Moreover, it is more difficult to create a forgiving environment for failures in a tall building than for a more modest structure: the onus is on the architect, and whoever commissioned him or her, to safeguard users of the building against hazards. This means both making the structure resistant to impact and ensuring that it can be cleared rapidly of occupants under difficult circumstances. This section will examine the impact on tall buildings of hurricanes, earthquakes and fires, and will consider the lessons learned in the fields of evacuation and medical response.

High-rise buildings and hurricanes
High-rise buildings are fully engineered structures: they receive individual attention from engineers and architects. However, little of such attention has been directed towards mitigating the effects of hurricanes. An

interesting example of what can happen was furnished by Hurricane Alicia, which struck the central business district of Houston, Texas, on 17–18 August 1983 (Savage et al. 1984). It damaged the glass cladding of buildings (Fig. 5.10), but not their structural members. There had been plenty of warning and time for evacuation, and hence personal injury was not a factor in the impact. Structures swayed under the duress of wind loading, but their frames were not damaged, only their glass cladding and windows.

The glass used in cladding modern high-rise buildings is annealed, heat strengthened and tempered. The strength of an individual window is increased by using double panes, more heat-strengthening or tempering,

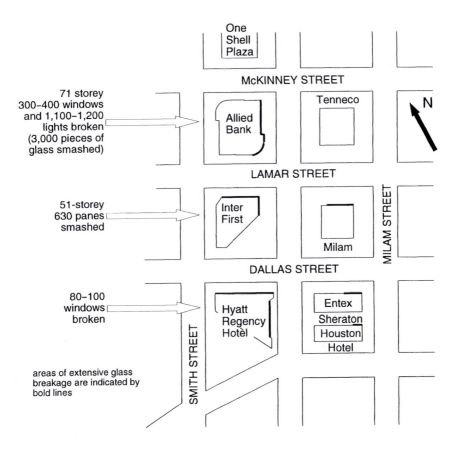

Figure 5.10 Effect of Hurricane Alicia (17–18 August 1983) on Houston, Texas (Savage et al. 1984: 55). Reprinted with permission of the publishers, National Academy Press.

or by using thicker plates. Wind loads on the surface of the glass produce stresses that interact with minute, widespread surface flaws to produce local nodes of stress concentration. If the stress is sufficiently high or concentrated, fracture begins at the node. Storms can also weaken glass over time, so that it breaks under the duress of subsequent, and perhaps less intense, storms.

The fracture patterns in glass during hurricanes are similar to those made by the impact of small missiles. Falling glass hits other glass panels, causing further damage: for example, in Houston the skylight panels of a hotel were damaged and falling glass impacted other windows in the wet, swirling atmosphere beneath.

The pressure field around the glass cladding of a building does not appear to fluctuate with the same frequency as the glass itself (if the latter is affected by wind loading). Damage tends to be concentrated on the windward edges, sides and angles of buildings. The windward elevation is a zone of positive pressure, while the sides and roof are generally areas of negative pressure, as is the leeward face.

The glass of modern high-rise buildings will fracture if wind pressures exceed design values on surfaces of the building. It may also suffer from missile impact by windborne debris and falling shards, which can either break windows or weaken them by causing craters or fissures to appear. Under wind stress the glass may break if it is badly installed, or if structural displacement causes excessive stress. Thus, at Houston in 1983, wind speeds did not exceed design values for the glass itself (most buildings having been tested in wind tunnels prior to construction), but damage was nevertheless widespread.

The pressure on buildings is increased when wind is channelled into the "street canyons" between structures (this is the so-called **Venturi effect**), or by the presence of other buildings nearby. Wind tunnel experiments may not be designed to model the wind loading caused by adjacent structures, especially if these are constructed after the building in question. Apart from the potential problems of structural loading, wind-borne debris can be funnelled at very high velocity by "street canyons". Most wind-borne missile damage occurs in the middle or lower zones of a tall building, whereas the highest parts usually have the strongest glass, as it is generally considered that these will suffer the highest wind velocities.

At Houston during Hurricane Alicia 80 per cent of glass breakage was caused by wind-borne debris. Pieces of falling glass from a tall building can become missiles in the wind, and their number can increase in the form of a "cascade effect" where there is a sequence of tall buildings. Glass that is not broken by flying debris may become pitted and scratched until it is too weak to resist further wind loading. Potentially dangerous situations can also be created by improper installation or poor maintenance of glass panes and stone or concrete façade panels. The former may become unstable and shift within the window frame or be overstressed where there is insufficient bite in the glazing system, if jamb blocks are missing or if gaskets are improperly installed. The latter may work loose if they are insufficiently anchored to the building's frame. In this respect, building designers at Houston did not analyze the rôle of structural displacement in overstressing buildings; the effect of this on panes of glass is termed **racking**.

Earthquakes and tall buildings
While hurricanes may damage tall buildings, earthquakes can destroy them. Seismic shocks pose critical problems for high-rise buildings, as a result of the high load factors and substantial inertia that ground shaking can generate. If a first main shock weakens a tall reinforced concrete building, it may collapse if it is then unable to absorb the energy of a second shock. Irregular structures will distribute their loads badly during tremors and may fall apart. It does not help to have weak zones, such as open, unstrengthened ground floors (known to architects as the "soft storey"). Hence, design needs above all to be even, such that the unit performs as a whole, not as a series of parts, and that no zones are weaker or stronger, stiffer or more flexible, than others.

The abruptness of earthquakes poses special problems of evacuating occupants, or protecting them until they can leave safely.

Building evacuation
Although computer models have been used to design evacuation procedures, there have been few investigations of the perceptual and social aspects of evacuation. It is not, of course, an easy topic to study, as it depends on a specific set of extreme circumstances, whose occurrence is relatively rare. One published study, however, refers to the rapid evacuation of a six-storey office block that was badly damaged by the 1979 Imperial Valley, California, earthquake. The east end of the building buckled and settled as the supporting columns cracked (Arnold et al. 1982).

Most of the occupants were sitting at their work stations when the earthquake struck, although some were in corridors, or standing (see Figure 5.11). During the impact one third took refuge under desks, one third did not move and half of the remaining people sheltered in doorways. The reasons for selecting a particular action included: the result of previous drills; reasoning; advice from a co-worker; or past experience of earthquakes. In terms of a second action, 56 per cent of people left the building spontaneously as soon as the tremors had ceased. This seemed the sensible action to most occupants, and the remainder

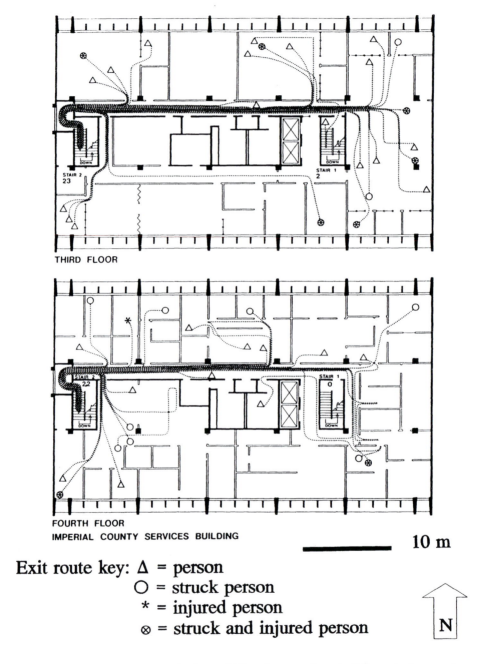

THIRD FLOOR

FOURTH FLOOR
IMPERIAL COUNTY SERVICES BUILDING

10 m

Exit route key: Δ = person
O = struck person
* = injured person
⊗ = struck and injured person

N

Figure 5.11 Evacuation of the middle floors of a building during and immediately after an earthquake (Arnold et al. 1982: 211).

were ordered to evacuate by floor monitors. Of the occupants, 79 per cent followed the pattern of a previous drill and left the building by its west stair (the east stair, being close to the elevator shaft, was felt to be more vulnerable to bomb attack and hence had not been used in drills). Investigators concluded that the instructions given out in evacuation drills should not be too complex, as people are not in a state of mind to evaluate alternatives directly after an earthquake impact, which is a time when familiar actions are dominant. Computer modelling of evacuation flows is a feasible technique, as the evacuation of a tall building involves predictable flows of people along well-defined passageways (Watts 1987, Weinroth 1989). Such models should, wherever possible, include a socio-perceptual component.

Fire in high-rise buildings

Although it is not a natural hazard, fire deserves special consideration because it is probably the principal risk to a tall building (it may, of course, result from the impact of an earthquake or other natural event). Examples from Sao Paulo, Brazil, in the 1970s show that tall buildings can be entirely consumed by fire in less than an hour if they contain flammable materials such as plastics. Considerable loss of life may then ensue if elevators jam, and there are no automatic sprinkler systems, interior fireproof stairwells or exterior fire escapes (Foster 1980). In such cases helicopter rescue may be practicable, but only if there is a landing space on top of the building. However, in a high-rise fire which occurred in Lima, Perú, in 1977, updrafts of air prevented helicopters from landing on the building. Expandable or hinged external fire escapes have been designed, but have not been adopted generally as standard equipment, and are virtually unknown in the Third World.

High-rise buildings and medical emergencies

The Director of Emergency Services at Sunrise Hospital in Las Vegas, Nevada, has argued that high-rise disaster poses a new threat to society, and that hospitals have not developed adequate emergency procedures to cope with it (D'Amore 1982). Two events in Las Vegas provided the experience from which this observation was drawn: on 21 November 1980 the MGM Grand Hotel caught fire, killing 85 people and injuring 700. Then, on 10 February 1981, the Las Vegas Hilton burned, causing nine deaths and 300 injuries. During the next 15 months, four more high-rise hotel fires occurred in various parts of the world, killing more than 100 people.

New disaster responses are required to counteract this threat and they can be stated with respect to four locations: the scene of the disaster, the receiving hospitals, secondary treatment centres and mortuaries. These places should be connected by special telephone links, helicopter flight plans and ambulance routes. Each location should be able to avail itself of call-up and alert procedures specific to high-rise disasters.

At the scene of the disaster, triage areas (see Ch. 7) should be established at the main exits to the building, and triage vehicles should be sent to them with oxygen masks and tanks and resuscitation equipment. Only one physician is required at each exit and he or she should use triage tags which are specific to the injuries likely to be caused by high-rise disasters (smoke inhalation and burns in the case of fire, crush injuries in the case of structural collapse). Buses should be organized in order to transport the lightly injured to secondary triage points.

The receiving hospitals should be ready for large numbers of victims. Specialists should be called up, including experts on pulmonary ailments, burns or crush injuries if it is necessary to treat the victims of smoke inhalation, burns or structural collapse, respectively. A special hospital disaster plan should be created and integrated into the main emergency medical plan. The secondary treatment centre should also have a special emergency and triage plan. In the case of fire disasters, it should be supplied with arterial blood gas capability. Finally, the mortuary requires specific plans for coping with large numbers of victims, including a temporary morgue and special autopsy arrangements to ascertain the causes of death. It also needs a specific plan for the notification of next of kin and for the positive identification of bodies.

Throughout the disaster and its aftermath there should be a careful planning review. The only authorities capable of changing emergency medical plans should be the emergency medical services themselves, not administrative authorities. It has also been recommended that an organization be set up to look into the safety of high-rise buildings, and although fire is the main preoccupation, other hazards could profitably be considered.

Dam disasters

Like the high-rise building, the reservoir dam is another potential source of concentrated vulnerability to natural disaster. Dams retain runoff for domestic and industrial supplies, flood control, power generation, groundwater recharge or irrigation. They may be constructed for a single purpose or for a combination of several uses, including recreational ones such as boating and fishing, and scientific ones, such as wildlife sanctuaries. In this context, it is notable that some uses may be mutually exclusive: for instance, reservoirs kept half empty in order to trap flood discharges may be of limited use for water supply or power generation. Their utility for their intended purpose must be balanced against their impact on the heritage and aesthetics of the local landscape, and on its vegetation, wildlife and habitat. They must also be considered in terms of the hazards which they create, and the way they interact with natural hazards such as landslides and earthquakes. The risks posed by large, artificially created bodies of water, and the dams that impound them, represent a point of interaction between man-made and natural hazards (Wahlstrom 1974).

There are five main types of dam. **Earth dams** and **rockfill dams** are triangular in shape and built of compacted sediments and boulders. By virtue of their relative cheapness and versatility, they are the most common of the five types. **Gravity dams** rely on the weight of concrete or dressed stone to support the structure, while **concrete arch dams** deflect the pressure of impounded water onto the adjacent foundations and abutments. **Buttress dams** add strength to the barrier through supports on its downstream side. Fewer than 10 per cent of dams are constructed of concrete. On a world scale, more than 50,000 dams exceed 15 m in height and about 40 exceed 180 m (Clark 1982).

The nature of hazards
The hazards associated with reservoir impoundments vary from the sudden (dam failure) to the gradual (silting problems), to the unexpected (reservoir-induced seismicity). They represent a point of overlap between man-made and natural hazards. The main problems are as follows:

(a) pollution of stored water, usually (but not always) by sediment;
(b) loss of storage capacity by siltation or leakage through the reservoir bed or banks, or through the dam itself;
(c) seiching during earthquakes, windstorms or hurricanes;
(d) surges created by landslides into the stored water, which may cause water waves to overtop the dam;
(e) failure of the dam as a result of design or construction errors, or accelerated leakage;
(f) faulting or other forms of instability in the rock formations beneath the dam foundations;
(g) overtopping, due to failure of the spillway or outlet structure;
(h) subsidence or uplift of the ground, leading to loss of free board and diminution of storage capacity;
(i) adverse modifications of environmental quality, such as the creation of minor-to-moderate earthquakes during filling of the reservoir.

Catastrophic failures are rare, but not excessively so (see Table 5.4).

Table 5.4 Resumé of notable dam failures.

Date built	Location	Reason for failure	Date failed	Number of deaths
1791	Puentes Dam, Murcia Province, S.E. Spain	Foundation instability and excessive rainfall	1802	608
1853	S. Fork Reservoir S.W. Pennsylvania	Spillway failure led to overtopping and structural failure	1879	2,100
1865	El Habra Dam, Algeria	Leakage in poorly constructed gravity dam	1881	209
1840–53	South Fork Dam, Johnstown, Penn.	Spillway failure	1889	2,209
1878–81	Bouzen Dam, France	Construction faults	1895	100
1911	Austin Dam, Pennsylvania	Poor construction of concrete gravity dam	1911	80
1924–6	St Francis Dam (near Los Angeles, California)	Bedrock and foundation failure	1928	>500
1935–7	Mono Creek, Santa Barbara, California	Siltation rapidly reduced reservoir capacity to zero	1938–9	0
1952–4	Malpasset, France	Increased permeability of faulted gneisses	1959	421
1960	Konya, Turkey	Karstic streamflow bypassed 0.043 km^3 reservoir while it was being filled	1960	0
1951	Baldwin Hills Reservoir, near Los Angeles, California	Leakage and structural failure consequent upon fault creep and subsidence under foundations	1963	5
1957–60	Vajont Reservoir, Piave River Valley northern Italy	Floodwave caused overtopping after landslide into reservoir: dam remained intact	1963	1,925
c.1962	La Penna, near Florence, Italy	Overtopping and release of floodwave into the River Arno	1966	12
1967	Koyna, India	Reservoir filling induced catastrophic earthquake; deaths were caused by earthquake	1967	177
?	Van Norman Dam, San Fernando Valley, California	Slumping at rim of earth dam almost caused failure during 1971 San Fernando earthquake	1971	0
?	Coal spoil dam, Buffalo Creek, West Virginia	Structural failure of dam materials, caused by bad design	1972	118
1976	Teton River, Snake River Basin, Idaho	Leakage led to structural failure of rolled earth dam	1976	11
1978	Machu, W. India	Structural failure when discharge reached twice design flow	1979	>1,000
?	Val Stava, Trentino, northern Italy	Breaching of earth dam, followed by catastrophic mudflow	1985	264
1987	Adda Valley, northern Italy	Damming of river caused by landslide threatened to breach disastrously	1987	58 (in landslide)

More than 2,000 such events have been chronicled in human history, and the 10 per cent of them that have occurred in the twentieth century have killed more than 8,000 people. Between 1918 and 1958 33 dams failed in the USA alone, of which five caused major disasters with the loss of 1,680 lives in all (Gruner 1963, Johnson & Illes 1976, Leonards 1987). From 1959 to 1965 nine major dams failed in various parts of the world, while in 1976 six failed, four of which produced major disasters with a total of more than 700 deaths. A study of 308 dam failures attributed 40 per cent to foundation failures, 23 per cent to inadequate spillway capacity and the remaining 37 per cent to poor design and construction, site inadequacies, subsidence and earthquakes (Gruner 1963).

In the context of dam failure, seepage must be distinguished from leakage. The former involves slow losses of water which are great only in the long term, while the latter consists of abnormally large escapes of water from fissures, breaches, natural pipes or permeable terrain. In earth dams, or those with a core of compacted sediments, uncontrolled leakage may result in an exponential increase in the rate of erosion of the materials of which the dam is constructed. At best this will deplete the storage capacity of the impoundment and increase discharge down stream, while at worst it could lead to weakening and catastrophic breaching of the entire structure. Moreover, leakage through the foundations of a dam may weaken it and lead to collapse. This may occur as a result of the weight, or **hydraulic head**, of water being forced through the leak and the abnormally high rate of shear of water running against the surfaces of the breach. Leaks can often be stopped by grouting (or injection with asphalt, bitumen, cement or clay), while they can be prevented by making surfaces impermeable and ensuring the homo-geneity of the structure and its foundations. Leakage through soil pipes has caused more than a hundred small earth dams to fail in western Australia.

Dam safety

The design of a safe dam (ASCE 1967) involves locating the structure where there are no active faults or possible landslides. Thus, the local fault geology, patterns of regional seismicity, and slope stability at the dam site and along the margins of the reservoir must all be examined carefully. The structure should be designed to allow for differential movements, seiches, seismic activity and shock waves, and to resist the stresses caused by these impacts. This will involve using wide sections which tend not to break or materials which tend to self-heal if cracked or penetrated.

Over-designing (for example, against damaging earthquakes for which a long return period is expected) is usually costly and uneconomical, but is one means of ensuring public safety. Both elastic and permanent displacements must be kept within a safe range, the former to avoid seiching and the latter to prevent overtopping. If all hazards cannot be countered in the design, then risks can be analyzed in order to assess the likelihood and expected cost of failure.

Risk analysis applied to reservoir dams
In the 1970s the US Federal Government began the construction of a double-arch reinforced concrete dam at Auburn, 56 km upstream of the city of Sacramento (pop. 260,000) in California. Geological surveys identified five rock types, various shear zones and nine faults at the site of the dam. Construction went ahead assuming the faults to be inactive, but the period of inactivity was unknown. In 1975 an earthquake occurred with epicentre 80 km from the site. An independent investigation was commissioned, and the consultants' report suggested that there might be a 1–10 per cent probability of movement on any of the faults (Carter 1977). It also argued that the type of dam was inappropriate for the geological conditions of the site. The government then spent $10 million on seismic studies and commissioned a risk analysis (Duffield 1980).

The cost of failure of the Auburn Dam once built has been estimated by risk analysis, in which the value of a human life is considered (in purely economic terms with no ethical connotations) to be $200,000 in lost income and social and medical payments (Mark & Stuart-Alexander 1977). The population at risk comprises the 260,000 inhabitants of Sacramento, and the risk of dam failure is once in 30,000 years, which is the estimated recurrence interval of fault rupture at the site:

$$\frac{\$200{,}000 \text{ per life} \times 260{,}000 \text{ lives}}{30{,}000 \text{ year estimated recurrence interval of fault rupture}} = \frac{\$52{,}000 \text{ million}}{30{,}000 \text{ years}} = \frac{\$1.7 \text{ mn}}{\text{year}}$$

A more realistic estimate is obtained if the lifespan of the dam is taken into consideration:

$$\frac{\$200{,}000 \text{ per life} \times 260{,}000 \text{ lives} \times 2\% \text{ chance of failure/100 yr}}{100 \text{ year expected life of dam}}$$
$$= \frac{\$10.4 \text{ million}}{\text{year}}$$

This is the overall cost, assuming the worst possible conditions, as spread over the 100-year life of the dam. However, in reality, failure would come at once, leading to a peak in costs, followed by prolonged costs related to the long-term payments associated with bereavement (e.g. pensions). These would diminish steadily over time as a result of inflation.

Some notable dam failures

The following examples indicate the range of problems and impacts caused by dam failure, including the often subtle relationship between the technological hazards created by reservoirs and their dams, and the natural hazards which they may provoke or which may affect them.

The St Francis Dam, which was 84.5 m high and 215 m wide, was located 73 km northwest of Los Angeles, California, in San Francisquito Canyon, a tributary of the Santa Clara River. It failed on 12 March 1928, two years after its construction, with the loss of more than 500 lives. The collapse of the structure released 46.5 million m^3 of water in a wave with an estimated discharge of 11,000–14,000 m^3/sec. The dam had been built on a fault, of a kind that engineering geologists would recognize easily, and failure was caused not by movement of this fracture but by seepage along it, which accelerated into erosion of the conglomerates on either side of the fault, which rapidly undermined the structure. As a result of this disaster, the State of California passed legislation to make geological site survey compulsory before dams can be constructed.

The Baldwin Hills Reservoir was built in 1951 at a site located 12 km southwest of Los Angeles; it failed on 14 December 1963, releasing 1.06 million m^3 of water onto a residential area, where $15 million of damage was caused and five people were killed. The impoundment consisted of a rolled earth dam 50 m high and was built on a site at which there are several minor faults. Warping and uplift had deformed the surface during the Late Pleistocene, while subsidence of up to 3 m had occurred nearby as a result of fault movement during the 1933 Long Beach earthquake and oil withdrawal over the period 1924–63. The latter was responsible for about 1 m of subsidence at the site, and in the vicinity for earth cracks that were 800 m long. Essentially, the subsidence led to fault movement, which caused dislevelment of the ground and overstressing of the dam. A small landslide which occurred during construction of the dam in 1950 had been inexpertly stabilized with cement and clay grout. Failure eventually occurred after leakage had begun along one of the fault lines and progressively accelerated over a period of 4½ hours into a major V-notch. Throughout the life of the reservoir water had leaked into the fractured zone under its foundations, developing pipes and cavities there. Failure was consequent upon the collapse of the structure along the eroded fault line that ran beneath it (James 1968, Castle et al. 1973).

The Teton Dam in Idaho failed on 5 June 1976 when the reservoir behind it was being filled for the first time (Boffey 1977). Some 302.8

million m³ of water flooded out of an enlarging crack in this structure, which was 94 m high and 915 m long. Eleven people died, 25,000 were made homeless and damage amounted to between $400 million and $1,000 million, according to different estimates. An independent panel of experts convened by the US Federal Government to enquire into the disaster blamed poor engineering work by the Federal Bureau of Reclamation. Water penetrated a grout barrier in fractured rock and eroded the silty core of this large earth dam.

The Vajont Dam (Fig. 5.12) was constructed over the period 1957–60 in a tributary valley of the River Piave in the Southern Alps of Trentino Alto-Adige Region in Italy. At 276 m it was then the world's second-highest, double-arch reinforced concrete dam. The alteration of ground-water conditions at the base of one of the reservoir's sideslopes led to a

Figure 5.12 The Vajont Dam in the northern Italian Alps. Behind the concrete arch can be seen the toe of the 1963 landslide, which sent a wall of water 70 m high over the lip of the dam and killed 1,925 people.

landslide of 700,000 m³ of rock into the reservoir as it was being filled on 4 November 1960. Over the succeeding three years, rock creep occurred in the brecciated dolomitic limestones interbedded with clays on the southern flank of the valley, whose strata dipped in the direction of inclination of the slope (Fig. 5.13). Initially, the rate of creep was 1 cm/week, but gradually it accelerated to 1 cm/day, and finally to about 80 cm/day. Then, on 9 October 1963 at 22.39 hrs, 275 million m³ of rock slid into the 115 million m³ of water in the reservoir, which was about two-thirds full. The sliding mass travelled as a flexible sheet of undisturbed rock formations (Fig. 5.13) at maximum speeds that have been variously estimated as 90 and 250 km/hr. A water wave was created which overtopped the dam by 70 m. The Piave Valley was flooded in only 7 minutes, killing 1,925 people. Debris climbed 260 m up the valley sides and spread 400 m wide some 1,800 m down valley, obliterating the town of Longarone (pop. 10,000), and the villages of Pirago, Villanova, Rivalta and Fae. The vertical fall of 25 million m³ of water 500 m over the lip of the dam produced energy equivalent to 170,000 kW/hr.

Essentially, the geological conditions were not suited to the construction of a reservoir, but the utility company responsible for the dam failed to act on the warning signs with sufficient decisiveness. Moreover,

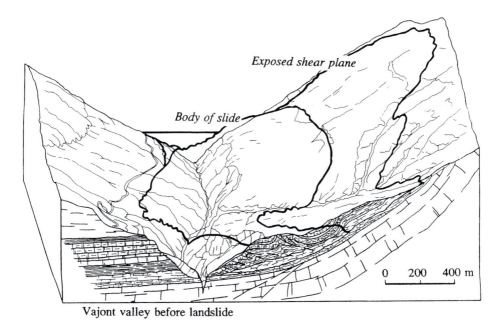

Vajont valley before landslide

Figure 5.13 Cross section through the Vajont Valley as it was in 1960, with the shear plane and mass of the 1963 landslide superimposed on it (after Selli & Trevisan 1964, Figs 9a & b).

geologists and engineers differed over the likelihood of a catastrophic slide (Müller 1968, Chowdhury 1979, Trollope 1981). However, the dam resisted forces that exceeded by eight times those for which it was designed, thus proving the durability of double-arch concrete spans across deep, narrow valleys. In conclusion, the disaster could have been avoided by the timely application of several measures, including slope drainage and stabilization of the slope foot. However, although it is easy to recognize this with hindsight, the necessity was not so evident before the disaster.

In the succeeding 25 years, two other disasters were to occur in the Italian Alps in connection with impoundments of water. On 19 July 1985 in the Stava Valley (part of the Dolomites), the rolled earth walls of two settlement tanks that formed part of a fluorite mining enterprise collapsed (Alexander 1986). A mudflow occurred at high velocity for about 5 km down a valley full of tourists, and 264 people were killed. Although the relief effort was extremely rapid and comprehensive, there were few survivors. Inadequate inspection of the safety of this two-tiered earth dam was blamed. In Valtellina, to the northeast of Val Stava, in July 1987, a major high-velocity rockslide occurred, obliterating two villages from which all the inhabitants had not been evacuated, of whom 43 died. The debris impounded the River Adda, creating a lake of 6 million m^3 of water, which threatened to breach the landslide barrier catastrophically and flood the valley. Some 25,000 people were evacuated down stream and for months frantic efforts were made to drain the lake and stabilize the debris tongue. This was finally accomplished without further fatalities (Alexander 1988, Costa 1991).

Reservoirs and seismicity
Seismicity resulting from tectonic forces may constitute a significant risk to reservoir dams. During the 1971 San Fernando Valley earthquake in southern California the tremors provoked a liquefaction flow on the flanks of the Van Norman Dam, a rolled earth structure 60 m high. At the time the water level was low and overtopping was narrowly avoided, but 80,000 people had to be evacuated from the downstream area where they had been threatened by catastrophic flooding. Had the tremors continued, more of the rolled earth and fill of the dam might have liquefied and total collapse might have ensued. As it was, the liquefied material flowed more than 100 m (Grantz et al. 1971).

Seismicity induced by the creation of reservoirs is discussed in Chapter 2.

Reservoir sedimentation
The products of erosion from upstream areas tend to settle out in the calm water of a reservoir and to shorten its design life by reducing the storage capacity. The reduction in gradient and enlargement of width

which reservoirs cause result in slower velocity of flow and hence fewer buoyancy forces maintaining transported particles in movement. The longer the residence time of water, the greater the amount and velocity of settling that takes place. If the ratio of reservoir capacity to mean annual inflow of water is high (as might be expected of ambitious schemes to impound the discharge of relatively arid areas), the dam will trap a high proportion of the solid load that feeds into it. This is known as its **trap efficiency** (Brune 1953). But studies have found the loss of capacity to be up to 15 times higher in small reservoirs (with capacities of less than 0.12 km^2) than in larger ones (see Table 5.5, Guy & Ferguson 1962).

In this manner, New Lake Austin, on the Colorado River (USA) lost 95.6 per cent of its capacity in only 13 years. Moreover, as a result of sedimentation, the Anchicaya Dam in Colombia, which was completed in 1955, could no longer store water or generate electricity by 1964. A study completed by the US Geological Survey in 1941 suggested that by the year 2000, 39 per cent of the reservoirs that then existed in America would largely have been filled with sediment. More recent data suggest that only 54 per cent of American reservoirs will function as such for more than a century, and 21 per cent will be in use for less than half that period (Dendy 1968).

In some cases bottom-hugging density currents can be deprived of part of their load of settling particles by building a pre-silting tank immediately up stream of the reservoir. Dredging tends to be a very expensive option, but in some cases the reservoir can be flushed of some of its sediment when scour occurs after opening the sluice gates wide. However, much reservoir sedimentation is not reversible. Its counterpart is **clear-water erosion**, which signifies the increased erosive capacity down stream of a river that has been deprived of its solid load by reservoir sedimentation (Chien 1985, Williams & Wolman 1984). Reservoirs that through being filled with sediment are not able to store excess floodwaters

Table 5.5 Sedimentation of 968 reservoirs (after Dendy 1968: 137).

Reservoir capacity (km^3)	Total initial storage (km^3)	Total storage depletion (km^3)	(%)	Average per reservoir per year (%)
0–8.11	556	145	26.3	3.41
8.11–81.1	6,650	1,387	20.9	3.17
81.1–811	78,706	13,158	16.7	1.02
811–8,110	396,087	41,440	10.5	0.78
8,110–81,103	3,400,000	299,097	8.8	0.45
81,103–811,030	14,800,000	514,393	3.5	0.26
811,030–8,110,000	31,000,000	1,090,000	3.5	0.16
Total	49,700,000	1,950,000	3.9	1.77

Length of record: 11–23 years.

may increase rather than abate the flood risk. As urban construction can cause a very substantial increase in the rate of sedimentation it should not be carried out up stream of the reservoir without appropriate measures to minimize the impact on the solid loads of streams. Indeed, in general terms, before a new reservoir is filled for the first time, slopes that show signs of instability should be carefully stabilized throughout the contributing catchment.

The ability of dams to trap sediment can, however, be utilized to good effect. Small sediment-trap basins on the urbanized hillsides of southern California trap an average of 2,000 $m^3/km^2/yr$ (but with a range of 500–25,000 $m^3/km^2/yr$). Provided that they are properly maintained, these traps are able to detain the raw material of what would otherwise be highly destructive mudflows. Periodic dredging, though, is very expensive.

Landslide dams
Natural dams created when landslides block rivers can be both spectacular and highly dangerous (Adams 1981, King et al. 1987). For instance, the earthquake-induced avalanche of 1911 at Usoy in the former USSR involved up to 2,500 million m^3 of rock and dammed the Murgab River to a height of 550 m. The Deixi landslide dam on the Min River in central China failed in 1933 and led to a catastrophic flood that killed 2,423 people. About 90 per cent of landslide dams are initiated by heavy precipitation or earthquake shaking, while volcanic eruptions are responsible for 8 per cent and processes such as devegetation and slope undercutting for 2 per cent.

Costa & Schuster (1988) classified documented examples of landslide dams into six types:
(a) Type I dams are caused by falls and slumps. They are small and shallow and do not reach right across the valley floor. In a worldwide survey of 184 landslide dams, they constituted 11 per cent.
(b) Type II dams result from avalanches, slumps and slides. They are large enough to span the entire valley floor and are much more hazardous than the previous category. They constituted 44 per cent of the sample of 184 dams.
(c) Type III dams are caused by debris flows or avalanches and are often the largest such movements. They not only span the valley floor (and perhaps part of the opposing sideslope), but also extend up and down valley for considerable distances. They constituted 41 per cent of the sample. Such landslides can not only create large, hazardous lakes, but may sometimes also block tributary valleys.
(d) Type IV dams result from falls, slumps, slides and avalanches of rock, soil and debris. They involve simultaneous movements from either side of the valley which either meet head-on or interdigitate, and they are rare (less than 1 per cent).

(e) Type V dams result from falls and avalanches and are characterized by multiple lobes from the same general movement, which cause multiple blockages of the river and corresponding impoundments of water. Like the last category they are rare, but they can be very hazardous if the valley is narrow and the sliding mass is voluminous enough to form a high, unstable dam.

(f) Type VI dams are the result of slumps or slides and have failure surfaces that extend under the stream to emerge on the opposite side. They constituted only 3 per cent of the sample.

Most landslide dams fail rapidly after formation, some within a day. The main factors that determine how long such a dam will last are the rate of river flow (hence the speed at which water accumulates behind the impoundment), the size and shape of the dam and its resistance to pressure, infiltration, undermining or other forces. Flooding can occur up stream as a result of the backup of impounded water, or down stream upon failure of the dam. The most common method of stabilizing dams has been to construct spillway channels across the debris (Fig. 5.14). Other methods include building tunnels through the material or bypass channels around it. Such work is often hazardous and has to be carried out rapidly before the dam yields catastrophically of its own accord.

Figure 5.14 Engineering works were rapidly constructed on the River Adda in the Italian Alps to drain and stabilize the natural dam created by the landslide of July 1987, which killed 58 people

In summary, world experience of dam hazards offers a sorry catalogue of failures, but a wealth of experience on which to base future design investigations. The price of safety is care in construction, strict standards of maintenance and constant vigilance.

References

Adams, J. 1981. Earthquake-dammed lakes in New Zealand. *Geology* **9** 215–19.

Alesch, D. J. & W. J. Petak 1986. *The politics and economics of earthquake hazard mitigation*. Boulder, Colorado: Institute of Behavoiural Science, University of Colorado.

Alexander, D. E. 1986. Northern Italian dam failure and mudflow, July 1985. *Disasters* **10**, 3–7.

Alexander, D. E., 1988. Valtellina landslide and flood emergency, northern Italy, 1987. *Disasters* **12**, 212–22.

Anagnostopoulos, S. A., J. Petrovski, J. G. Bouwkamp 1989. Emergency earthquake damage and usability assessment of buildings. *Earthquake Spectra* **5**, 461–76.

Arnold, C., R. Eisner, M. Durkin, D. Whitaker 1982. Occupant behaviour in a six-storey office building following severe earthquake damage. *Disasters* **6**, 207–14.

ASCE 1967. *Design criteria for large dams*. New York: Committee on Large Dams, American Society of Civil Engineers.

Baldwin, J. E., II, Donley, H. F., Howard, T. R. 1987. On debris flows/avalanche mitigation and control, San Francisco Bay area, California. In *Debris flows/ avalanches: process, recognition and mitigation*, J. E. Costa & G. F. Wieczorek (eds), 223–36. Boulder, Colorado: Geological Society of America.

Boffey, P. M. 1977. Teton dam verdict: a foul-up by the engineers. *Science* **195**, 270–2.

Bolt, B. A. 1988. *Earthquakes*. New York: W. H. Freeman.

Borg, S. F. 1983. *Earthquake engineering: damage assessment and structural design*. New York: John Wiley.

Brune, G. M. 1953. Trap efficiency of dams and reservoirs. *American Geophysical Union, Transactions* **34**, 407–18.

Burby, R. J., B. A. Cigler, S. P. French, E. J. Kaiser, J. Kartez, D. Roenigk, D. Weist, D. Whittington 1991. *Sharing environmental risks: how to control governments' losses in natural disasters*. Boulder, Colorado: Westview Press.

Carter, L. J. 1977. Auburn dam: earthquake hazards imperil $1 billion project. *Science* **197**, 643–7.

Castle, R. O., R. F. Yerkes, T. L. Youd 1973. Ground rupture in the Baldwin Hills: an alternative explanation. *Association of Engineering Geologists, Bulletin* **10**, 21–46.

Changnon, S. A., Jr., 1972. Examples of economic losses from hail in the United States. *Journal of Applied Meteorology* **11**, 1,128–37.

Cheng Yong, Kam-ling Tsoi, Chen Feibi, Gao Zhenhuan, Zou Qijia, Chen Zhangli (eds) 1988. *The great Tangshan earthquake of 1976: an anatomy of disaster*. Oxford: Pergamon.

Chien, N. 1985. Changes in river regime after the construction of upstream reservoirs. *Earth Surface Processes and Landforms* **10**, 143–60.

Chowdhury, R. N. 1979. *Slope analysis*. Amsterdam: Elsevier.

References

Clark, C. 1982. *Flood*. Alexandria, Virginia: Time-Life Books.

Coates, D. R. 1985. *Geology and society*. New York: Dowden & Culver, Chapman & Hall.

Coburn, A. W., A. Pomonis, S. Sakai 1989. Assessing strategies to reduce fatalities in earthquakes. *International Workshop on Earthquake Injury Epidemiology for Mitigation and Response*, July. Baltimore, Maryland: The Johns Hopkins University Press.

Cooke, R. U. 1984. *Geomorphological hazards in Los Angeles*. London: Unwin - Hyman.

Cornell, C. A. 1968. Engineering seismic risk analysis. *Seismological Society of America, Bulletin* **58**, 1,583–606.

Costa, J. E. 1991. Nature, mechanics and mitigation of the Val Pola landslide, Valtellina, Italy, 1987–1988. *Zeitschrift für Geomorphologie* **35**, 15–38.

Costa, J. E. and R. L. Schuster 1988. The formation and failure of natural dams. *Geological Society of America, Bulletin* **100**, 1,054–68.

D'Amore, R. D. 1982. The high-rise fire distaser: a plea for a new disaster response. In *Mass casualties: a lessons learned approach*, R. A. Cowley, S. Edelstein, M. Silverstein (eds), 257–61. Washington, DC: US Department of Transportation.

Dendy, F. E. 1968. Sedimentation in the nation's reservoirs. *Journal of Soil and Water Conservation* **23**, 135–7.

Dobry, R., I. Oeis, A. Urzua 1976. Simplified procedures for estimating the fundamental period of a soil profile. *Seismological Society of America, Bulletin* **66**, 1,293–321.

Driscoll, D. J. 1984. The legal implications of earthquake prediction. In *Earthquake prediction*, T. Rikitake (ed.), 849–56. Tokyo: Terra Scientific Publishing & Paris: UNESCO Press.

Duffield, J. W. 1980. Auburn Dam: a case study of water policy and economics. *Water Resources Bulletin* **16**, 226–34.

Dynes, R. R. & E. L. Quarantelli 1976. Community conflict: its absence and its presence in natural disasters. *Mass Emergencies* **1**, 139–52.

Fielden, B. M. 1980. Earthquakes and historic buildings. *Proceedings of the Seventh World Conference on Earthquake Engineering* **9**, 213–26.

Foster, H. D. 1980. *Disaster planning: the preservation of life and property*. New York: Springer.

Furbish, D. J. & R. M. Rice 1983. Predicting landslides related to clearcut logging, northwestern California, USA. *Mountain Research and Development* **3**, 253–9.

Geipel, R. 1990. *The long-term consequences of disasters: the reconstruction of Friuli, Italy, in its international context, 1976–1988*. Heidelberg & New York: Springer.

Godschalk D. R. & D. J. Brower 1985. Mitigation strategies and integrated emergency management. *Public Administration Review* **45**, 64–71.

Grantz, A. et al. 1971. *The San Fernando, California, earthquake of February 9, 1971*. US Geological Survey Professional Paper **733**, 1–254.

Green, N. B. 1987. *Earthquake resistant building design and construction*, 3rd edn. New York: Elsevier.

Gruner, E. 1963. Dam disasters. *Proceedings of the Institution of Civil Engineers* **24**, 47–60.

Guy, H. P. & G. E. Ferguson 1962. Sediment in small reservoirs due to urbanization. *American Society of Civil Engineers, Proceeding; Journal of the Hydraulics Division* **88**(HY2), 27–37.

Guy, H. P. & D. E. Jones 1972. Urban sedimentation in perspective. *American Society of Civil Engineers, Proceedings; Journal of the Hydraulics Division* **98**(HY12), 2,099–16.

Hageman, R. K. 1983. An assessment of the value of natural hazard damage in dwellings due to building codes: two case studies. *Natural Resources Journal* **23**, 531–47.

Hammond, D. J. 1989. A course in structural aspects of urban heavy rescue. *International Workshop on Earthquake Injury for Mitigation and Response*, July 1989. Baltimore, Maryland: The Johns Hopkins University Press.

Harriss, R. C., C. Hohenemser, R. W. Kates 1978. Our hazardous environment. *Environment* **20**, 6–15, 38–40.

Havelick, S. W. 1986. Third World cities at risk: building for calamity. *Environment* **28**, 6–11, 41–5.

Hays, W. W. 1986. The importance of post-earthquake investigations. *Earthquake Spectra* **2**, 653–68.

Huffman, J. 1986. *Government liability and disaster mitigation: a comparative study*. Lanham, Maryland: University Press of America.

Hughes, R. 1981. Field survey techniques for estimating the normal performance of vernacular buildings prior to earthquakes. *Disasters* **5**, 411–15.

James, L. B. 1968. Failure of Baldwin Hills reservoir, Los Angeles. *Engineering geology case histories no. 6*, 1–11. Boulder, Colorado: Geological Society of America.

Jennings, P. C. (ed.) 1980. *Earthquake engineering and hazards reduction in China*. Washington, DC: National Academy of Sciences, National Academy Press.

Johnson, F. A. & P. Illes 1976. A classification of dam failures. *Water Power and Dam Construction* **28**, 43–5.

Key, D. 1988. *Earthquake design practice for buildings*. London: Thomas Telford.

King, J. P., I. Loveday, R. L. Schuster 1987. Failure of a massive earthquake-induced landslide dam in Papua, New Guinea. *Earthquakes and Volcanoes* **19**, 40–7.

Kunreuther, H. 1974. Economic analysis of natural hazards: an ordered choice approach. In *Natural hazards: local, national, global*, G. F. White (ed.), 206–14. New York: Oxford University Press.

Lagorio, H. J. 1990. *Earthquakes: an architect's guide to nonstructural seismic hazards*. New York: John Wiley.

Leonards, G. A. (ed.) 1987. Dam failures. *Engineering Geology* **24**, 1–612.

McNutt, S. R. & R. H. Sydnor 1990. *The Loma Prieta (Santa Cruz Mountains), California, earthquake of 17 October 1989*. Sacramento, California: Department of Mines and Conservation.

Mark, R. K. & D. E. Stuart-Alexander 1977. Disasters as a necessary part of benefit–cost analysis. *Science* **197**, 1,160–2.

Meli, R. & J. A. Avila 1989. The Mexico earthquake of September 19, 1985: analysis of building response. *Earthquake Spectra* **5**, 1–18.

Minor, J. E. 1984. The mode of construction as a determinant of the extent of tropical cyclone damage to housing. *Ekistics* **51**, 463–9.

Mitchell, J. K. 1990. Natural hazard prediction and response in very large cities. *Proceedings of the IDNDR Workshop on Prediction and Perception of Natural Hazards*. Perugia, Italy: WARREDOC Center.

Müller, L. 1968. New considerations on the Vajont slide. *Rock Mechanics and Engineering Geology* **6**, 1–91.

Nakabayashi, I. 1984. Assessing intensity of damage by disasters in Japan. *Ekistics* **51**, 432–8.

References

Nakano, T. & I. Matsuda 1984. Earthquake damage, damage prediction and countermeasures in Tokyo, Japan. *Ekistics* **51**, 415–20.

Nilsen, T. H. et al. 1979. *Relative slope stability and land-use planning in the San Francisco Bay region, California.* US Geological Survey Professional Paper 944, 1–96.

Osteraas, S. & H. Krawinkler 1989. The Mexico earthquake of September 19, 1985: behaviour of steel buildings. *Earthquake Spectra* **5**, 51–88.

Page, R. A., J. A. Blume, W. B. Joyner 1975. Earthquake shaking and damage to buildings. *Science* **189**, 601–8.

Park, Y. J., A. H.-S. Ang, Y. K. Wen 1987. Damage-limiting aseismic design of buildings. *Earthquake Spectra* **3**, 1–26.

Rantz, S. E. 1970. Urban sprawl and flooding in southern California. *US Geological Survey Circular* 601B, 1–11.

Reitherman, R. 1985. *Reducing the risk of non-structural earthquake damage: a practical guide.* Washington, DC: Federal Emergency Management Agency.

Savage, R. P. et al. 1984. *Hurricane Alicia: Galveston and Houston, Texas – August 17–18, 1983.* Washington, DC: National Academy Press.

Scawthorn, C., Y. Yamada, H. Iemura 1981. A model for post-earthquake urban fire hazard. *Disasters* **5**, 125–31.

Seed, H. B. & S. D. Wilson 1967. The Turnagain Heights landslide, Anchorage. *American Society of Civil Engineers, Proceedings; Journal of the Soil Mechanics and Foundations Division* **93**(SM4), 325–53.

Selli, R. & L. Trevisan 1964. Caratteri e interpretazione della frana del Vajont. *Giornale di Geologia* **32**, 7–123.

Steinbrugge, K. V. 1982. *Earthquakes, volcanoes and tsunamis: an anatomy of hazards.* New York: Skandia America Corporation.

Sugg, A. 1967. Economic aspects of hurricanes. *Monthly Weather Review* **95**, 143–6.

Tank, R. W. (ed.) 1983. *Environmental geology.* New York: Oxford University Press

Thiel, C. C. & T. C. Sutty 1987. Earthquake characteristics and damage statistics. *Earthquake Spectra* **3**, 747–92.

Tiedemann, H. 1984. Quantification of factors contributing to earthquake damage to buildings. *Engineering Geology* **20**, 169–80.

Tobriner, S. 1984. A history of reinforced masonry construction designed to resist earthquakes, 1,755–907. *Earthquake Spectra* **1**, 125–50.

Trollope, D. H. 1981. The Vajont slide failure. *Rock Mechanics* **13**, 71–88.

Vice, R. B., H. B. Guy, G. E. Ferguson 1969. Sediment movement in an area of surburban highway construction, Scott Run Basin, Farifax County, Virginia, 1961-1964. *US Geological Survey Water Supply Paper* 1591E, 1–41.

Wahlstrom, E. E. 1974. *Dams, dam foundations and reservoir sites.* Amsterdam: Elsevier.

Watts, J. M. 1987. Computer models for evacuation analysis. *Fire Safety Journal* **12**, 237–45.

Weinroth, J. 1989. A model for the management of building evacuation. *Simulation* **53**, 111–19.

Williams, G. P. & M. G. Wolman 1984. *Downstream effects of dams on alluvial rivers.* US Geological Survey Professional Paper 1,286, 1–83.

Wolman, M. G. 1967. A cycle of sedimentation and erosion in urban river channels. *Geografiska Annaler* 49A, 385–95.

Wolman, M. G. & A. P. Schick 1967. Effects of construction on fluvial sediment, urban and suburban areas of Maryland. *Water Resources Research* **3**, 451–64.

Wood, R. M. 1981. El Asnam earthquake: notes from the fault line. *Disasters* **5**, 32–5.

Select bibliography

Alexander, D. E. 1989. Urban landslides. *Progress in Physical Geography* **13**, 157–91.

Ambrose, J. & D. Vergun 1985. *Seismic design of buildings*. New York: John Wiley.

Appelbaum, S. J. 1985. Determination of urban flood damages. *Journal of Water Resource Planning and Management* **111**, 269–83.

ASCE, 1976. *The evaluation of dam safety*. New York: American Society of Civil Engineers.

Bertero, V. V. 1986. Lessons learned from recent earthquakes and research, and implications for earthquake-resistant design of building structures in the United States. *Earthquake Spectra* **2**, 825–58.

Bertero, V. V. & E. Miranda 1991. Evaluation of the failure of the Cypress Viaduct in the Loma Prieta earthquake. *Seismological Society of America, Bulletin* **81**, 2,070–86.

Chandler, T. J., R. U. Cooke, I. Douglas 1976. Physical problems of the urban environment. *Geographical Journal* **142**, 57–80.

Coates, D. R. (ed.) 1976. *Urban geomorphology*. Boulder, Colorado: Geological Society of America.

Coburn, A. W. & Spence, R. J. S. 1992. *Earthquake protection*. Chichester, England & New York: John Wiley.

Davis, I. 1983. Disasters as agents of change? Or: form follows failure. *Habitat International* **7**, 277–310.

De Alba, P., H. B. Seed, E. Retamal, R. B. Seed 1988. Analyses of dam failures in 1985 Chilean earthquake. *Journal of Geotechnical Engineering* **114**, 1,414–36.

Durkin, M. E. 1984. Improving seismic safety in unreinforced masonry buildings. *Ekistics* **51**, 553–7.

Durkin, M. E. 1985. The behaviour of building occupants during earthquakes. *Earthquake Spectra* **1**, 271–84.

Durkin, M. E. 1989. The rôle of building evaluation in earthquake hazard reduction. In *Building evaluation: advances in methods and applications*, W. F. E. Preiser (ed.), 67-80. New York: Plenum.

FEMA, 1988. *Rapid screening of buildings for potential seismic hazards: a handbook*. Washington, DC: Federal Emergency Management Agency.

Gupta, H. K. & B. K. Rastogi 1976. *Dams and earthquakes*. Amsterdam: Elsevier.

Hollis, G. E. 1975. The effects of urbanization on floods of different recurrence intervals. *Water Resources Research* **11**, 431–5.

Judd, W. R. 1974. Seismic effects of reservoir impounding. *Engineering Geology* **8**, 1–212.

Komura, S. & D. B. Simons 1967. River bed degradation below dams. *American Society of Civil Engineers, Proceedings; Journal of the Hydraulics Division* **93**(HY4), 1–14.

Lennis, G. 1980. *Earthquakes and the urban environment* (3 vols). Berlin: The Chemical Rubber Company.

McCuen, R. H., B. M. Ayyub, T. V. Hromadka 1990. Risk analysis of debris-basin failure. *Journal of Water Resources Planning and Management* **116**, 473–83.

Mahr, T. & J. Magot 1985. Devastation of the environment by landslides activated by construction. *International Association of Engineering Geologists, Bulletin* **31**, 81–8.

Müller, L. 1964. The rock slide in the Vajont Valley. *Rock Mechanics and Engineering Geology* **2**, 148–228.

Newmark, N. M. & E. Rosenblueth 1971. *Fundamentals of earthquake engineering*. Englewood Cliffs, New Jersey: Prentice-Hall.

Parker, D. & E. Penning-Rowsell 1982. Flood risk in the urban environment. In *Geography and the urban environment*, D. T. Hubert & R. L. Johnston (eds). New York: John Wiley.

Pomonis, A. 1990. The Spitak (Armenia, USSR) earthquake: residential building typology and seismic behaviour. *Disasters* **14**, 89–114.

Reitherman, R. 1985. A review of earthquake damage estimation methods. *Earthquake Spectra* **1**, 805–49.

Rose, D. 1978. Risk of catastrophic failure of major dams. *American Society of Civil Engineers, Proceedings; Journal of the Hydraulics Division* **104**(HY9), 1349–51.

Simpson, D. W. 1976. Seismicity changes associated with reservoir loading. *Engineering Geology* **10**, 123–50.

Steinbrugge, K. V. & S. T. Algermissen 1990. Earthquake losses to single family dwellings: California experience. *US Geological Survey Bulletin* 1939A.

Tarry, W. I. 1980. Urban earthquake hazard in developing countries: squatter settlements and the outlook for Turkey. *Urban Ecology* **4**, 317–27.

Terwindt, J. H. J. 1983. Prediction of earthquake damage in the Tokyo Bay area: a literature survey. *GeoJournal* **7**, 215–27.

US National Research Council 1985. *Safety of dams: flood and earthquake criteria*. Washington, DC: National Academy Press.

Wiegel, R. L. (ed.) 1970. *Earthquake engineering*. Englewood Cliffs, New Jersey: Prentice-Hall.

Wood, W. J. & G. W. Renfro 1960. Trap efficiency of reservoirs and debris dams. *American Society of Civil Engineers, Proceedings; Journal of the Hydraulics Division* **86**(HY2), 29–87.

CHAPTER SIX

The logistics of planning and emergency action

Various practical means of conceptualizing disaster are reviewed in this chapter, in a sequence that progresses from early preparedness, through short-term reactions, to long-term aftermaths. First, hazards and their impacts can both be mapped as static geographical distributions or dynamic spatial processes. When the scale is large, satellite remote sensing can sometimes give a remarkably complete and concise picture of events as they occur, offering mappable data, especially when automated cartographic techniques are used. A map or remotely sensed image is essentially a form of low-level explanatory model. As shown here, other, more sophisticated kinds of model can be used to simulate the dynamics or effects of disaster, including the social processes which it entails. Models have also been made of the warning process, which blends social organization with the deployment, use and perception of modern technology. But warning itself must be a part of planning to reduce the impact of disasters, a process that should involve the mass media as a means of disseminating useful information. Hence the rôle of the media in disasters is examined here. Particular attention is then given to how, when disaster strikes, the emergency period is planned for. This is followed by an examination of the dual problems of providing temporary shelter for survivors and permanent reconstruction in disaster areas.

Disaster and hazard mapping

Any spatial aspect of disasters and hazards can be mapped, providing there is sufficient information on its distribution. In this respect the aspects most often plotted are the pattern of manifestations of the geophysical agents that cause disasters, the spatial distribution of their impacts and the distribution of human vulnerability. However, it is also possible to map emergency response and the viability of routeways and lifelines after disaster, although the potential contribution to disaster management of these methods has not yet been realized.

One of the principle goals of mapping is to define spatial units in such a way as to minimize differences within each unit while at the same time maximizing the differences between adjacent units. Maps are, by and large, a static representation of reality, and one that represents dynamic processes only very imperfectly. This limitation is particularly important in disaster and hazard mapping, as there may be little spatial information on the processes that cause the impact, such as the location of active faults or of areas which contribute to flood runoff in a particular drainage basin.

One way of increasing the power, potential and flexibility of mapping is to create a Geographical Information System (GIS) (Gupta & Joshi 1990). This is a form of electronic spatial library that involves data input, editing, verification, management, analysis, overlaying, comparison, output and display. Spatial statistics and spatial–statistical comparisons of data sets can be carried out. It is thus very useful as a low-cost form of seismic hazard assessment: geological information can be compared with vulnerability data, which can be related to seismic attenuation for hypothesized events, and so on. Remote sensing input can be derived from gravimetry, aeromagnetometry, Landsat or synthetic aperture radar (Ehlers et al. 1989).

Scales of measurement

There are four scales of measurement, each of which has a different relevance to disaster mapping. The simplest is the **nominal scale**, which involves mere presence or absence of a phenomenon or characteristic. It can, for instance, be used to keep a tally of the location of landslides of a particular type, or map the location of households that have purchased earthquake insurance. At a more detailed but still only semi-quantitative level, the **ordinal scale** is one of ranked categories, such as bad, mediocre, good and excellent, or slight, moderate and severe. It can be used to represent the relative intensity of a phenomenon, but in some cases the ranks may be assigned on the basis of subjective evaluation, rather than rigorous quantities. The quantitative **interval scales** consist of relative gradations according to a fixed interval, but with no absolute zero. For example, temperature is measured relative to an arbitrary starting point. These scales are of limited use in hazard mapping and are overshadowed by **ratio scales**, which involve relative gradations on a fixed interval relative to an absolute zero. Length is measured in this way, and mapping on such scales involves precise measurements and usually also the need for abundant data. The choice of which scale to use, and what level of sophistication to aim for, will be made on the basis of the availability of appropriate data, the resources of time, manpower and funds available to carry out the project and how detailed are the requirements of the eventual map users. At some point, there will be a trade-off between need for detail and for simplicity and legibility.

Types of map used in disaster studies

A wide variety of map types is used in the study of hazards and disasters. Little can be achieved without **base maps**, which furnish representations of topography and the most essential elements of human geography (such as political boundaries, highways and the location of settlements). On these, representations of the spatial aspects of hazards can be superimposed. **Thematic maps** include tectonic, lithostratigraphic or soil maps, as well as hydrological and geotechnical compilations. They furnish basic locational information concerning hazards.

A special type of thematic cartography, **hazard mapping**, indicates the danger present at any of the locations represented. The method defines the spatial pattern of past disasters (as these are expected to be a guide to future impacts) and the distribution of vulnerable buildings and structures, or of populations at risk. Co-variation from place to place in the intensity of both hazards and vulnerability can be depicted by **risk mapping**, which quantifies the hazard in terms of potential victims or damage. The representation of actual impacts, **disaster mapping**, is something of a contradiction in terms, given that cartography should always be methodical, precise and rigorous, which are qualities which the chaotic and tempestuous atmosphere of disasters tends not to foster. Thus, the vast majority of maps in this field relate to past events or impacts, not to crises as they evolve.

Maps of the physical nature of terrain can be constructed in a variety of ways in order to promote the identification of hazards and decision-making to mitigate them. For example, there is a hierarchy of four orders of engineering geology map (Mathewson & Font 1974). **First order (observational) maps** involve assessing each factor in terms of its impact upon the community. They show the technical results of geological studies in the form of hazards to settlement or lifelines. **Second order (engineering) maps** are used by engineers who have to construct the structural response to the hazard or damage. **Third order (interpretive) maps** show hazards and risk zones in terms of which areas are suitable for occupance or for structural or other modification of the risk. They are based on the first and second order maps prepared by and for geologists and engineers, but are for use by the general public. Lastly, **fourth order (planning) maps** show recommended land uses and thus help guide the formulation of policy.

Geological maps portray and interpret the distribution and structural relations of rock units that crop out at the Earth's surface and the sediments, or drift deposits, that mantle them (Blikra 1990). The maps are sometimes differentiated into those that deal with crustal units and those purely concerned with the sediments deposited upon the underlying bedrock. Both types of map are intended to elucidate spatial and chronological aspects of the stratigraphic column (the sequence of rocks and deposits that constitutes the outer part of the Earth's crust). They can often be used to clarify and identify earth materials, geological hazards (such as landslides and volcanic eruptions) and natural resources.

Geomorphological maps are a particularly useful means of creating an integrated picture of the natural land surface and its hazards (erosion, floods, landslides, subsidence, and so on). Such maps may be static, if they involve merely classifying land form, or dynamic, if they include some element of process (Brunsden et al. 1975). They rely on the correct interpretation of features as single, undifferentiable units (when boundaries in the natural landscape may be indistinct) and the correct understanding of the origin and material composition of each feature. Moreover, according to Doornkamp et al. (1979), the whole geomorphological system must be understood if the correct significance is to be attributed to any one element of it.

Most geomorphological maps are slope maps, with special symbols for ridges, crags, channels, and so on (Kertesz 1979). Although these are not essentially diagnostic, they can be a useful means of indicating the location of hazards, if some knowledge of processes is incorporated into them. This may include the critical angles above which slopes will fail, and the depth of weathered rock or soil liable to contribute to erosion, sedimentation or mudflows. Maps that emphasize landslides or erosion may be useful after other types of impact, such as floods or earthquakes, but there is always a danger of regarding pre-existing and normal "seasonal" features as having been induced by a particular disaster. More generally, it is essential to identify correctly the timescales which govern the evolution of the geomorphic processes which are to be mapped.

Geomorphological map making involves survey of literature and use of base maps, or other thematic maps, and use of aerial photographs. Ground survey is usually carried out at scales of 1:2500, 1:10,000 or 1:25,000 depending on the area to be covered in a given time or at a given level of detail. Field investigation must be used to verify or expand upon the information derived from aerial photographs or satellite images. From the point of view of engineers and engineering geologists, geomorphological maps can furnish information on how natural landforms and processes affect a site, how developing or otherwise altering the site will affect its geomorphology, how sensitive geomorphic processes are to changes induced by man, and to what extent landscape processes can be predicted, managed and controlled. The geomorphic maps thus form part of a wider endeavour to understand the risks associated with development and settlement of hazardous sites.

Soil maps are usually intended to represent pedological catenas, or the variation and change in soil characteristics along spatial traverses. Soil units are defined on the maps as areas in which there is a reasonable degree of uniformity or homogeneity in pedological characteristics. These are classified according to the type of soil (its origin, depth, colour, texture and particular horizonation), slope form and angle, the soil's parent material (bedrock or drift deposits), and other characteristics. Hazards come within the purview of the last of these: specialized pedological maps can represent factors such as a soil's shrinking and swelling capacity, its contamination with precipitated salts or its erodibility. They can also act as a guide to past flood deposits (Cain & Beatty 1968).

Both erodibility and erosivity can be represented on maps in the form of isohyets of the intensity of the phenomenon. The latter requires considerable amounts of data and calculation if any form of detailed spatial representation is to be achieved. Similarly, current erosion rates can be shown as isohyets for different soil losses in t/km^2, which can be related to causative factors, including rainfall intensity, slope angle and strength of sediments. When all of these factors are combined into an index and represented spatially, the result is an erosion potential map, in which the isohyets are intended to predict future distributions of soil loss on the basis of known susceptibilities (Whitlow 1986).

Many other forms of thematic map have been devised for particular natural hazards. Tsunami run-up and hurricane storm surge probabilities are often represented on maps of coastlines as a series of sectors in which the probabilities or recurrence intervals of particular levels of inundation are shown. Volcanic hazards or the effects of eruptions can be mapped in terms of the distribution of ashfall, the tracks of lahars or lava flows or the extent of lateral blasts and nuées ardentes (Booth 1977). Thermal emissions can be mapped for any given time on the basis of infra-red remote sensing, and geodetic maps can be created to show tilting, bulging or subsidence caused when magma rises beneath the surface (Rothery et al. 1988).

Seismic phenomena are among the most extensively mapped natural hazards. The basic cartographic forms are the maps of known epicentres (in which the symbols used to depict them are usually proportional to the magnitude of the earthquake) and isoseismals of the intensities of certain events (**macrozonation**; see Ch. 1). These can then form the basis of more sophisticated maps. Seismic risk maps use the accumulated evidence of past earthquakes to depict the pattern of future hazard (Fahmi & Alabbasi 1989, Muñoz 1989). One form of this is a map of recurrence intervals of damage to a certain intensity (e.g. $MM_I = VIII$), or a map of probable maximum accelerations in hard rock, expressed as a proportion of the gravitational constant, $g = 9.81$ m/sec^2 (see Fig. 2.7). Both maps are useful for determining where to concentrate expenditure on the seismic upgrading of existing buildings and what parameters to include in building codes for future anti-seismic construction.

At the local scale, microzonation may be necessary in order to express the detailed picture of variations in risk.

Microzonation

According to Varnes et al. (1984) the term **zonation** signifies division of the land surface into areas and ranking these according to degrees of actual or potential hazard. The technique of **microzonation** is designed to show the spatial variation of risk, including how to define acceptable risk in spatial terms, where to concentrate resources and where to send emergency services during critical periods, such as an annual rainy season (for floods and landslides) or a spring thaw (for avalanches). In the long term, land use should adjust to the recommendations provided by microzoning in order to ensure that impacts and settlement or communications patterns are as little coincident as possible. Microzoning maps are very expensive to produce, but the benefit:cost ratio of planning based on microzonation is high (it may be 3:1 for landslides, 5:1 for volcanic eruptions and earthquakes, and 1.3:1 for flooding).

Foster (1980) distinguished between three approaches to hazard mapping as a form of microzonation, depending on whether the hazards mapped and uses of the map are single or multiple. **Single hazard–single purpose mapping** is the cheapest and most rapidly executed strategy. It is appropriate where only one hazard is likely and impacts are expected to be simple rather than complex (for example, landslide risk in a suburban residential area), or where detailed information is required on only one aspect of the local natural hazards situation. There is a need to combine knowledge of past impacts (which is what are usually mapped) with theoretical predictions of future ones: for example, areas likely to be inundated by tsunamis are usually mapped according to the location of past impacts, but modelling of run-up processes could help give a better theoretical idea of future patterns of destruction. It is important not to base such maps on untenable assumptions: for example, that every slope above a certain angle is liable to fail, or that every stream draining a certain minimum area is a floodway.

Single hazard–multiple purpose mapping is appropriate where the risks created by a particular type of impact pertain to more than one activity: for example, risks to housing, manufacturing activities and transportation routes (e.g. Fig. 6.1). Microzoning of seismic vulnerability will take into account physical variables such as proximity to epicentres, probable acceleration levels and the effects of strong motion on soils and topography. It will also incorporate aspects of the human landscape, such as construction type to be permitted at each location, the likely nature of seismic impacts on structures and the logistical connotations of earthquakes. Hence, such maps show the varying intensity of and response to geophysical processes (e.g. Figure 6.2).

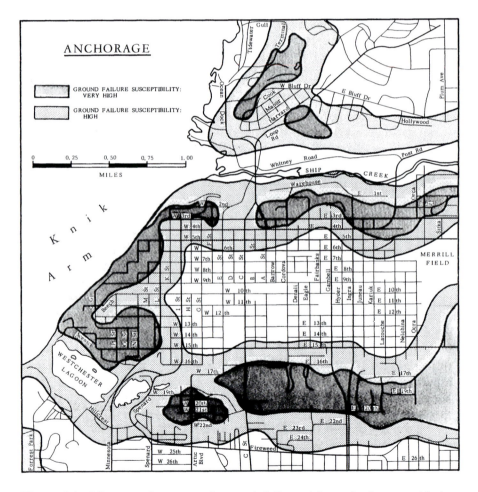

Figure 6.1 Microzonation map of ground failure risk at Anchorage, Alaska (Steinbrugge 1982: 81).

380

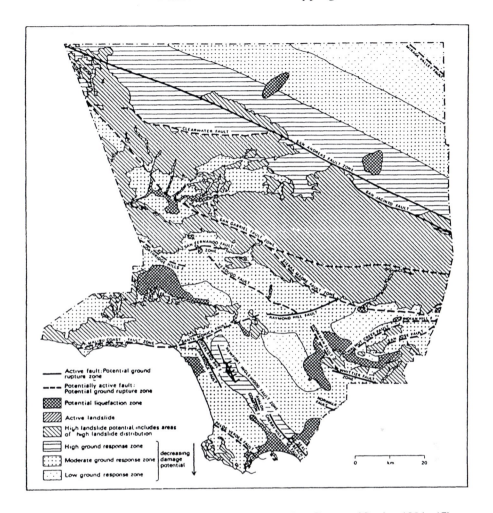

Figure 6.2 Seismic zoning map of Los Angeles County (Cooke 1984: 17).

Lastly, **multiple hazard–multiple purpose mapping** is the form of microzoning most useful to planners, but it is also the most time-consuming and expensive of the three approaches. Computerization, such as the creation of a Geographic Information System, may help handle the large amount of information needed, but it can also lead to artificiality. A detailed map of hazards to be expected in a defined area can be used as input to a detailed risk analysis and assessment of the spatial variation in loss potential. Both the hazard and the human response (in terms of damage and losses) must be quantified. The expected magnitude of each hazard must be weighted in relative terms, and the rôle of secondary hazards (such as seismic landslides) must be investigated. Total risk will

381

consist of some combination of hazard probabilities, in which timescales will be critical in terms of the arrival time of each hazard, its frequency and the number of impacts per unit time.

One example of single hazard–single purpose microzoning is the survey of seismic and post-earthquake fire damage potential carried out for Tokyo (Nakano & Matsuda 1984). First the city was divided into 2,300 blocks of 500 m^2. Secondly, basic assumptions were made about the magnitude and timing of the earthquake. Thirdly, liquefaction potential and ground response characteristics were investigated. Fourthly, a census was carried out of wooden houses. Next, the demographic characteristics of each block were investigated (population density and age distribution, traffic volume and infrastructure capacity, etc.). Then, the flammability of blocks was assessed: the spreading velocity of urban fire in Tokyo appears to be a function of wind velocity, building density and ratio of wooden buildings to other structures. Finally, the distance from each block to open-space refuges was measured. The results indicated the relative vulnerability of different buildings to collapse or burning, the probability that fire will spread and the danger posed by each block. A five-point damage potential scale was used in the eventual map of hazards.

But as microzoning is costly and slow, it is rarely carried out before disaster and, in fact, has been utilized mostly after earthquakes, landslides and floods when there is greater political consensus about the need for such studies. In the aftermath of a disaster the planning authorities should ensure that there are sufficient resources and there is adequate time to complete a microzonation study before reconstruction takes place. They should also decide on what level of total risk is acceptable for individual locations that have had their hazardousness mapped under a microzonation programme.

An example of flood hazard mapping
The following example illustrates the various components of a comprehensive scheme of single-purpose risk mapping. It is taken from the British cartographic provision for flood hazards, which was instituted as a legal requirement of local authorities in 1973 (Parker 1981, Handmer 1987).

First, maps must be drawn up at a scale of 1:25,000 to depict the location of floodable rivers, including both present channels and canals and proposed modifications to the fluvial network. Pumping stations, dams and locks should be included, as should discharge measuring stations and flood monitoring points. The maps must also show the jurisdiction of the various local and regional authorities within the drainage basin.

Secondly, overlays must depict areas likely to be subject to inundations deeper than 1 m, or which are permanently at risk from flooding. Separate overlays should chart the directions of water flow during flooding, the probable velocity of flow (where this can be estimated), and the location of flooding likely to be caused by blocked drains. They should also show floodplains, areas which lack adequate drainage or outflow, areas of excessive overland flow (actual or future) and zones of land subsidence. Vulnerable flood defences must be shown, as must buildings and structures at risk from flooding. Special symbols are used for those buildings which have been flood-proofed. More generally, land use and level of development of floodplains are mapped on the overlays, as are proposed and existing structural defences against floods. Finally, estimates of the duration of flooding should be incorporated into the overlays.

To accompany the maps, tables must be compiled of population size, the numbers of properties at risk and estimates of damage expected in future floods. Costs should be tabulated for warning and forecasting measures, the adaptation of structures and appropriate planning measures. Lastly, the appendices of the flood study must list the criteria used to derive cost estimates.

Remote sensing and disasters

Remote sensing constitutes a versatile and varied range of techniques, whose full potential for the study and management of disasters has not yet been realized. It has potential, and often untapped, applications to the study of accelerated erosion, desertification, drought, floods, forest fires, landslides, severe storms (including hurricanes and tornadoes), seismic activity (fault location and the more evident aspects of devastation caused by earthquakes), and volcanic activity (for example, heat emissions and atmospheric pollution). Moreover, the environmental impact of certain technological disasters can be monitored, such as algal growth ("red tides") associated with the eutrophication of coastal seas and the destruction of vegetation by the spread of catastrophic pollution.

The range of techniques known as remote sensing entails acquiring information about the ground surface from a distance without physical contact. This can involve ground-based sensors, such as infra-red telescopes, or those based in aeroplanes or satellites. Various terrestrial and atmospheric phenomena can be sensed remotely. One of the most common is **electromagnetic radiation (EMR)**, which is emitted by or reflected from almost all surfaces or objects. This includes heat emission or reflection: hence, albedo is useful for the study of erosion, desertification, devegetation, drought and sedimentation; geothermal heat for the study of seismic zones and volcanic activity; and man-made heat emissions relate to concentrations of environmental pollution. In a different part of the EMR spectrum, light emission facilitates visual pictures or images of light distribution. **Force fields** such as gravitational variations or the magnetism of the Earth's crust can be sensed remotely, as can the **mechanical vibrations** (generally acoustic or seismic) emitted or transmitted by the Earth's surface or objects on it. Finally, **radar** can be used to penetrate cloud cover and pick up geological lineaments such as fault breakages.

The hardware of remote sensing

Images which are useful for disaster and hazards studies are obtained from ground-based instruments, aerial surveys or space. Ground-based sensing is carried out by **activated sensors** (which operate automatically; for example, a flash flood sensor activated by rising water levels) or by **manual sensors** (for example, an infra-red telescope for observing volcanic heat emissions).

Aerial images may be either vertical or oblique. Oblique images can sometimes give a better three-dimensional effect than vertical ones, but the latter give more objective information and are much more useful for mapping. Panchromatic vertical aerial photographs taken with overlap have sufficient difference in aspect to produce a three-dimensional stereographic effect when viewed as pairs, and this aids recognition of surface features. Aerial images are divided into high-level (taken from altitudes of 5,000–10,000 m) and low-level (obtained at lower elevations, perhaps a few hundreds of metres).

There are four main types of space-based remote sensor. **Communications satellites** are used for video, radio or telephone transmission (see section on emergency management). **Earth resources satellites** include Landsat (formerly ERTS) modules, which orbit the Earth elliptically between 50°N and 50°S with a 16-day repeat cycle. Likewise, the French SPOT (*Système pour l'observation de la terre*) satellite orbits on a 26-day cycle and covers a 60-km swathe. **Geophysical and geodetic satellites** are used for accurate global surveying and sensing of physical variables such as the Earth's magnetic potential. In addition to conventional sensors new geophysical satellites will carry microwave altimeters having a vertical sensitivity of 5–10 cm, which could help monitoring of storm surges. Similarly, using the Global Positioning System, satellite-borne lasers with a sensitivity of 2 cm could help measure ground deformation associated with volcanic eruptions or impending earthquakes. Finally, **sea satellites** such as SEASAT and meteorological satellites such as the GOES, NOAA and GMS families are designed to sense, respectively, marine and atmospheric conditions. Besides orbiting satellites, there are geostationary ones, which rotate with the Earth and can supply pictures of the same area at intervals of fractions of an hour.

Landsat and similar satellites collect information and relay it to receiving platforms on Earth in the form of picture elements, or **pixels** (Curran 1985). Each pixel registers the intensity of EMR at the wavelength being sensed, and the image is made up of a mosaic of many millions of pixels. Landsat produces a square image of an area which is 185 km on each side (34,225 km^2) and is known as a **scene**. The sensor on SPOT is pointable and captures scenes which are 60 km square. Pixels are 67 m square in the older forms of Landsat sensor, namely the multi-spectral scanner (MSS) and return beam videcon (RBV), giving an area of 0.45 ha. Later MSS data involve pixels that are 30 m wide, and the newer thematic mapper (TM) of Landsats 4 and 5 has reduced the size of pixels to 20 m (10 m in the case of panchromatic data relayed by SPOT). New satellites are planned by India, Japan, the USA and the Commonwealth of Independent States with resolutions that vary from 3.6 to 30 m, but this increases correspondingly the amount of data to be manipulated.

Although about half of all satellite passes fail to obtain any scenes, as many forms of sensor cannot pierce cloud cover, there is a good chance that orbiting satellites will pass during the critical phase of an extreme geophysical event, and the virtual certainty that geostationary modules will transmit useful images back to Earth. Tradeoffs occur between radiometric, spatial and temporal resolution: the faster a Landsat-type satellite moves (e.g. 24,000 km/hr), the more rapidly it must acquire data, the less time it will have to spend on sensing a particular location over which it passes, and therefore the less sensitivity there will be in the image. Alternatively, it must circulate at lower altitude and thus sacrifice breadth of field.

Images must be filtered digitally to **geocorrect** for distortion caused by the Earth's curvature and to reduce the effect of interference. During filtering, the brightness, colour or tone of the individual pixel is altered on the basis of its relationship with its neighbours. Low-pass filtering smooths out detail, while high-pass filtering amplifies it. The angle at which the sun strikes the land surface may influence the resolution level and degree of definition of the image. Filtering can help correct this and also compensate for the effect of cloud cover, which obscures the image. Digital methods of **contrast stretching** enhance the image by increasing diversity in the brightness of pixels to cover the entire dynamic range rather than only those tones sensed during the collection of the raw data. **Edge enhancement** emphasizes pixels which differ from their immediate neighbours and can thus help to increase contrast within the processed image. The aim of these methods is to enhance the **signature**, or unique signal response, which differentiates what one wishes to sense from all other phenomena on the image.

In view of the cost of computer time and vast amounts of data involved in an image composed of millions of pixels, processing tends to be expensive. However, it is versatile: images can be used to produce conventional maps of large areas or rapidly changing phenomena, or the end product can be portrayed statistically. Thus, indices or principal components can be used to relate characteristics and qualities of the land surface to terrestrial processes.

According to Walter (1990), there are three phases of disaster management with which satellites can help: prevention, preparedness and relief. Both cause and effect can be observed; thus a situation can be monitored as it develops, or its consequences can be assessed afterwards. The remote sensing implications are different for sudden onset and creeping disasters. In the case of the latter, monitoring by comparison of satellite images has proven particularly successful. For disaster studies, two factors in remote sensing are critical: the **resolution level** of the image, or in other words its detailed information content, and the repetitiveness of image transmission. Disasters tend to be dynamic situations and their evolution needs to be followed closely.

Limitations of remote sensing
The characteristic to be sensed must produce a valid signature, or unambiguous sign, on the image. As different features respond to different EMR spectra, much thought may need to go into the processing of terrestrial satellite images in order to be able to distinguish clearly between, say, flooded areas and perennial wetlands. Assumptions about the extent or intensity of sensed phenomena must be checked in the field in order to establish **ground truth**.

The individual pixel is of no particular use, and thus the resolution level of satellite photographs (and hence what they can depict) is limited by the contrast between pixels. Small, linear or variegated patterns are not necessarily visible on Landsat images. Moreover, few Landsat or SPOT images that have more than 10 per cent cloud cover are useful, and the best images from Earth resource satellites are obtained in dry, cloudless weather.

Clearly, Landsat images obtained from an altitude of 35,000 km and aerial photographs taken at 3,500 m above ground pertain to different levels of detail. The difference is often one of reconnaissance and detailed mapping. In most cases, air sorties cost too much to be carried out on a very repetitive basis, and they cannot cover the area surveyed routinely by satellites. Combinations of methods are thus needed – also combinations of frequencies when using satellite data, as many phenomena show up best when EMR wavebands are combined in a false-colour composite image. Hence the monochrome, infra-red and visible spectrum wavebands have complementary rôles.

Remote sensing in disaster situations

Using remote sensing techniques, large and inaccessible areas can be surveyed rapidly after disaster has struck them. Satellite images can be used to assess the passage of storms or the extent of flooding (Brown et al. 1987), while aerial photography can provide information on landslides or damage to structures (Maruyasa 1984). In many cases, however, the most diagnostic effects of disaster may not be visible from the air, as in the case in which a settlement has been rendered uninhabitable by an earthquake, but full-scale structural collapse there is limited. Such information can also be ambiguous as to the nature of cause and effect and should always be followed up by a full-scale field visit. In any case, fog or cloud blankets can invalidate the image by obscuring the damage or other effects, and they are common after meteorological disasters. Nevertheless, aerial photograph sorties are usually flown after geological disasters such as earthquakes and volcanic eruptions, and they generally give a good basis for directing emergency services to points of maximum impact, especially as they may indicate which transportation routes are functional and which severed.

As the following sections show, different kinds of disaster benefit in a variety of ways from different remote sensing techniques.

Meteorological and hydrological disasters

As they develop, hurricanes produce a distinct radar and satellite image in the form of a hook-shaped cloud funnel, which represents the top of a circular depression. Nascent hurricanes can be classified on the Dvorak scale (see Ch. 3). Some die out, while others grow to the proportions of major storms, and the direction of development is difficult to predict, as one is looking only at the "exhaust" production of the hurricane "heat engine" (the outflow of air and water vapour aloft), not at its inputs. Geostationary satellites (such as GOES-2, GOES-3, GMS-1 and Meteosat) give global coverage of weather conditions between 50°N and 50°S at half-hourly intervals. This is particularly useful in monitoring hurricanes as they move towards landfall (their most dangerous and destructive stage), at which point coastal radar stations will be tracking them continuously.

Since 1982, the USA has used a Centralized Severe Storms Information System (CSSIS) based at the National Severe Storms Forecast Center at

Kansas City. It receives data from geostationary satellites in order to determine the atmospheric conditions which give rise to severe thunderstorms and tornadoes. Using the CSSIS, the percentage of tornado watches ending in the appearance of at least one tornado has risen from 45 to 53. Doppler radar, which measures the velocity of moving raindrops, has also had a dramatic impact on the detection of tornadoes and the elimination of false alarms (see Ch. 3).

Where cloud cover does not inhibit remote sensing, satellite images can record flooding as it occurs, using, for example, Landsat Multispectral Scanner band 7 (0.1–1.1 µm) or AVHRR channels 1 (0.58–0.68 µm) and 2 (0.725–1.1 µm; Ali et al. 1989, Rasid & Pramanik 1990). From such images it may be possible to identify the storms that produce floods, map the extent of inundation, map the varied characteristics of floodplains, and monitor flood water drainage. Runoff into lakes can be detected, as can the crest of a floodwave as it passes down a river. If appropriate criteria can be developed and verified by inspection on the ground, floodplains can be delineated rapidly and inexpensively using the near infra-red and red bands of the electromagnetic spectrum (Harker & Rouse 1977). As plants retain moisture, the near infra-red band can record the maximum extent of flooding up to a week after it has occurred. However, cloud cover associated with flood-producing storms is a major problem for Landsat-type sensing of floods, although, for instance, meteorological satellites help define the developing atmospheric conditions that enable the forecasting of thunderstorms which are sufficiently intense to cause flash floods. Finally, basic black-and-white aerial photography can be used to define the limits of flooding (Blyth & Nash 1980), but neither this nor satellite imagery will give the depths of floods.

Avalanche hazard maps can be produced accurately using aerial photographs (Walsh et al. 1990). These enable avalanche tracks up to 200 years old to be delineated according to topographic features, aspect, slope steepness, the presence of fallen rock and soil debris and the composition and age structure of vegetation. Landsat positive prints are not useful for this, because the image quality is degraded, but satellite data in digital form are useful because there is high contrast between snow-covered areas (with high albedo) and adjacent forested areas (where albedo is low). Ground truth must always be sought, and unconfined avalanches, above all, require supplementary information in addition to remotely sensed data. Landsat and SPOT could at least be used for rapid reconnaissance mapping of avalanche hazards, especially if other data are lacking (Francis & Wells 1988). Landsat images can also help trace the seasonal melting of snowpacks, which, if it is rapid or severe, may be a prelude to avalanches or severe flooding (Hall & Martinec 1985).

The Food and Agriculture Organization considers satellite information very useful for monitoring agricultural disasters such as drought. Satellites can be used to sense the condition of both the atmosphere (the configuration of severe storms or occurrence of prolonged clear weather) and the Earth's surface. Data may be reliable, routinely collectable and relatively cheap to obtain. Moreover, remote sensing can serve as a means of checking reports of drought or other agricultural disasters in remote areas. Drought causes changes in soil moisture, surface waters and vegetation (which undergoes a loss of greenness on digital images), all of which are reflected in albedo (Teng 1990). Worldwide, satellite image processing is combined with assessment of historical data on climatic variations, crop condition indices and other information to give a potential 3–6 month forecast of drought in any of 400 agroclimatic risk zones.

Geological disasters
There is some potential for using terrestrial satellite data to detect and monitor soil erosion. Variations in the organic material content of soil can be detected on the MSS colour infra-red, visible red and thermal infra-red wavebands (Frank 1984). Soil moisture is detectable on the near infra-red black-and-white and colour bands, and the thermal infra-red spectrum. Topography is shown on the colour infra-red and panchromatic black-and-white red band, and to a lesser extent on most other MSS wavelengths. Some erosional features such as gullies and geological lineaments such as faults can be detected on the Landsat MSS thermal IR band (10.4–12.5 µm) and the SPOT colour IR band (0.5–0.89 µm).

A semi-quantitative approach to factors which contribute to erosion potential is possible using satellites, but no single signature or band shows up soil erosion unequivocally. Both albedo and texture measurements can be used to characterize the relative condition of geomorphic surfaces. But in semi-arid lands (which tend to have the greatest erosion potential) albedo only distinguishes eroding areas well during the dry season, when vegetation cover does not obscure them. Moreover, erosional patterns and textural features are not very directly related. However, contrasts in both albedo and texture can be enhanced by high- and low-pass filtering.

Landsat is only useful for the preliminary mapping of landslides at a large scale. Terrain mottling and banding show up on frequency band 7, while band 5 shows the vegetational changes sometimes associated with landslides (Sauchyn & Trench 1978). After a landslide disaster, low-level aircraft sorties can be helpful, not merely for standard aerial photographs, but also thermal, black–and–white and colour infra-red images (these were used to great effect, for example, after the 1971 St Jean-Viannay quick-clay landslides in Quebec, in which 18 people died). Such sorties can help in the planning of ground surveys and the detection of groundwater seepage zones, while, if repeated, they can act as a monitoring method (Penn 1984). Groundwater conditions can best be assessed by thermal infra-red imagery collected shortly before sunrise, and by black-and-white infra-red images. Similarly, colour infra-red photography can help locate seepage zones by indicating high concentrations of surface moisture or areas of hydrophytic ("water-loving") vegetation.

There is considerable potential for remote sensing as a means of monitoring desertification (Walker & Robinove 1981). Standard and enhanced Landsat images can be used to make inventories of land use, and to monitor changes in vegetation, soil, soil moisture, surface water and land use. The repetitiveness of satellite images makes them particularly suited to temporal comparisons, for example of migratory sand dune belts, of present and past drainage and irrigation systems, or of the transport of dust in the wind. Temporal variations can be measured quantitatively on digital tapes, although the process of computer enhancement of the tapes tends to be very expensive. Terrain brightness, albedo and green indices can all be used to assess land degradation, as many of the phenomena which lower the productivity of land systems in arid areas come out brighter on remotely sensed images. Much debate has taken place about what phenomena, and above all what changes, are being observed, but there are clearly dramatic contrasts in the productivity of land. Some of these are related to political boundaries reflecting varied intensities of land conservation, for example between Botswana and Namibia, the Negev and Sinai, and the Orange Free State and Lesotho. The fertile corridor of the Nile stands out particularly sharply against the surrounding desert sands.

Volcanic eruptions are receiving increasing attention from remote sensing specialists (Matson 1984, Bianchi et al. 1990, Adams et al. 1991). In general, satellite images do not have resolution levels below 120 m, but aerial surveys are too expensive to use in routine volcanic monitoring. Hence, most volcanoes are badly documented, and measurement is usually carried out after, rather than before, an eruption begins. This means that the effects are more widely monitored than the precursors. Nevertheless, there is ample scope for predicting many eruptions using an array of techniques that includes various methods of remote sensing.

One of the first volcanic phenomena to be sensed remotely was heat emission (Rothery et al. 1988). Besides helping to forecast eruption, knowledge of volcanic heat emissions is also useful to scientific studies of volcanism in general. Although not all eruptions are preceded by substantial changes in ground temperature, the event is usually preceded by the introduction of hot magma into a chamber beneath the volcano and there may be a detectable surface heat anomaly, especially with slow-moving andesitic magmas. A thermal map of the volcano can be produced from aerial, satellite or ground-based infra-red sensors, using the 3–5 μm and 8–14 μm wavebands. In this respect, the best data are obtained from aerial surveys that use scanning instruments that create thermal images. Thermal Infra-red Multispectral Scanners (TIMS), which sense in the 8–12 μm band, can distinguish heat sources and the relative age of young lava flows, thus giving an idea of the magnitude and frequency of eruptions. The Thematic Mapper Simulator (TMS) senses the visible and three infra-red spectra, and can provide synoptic information on very hot volcanic features which emit electromagnetic radiation in the 1–2 μm band. Erupted products can be delineated by Landsat, although aerial photographs are often sufficient (Mouginis-Mark et al. 1989).

Weather satellites can be used to monitor patterns of the emission and transmission of volcanic dust in the stratosphere. As it can be invisible to pilots' vision or on-board radar, volcanic ash is a severe hazard to aviation. It can cause abrasion and corrosion of aircraft windscreens and fuselages and flame-out stoppage of jet engines, which may ingest and fuse ash particles (Kienle et al. 1990). Yet volcanic plumes can be detected by the Advanced Very High Resolution Radiometers (AVHRRs) carried aboard NOAA weather satellites, which provide coverage of much of the Earth every 12 hours. Clouds of volcanic ash appear on band 5 (11.5–12.5 μm) and at night on thermal infra-red band 3 (3.55–3.93 μm). Bands 4 and 5 together (10.3–12.5 μm) show up the contrast between normal cloud cover and clouds rich in ash particulates. Contrast can be enhanced by assigning "false" colour tones to particular ranges of values on the digital image.

In fact, weather satellites are proving extremely versatile in terms of rapid volcanic monitoring (Malingreau & Kasawanda 1986). The Total Ozone Mapping Spectrometer (TOMS) carried aboard Nimbus 7, for example, can use the 0.3–0.335 μmm waveband to map the atmospheric distribution of sulphur dioxide emitted during eruptions. GOES satellites can give information on the changing location of ash plumes over time, while radar sensors, which respond to wavelengths of 3–70 cm, can monitor the volume and position of lava flows.

The rôle of satellites in volcano monitoring is set to expand. The Earth Observation System (EOS) scheduled for launch in the late 1990s will carry ultraviolet, visible, infra-red and microwave sensors: an 11-year investigation is planned, using EOS as a synoptic volcano eruption and hazard monitoring system (Mouginis-Marr et al 1991). At the same time, a new Orbiting Volcano Observatory (OVO) will provide resolution of a few metres and will enable gas emissions, thermal anomalies and changes in volcanic landforms to be assessed accurately, especially as microwave and radar sensors can obtain an image of surface conditions under cloud cover.

The map and the remotely sensed image can be considered as elementary models of geographical distributions, which they simplify to their essential elements. But as the following section shows, a wide range of modelling techniques can be applied to the study and management of disasters.

Models of disaster and their practical uses

Simulation models can be used to study and predict the probable effects of disaster so that action can be taken to avoid the worst consequences. If necessary, this action can be built into the model in order to see whether it will be effective in reducing losses caused by the disaster. A model is a simplification of reality which may take one of several forms, for example, scale, analogue or mathematical (digital or conceptual). Any of these methods can be used to test the effects of disaster without the harm involved in the real event.

Scale models

Scale models use physical hardware. Landslides, floods, mudflows, reservoirs with dams, tornadoes, earthquakes and tsunamis can all, to a greater or lesser extent, be reproduced in scale ratio (with respect to the original) in the laboratory. The causes of disaster impacts can be investigated, and structural adjustments tested before they are constructed in the real situation (for example, simulating a storm and flood and, interactively, a set of flood barriers). The aim is to see where the worst impacts are and how best to design engineering structures to mitigate them (Lang & Dent 1980).

One good example of the investigative use of physical simulation is the model built after the mudflow disaster of 1966 at Aberfan in South Wales (see Ch. 4). It was possible to simulate the effect of high slope angle on the stability of the material that slid and flowed, high water pressure that arose in sediment pores after heavy rainfall, and the development of internal shear planes as a result of the pattern of dumping colliery waste on the tip.

In Japan an earthquake simulator has been built capable of subjecting a 200-ton model structure to a vertical acceleration of up to 1 g (9.81 m/s^2), which allows precise prediction of the impact on buildings of earth tremors. Models of ports, harbours and estuaries have been built as large as 6,000 m^2 (e.g. for the San Francisco Bay). Dangers to shipping, salt-water infiltration into neighbouring land and river sedimentation patterns can be studied. The model of Hilo Harbour, Hawaii, has been used to help design structures that were built to provide protection against tsunami impacts. The construction of the Libby Dam in Montana involved prior work with a model at 1:120 scale, which was able to simulate hazards of landsliding into the Libby Reservoir of what, in a real situation, would amount to 4,300,000 m^3 of rock.

Laboratory models of tornadoes use a variety of fluids – liquid and gas – to study airflow patterns, latent heat release, vortex creation and the effects of gravity on airflow (Turner & Lilly 1963). In a rather gentler capacity, rainfall simulators are much used in the study of soil erosion. The duration and intensity of rain can be controlled; drop size, impact velocity and the kinetic energy of the storm can be varied. And, finally, water and sediment can be collected at the base of the experiment to determine the infiltration capacity and erosion potential of particular soils. Portable simulators can be used on soils in the field, or relatively undisturbed samples of soil can be taken into the laboratory for use under larger but immobile equipment (Yoo & Molnau 1987).

Electrical analogue models
The flow and regulation of electrical current can be used to model water flow, heat flow, traffic flow, the spread of fire and other diffusion patterns. Analogue computers of this kind have largely been supplanted by their digital equivalents, as the manipulation of numbers allows more flexibility to be exercised than is possible with circuits. But analogue models can still be used profitably to help identify breaks in complex networks, such as water mains or electricity grids, and indicate how to avoid losing the supply.

Digital computer simulation
Simulation using mainframe, mini- or microcomputers is often capable of providing the essential overview on which preparedness plans can be based (Carroll 1983, Sullivan & Newkirk 1989). It can help forecast when and where hazards will strike, and which roads, hospitals, services and utilities will be rendered inoperative by a disaster of a given magnitude. It can be used to help establish danger zones and thus aid warning procedures and evacuation plans, and it can help plan evacuation efficiently. Thus, Belardo et al. (1983) designed a computer simulation to train emergency managers to assign resources in the order in which they become available and in the smallest possible transport time. Moreover, Weinroth (1989) used computer modelling to simulate building occupance, exit time and rate and mode of evacuation during an emergency. Evacuation policy was modelled in the form of instructions to evacuees, and delays and bottlenecks were built into the simulation.

In 1969 computer models discerned heavy snowpacks and predicted large-scale spring flooding in Missouri. Having forecast the location, timing and magnitude of the floods, $36 million were spent on preparation for the event and it is estimated that $250 million of damage was prevented (a benefit:cost ratio of 7:1). Structural modifications built at this time have helped protect against subsequent floods.

Some of the most elegant computer simulation models have combined physical and human variables to simulate the eventual outcome in terms of damage, losses and disruption. The flood simulation shown in Figure 6.3 does this by inputting random numbers (which represent "chance" factors) selected from statistically reasonable distributions of values. Digital outputs, in the form of matrices, can be plotted for selected iterations to give a visual impression of the simulation outcome. In Figure 6.4 wave travel patterns are shown (with a vertical exaggeration of 2 million in order to make them visible) for a tsunami generated in Prince William Sound, Alaska.

One possible technique of computer modelling of disasters (Foster 1980) starts with an initial analysis of the hazard or threat. Next, a model is developed that produces a spatial pattern of values with contours that correspond to the size, shape and configuration of observed patterns. The distribution of these will be controlled by the magnitude of the disaster event and the various influences that modify it. Hence, there are three types of input to the computerized disaster model. First, the **natural event generator** creates a data bank which represents the frequency and magnitude of a given event according to geographical location. It will utilize records of the magnitude and frequency of past geophysical events that have caused disasters. The **populations at risk** in each area form the second source of data, fed in according to density and geographical area, as defined by census returns. Economic data can also be used here. A third input, to be fed in with the other data, is constituted by the **vulnerability of the populations at risk** to a given severity of event. This will require surveying the susceptibility to damage of structures located in the predicted impact area (for example, the inner floodplain in the case of a riverine flood hazard).

Once the methodology is defined and the input data have been collected, losses are simulated. The result is a matrix of impact or intensity values and a loss-propensity matrix. This latter, derived from both field survey of present conditions and records of past disasters, tells us what losses to expect from an impact of a given size. The simulated impact of disaster is then expressed as a matrix, which is multiplied by a matrix of vulnerabilities to loss in order to yield the pattern of outcomes, location by location. The use of different ranges of input values allows the losses to be assessed for a variety of sizes of disaster impact, and also allows various loss-reduction strategies to be tested. Finally, the results of simulations must be tested against the record of past disasters, in order to ensure that they are realistic and probable.

Computer simulation has several drawbacks. Only a limited timespan can be covered, many simplifying assumptions must be made and the

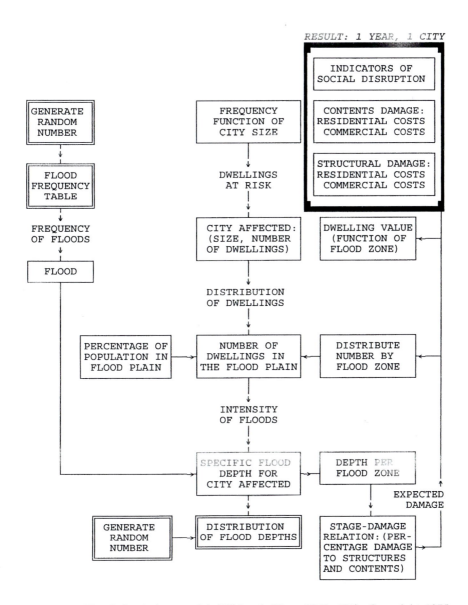

Figure 6.3 Flood simulation model (White & Haas 1975: 131). Copyright 1975 by MIT Press.

input data tend to be crude by comparison with the real environment. Unless the model is specifically made to cope with secondary hazards, these tend to be overlooked in the simulation. Apart from these limitations, there are two main types of constraint upon computer

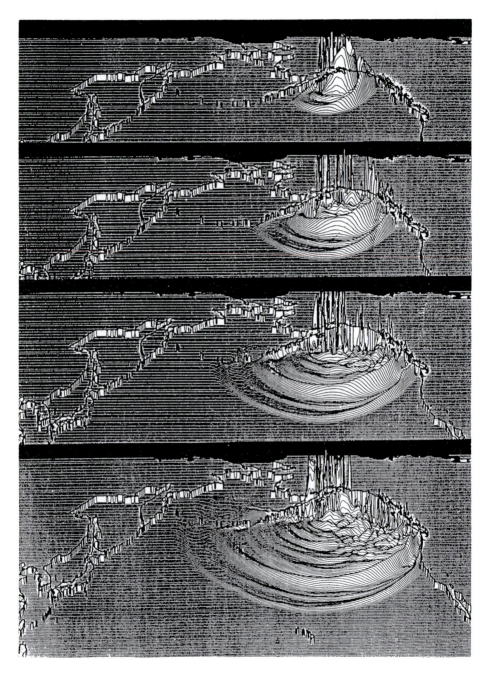

Figure 6.4 Output from a numerical simulation of tsunami travel times (*Earthquakes and Volcanoes* 1989).

simulation: the climatology, seismology, etc. (as appropriate), of the region; and the real distribution of events, which may in part be random.

The scenario method

One type of disaster modelling that does not involve numerical calculations is the scenario method (Borchardt 1991). It is instead an explication, or exploration, of possibilities, which sets up a logical sequence of events in order to ask how, starting from a given condition, alternative futures might evolve. Each point of fundamental change in the system serves as a juncture at which new pathways can be explored. Hence it is a way of answering the question "What do we expect the consequences of a given decision to be?" Its usefulness lies in its ability to provide insight into decisions needed for preventing, diverting or encouraging the evolution of the social system at specific points in time.

The method begins with the establishment of base conditions, involving a historical review of a hazard at a particular place, and the choices of adjustment that were considered and adopted or rejected. To these are added the environmental considerations that represent external forces. These include factors that effect the hazard within the community, such as urban renewal and transportation programmes, and those that impinge on it from outside, such as insurance, taxation and hazard management at the national level. The external forces can be changed during the scenario to affect its progress.

The scenario evolves from its base conditions under constraint from the external forces. One main objective of writing progressions and sequences of events in this way is to explain the impact of change in any one element on other aspects of the natural hazard system. A principal advantage of the method is its ability to "freeze" time, to enable the writer to look in detail at how events have progressed up to a particular point in the sequence (for example, how reconstruction costs have risen). Often, however, the scenario method is used for "backcasting", and hence answers the question "What would have happened if. . .?" For instance, the 1972 flood at Rapid City, South Dakota, which killed 231 people and caused $18 million in damage, was later "backcasted" by scenario, in order to determine what could have been done to prevent the impact, and what lessons could be learnt for the future.

The Delphi technique

The Delphi technique, so-called after the Oracle at Delphi in Ancient Greece, is based on questionnaire survey of a panel of experts. Questionnaires are analyzed on the spot and the results fed back to the panel, after which a new questionnaire is given out and the process repeated as many times as is necessary. In this way one of the common failings of group discussions is avoided, that of the dominance of the group by one or more of its members. Moreover, the information supplied to the panel can be controlled in order to improve the efficiency of its thinking, such that extraneous or misleading information is excluded.

Apart from the value of the information collected in the surveys, this method may encourage the members of the panel to accept disaster mitigation strategies and may also help to increase awareness of the risks posed to the local community by natural disaster.

Games simulation

Games simulation is a functional method in which the dynamic processes of society are acted out in a highly simplified form, consisting usually of a board game (Belardo & Wallace 1989). The use of dice or other probablistic methods allows chance to have a hand in the results and ensures that the outcome of the game is not fixed. Oil tanker spills, air pollution episodes and disaster relief management exercises have all been simulated in this manner. The advantage is that the problem definition is made explicit, as are the rôles played and strategies adopted by the players. The exercises must be plausible, probable emergencies, and thus can benefit from the prior use of other simulation methods, such as scenarios and computer methods, which could be used to provide appropriate background data.

Games simulation involves various components and stages (Foster 1980: 155). First, the problem (i.e. the disaster) must be defined, as must the scale of game, the time factor involved and the nature of the input data. Groups, umpires, structure, laws, goals and rôles must be established, and their interactions, strategies, policies and tactics formulated. Once this is done, information is given and generated, after which play can begin. During the simulation, events will take place according to predictable patterns tempered by chance factors, and thus the game will develop over time. In order to keep track of this process, accounting procedures can be used, such that information is compiled on, for example, the number of dead and injured in the disaster and the magnitude of losses and gains. Techniques of communication among the participants can also be monitored and studied. Finally, once a halt is called, the simulation should be subjected to a thorough evaluation.

The method has various purposes, including its use as a learning exercise for the understanding of events, decision-making processes and rôles, for exploring and solving particular problems and for simplifying and abstracting the complexity of a real disaster in order to understand it better. Techniques of communication used during the game may include written presentation, oral discussion and audio-visual methods, such as the games board. The participants will assume rôles played in a real disaster, either singly or as members of groups which act in a concerted manner. Finally, the outcome will involve simulating the magnitude, extent, effects and response to a real disaster.

This method is not suitable for modelling physical hazards, but it can be used as a training procedure to teach participants in situations of risk about behaviour, community actions and the functioning of committees and relief and management organizations. However, despite such versatility, the technique lacks the sense of total involvement that disaster inevitably brings. As many actual difficulties can easily be overlooked in the simulation, it should be carried out in conjunction with field exercises.

Field exercises
Field exercises are a means of ensuring the functionality of relief operations without the hazard of a real impact or the ensuing risk of increased physical injury associated with badly run operations. The method can be very useful as a means of ensuring that the timing of complex logistical movements is correct. It can also highlight particular problems, such as the inability of a certain hospital to deal with large numbers of casualties. Such exercises are often used to stimulate publicity for a particular hazard, usually by involving the press, media and community leaders. The simulated emergency must be clearly identified as a mock-up, in case it is mistaken for, or overtaken by, the real emergency.

Conclusion
The models described above can be used in a complementary manner, as part of a complete strategy designed to reduce hazards and educate officials and the public. Scale and computer models can be employed to determine the physical characteristics of the hazard and to predict when, where and how the next disaster may occur. Computer simulation can then be used to forecast damage and casualties. Less precise methods, such as the scenario and Delphi techniques, can be employed to forecast less tangible threats, such as failure of co-ordination of the relief effort; and, finally, games simulation and field exercises can be used to teach the community to react efficiently to the threat. If used intelligently and creatively, these methods will encourage proper land use and emergency planning and improve safety through design of better environments for living and better emergency response systems.

We now turn from the more general processes of monitoring and mitigating hazards to specific aspects of the logistics of emergencies. Because of the opportunity to save lives and reduce damage that it offers, warning is one of the most vital of these aspects. But, as will be demonstrated, it should not be considered in isolation from the social and organizational processes that form its main context, including planning for and managing disasters.

Disaster warning

Volcanologists distinguish between a **factual statement**, which describes current conditions but does not anticipate future developments, a **forecast**, which is a comparatively imprecise statement about future events, and a **prediction**, which usually covers a shorter time period than a forecast and specifies more clearly what is likely to happen (Swanson et al. 1985). Warning is distinct from each of these and usually constitutes *a recommendation or order based on a prediction*. The process of creating and issuing a warning involves complex interactions between physical, technological and social systems, whose operation must be carefully co-ordinated in order to achieve a satisfactory result. The objective of disaster warning is to avoid death and injury and, where possible, to achieve a reduction in the scale of devastation.

Warning systems designed to alert the public or community almost always involve in an active rôle the members of regional or local government. For example, once a tsunami has been identified or predicted in the Pacific Basin, the Pacific Tsunami Warning System (PTWS), based at Honolulu in Hawaii, will alert each of its member countries in the Pacific area hours in advance of the arrival of the waves (see Ch. 2). But the warning must then be transmitted to the public and emergency services by the government of the country in question, which has sole responsibility for this task. In general, whatever the size, scope and effectiveness of a particular warning system, it will probably involve co-operation among a variety of organizations and individuals.

Once a disaster impact has been predicted, warning can be divided into two distinct components: recognition of the danger and avoidance of it, or other action taken to minimize the impact (McLuckie 1973). Response and readiness tend not to follow smoothly from the identification of danger, but occur instead as a series of discrete jumps. Malfunctions in the warning process will manifest themselves as inefficiency in response, failure to inform people at risk or inability to mitigate avoidable consequences of the impact. There is also a potential for false alarms. Hence it is important to learn from the practical use of warning systems and to operate them with a high degree of self-awareness.

The effectiveness of warning

In the industrialized countries in recent decades the balance has tended to shift from large-scale loss of life to increased property damage. The impact of hazards has thus not decreased in severity, but reductions in mortality (hence increases in personal safety) can be attributed to improvements in warning and evacuation measures.

Several factors influence the effectiveness of warnings (Mileti & Sorensen 1988). For example, a high frequency of event may encourage a community to prepare effective mitigation measures. Experience of how to recognize and deal with the hazard will accumulate and be utilized regularly, while plans are unlikely to become stale and obsolete. On the other hand, a series of near misses or light impacts may generate complacency, which will be counter-productive if the area runs a risk of occasional severe impacts.

Secondly, warning systems must be designed to function after the initial impact has occurred, and to give details of the exact physical consequences to be expected, including those relating to secondary hazards, such as post-earthquake landslide or fire. Thirdly, speed of onset and length of forewarning must be taken into account when designing a warning system. Many sudden-impact disasters allow little time for the issue of warnings; hence reaction to them must be prompt, and more time-consuming adjustments are thus ruled out. Long warning time may lead to apathy rather than optimal public response. In this respect, one can distinguish between three types of impact: rapid or sudden, such as earthquakes; those involving gradual build-up, such as drought; and those which strike repeatedly, such as a succession of tsunami waves.

Effective warning messages will include some assessment of how long the emergency will last, as its duration will affect the nature of both warning and response. For instance, the public may need to be prepared for a long drawn-out or repeated impact. The messages must also include an assessment of the scope of the impact: the scale of damage, the geographical area to be affected and the number of personnel needed to participate in the subsequent relief operations must be predicted. Prior knowledge of the likely scope of impact would enable the authorities to negotiate agreements for reciprocal assistance from communities outside the impact area, which could then be included in the warning process.

If warning is to be adequate, the destructive potential of the disaster must be predicted: thus, a hurricane may cause much damage but little loss of life (if evacuation is carried out successfully), whereas a drought may kill many people but not damage property. Between these two extremes there are many combinations of destructiveness and there is considerable variation in the gross predictability of disaster types. For instance, the vast majority of earthquakes are not yet predictable; the impact of tornadoes and hurricanes is only partially foreseeable, but that of floods and avalanches caused by snowmelt may be largely amenable to forecasting. In this respect, a hazard that has a high level of gross controllability may engender reluctance to warn and evacuate people. This is often the case with reservoir hazards, which are controllable at least potentially, although none the less dangerous. Warnings seem more likely to be issued and responded to if there is no way of controlling the impact of the hazard.

In developed societies there is a dichotomy between an individual's responsibility for his own safety and collective responsibility for public safety. If the balance is wrong, then there will be some form of abnegation of responsibility, either at the personal or the institutional level. Government officials and scientists are presumably professionals trained in appropriate aspects of hazard management, but all aspects of personal security cannot be left in their hands; and, in fact, individualism in evacuation behaviour and the public's response to specific emergencies suggest that there is a general awareness of this fact.

In any case, it is clear that as a result of the variety of individual perception, the warning process is a very approximate one. In considering the rôle of information in inducing people to take immediate action to prevent the worst effects of disaster, Sims & Baumann (1983) were only able to conclude with the following highly tentative statement: "Information may lead to behaviour change . . . under highly specific conditions . . . if properly executed [i.e. delivered to recipients] . . . with specific targets."

In addition, failure to predict or forecast will inevitably lead to failure to warn. For example, the British Meteorological Office failed to predict the storm of 15–16 October 1987, in which central pressure fell to 958 millibars, winds rose to 62 knots and gusted to 90 knots, 18 people were killed and 25 million affected. An enquiry recommended increases in the density, frequency and areal extent of weather data collection, in the use of new technology (including microwave sensors borne by satellites) and a doubling of the frequency with which computer forecasting models are run each day. The procedures for impressing a warning on the public also needed clarification and improvement (*Weather* 1988). However, Mitchell et al. (1989) attributed part of the failure to the contemporary political context: a stock market crash and uncertainties about local government reorganization both tended to distract official and public attention from the storm.

Steps in the warning process

Foster (1980) identified the following steps in the creation, development and application of a warning system (see also Fig. 6.5):

(a) *Recognition by decision-makers of the possibility of danger.* A disaster plan must identify all hazards likely to threaten a community. Public awareness of, and belief in, the existence of specific hazards is vital.

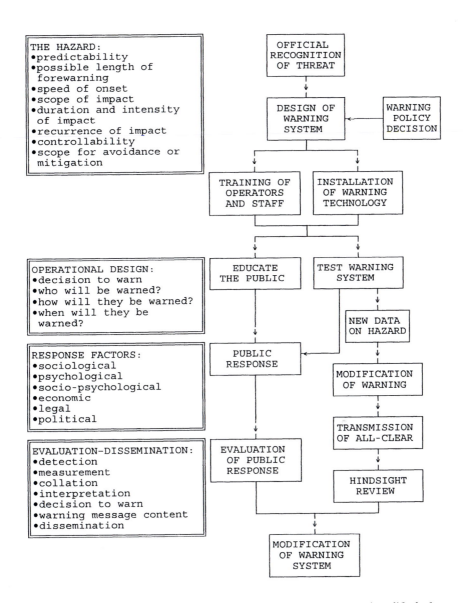

Figure 6.5 Design and testing of a natural hazard warning system (modified after White & Haas 1975: 184–5, Foster 1980: 172).

(b) *Design of a system to monitor changes in a particular hazard and issue warnings.* If the hazard is predictable in the first place, the warning system must be integrated with the system of prediction, such that when the danger suddenly grows too great a warning is issued. Like a

403

chain, the warning system is only as good as its weakest link. There must also be a complete chain of command, which establishes who has responsibility for each aspect of or stage in the warning process, and which includes provision for substitution of participants who fail to report for work. Assistance with designing and implementing warning systems can usually be obtained from research institutions.

Warning systems need to be scrutinized by legal experts before being implemented. Failure of a local authority to warn against predictable hazards may be considered as criminal negligence or abnegation of legal responsibility. On the other hand, false alarms may make the authority liable for the costs incurred by the disruption and curtailment of commerce and industrial production.

(c) *Installation and operation of a warning system.* It is possible to distinguish between the technological and social aspects of warning (Christensen & Ruch 1980). Hence, instrumental networks may be used as much for warning as for prediction of geophysical events. The network of devices used to sense the hazard needs to be sufficiently dense to be able to detect it efficiently, and there is a distinct advantage in using automatic equipment (for example, recording streamgauges that telemeter data to base). As there is a risk that vandalism, insurance fraud or simply the impact of the event being monitored will render the equipment inoperative, a back-up system would be helpful.

(d) *The decision on who to warn, about what, and how.* It stands to reason that any disaster warning system must prevent more damage than it causes! When impact is uncertain, officials responsible for warning may face a dilemma: making a probablistic statement about the likelihood of impact may be much less successful in stimulating the public to act on the warning, but may also be considerably more realistic than giving out information which is inaccurate because it is categorical. In fact, forecasts and risk estimates are often made with an unwarranted degree of confidence in their veracity. But on the other hand, messages must be specific in terms of the consequences of impact and the action to be taken. A sense of danger must be conveyed to the public, by giving information that is as specific and as reliable as possible.

Particular groups among the population may require special warnings. Elderly people, infants and children, handicapped people and certain ethnic groups may need to be informed and assisted in particular ways. Hence it must be remembered that the general population is far from homogeneous.

(e) *Public perception and education.* It is not easy to ensure optimal public response to disasters, for people differ in terms of age, intelligence, health and attitudes. They tend also to be stimulated differently depending on who they are, whom they are with and what they see. Moreover, the impact area may contain occasional visitors, such as tourists, who are much less aware of the hazard than are permanent residents. Hence, although many people may hear a warning message, all of them will believe different things, and the best level of success which can be hoped for is that the differences are minimal. Education programmes that aim to sensitize the public to hazards and emergency procedures can help ensure this, but should take account of the fact that people tend to respond to warnings on the basis of how what they hear encourages them to behave.

Social research indicates that people confirm disaster warnings by observation, waiting to see whether there is evidence that they may be true (the initial response may be disbelief, the so-called **normalcy bias**; Drabek 1986: 72), and, to a lesser extent, by contacting the authorities or asking friends, relatives and acquaintances (Drabek et al. 1969). Public information campaigns about warning or the risk of disaster are only likely to succeed if they assume that the public will only be marginally interested in them. Their message and objectives must be highly specific, and the target population must be considered in terms of its habits, interests and life-styles (Drabek 1986: 364).

To ensure rapid and appropriate response, warning messages must be intelligible (sirens, for example, do not indicate the nature of the emergency). Moreover, they must not be given in a form that is easily discarded, as might be the case if they are simply printed in newspapers and read out on television broadcasts. When information is distributed, its impact must be assessed in order to know what proportion of the public is aware of the hazard and of the appropriate responses to it. In one notorious case (the River Thames flood warning exercise of 1981 in the City of London) a costly and extensive publicity campaign came to nothing: social surveys after the event revealed that 50 per cent of respondents did not hear the warning, and of those who did, 30 per cent did not understand what it meant and 60 per cent did not know what to do (IDI 1981: 44).

(f) *Monitoring and managing the event as it develops.* In detecting the danger, the need for rapidity is pitted against the need for accuracy. Psychologists argue that, when the information received by decision-makers is conflicting, they have an innate tendency to accept that which is least threatening. Hence, not all accurate information on impending disasters is readily believed. If the public fails to respond adequately to messages broadcast to them, these can be modified and transmitted again. Thus the information flow must be two-way. Repeated warning may convince those waverers who did not intend to act on the initial messages.

As the emergency progresses, warning messages should be adapted to the change in conditions brought by the dynamics of the impact. Once this is over, all-clear messages can be transmitted, so that people may return to the disaster area. Repetition of warning messages and bulletins can ensure that they do not attempt to do so before the danger has passed, i.e. that they are aware of the need to wait for an all-clear message.

(g) *Reviewing, testing and modifying the system.* When there is little danger the warning system must be tested to ensure the functionality of both hardware and personnel. There is little substitute for experience in stopping people from making errors under stress, and hence disaster warning systems tend to improve with use. This is all the more true in that tests and actual disaster events show how such a system could be modified, if its functioning could be improved.

It is important to keep detailed records of the messages broadcast, the timing of events and the response to (or in other words the effect of) the warnings. These can later be used for a debriefing review, in the light of which the warning system can be revised. Hence the lessons of present disaster can provide a basis for upgrading the capacity of a warning system, which, of course, must then be tested again before the next disaster strikes.

Clearly, the design of warning systems forms part of the wider issue of planning for disasters, and their operation fits into the general question of emergency management.

Planning for disasters

In the words of disaster researcher Harold D. Foster, "Victims trapped beneath fallen buildings, on the roofs of burning high-rises, or caught in rapidly submerging residences cannot afford the luxury of bureaucratic procrastination" (1980). Hence, inefficiency in disaster planning may have as a consequence numerous avoidable deaths and injuries.

Unusual needs created by the disaster have been classified by sociologists into **agent-generated demands**, which are created by the geophysical event, and **response-generated demands**, which are created by the damaged social fabric as it responds to the event. The former include warning, search and rescue, restoration of essential services and the protection of community order, while the latter include communication, co-ordination and command (Quarantelli 1981a). The level of unfulfilled demands can be lessened by adequate pre-impact preparations, especially if the interval between identifying the hazard or forecasting the event and the moment of impact is long enough to permit them to be substantial.

When impacts are likely to exceed the capacity of normal organizations and services to cope, a disaster plan will be necessary. It will need to provide for prior authorization and fast authorization of decisions regarding emergency services, and to vest authority in a co-ordinator, head or chairman who will direct the plan and its participants. A word of caution must be interjected here: one of the most common failings of hazard management has been the tendency to make massive attacks on secondary problems rather than dealing with the most important hazards first (Fischhoff et al. 1978).

The process of disaster planning can be divided into four main themes (see Petak 1985):

(a) **Mitigation**: the planned reduction of risks to human health and safety. This may involve modifying the causes or consequences of the hazard, the vulnerability of populations or the distribution of losses.
(b) **Preparedness**: the development of a plan and reciprocal agreements with other jurisdictions to reduce loss of life and damage by prompt intervention. This category includes the preparation of a plan and the establishment of warning systems, training programmes and public information services.
(c) **Response**: the provision of emergency relief and assistance when it is needed and the maintenance of public order and safety. Evacuation, the distribution of primary necessities and the mobilization of emergency services come under this category.
(d) **Recovery**: the provision of support during the aftermath of a disaster, so that community functions can quickly be made to work again. Such activities can be divided into those connected with short-term restoration and those that facilitate long-term reconstruction.

Sociologically, the degree of acceptance of disaster planning varies with socio-economic status and frequency of experience of disasters (Drabek 1986: 24). With regard to the latter, both mitigation and policy formulation tend to occur in the wake of particular disasters, when public opinion is favourable. But with time official and public interest in the problem diminishes. For instance, in the United States there is clear evidence that the funding of seismic and volcanological research waxes and wanes with the current prominence of earthquakes and eruptions both at home and in nearby countries (but not further afield).

In most countries, the responsibility for disaster preparedness and relief is likely to be shared among various levels of government (national or federal, state or provincial, county, municipal, etc.) and neighbouring jurisdictions. May (1985) identified four possible ways in which these are likely to work together. In the **regulatory mode**, legislation is promulgated and reinforced jointly between the various bodies. In the **mobilization mode**, the local or regional government will bear the main responsibility for addressing the crisis, assisted by the national or Federal

Government. The **collaboration mode** involves greater co-operation between the levels of government than under the mobilization mode, and will be applicable to problems whose solution requires a more sophisticated approach or a greater input of resources. Finally, the **"degenerative" collaboration mode** implies that some partners in the joint effort have withdrawn their support in part or entirely.

The speed of recovery depends on the degree to which local government has motivation, competence, knowledge of what needs to be done and the political ability to do it (Rubin & Barbee 1985). Civil authorities essentially have a choice among three strategies for risk reduction. First, the behaviour of individual citizens can be controlled by regulations, penalties and incentives. Secondly, public resources can be used to combat the hazards (for example by engineering works) and, thirdly, services (such as education programmes) can be provided to assist with adjustment to hazards.

Despite the extra costs involved, multiple rather than single hazard management is advantageous for several reasons. First, it enables the full risk and vulnerability of a particular locality to be tackled. Secondly, it permits linkages between hazards to be mitigated. Thirdly, it allows for economies of scale in tackling them, and, finally, it involves the full range of organizations involved in hazard mitigation and favours collaboration between them.

In synthesis, Dynes & Quarantelli (1975) proposed the following nine models of disaster planning (none of which, however, has been verified with sociological data):

(a) The **maintenance model** emphasizes the preservation and use of accumulated resources (facilities, supplies and money).

(b) The **military model** uses the armed forces as the principal source of aid in civil emergencies.

(c) The **disaster expert model** emphasizes professional opinion and services within the community.

(d) The **administrative staff model** stresses organizational skills.

(e) The **derived political power model** emphasizes that it is necessary to co-ordinate emergency plans, but does so through the authority of one particular leader, such as a city mayor.

(f) The **inter-personal broker model** highlights the rôle of liaison between emergency workers in different organizations.

(g) The **abstract planner model** stresses planning based on various contingencies and scenarios.

(h) The **community educator model** places emphasis on the need to overcome public apathy with regard to disaster planning and mitigation.

(i) The **disaster simulation model** stresses the rehearsal of disaster plans.

It is unlikely that any of these models can be applied to a real emergency with the complete exclusion of all the others.

Decision-making and disasters

During the community emergencies which disasters cause there are likely to be acute shortages of relief supplies, equipment and trained personnel. It will be necessary to make decisions rapidly, but there will be little time in which to do so.

In the normal processes of decision-making in government and civil administration, problems are defined, objectives are ranked in order of priority, criteria are established that can be used to guide the decisions and constraints upon the various options are recognized (Foster 1980). Alternative strategies for achieving a particular objective are evaluated and their consequences estimated. Before decisions become final there may be a need for consultation and, in a democratic society, they may need to be ratified by other organs of government or after discussion with interested groups or the public at large. Any conflict of interests is likely to cause delay until it is resolved; indeed, objection and conflict can be used as a delaying tactic by people opposed to a particular decision (Heffron 1977).

As inability to make decisions with sufficient rapidity can lead to inferior performance in disaster situations, the disaster plan should make as many relevant decisions as possible before the impact happens and, having identified which decisions must be made in the heat of the moment, streamline the procedure for making them. In order to do this, some powers will have to be suspended during the crisis and others augmented.

Finally, the importance of emergency training cannot be overestimated (Toulmin 1987, Daines 1991). Inadequate education of emergency managers tends to lead to several characteristic problems, among them inability to estimate real costs and compile budgets, to forecast problems, to administer personnel or to programme activities in their proper sequence.

The disaster plan

The co-ordinator of emergency services should ensure that a disaster plan is drawn up, made known to all participants and tested before it has to be used (Herman 1982). The plan must have the full support of senior members of the government or other organization for which it is being prepared. Figure 6.6 gives an example of the structure of command assumed by a typical disaster plan and the organizations involved in it.

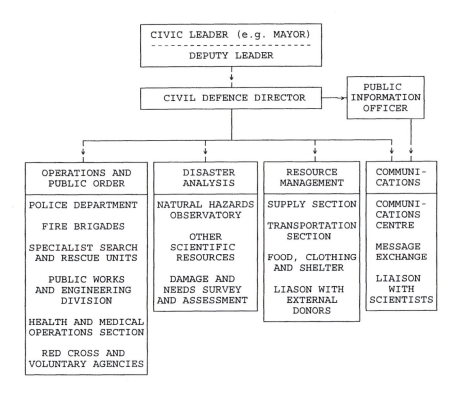

```
┌─────────────────────────────────────┐
│    CIVIC LEADER (e.g. MAYOR)         │
│    -----------------------------     │
│          DEPUTY LEADER               │
└─────────────────────────────────────┘
```

OPERATIONS AND PUBLIC ORDER	DISASTER ANALYSIS	RESOURCE MANAGEMENT	COMMUNI-CATIONS
POLICE DEPARTMENT FIRE BRIGADES SPECIALIST SEARCH AND RESCUE UNITS PUBLIC WORKS AND ENGINEERING DIVISION HEALTH AND MEDICAL OPERATIONS SECTION RED CROSS AND VOLUNTARY AGENCIES	NATURAL HAZARDS OBSERVATORY OTHER SCIENTIFIC RESOURCES DAMAGE AND NEEDS SURVEY AND ASSESSMENT	SUPPLY SECTION TRANSPORTATION SECTION FOOD, CLOTHING AND SHELTER LIASON WITH EXTERNAL DONORS	COMMUNI-CATIONS CENTRE MESSAGE EXCHANGE LIAISON WITH SCIENTISTS

The box structure also includes: CIVIL DEFENCE DIRECTOR and PUBLIC INFORMATION OFFICER.

Figure 6.6 Organization of emergency operations (after White 1975: 67).

Some estimation of the lead time between prediction and impact should be inherent in the disaster plan. If days or weeks fall between identifying an impending extreme event and its occurrence, a good plan will provide for readying personnel and equipment, and putting into effect transient mitigation measures (such as sandbagging against floods, or digging diversion channels for lava flows). Supplies can be stockpiled, the maintenance of vehicles and equipment accelerated and the leave of key personnel cancelled. In this way a high degree of readiness can be achieved in a relatively short space of time.

The disaster plan must make provision for detecting and dealing with threats and thus should contain a mechanism for continuously monitoring both the hazard and the progress of the emergency. In this context, Table 6.1 summarizes the aspects of disasters relevant to management decisions. By virtue of its individual mix of characteristics, each disaster will present unique opportunities for planning, and these should be evaluated before the emergency arises. Obviously, the results of the assessment should be fed back into the decision-making process.

410

Table 6.1 Aspects of disasters pertinent to their management.

- *Physical occurrence:*
  Probability
  Frequency
  Transience (duration)
  Physical magnitude
  Energy expenditure
  Physical effects: direct, indirect and secondary
  Area affected: directly and indirectly
  Degree of spatial concentration or ubiquity
  Volume of products (e.g. lava, floodwater)

- *Predictability:*
  Short-term (for avoiding action)
  Long-term (for structural and non-structural adjustment)

- *Controllability:*
  Can physical processes be modified?
  Can physical energy expenditure be reduced?
  Can effects be mitigated?
  Can effects be modified?

- *Socio-cultural factors:*
  Belief systems inherent in societies
  Degree of knowledge of risk
  Complexity of social system and its constituent groups

- *Ecological factors:*
  Environmental damage propensity
  Environment compatibility of mitigation measures

The plan should aim to preserve both public safety and community order. Although looting is seldom a problem (unless before the disaster there were strong preconditions for it), it will be necessary to restrict access to damaged or hazardous areas by cordoning off their perimeter. Property must be guarded, traffic directed and dangerous areas patrolled. Points where there is a high risk of secondary hazards (for instance, power stations and dams), must be examined and supervised.

The rôle of the armed forces poses something of a dilemma for civil emergency planners (Anderson 1969, 1970). Thanks to its capacity for rapid action, stockpiling of emergency resources and logistical abilities, the military has great potential for action during disasters. It tends to be strong on logistics and communications, well disciplined and well organized. However, military units may adopt too authoritarian an attitude to the problems of survivors and could conceivably deprive local people of initiative, if the solutions that they impose are too strongly regimented.

Although local self-sufficiency should be one of the goals of the emergency plan, reciprocal agreements for mutual assistance can be concluded with neighbouring communities, regions or states. These provide useful reinforcement in times of shortage and also distribute the experience of coping with a real emergency. Thus, expertise is pooled and help will be supplied with a minimum of delay. However, the effectiveness of reciprocal agreements will depend heavily upon how well and how quickly needs are communicated to the reciprocal donor.

Call-up and rendezvous

The success of the disaster plan will depend on its chain of command; like all chains, this will only be as strong as its weakest link. Assembling the command structure will begin with calling participants to their posts. There should be a list of emergency staff, with their addresses and telephone numbers. The list may include specialized personnel, such as skin-divers, skiers or owners of earth-moving equipment, whose services may be useful in particular emergency situations. For the principal participants a parallel list of substitute personnel should be compiled, in case the first individual to be contacted is incapacitated or unavailable.

Emergency personnel can be called up in several ways. For example, a telephone fan-out system can be used, in which person A calls persons B and C, B calls D and E, C calls F and G, and so on. There should be cross-checking to avoid a break in the chain caused by the absence of any participant. Where the resources are available, it is possible to design a computerized means of telephoning participants automatically and using a prerecorded message to tell them what to do next. Where telephones are not available, or are out of action, the fan-out system can be activated by person-to-person journeys.

Alternatively, a courier service can be used to alert participants. However, this is a laborious and time-consuming method which, like the manual fan-out system, may subject the message-bearer to danger en route. Until participants have reported for duty, the operations centre is unlikely to know what level of success the courier has had in his mission. Some of these problems can be avoided by arranging that in the event of impact participants in the disaster plan rendezvous at a prearranged meeting point. This system is feasible if one can be sure in advance that the meeting point is the right one for the particular circumstances, that it has not been seriously damaged by the impact and that a rendezvous there will not put individuals in danger.

The operations centre and command process

In its command structure the disaster plan should aim to avoid ambiguity and duplication, especially that which occurs between competing and parallel organizations. In comparison with countries where much power resides at the local or regional level, highly centralized societies tend to involve fewer people in decision-making, especially among the lower echelons, and to delegate a portion of the command structure to military authorities. As civil and military authorities usually have separate command structures, the plan should determine who is subordinate to whom. It should also ascertain whether charities and voluntary agencies are to be permitted to act on their own authority, or whether they are responsible to government.

Centralization is the key to the effectiveness of command operations in a sudden impact disaster (Drabek 1986: 33). Therefore, the disaster plan should designate a control centre or **command post** (Quarantelli 1981b). This must be located in a structurally sound building, which is outside the area of highest risk, but is sufficiently near to it to be accessible. It should also have good and robust communications, and hence function as a nerve centre for operations, and one that is unlikely to become damaged or isolated during the emergency. In certain types of emergency, a subsidiary command post will be set up on the periphery of the impact area, perhaps in a vehicle or mobile shelter, from which commands emanating from the central post will be relayed to emergency personnel.

Each group and organization that participates in an emergency will view the situation in terms of its own perception, experiences, objectives and ability to respond. The command post must co-ordinate these varying capabilities and must acquire sufficient information about the situation to be able to channel emergency relief to where it is needed and relay instructions to field personnel.

Communications are vital and must be supported and strengthened under the provisions of the disaster plan. Normal channels of information must be restored and augmented by emergency channels, especially as "communications overload" is a common feature of the immediate aftermath of natural disasters, as in the case of public telephone systems, which tend to function badly under the duress of vastly increased traffic. Hence, channels must be reserved exclusively for the emergency services.

Public information must also be sorted, checked and disseminated with care. If contact and liaison centres give out too little information the public is apt to believe that the truth is being kept secret by the authorities. On the other hand, when too much information is disseminated it may be interpreted too creatively, leading to unjustified rumours. In this context, authorization to make official statements to the news media should be limited to responsible leaders who are supplied with all the information that they need in order to understand the situation.

Evacuation and the disaster plan
Where warning time permits, evacuation is often the best way of ensuring public safety under the duress of a disaster. In the event, unexpected flash flooding may simply involve getting people to climb the sideslopes of valleys in order to escape the waters. But in general, moving large numbers of people from a threatened area is a complex process which

requires considerable planning (Perry et al. 1981, Lindell & Perry 1991). Reception centres must be organized where the risk of their destruction is as low as possible, and they should have adequate structural integrity to survive an unforseen geophysical impact. They should be stocked with food, clothing, bedding, medical supplies and communications facilities which can collect and relay information on evacuees. They should also be as accessible as possible to the evacuees, and clearly identifiable as the proper point of arrival. For greater efficiency, the centres can be devoted to several uses. For example, in the southern Indian states of Andhra Pradesh and Tamil Nadu cyclone shelters built by the government are used as community centres, clinics, rest homes and schools in the intervals between storms.

The disaster plan should ensure that there is time to complete evacuation before the impact, as in many cases people will be more vulnerable when they are on the move than when they are at home (Sorensen 1991). Care should be taken to ensure that all people to be evacuated receive and obey the necessary instructions. Although the type of evacuation is likely to differ with the kind of disaster, in all cases evacuees should be channelled into areas of progressively lower risk. Their route should not lead them across zones that are highly unsafe, and hence, several evacuation routes may be needed or routes may vary for different kinds of hazard. Alternative routes should be planned, and measures should be designed to rectify and avoid bottlenecks. For example, the rapid evacuation of highly urbanized barrier islands, which are faced with an imminent storm surge, may involve vehicle breakdowns on the bridges and causeways that connect the islands to the mainland. Swift intervention will be necessary to clear the roads.

The disaster plan should also make special provision for the evacuation of particular groups; not merely infants, pensioners, the sick and the handicapped, but also the inmates of prisons, who must be kept securely detained at all times.

Figure 6.7 shows the flood risk areas, evacuation routes and evacuee reception centres for the Italian city of Florence. The location and severity of flooding are indicated by the record of the 1966 flood disaster (see Ch. 3). The reception centres are public buildings and the evacuation routes are chosen to enable people to reach high ground as quickly and efficiently as possible, using the widest, least cluttered roads. In order to minimize bottlenecks, emergency relief is planned to arrive on different roads to those that carry the evacuees. The map is published regularly in successive editions of the city's telephone guide and street maps. But at the time of writing the evacuation plan has yet to be tested, and there are no data on the extent to which Florentines are aware of it.

Further observations on evacuation are given below in the section on emergency management.

414

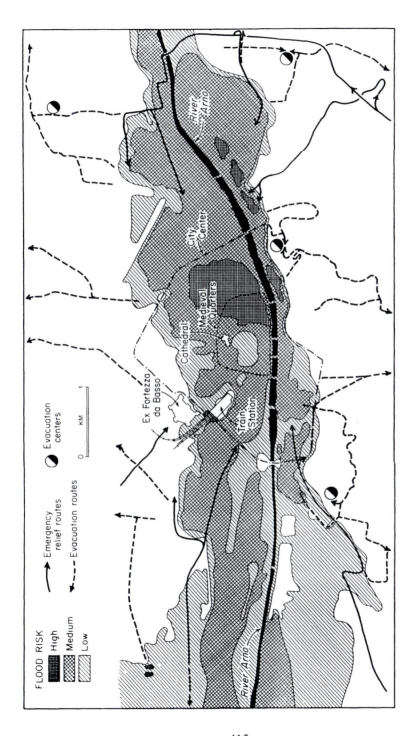

Figure 6.7 Flood hazard evacuation map of Florence, Italy (after *Tutto Città-Firenze* 1990).

415

Concluding remarks

The disaster plan will need to be tested carefully. This can be done in one of several ways (see the section on simulation models above). A mock emergency can be mounted as a form of field exercise, which is a good, if expensive, way of testing whether logistical arrangements will actually function. Alternatively, the functioning of the plan can be thought through as a scenario. In any event, the plan must be constantly maintained and updated. If it contains out-of-date information, such as obsolete names and telephone numbers, then it will not function efficiently.

Finally, political developments are often highly salient to disaster management. With regard to public administration and natural catastrophe, Olson & Nilson (1982) classified politics into participatory, specialist, pluralist and elitist forms, and public policy into distributive, constituent, regulative and redistributive. They noted that political culture, expressed as power relations, may inhibit decision-making by preventing a latent issue, such as a disaster mitigation strategy, from becoming a question for decision. Furthermore, Mitchell et al. (1989) argued that political issues may on occasion overshadow the impact of natural disaster.

Managing hazards and disasters also involves managing the process of information dissemination. The success of risk mitigation or emergency relief may depend on the quality of news broadcast or published, and hence the media often play a crucial rôle in disasters: their influence may be positive or negative, depending on what information they receive and how they interpret it.

Disasters and the mass media

After disaster has struck, news comes from four main sources: reporters sent into the field; journalists based at the nearest centre of government; press agency syndicated reports; and official communiqués. Newspapers, television and radio will be primary sources of information for survivors, the public and, to a certain extent, the suppliers of national or international aid (who will also be influenced by the opinion and solidarity of the general public, which tends to be sensitive to the quality of news received). But although the news media can motivate the public to contribute to relief appeals (Phillips 1986), the agents of publicity can, to a certain extent, turn disaster relief on and off like a tap by highlighting, minimizing or ignoring the plight of survivors as they please (Scanlon et al. 1978).

While it is powerful, the process of communication between the media and the public is a very imprecise one. Although radio and television may be the principal sources of information for most people, few individuals end up with a clear idea of the risks entailed by the events being reported. Moreover, reports of disaster in foreign countries are seldom translated by the public into practical lessons for safety at home.

The process of gathering and relaying information involves several potential problems. First, the visit of the reporter to the area may be very brief, and he or she may have a weak understanding of the local language, culture or geography. Details may be gathered in a great rush, if the journalist cannot be spared from the next assignment, and names and nouns may be misspelt. News about the disaster obtained from local participants is likely to be filtered by their perception, as well as by that of the reporter. Alternatively, the news may be conditioned by the "reflected glow" of those who are not directly affected by the event, such as government officials, whom the reporter has interviewed.

Hence, unless the job is well done by a journalist who is highly professional, objective and sensitive, there is a strong risk of bias or misperception, or at least that the report will tend to be brief and uninformative, its details gathered in a tearing hurry. Bias is a particularly thorny problem in international news (Alexander 1980). Unfortunately, reporters often tend to distort the reality of the situation to the angle judged to be of most interest to the public for whom the story is destined. In many cases prejudice is evident, as in the portrayal of national stereotypes; on the other hand, reports which are heavily flavoured with nationalism are unlikely to be any more objective. Moreover, journalism dictates that every day the front pages of newspapers, or the headlines of television and radio broadcasts, must differ from the previous day's offerings. Thus the disaster may be dropped long before it ceases to be news.

Other negative aspects of reporting include the risk that massive publicity will increase the "convergence reaction" (see Ch. 9), thus bringing large influxes of people unnecessarily into the disaster area, including hordes of reporters and television crews. Transportation, accommodation, communication and information systems can be over-loaded by the demands of the media at a time when they are needed for search and rescue, while myths and rumours can be increased by biased reporting. The credibility of authorities responsible for disaster relief can be affected by negative reports in the media, especially as these will exert intense pressure on the authorities to supply information at a time, shortly after the impact, when the situation may be far from clear.

The relationship between scientists and news reporters deserves special consideration (Burkhart 1987). In simple terms, a scientist working on hazard prediction or monitoring should be a source of reliable information and advice which the reporter can readily obtain. If the relationship is a good one, valuable information can be relayed to those who need it and the public can be made fully aware of efforts to monitor and mitigate the hazards. Moreover, reliable scientific information obtained from original sources reduces the risk that rumour and exaggeration will be acquired second hand from dubious sources. It may also convince the public that no vital facts are being withheld from them.

But the situation is easily complicated. Any dissent and controversy among scientists may lead to confusion in the press and loss of confidence on the part of the public. For the sake of newsworthiness reporters may overdramatize scientific conclusions and thus distort them. News staff are not always able to distinguish the bona fide scientist from the credible charlatan, whose principal aim is to gain publicity for himself or his ideas, especially as the two categories are not always mutually exclusive! Finally, although news must, of course, be obtained in a timely way, frequent interruptions by reporters can hinder and distract scientists at work.

However, there is a positive side to news reporters. If honest attempts are made to involve them in the disaster relief operation they will usually co-operate by utilizing information in a responsible way (Burkhart 1991). The emergency manager must first understand the media: for example, radio needs information quickly, television requires a startling visual impact and sense of immediacy, while newspapers require deeper knowledge and more detailed graphics. All reporters will require detailed information on casualties, damage, the cause of the disaster, search-and-rescue work in progress and the likelihood of future impacts. Moreover, in a serious disaster, coverage is likely to be much more extensive than for most workaday events: long, live transmissions and large newspaper supplements are the rule. Hence, by careful – but honest – management of the media rumours can be stopped, confusion avoided, the public educated and the emergency workers kept free to do their jobs properly. This may involve setting up a pool of accredited journalists, including the major telegraphic services (news agencies) close to the emergency management command post (Scanlon et al. 1985). This is one way to keep the number of journalists down to manageable levels, given that, in the developed world, a major disaster may lead to the arrival on the scene of more than a hundred journalists within a very few hours. Thus, disaster managers cannot afford to ignore the media.

Emergency management

According to the sociologist Gary Kreps (1983), the essence of disaster is that the physical and technological foundations of social regularity are drastically altered. As natural disasters tend to be sudden and short-lived, so also are the organizational responses. Therefore, to meet new and transient demands caused by natural catastrophe, social organizations that are normally static must adapt or be replaced by more effective units.

Information

The first requirement of emergency management is information, which will be the one commodity that every participant in the emergency needs (Guha-Sapir & Lechat 1986). But the flow of news and data may be disrupted or interrupted immediately after a disaster. Hence, the services of information collection and dissemination must be restored and perhaps augmented. Thus, it will be necessary to co-ordinate and collect information, disseminate it to the organizations which need it, and ensure that it is not misinterpreted (Comfort 1991).

Information collection should take several forms in a known hazard area before disaster strikes. These include hazard mapping and microzonation, or in other words, spatial assessment of risk. Also, data should be gathered on the frequency and magnitude of geophysical phenomena, and on patterns of socio-economic behaviour, in order to design prediction and warning systems. Vulnerability should be analyzed, beginning with the collection of data on the populations, buildings and structures which are at risk: data on past casualties and damage, and on responses to past disasters, can be used to help design better mitigation measures. Equipment and supplies should be stockpiled, maintained and an inventory of them drawn up. This includes items such as earth-moving equipment, pumps, beds and bedding, cooking equipment, rations and medicines. Care should be taken to maintain the functionality of all such stocks. Finally, emergency plans should be created and updated, and the people who will have to participate in them should be educated and re-educated in the relevant procedures (see the section on disaster planning above).

Information collection should, of course, not cease when disaster strikes, but the character of the information required will change as will the means of collecting it. The objectives are now to develop an overall system of accounting for damage and the assessment of preliminary needs, and to define the relationship between these and what will be required in order to facilitate long-term development. For example, after an earthquake damage to housing must be surveyed in order to know how many dwellings are no longer habitable; immediate and temporary forms of shelter must be organized and arrangements made to draw up long-term plans for reconstructing homes.

In general, post-disaster information collection involves the following categories. First, the scope of the impact must be determined in terms of deaths, injuries, homelessness and damage to housing, infrastructure and public utilities. The size of the impact area must be determined and so must the location and accessibility of affected places. Then needs must be assessed, including burial, health care, surgery, disease surveillance, shelter, heating, water supply and supplementary feeding. Simultaneously, existing local resources should be surveyed in order to determine the increment in supply that is required. Similarly, relief efforts that are already underway should be evaluated so as to estimate how far they must be supplemented. Next, continuing and future threats to public safety and the social system must be identified. And finally, information should be gathered that will be of use in the management of future disasters, and relevant observations should be fed back into preparedness plans in order to improve them. As needs are often more than met, it is, unfortunately, often necessary to stave off unwanted assistance (e.g. useless goods, unorganized volunteers and superfluous analysts), which would use up vital resources such as food, accommodation and the valuable time of relief managers. Proper information management can help reduce this problem to a minimum.

Emergency managers should inform the public by setting up an information service on hazards, victims, and so on, and by educating people about how to cope with hazards (through meetings, broadcasts, etc.). But they should also listen to public opinion and collect facts from members of the public. This can be formalized in the setting up of citizens' advisory groups and consultative councils. Hence, information flow should be a two-way process in disasters (UNDRO 1979).

Information technology in disasters
As information is a prime requirement in disaster situations, so modern information technology has risen to prominence, for on it the storage, manipulation and transmission of information depend. Hence, in the late twentieth century the satellite and the microcomputer have become central to international disaster management (Toigo 1989).

In hurricane disasters more than half the ground-level telecommunications (telephone wires, electricity pylons, radio masts, and so on) may be damaged beyond use, as happened in Jamaica during Hurricane Gilbert in 1988 (see Ch. 3). In North America a portion of the reserve radio bands, 806–890 Mhz, has therefore been reserved for a land-based service of mobile satellite communications, which can be mounted on the roof of an ambulance or other vehicle and which use the Experimental Applications Technology satellites developed by NASA (Congressional Research Service 1984). Moreover, the United Nations Disaster Relief Organization (UNDRO) has developed a cheap, portable beacon capable of transmitting a short, coded message about a disaster situation to polar-orbiting satellites and then to ground receiving stations in the USA and France. The information should reach the UNDRO headquarters in Geneva within two hours of being sent, but at present there is no way for the user to verify that transmission has been successful.

Communications satellites generally follow an equatorial orbit and supply broad band transmission of voice, video and digital data. Fifteen of them, with 2,100 international communications pathways, serve 117 member countries of the International Telecommunications Satellite Organization (Intelsat). When a disaster occurs, the system can be a vital means of communication if the Earth station facility remains intact. The weak link is represented by ground communications to the transceiving station: but medium-range UHF radio-telephones tend to be relatively resistant to floods, hurricanes and earthquakes. However, only 25 among 150 developing countries have telecommunications adequate to cope with disasters when they occur near major cities, and few if any are equipped to use robust, satellite-based radio-telephones to cover impacts in remote areas. Here, there is scope for the use of portable Earth stations with antennae 1.2–2.4 m in diameter, which could be deployed more widely (at major airports, for example) and prequalified for access to Intelsat satellites. At US$50,000–$75,000, their cost is relatively low. Finally, problems of economic feasibility are associated with the use of commercial television satellites for emergency communications: it has been estimated that pre-emption of satellite programming for uses connected with a specific disaster could cost between US$200,000 and $1 million in lost revenue.

Microcomputers can be particularly useful aids to rapid decision-making under stress. According to Belardo et al. (1984), they are gradually being accepted as practical tools in unusual situations such as disasters. Various algorithms have been developed for emergency management using personal computers (Fig. 6.8). Basically, the user interacts with the environment in which disaster relief is taking place by manipulating data in order to design the best possible response under the constraints imposed by actual circumstances and events. There are usually five main subsystems (cf. Fig. 6.9). The first involves **data input, storage and display**, as the systems will work only if updated periodically with inputs of relevant information. The second concerns **data management**, as large volumes of information must be made both accessible and comprehensible to the user. Thirdly, **data analysis** enables the information stored in the computer to be presented in synthesized form, for example as a map, a graph or summary statistics. Next, **simulation modelling** permits current data to be turned into a prediction of what will happen in the immediate future. And finally, in order to make the computed information accessible and understandable, **display technology** must be coupled with software for data entry and for interrogation of the system by its users.

Personal computers, and the software that runs on them, have advantages of cheapness, flexibility and ease of use. In disasters they can be used for the following tasks: to monitor events; display information in

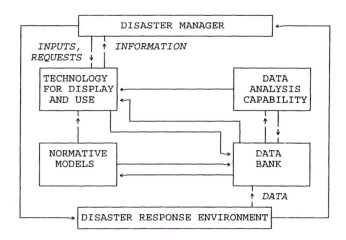

Figure 6.8 Use of microcomputers in disaster management (after Belardo et al. 1983: 32).

order to aid decision-making; keep a record and inventory of resources, events, casualties and damage; notify emergency personnel or dissemi-nate a warning automatically; or simulate the consequences of hypotheti-cal decisions, which can then be made if the outcome is judged satisfactory. As they are so easily modified, maps displayed on computer screens are more versatile and dynamic than hard-copy versions. They can be used, for example rapidly to identify schools, homes and hospitals and their special needs during the emergency, or to locate the most appropriate routeways to be used by the emergency services. Carrying out such tasks by computer reduces the administrative burden on emergency managers and increases the speed and accuracy of decision-making.

Emergency evacuation
If carried out properly on the basis of prior planning, evacuation is one of the best means of ensuring public safety during disasters. It can be divided into six time components. First, **decision time** elapses between detection of the threat and decision by a competent authority to evacuate. Secondly, **notification time** is required in order to inform and instruct all evacuees. Thirdly, people require some **preparation time** before begin-ning the evacuation, while, fourthly, **response time** is the interval between departure and arrival outside the danger area, or in the designated evacuation centre. Next, **verification time** is required in order to account for the whereabouts and safety of evacuees. Finally, a **return** will have to

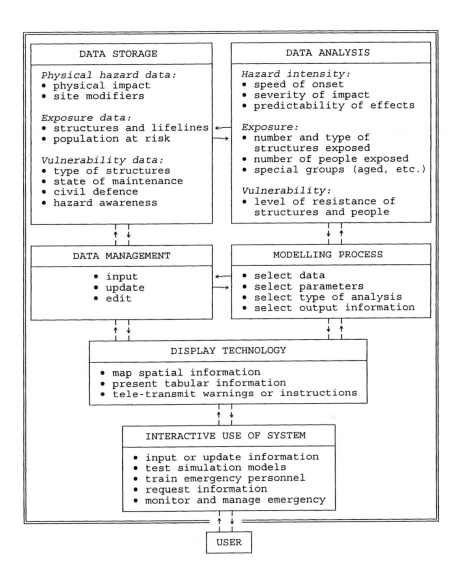

Figure 6.9 Decision support system for hazard mitigation (after Berke & Stubbs 1989: 105).

be organized when the emergency is over (Stallings 1991). The process of evacuation may serve different functions depending on when it is carried out. According to Perry & Meredith (1978: 169), short-term pre-impact evacuation has been termed **preventative**; long-term pre-impact evacuation

423

can be described as **protective**; short-term post-impact withdrawal involves **rescue**; and long-term post-impact evacuation is linked to **reconstruction**.

Generally, enforced evacuation is inefficient, but voluntary removal needs incentives. Plans must be made and publicized well in advance, and must be enforced by officials during the movement. Most evacuees seek confirmation of an evacuation order from neighbours, family members and officials. In fact, kinship groups should not be separated during evacuation or attempts will be made to return to the evacuation area. Finally, property must be safeguarded, as evacuees tend to be worried about the security of what they have left behind.

Volcanic risk in the Philippines offers one good example of careful evacuation planning (UNDRO/UNESCO 1985). The area around an active volcano is divided into a permanent danger zone (within 6 km of the summit, where settlement is not allowed), a high-risk zone (within 8 km), a probable danger zone (within 10 km) and mudflow (lahar) risk zones. Evacuation during volcanic emergencies is organized into four phases. In the first of these, geophysical signs of an impending eruption require mobilization of provincial civil defence authorities, preparation of evacuation centres and prohibition of entry into the danger zone. In the second, the early stages of eruptive activity require evacuation of people from the high-risk zone and instructing adjacent communities to prepare for evacuation. The third phase consists of the main eruption and may last as long as several months. The high-risk zone and selected areas of the probable danger zone are evacuated and there is an absolute prohibition of entry into these zones. Lastly, in the fourth phase eruptive activity quietens down and, it is to be hoped, ceases. Residents are allowed back first to the probable danger zone and then to the high-risk zone. Due to the continuing risk of secondary lahars (see Ch. 2), mudflow hazard zones are the last to be reoccupied.

In American disasters the majority of evacuees go to friends and relatives, and a minority go to second homes, hotels or public shelters (such as those organized by the Red Cross). The lower a person or family is on the socio-economic scale, the more likely it is that he or she will evacuate to a public shelter (Baker 1991), a form of haven that is normally used by only 15–25 per cent of evacuees (Drabek 1986: 126). Similar regularities are likely to be found in other countries, but studies are not numerous enough to confirm them (Cutter 1991).

Several important factors distinguish evacuation in developing countries from that in the industrialized nations. First, it tends to have a much greater impact on the local economy. Secondly, transport modes tend to be slower and more diverse, while routeways may more easily become cluttered with slow-moving vehicles (such as bullock carts). The poor, the uneducated and the subsistence farmers may be especially reluctant to leave their few possessions, if, that is, they understand the message at all. Also, distrust of authority may be endemic. Hence, there may be cases where evacuation is simply not a viable option or at least unlikely to be entirely successful.

There is some evidence that the people most willing to evacuate in developing countries are transient visitors, and that residents prefer to stay put, if they can. During the eruption of the Mexican volcano Paricutin, the richer, better educated residents of nearby villages were the first to leave, as they tended to have more ambition and better opportunities. Older and poorer people, in particular, feared the aculturation and exploitation that seemed to go with departure from home and lands. In several cases, people have been evacuated (willingly or otherwise) *en masse* from small volcanic islands (in the Tonga and Vanua Atu groups, and at Tristan da Cunha, for instance). But the prevailing mood has been to return as soon as possible, although small proportions of the groups have been attracted by the more up-to-date conditions in the place of evacuation.

In all societies, evacuation lasting more than, at the most, a few days is often inhibited by the strong desire of families to remain close to their homes or lands, whether or not these have been rendered useless by damage. Owner-occupiers may fear that absence will lead to dispossession and loss of title, while renters may fear eviction. Finally, it should be noted that the evacuation of hospitals poses very special problems (Drabek 1986: 36), as it may be vital not to interrupt medical care (especially for babies in incubators and patients undergoing intensive care).

The emergency period
After sudden-impact natural disasters short-term responses may last from a few hours to more than a week, depending on the type and seriousness of the impact (Drabek 1985, Drabek & Hoetmer 1991). Figure 6.10 shows the range of possible exigencies during the first nine days after two kinds of sudden impact disaster, earthquakes and floods. In both cases, operations such as evacuation and search and rescue dominate the early period, while measures such as sanitary provision dominate the agenda for subsequent days.

The first problem is to recognize the scope of the emergency, given the likely disruption of populations, transportation, organizations and information flow. Hence reconnaissance must be undertaken immediately. This can often be accomplished relatively cheaply by aerial survey, the exceptions being at night, in bad weather (especially when there are clouds) and during volcanic eruptions that send large volumes of ash into the atmosphere. The aerial view may be misleading if damage is in some

(a) Earthquakes

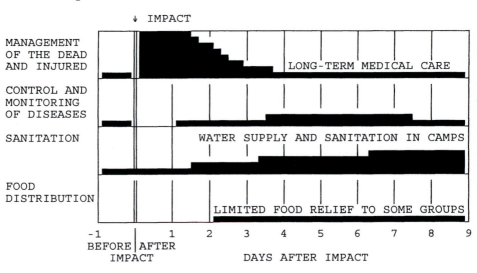

(b) Floods and sea surges

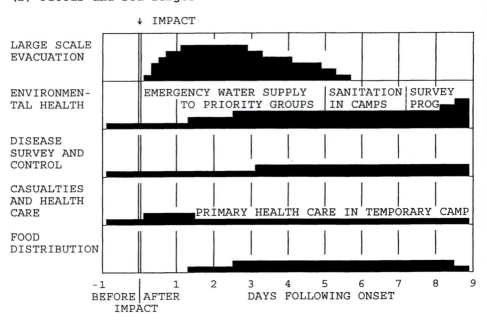

Figure 6.10 Changing needs and priorities after disaster (PAHO 1981: 15–16).

way concealed or if photographs do not provide the depth of field to be able to ascertain exactly what has happened.

Having established exactly where the impact has occurred, the first priority should be casualty management. Search and rescue will be vital life-saving activities to which the maximum attention, effort and promptness should be devoted. The functionality and capacity of hospitals and clinics should be assessed and those that have not been rendered unusable by the disaster should be supplied with emergency first-aid equipment. Special attention should be given to the evacuation, care and protection of vulnerable groups who have less than average ability to fend for themselves, including young, aged, infirm and handicapped people, especially if any of these groups has suffered the loss of some vital commodity such as heat or shelter.

Life-saving resources must be concentrated where they can be used effectively. When buildings have collapsed or been inundated with rock or soil, lifting and digging gear will be necessary, as will floodlights and generators for use at night. In floods, small powered boats will be needed. One limiting factor will be the speed with which stockpiled resources can be deployed. Columns of relief vehicles are by their very nature slow, and heavy plant and equipment cannot speedily be deployed unless there is some system for rapid loading and rapid transit in potentially difficult circumstances. Clearly, the overriding objective will be to rescue people alive.

In managing the population of a disaster area it is vital first to define the living victims and to distinguish them from interlopers, sightseers and other undesirable occupants of the impact zone. This is not necessarily an easy task during the chaotic moments that follow a highly destructive event. However, it should borne in mind that self-help will proliferate among survivors, including some mutual policing. Survivors will tend to share the remaining food stocks, shelter and transportation.

As has been emphasized elsewhere in this book, hasty and indiscriminate measures should be avoided, particularly during the emergency phase of the disaster. These include mass disinfection or vaccination, bulldozing rubble to stop the spread of disease, and the indiscriminate distribution of food, clothing, blankets and water purification tablets. In the aggregate, none of these measures is likely to have a particularly positive effect.

Towards a systematic and rational response
It is often convenient to consider the short term in sudden impact disasters as lasting about three days, and to divide it up into 12- or 24-hour periods. The disaster researcher Robin S. Stephenson has proposed a sequence of priorities for tackling the emergency in the most systematic and efficient manner possible (Stephenson 1981). It applies principally to

major earthquakes, but some of the elements will be relevant to other forms of disaster. According to Stephenson, during **the first 12 hours** after the disaster it is advisable to work through existing local organizations wherever possible. The boundaries of the affected area and the location of major damage should quickly be established, and continuing threats to survivors identified. Key points, such as hospitals and the worst damaged urban areas, should also be identified. Expert personnel, such as medical corps, engineers, police and firemen, should be moved into the area, and means of rapid transport, such as helicopters, should be mobilized. Local manpower and equipment should be assembled in order to begin clearing major road and rail links. Finally, steps should be taken to assess the viability of airports, including damage to runways, functionality of air traffic control, level of fuel and oil stocks, and adequacy of storage facilities for arriving relief supplies.

From **12–24 hours** search-and-rescue operations should begin in remote areas. A survey of water supplies to major urban areas should be initiated, and hyperchlorination used if there is a risk of vector-borne diseases being carried in the mains. A start should be made on assembling fuel stocks and repairing electricity supplies at the local grid level. Lastly, a casualty reporting system should be established based on hospitals and their contacts with local doctors and medical units.

From **24–48 hours** after the impact, remote areas should be sample-surveyed to assess their levels of damage and need. Resources should be shifted from the centre of the disaster zone to its periphery and from urban to rural areas. In order to accomplish this, route blockages should be identified (if this has not already been done by local personnel) and a start made on clearing them. The number of people requiring immediate emergency housing should be estimated. Message exchanges should be set up in parallel for both the general public and emergency services.

From **48–72 hours**, field kitchens should be deployed, if there is to be a mass feeding programme for survivors, and food generation should become a major priority. Tents, footwear and clothing should be transported to rural, remote and peripheral areas. Lastly, information centres should be set up and their telephone numbers publicized outside the disaster area (for example, in embassies and consulates abroad).

According to a League of Red Cross Societies' survey of 106 disasters that occurred between 1965 and 1974 (including floods, earthquakes, famines and hurricanes), the relative frequency of requests for different relief items was as follows: food 68 per cent, cash for local purchases 65 per cent, blankets 60 per cent, clothing 59 per cent, emergency shelter 54 per cent, medicines 45 per cent, vitamins 15 per cent, cooking utensils 13 per cent, and antibiotics 9 per cent. However, each disaster will generate a unique set of needs, and it would be difficult to assemble a standard package of disaster supplies.

Organizations and emergency management

The organizations which participate in post-disaster emergencies can be classified according to two schemes, the first of which relates to their changing functions (Bardo 1978):

(a) **Emerging organizations** include the local relief committee and *ad hoc* volunteer groups. They are created specifically for the occasion.

(b) **Expanding organizations** exist permanently but must increase their size in order to cope with the disaster. For instance, the Red Cross and local fire services absorb trained volunteers.

(c) **Adapting organizations** also exist permanently, but must change their functions in order to cope with the emergency. Local councils, for example, must form relief subcommittees.

(d) **Disbanding organizations** are those which are not relevant to the emergency and which will cease to function until it is over. They range from cultural organizations, such as choral societies, to bureaucratic bodies like motor vehicle registries.

The second means of classifying organizations is hierarchical:

(a) **World-level organizations** include those which are directly related to the United Nations, or under its jurisdiction. UNDRO (the Disaster Relief Office), UNHCR (the High Commission for Refugees) and the World Health Organization are typical examples.

(b) **Supranational organizations** can be divided into governmental bodies (such as the Organization of American States and the Centres for Disease Control) and non-governmental agencies, the voluntary charities, such as the Red Cross, Charitas, Oxfam, Christian Aid and the Salvation Army. Many are based in the main centres for relief activity: Geneva, London, New York and Washington. Most have a high reputation for professionalism and co-operativeness. Charities and aid organizations are accountable to donors and governments, but, curiously, are almost never responsible to the actual victims of the disaster.

(c) **Governmental organizations** include national and regional committees and special commissioners for disaster relief, usually with overall powers to co-ordinate emergency activities (in the USA state governors have this power). Local government organizations tend to be normal institutions (such as town councils), which have been adapted for the occasion as the disaster constitutes the only imperative demand upon their attention.

(d) **Non-governmental organizations** at the domestic level include academic research groups, national and regional voluntary associations, and Quasi-Autonomous Non-Governmental Organizations ("QANGOs") such as geological surveys and weather forecasting units.

The main organizational need is for an overall committee of co-ordination of the relief efforts, led by a figurehead and his or her office. In some countries this individual is given substantial powers, which may even include the ability to pass temporary laws or other forms of legally binding decree. The relief co-ordinator should be a high-level represent-ative of the national government, and his committee should include representatives of all the principal accredited politicians, experts and relief organizations.

At this point it is worth noting that the desire to help – or to be seen to be helping – can be one of the most debilitating phenomena during the aftermath of a major national disaster and may require special agencies to cope with it, fend it off or absorb it. For instance, *ad hoc* groups can be a major drain upon resources, requiring to be fed, housed and trained during the emergency. If they come from far away (perhaps even from abroad), they may add problems of unfamiliarity with local climate, customs, language and geography. The donor or volunteer is likely to be offended and uncomprehending if aid that is superfluous or misguided is refused. Hence, it may be wiser to accept it and direct it where it can do no harm than, instead, to risk the adverse publicity associated with outright refusal.

Several problems are commonly experienced with international aid. For example, assessments of what is required are too often geared to the total needs of the stricken area, whereas not everything will have been destroyed. Imported resources that duplicate those which are already available in an area are wasteful and tend to use time, manpower and storage facilities that could be better employed on activities such as search and rescue. In this respect, international agencies may need to be acquainted with what is being achieved at the local level, and what else is being imported, in order not to provide surplus supplies and assistance. The local response is often better than is supposed. What is needed is a pre-existing network of contacts and data-gatherers who can help international agencies when disaster strikes. It is often very good policy for the international agencies to send observers into the field to assess exactly what they can supply that is actually required. Another problem is that conventional data management procedures do not tend to work well during a disaster aftermath, and a more flexible approach will probably be needed. In particular, inappropriate or inadequate monitoring of creeping disasters (such as drought) may mean that the international agencies are taken by surprise when the impact of these becomes acute.

In synthesis, one hopes that the increasing internationalization of disaster relief will allow negative experiences to be shared, so that problems such as these can be avoided. To date, countries that have not participated strongly in global initiatives have tended to repeat errors made years before in other parts of the world.

The longer term aftermath

After the impact of the disaster and the subsequent emergency are over, and as rehabilitation and reconstruction begin, particular needs and problems will be associated with the continuing relief effort (Rubin 1991). These will tend to differ according to three consecutive stages (see the final section of this chapter). First, the **rehabilitation** phase is dominated by the need for continuing care of the remaining long-term victims of the disaster. Those who have not succeeded in finding their own food, housing and other basic necessities may need to have these provided. Victims should not be encouraged to depend passively on charity; therefore, it is often better to donate facilities such as cooking equipment, the raw materials of housing and the means of creating employment, rather than provide food, furnished accommodation and money. This is the period in which pre-existing problems such as crime, corruption and unemployment re-emerge and mingle with the problems created by the disaster.

The second long-term phase is that of **temporary reconstruction**. The structures built during this period will have a short design life and will include prefabricated dwellings, girder bridges and bracing and buttressing systems for damaged buildings. These will probably not be designed to resist future high-magnitude impacts and hence should not be *emplaced* until the danger has receded, and should be *replaced* before another impact is imminent or statistically probable.

The final phase is that of **permanent reconstruction**. Organization and experience are needed to administer the money allotted to this endeavour. First, any relief funds left over from the emergency and rehabilitation phases must be used wisely and not squandered on lavish or irrelevant projects. Moreover, damage mitigation measures and planning precepts should be incorporated into the reconstruction, by as much as costs will allow. Immediate needs can often be met using loans raised on the international money market, or by disposing of inessential assets. There is, however, a notable risk that efforts to achieve rational reconstruction will be diluted by the corruption that handling large sums of money so often entails.

Aspects of the US experience

At the national level, in the USA the agencies responsible for disaster relief disbursements, and to a certain extent for co-ordinating mitigation and aid, are the Federal Emergency Management Agency (FEMA) and the Federal Disaster Assistance Administration (FDAA), which have their headquarters in Washington, DC (Kreps 1990). Full-scale FEMA and FDAA intervention in an emergency is contingent upon the president making an official Disaster Declaration, which he does on average between 30 and 35 times a year. Disasters Declarations can also be made by state governors, in order to liberate state funds for relief. If the president declares a State of Emergency or a State of Serious Emergency for a particular state, then extraordinary powers will be vested in its governor to direct relief operations and take action to ensure public safety.

These measures are codified in Public Law 93–288, the US Disaster Relief Act of 1974. The Act also provides funds (up to $250,000 per state) to develop and revise predisaster prevention and preparedness plans, and even for the provision of psychiatric services for survivors. It provides for the establishment of Recovery Councils to carry out plans for reconstruction and development after disasters in such a way as to reduce vulnerability and avoid previous mistakes. The costs of such activity are estimated to be about half the average cost of damages.

In the USA, as elsewhere, one particularly demanding aspect of emergency management is the provision of shelter to large numbers of survivors who have lost their homes in disaster.

Emergency shelter

Several types of natural disaster (particularly earthquakes and hurricanes) create a sudden and massive demand for alternative housing as a result of the widespread destruction of vernacular buildings that they cause (Davis 1981). There is a lack of data on emergency housing problems and the effectiveness of solutions in the aftermath of disasters. But still a considerable portion of international aid money (to which must be added what countries spend on their own homeless) is disbursed on emergency shelter, even though donated shelter usually does not comprise more than 20 per cent of total housing provision.

The problem of shelter is universal and is most acute in Third World countries. In these, reconstruction tends to begin immediately, irrespective of government schemes. But this should not encourage governments to indulge in sweeping and immediate reconstruction planning, as carefully formulated indigenous solutions are likely to be the most effective and rapid ones, and will probably suit local needs best. Moreover, programmes of shelter provision should be (but often are not) linked with land tenure patterns. Emergency accommodation, on the other hand, tends to be made use of by survivors only as a last resort, when they have not succeeded in improvising shelters or obtaining help from relatives and friends. As mentioned in the sections on disaster planning and emergency management, compulsory evacuation and relocation schemes invariably meet too much opposition from the evacuees to be very successful and are often unsatisfactory from the social and economic perspectives.

Individual disasters can lead to solutions to the shelter problem as complex as they are costly. Nine examples are given in Table 6.2, together with data on their rates of occupance and use. In one of these cases, Managua after the 1972 Nicaraguan earthquake, ten forms of shelter (using five building systems) were supplied. These constituted 12,000 shelter units (excluding direct reconstruction) and were built at a cost of $4 million ($288 each, or $9.90 for every member of the predisaster population of the city). Yet such initiatives must be kept in their proper perspective: foreign aid rarely accounts for more than one fifth of the homes rebuilt after disasters in Third World countries.

A wide variety of types of shelter has been employed in disasters, largely because there is no agreed international policy on what to supply. Sadly, past efforts by the international relief community to solve the problem of the abrupt increases in homelessness which disasters cause have all too often been based on inaccurate assessments of needs: shelters that are highly inappropriate in style or culturally unacceptable have been supplied, as well as forms that were irrelevant to the needs of the displaced population (Aysan 1987).

Shelter as a process
In the words of one of the principal experts, Ian Davis (1978), shelter should be considered as a process and not an objective. Housing is often evolutionary, developing as the needs, objectives or prospects of its occupants develop; and therefore emergency shelter is also subject to processes of evolution, change and mutation over time (see Fig. 6.11).

Sudden-impact disasters lead to three possible housing situations, none of which is entirely exclusive of the others: homes may have survived in usable form, temporary lodging may be needed until permanent construction can house the victims or permanent reconstruction may be rapid enough to be the initial solution. In developing countries, if not elsewhere, survivors tend to have clear preferences regarding where to lodge after a disaster has destroyed their homes. These are, in order: the homes of friends or relatives, improvised shelters, converted buildings (such as schools and barracks) and official shelters (Davis 1978).

Where emergency shelter must be utilized, it can serve any of the following functions: temporary housing during periods of risk to permanent accommodation; protection against the elements (either hot or cold climate); protection of ownership or occupancy rights; emotional security and the need for privacy; storage of salvaged property; or the nodal point for receiving relief or starting reconstruction.

Solutions used in the developed world
In the industrialized world and the more accessible parts of developing countries prefabricated buildings are commonly deployed after disaster. A structure giving 40 m^2 of ground space can be transported on a flat-bed

Table 6.2 Use of donor shelters in selected disaster situations (after Davis 1978, UNDRO 1982).

Disaster	No. of houses damaged and destroyed	Number of homeless	Shelter units supplied	Dates of occupance	Use of shelter provided	Homeless in donated shelter	Evacuation policy?
Earthquake July 1963 Skopje, Jugoslavia	13,700 damaged, 15,766 destroyed	160,000	5,000 tents and 1,711 huts	Days 1–28 (tents), 28–115 (huts)	Tents 60%, huts 100%	10.3%	Yes
Earthquake March 1970 Cediz, Turkey	5,105 damaged 14,852 destroyed	90,000	25,000 tents and 400 polyurethane "igloos"	From day 6 for variable periods	100% at first, dropping to low levels	Unknown	No
Earthquake May 1970 Chimbote, Perú	59,800 damaged, 139,000 destroyed	500,000	12,400 tents, 1,298 huts & "igloos"	Day 60 onwards	Huts 30–60%, "igloos" 100%	17.6%	Yes
Earthquake Dec. 1972 Managua, Nicaragua	Damage total unknown, 50,000 destroyed	200,000	1,960 tents, 11,600 wooden huts, 500 "igloos"	Tents from day 2, huts from days 28–98, "igloos" day 150+	Tents 20–60%, huts 35–100%, "igloos" 45%	6.7%	Yes

Hurricane "Fifi" Sept. 1974 Honduras	Damaged 12,000 destroyed ≤15,000	≤350,000	10,000 tents, 601 wooden houses	From days 2–60	Unknown	Unknown	No
Earthquake Sept. 1975 Lice, Turkey	Damaged 8,450, destroyed 7,710	5,000	3,691 tents, 463 Oxfam "igloos"	Tents from day 14, "igloos" from days 60–90	Tents 90%, "igloos" 10%	Unknown	No
Earthquake Feb. 1976 Guatemala	Damaged total unknown, destroyed 384,762	1,600,000	10,000 tents, >50,000 huts	Variable	Tents low to high, huts high	Unknown	No
Cyclone Nov. 1979 Andhra Pradesh, India	Damaged and destroyed 150,000	250,000	Unknown number of simple thatched shelters	Days 30–90 until day 400 or later	Unknown	Unknown	No
Earthquake Oct. 1980 El Asnam, Algeria	Damaged 60,000 destroyed 80,000	400,000	15,000 tents, 20,000 prefabs	Tents up to 1 yr, prefabs 1 yr to present	Tents low at first then high	Unknown	Yes

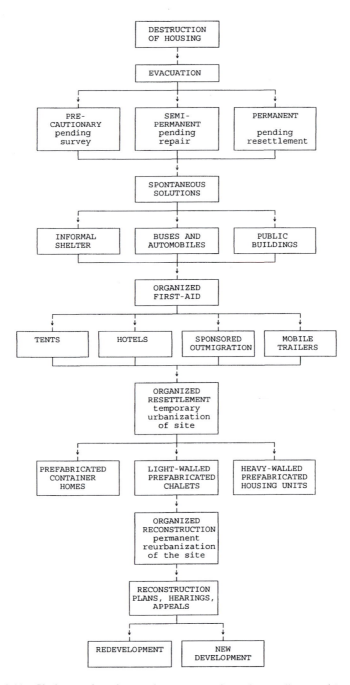

Figure 6.11 Shelter, rehousing and reconstruction after a disaster (the example of a developed country) (Alexander 1991a: 218).

truck to where it is needed. It may require a damp-proof concrete base pad, and may have simplified, flexible connections to electricity, water mains, sewers and bottled gas. Hence the large-scale use of prefabs requires considerable investment in temporary urbanization of the site, including access roads and retaining walls (Cavanagh & Johnston 1976).

When disaster strikes, mass evacuation of damaged buildings may be the only way to ensure public safety, although rubble should not be bulldozed away, as it can often be recycled into the reconstruction process. Thereafter, the temporary housing problem is immediate, serious and imperative, but amenable to various solutions (see Fig. 6.11). To begin with, redeployment of homeless survivors may involve accommodating them in hotels, mobile trailers (Fig. 6.12), holiday villages, public buildings (schools, offices, etc.) or even on ships and in trains. Some survivors will migrate out of the disaster area and others will find lodgings with relatives who have not lost their homes. On a simpler and more basic level, plastic sheeting is a cheap, easily transportable material that can provide some protection against the elements for very short periods, until something more durable can be organized.

Tents are a traditional remedy and they constitute one of the few types of post-disaster shelter that are regularly stockpiled – in their tens of

Figure 6.12 A survivor of the 1980 southern Italian earthquake, and the temporary home she was assigned pending the construction of long-term shelter.

thousands in some countries. They are not suitable for prolonged occupation or for use in freezing weather or monsoon rains, although their design should take into account the problems of sustained use and rigorous climates. Tents cannot resist high winds or even mild flooding, although they are easy to transport and quick to erect. Vacation-style tents tend not to function well and often puncture easily. Canvas army tents are more usually deployed and have the merit of being rugged. Lighter, cheaper tents made of woven plastics have been developed specifically for disaster relief and used in Southeast Asia and Malaysia (Greeman 1975).

At a slightly later stage, "monoblock" containers (Fig. 6.13) may be supplied to homeless families. These are oblong prefabricated modules that were originally developed as construction site offices and temporary building extensions. They are factory-assembled complete with glass-fibre insulation, plastic floors, interior partitions, stoves, bathrooms, and so on. Electricity connections are provided at head height, and flexible pipes are attached for water and sewer linkages. Monoblocks are designed to be equivalent in size to the standard road–rail–sea container, and have similar arrangements for lifting and transporting them. They can be positioned on jacks and hence do not necessarily require a base pad. In hot climates, however, they need an elevated second roof in order to insulate them from the heat of the sun.

As a more permanent and durable alternative, light-walled pre-fabricated dwellings (Fig. 6.14) are factory-assembled as panelling units and are bolted together, sealed and weatherproofed on concrete plinths at the site. As it is possible to transport them as separate panels, they can be quite spacious, but they tend to run a risk of rising damp and mould if the weatherproofing is not maintained. Their design life is usually about

Figure 6.13 Monoblock container of the kind used to house disaster survivors.

438

10 years, although many have the capacity to last longer. The physical appearance may be out of keeping with local architectural styles, and with the gravity of the situation, given that many were designed as holiday homes, with a very different clientele in mind.

If more substantial investment is to be made, the choice may fall upon heavy-walled prefabricated buildings, which use concrete panels instead of glass fibre and wood. They tend to last longer but are less easy to bring into position and erect, and may require special arrangements for transport and assembly. They were developed as permanent "system building", a cheap and rapid solution to housing crises with various causes. Problems tend to manifest themselves in the long term, as a result of weathering and inherent structural weaknesses not noticed in the design.

In providing shelter after disaster a balance must be struck between transience and permanence. Shelter which is too flimsy and poorly designed will neither protect nor satisfy its users: on the other hand, shelter which has an air of permanence may discourage full-blown

Figure 6.14 Light-walled prefabricated housing for homeless earthquake survivors at Potenza, southern Italy. Note the elements of regular settlement-landscaping and a one-way traffic system. Once occupied these huts rapidly acquired window boxes, ornamental porches and other trappings of permanence. There was thus a risk that their very longevity would slow down full-scale reconstruction.

reconstruction. If the temporary answer takes away all desire to restore normal patterns of town planning and traditional architecture, it virtually becomes the long-term solution. But it is rare that the provision of prefabricated shelters is based on sound principles of town planning. Sites tend to be laid out rapidly in very orderly patterns that resemble internment camps rather than good social space. Yet if the real buildings of urban centres are not restored, there is a risk that their functionality will not return and their populations will suffer frustration, discontent and loss of civic identity. Outmigration may result.

Hence, temporary shelter may introduce an extraneous note into local architecture (for example, the presence of wooden chalets in areas of traditional stone buildings; Alexander 1989). It may also create a problem when it wears out (Latina 1987), either of what to replace it with (stories abound of survivors living in prefabs for decades) or, if it has been abandoned in favour of permanently reconstructed buildings, what to do with the site.

Solving the emergency housing problem in developing countries

In the Third World most solutions to the emergency housing problem are indigenous and do not depart from local building traditions. This makes great sense when ordinary vernacular housing can be constructed cheaply and rapidly with abundant local materials (Fig. 6.15) and where there is no particular impetus to change in favour of newer and safer techniques (Kreimer 1979).

As part of international aid packages, donor countries have contributed to three approaches: advanced technology solutions, advanced techniques applied to local traditions in order to make them more efficient and solutions based entirely on local traditions. Donor solutions are a recent phenomenon related to the growth of rapid transportation, world relief programmes and the new spirit of global internationalism. Good solutions are based on careful study of factors such as practicability and logistics, local cultural and economic patterns, local climatic conditions and the performance of traditional housing.

Advanced technology has not proved to be the panacea for shelter problems after disaster in the Third World. At least 18 shelter systems are currently being manufactured and marketed by 21 companies in the developed world. These include several forms of polyurethane "igloos" designed by the German Red Cross (1,300 units were used variously in Perú, Nicaragua and Turkey over the period 1970–5) and by Oxfam (453 units were used in the Lice earthquake in Turkey in 1975; Greeman 1975). Although the igloos did prove useful in the early stages after disaster had struck, they suffered low rates of occupation and high rates of abandonment, which suggests that they had problems of functionality and cultural acceptability (Howard & Mister 1979). Other solutions have

Figure 6.15 Village building at Waimbasanga in Fiji. Traditional housing on the South Pacific islands is particularly vulnerable to hurricanes and floods, but can be reconstructed rapidly at minimal cost.

proved patently impractical, such as the Moss "air-drop" shelter, a parachute that becomes a parachute-shaped dwelling as soon as it reaches ground (Davis 1978). Advanced technology shelters tend to be relatively expensive to manufacture, transport and install, and if it is not possible to fit them in with local patterns of living, then they cannot be construed as an efficient solution to the shelter problem.

Advanced techniques applied to local materials tend to be less expensive than importing the shelter complete. For example, portable hydraulic rams have been designed to produce rammed-earth bricks for housing. These have been used by Oxfam after floods in Bangladesh. Elsewhere, Oxfam and Save the Children Fund have used "A"-frame housing, which was designed at Carnegie-Mellon University and relies on bamboo for the frame and thatch for the covering. The result is a relatively robust, spacious and well-ventilated dwelling suitable for parts of the warm tropics. After earthquakes in Perú and Nicaragua the relief agencies pioneered a form of construction known as "stack-sack", which uses sandbags filled with cement mix and then spiked with reinforcing bars before the cement sets. Lastly, "tilt-up" housing has been developed in response to hurricane damage in the Pacific Basin. Walls are constructed by pouring concrete into forms hollowed out of the earth at the site. When the concrete has set, the walls are tilted upright and fixed into position. This saves on the costs of constructing more elaborate moulds and panels, and transporting them to the point of erection.

The promotion of indigenous techniques simply involves help by the relief agencies to erect traditional local housing. In the villages of many developing countries this consists of wooden frames or posts, woven jute partitions coated with mud, and a corrugated steel roof. The total cost per unit is low, at about $100.

The experience with emergency shelter in developing countries

In developing countries, imported shelter is often never fully occupied and its use may decline by 70–75 per cent several months after the disaster. Thus, at Coyotepe in Nicaragua, where 360 tents were provided as temporary housing for the survivors of the 1972 earthquake, occupancy declined steadily from a peak of 60 per cent. The reasons for such lack of utilization vary from disaster to disaster. In the first place, the survivors' ability to improvise may be underestimated. For instance, up to 50,000 shelters were extemporized in Guatemala City within 24 hours of the 1976 earthquake (Davis 1977). Secondly, the number of people left homeless by the disaster may be overestimated, and local authorities may have deliberately exaggerated the toll of homelessness in order to obtain a greater proportion of attention or relief commodities. Concurrently, excessive volumes of aid may have arrived, and some donations may have duplicated others, while perhaps none of the aid matched up with local needs.

If relief planning is inadequate, units may have been located in the wrong place, such that they do not fit in with local patterns of activity. Similarly, unusual forms of housing may have been rejected for not conforming with local cultural predilections or economic activities. In this respect, hostility to multi-family units is almost universal. The only alternative form of high-density occupance is to house members of an extended family group together. In fact, survivors tend not to like being made conspicuous or set apart from people who did not suffer major losses in the disaster. If this happens, they may form a miserable underclass of people who are seen to be dependent on welfare. But whatever the advantages or drawbacks of temporary shelter, it is to be hoped that there will be a drift to more permanent forms of reconstruction, which should be substituted gradually for the shelter. Where the latter is not abandoned completely, it may be occupied by groups who are even lower down the social scale than the survivors of the disaster, such as recent immigrants or refugees who have not been settled.

Thus, after many disasters there has been a strong tendency to exaggerate the number of shelter units needed. There has also been a reluctance to apply the skills of town planners to the design of communities using shelters. In most cases, survivors will have a strong desire to regain their kinship and neighbourhood structures, and they will not tolerate being resettled at random or split away from family groups. Hence, shelter communities should attempt to preserve as much of the pre-existing social structure as possible, especially as survivors will need a web of mutual support in order to cope with difficult times during the aftermath.

There is an important time factor in solving the demand for shelter after disaster. Tents and other short-term solutions can usually be organized overnight, but large-scale camps of prefabricated buildings may take 8–10 weeks to plan, construct and assign units to families. Potential problems include financing the purchase of units and paying the labour required, the logistics of transportation and the difficulty of acquiring land. In the end, permanent housing may not be much more difficult to organize.

There are also important considerations of cost. Technological solutions manufactured in industrialized countries and transported to developing nations may be poor value for money (at perhaps $4,000–7,500 per family unit for a prefab), involving expensive imported materials, high Western labour costs and high transportation costs. Indigenous solutions tend to be cheap by comparison, but cannot be made to rely on products from the developed world that are either unavailable or too costly for the average Third World survivor. In most cases multiple units reduce per capita costs dramatically, but may be unpopular with their occupants.

A valid approach
The following criteria form a logical approach based on experience gained in Third World disasters (Davis 1978). No perfect or absolute solution exists, and the best compromise will depend on local cultural, climatic and economic factors and local materials. Novel solutions should be avoided, as they are likely to be rejected for cultural or functional reasons, and a "tried and tested" approach is more reliable. However, shelter needs to be provided within a week, not two months, and hence, if a durable solution is being worked out, an interim measure must be designed as well. But despite this, people are rarely so shocked and debilitated by the impact of natural disaster that they do not rebuild their own homes if they can (the outbreak of famine is an obvious exception). So it is important to bear in mind that reconstruction is likely to begin immediately after the crisis phase has abated.

It should be borne in mind that donor housing may be too costly, may generate little employment, may be unacceptable to the occupants and may take too long to deploy. Hence, local resources should be used wherever possible, as they are immediately available and as their use generates local employment. Thus, labour-intensive solutions should be encouraged, for the community generally needs work after a disaster. Furthermore, although shelter must not be too costly for its intended recipients, it is better to sell than to give (even renting preserves the survivors' dignity).

However ingenious they are, new construction methods often have little impact upon practices common in Third World countries. It may be better to identify the primary building materials and purchase them in bulk for sale or distribution to survivors. Thus, Oxfam made one of its largest allocations of funds for the distribution of corrugated steel sheet (known locally as *lamina*) in Guatemala after the 1976 earthquake (Bates et al. 1979). This can be used as a roofing material on very traditional forms of housing (in some cultures it is even a form of status symbol to have a house roofed with corrugated steel sheets). However, local communities should never be encouraged to depend too much upon supplies which may eventually cease.

Hence, in many Third World situations it often makes sense to do no more than simply provide materials, advice and encouragement so that survivors can rebuild their own homes according to traditional styles. The exception is when the occasion can be used to make traditional dwellings safer – for example, to render them anti-seismic. But even rural refugees in Africa are quite capable of using their skills and manpower to reconstruct traditional shelter with local materials. This enables relief monies and resources to be devoted to other uses, or more rapid progress to be made where aid and assistance are lacking.

In summary, although the problem of housing the world's poor and afflicted is probably insuperable, it can and should be tackled and, despite the mistakes of the past, useful results can be achieved. Shelter is part and parcel of the wider problem of achieving a sound, durable and safe reconstruction.

Recovery and reconstruction after disaster

There is little sign that in the modern era natural disasters have led to the wholesale abandonment of damaged towns and villages. Such settlements form part of economic systems and regional demographic trends, whose growth may represent a strong stimulus in favour of reconstruction, while the psychological entity known as "community" seems well able to survive physical devastation. In any case, relocation of a community after disaster will only succeed if there is full participation of all its members in the planning and decision-making process, if political leaders are sensitive to individual needs and if cultural and ethnic factors are taken into account (Drabek 1986: 302). Yet even when there is no move to raze damaged settlements, disaster may hasten changes that are occurring more slowly as a result of other forces, such as political developments, economic achievements, social expectations, and the quality of life.

The Kates and Pijawka model

Robert W. Kates & David Pijawka (1977) have argued that the reconstruction process typically occurs in four stages: an emergency period, a restoration phase, a replacement reconstruction period and one of developmental reconstruction. On examining recovery after earthquakes in the USA and Central America, they found that the phases overlap rather than follow each other in strict succession.

The first stage of the aftermath is the **emergency period**, in which the disaster plan is in operation, and emergency actions, such as search and rescue, take precedence. Normal social and economic activities are suspended or greatly modified while the situation is sorted out with regard to casualties, survivors and destruction. It may be imperative to clear rubble or wreckage if these are hazardous or if they obstruct search-and-rescue operations. Widespread wreckage is very costly to clear and the chances of doing so may depend on the availability of machinery, fuel, manpower and landfill sites. In fact, rubble clearance tends to destroy both salvageable materials and landmarks. Not only may this delay reconstruction, but the very presence of the heavy plant involved in demolition and clearance impedes the planning and execution of reconstruction.

The emergency phase may last from days to weeks, depending on the scale and type of disaster and the preparedness and capacity to respond of the affected community. It ends when major infrastructure has been cleared and basic services have been restored: this may involve temporary bridges or connections, standpipes from water supplies, and so on. Search and rescue will have ceased and mass feeding programmes been reduced to a minimum. Immediate shelter needs will have been met and precarious structures buttressed or braced.

In the 1972 Nicaraguan earthquake this phase lasted 3–4 weeks, while after the 1964 Alaskan tremors it lasted about seven days. Contemporary accounts suggest that in the 1857 southern Italian earthquake the emergency phase was still in full swing in the remoter settlements 10 weeks after the disaster.

The **restoration period** is characterized by repairs to utilities and to commercial, industrial and residential structures. Buildings that cannot be salvaged will be demolished (this will already have happened if they pose a particular threat to public safety), and efforts will be made to turn others back to normal uses. Survivors who have left the disaster area return during this period, if they ever come back. In societies with a large resource base restoration may only take a few months (two in Alaska after the 1964 tremors and nine in Nicaragua after the 1972 earthquake). However, to ensure that transportation and utilities are functioning will be a complex and laborious process, especially if damage has been widespread.

The **reconstruction-replacement period** is one in which capital stocks are rebuilt and the economy of the area recovers to predisaster levels. This may take some years, and the phase ends when population has returned to its former levels, and losses in jobs, housing and services have been rectified.

Finally, during the **phase of developmental reconstruction** building will not merely commemorate the disaster but also mark the further development of the area (Pantelic 1991). For example, over the period 1915–29 a monumental civic centre was built in San Francisco, which commemorated the recovery from the earthquake of 1906 and the general improvement of the city's facilities.

Both the general principals and the details of the Kates and Pijawka model have been criticized. Although in the developed world reconstruction can take months or years, in the world's poorest countries it may instead last for days, such is the paucity and simplicity of materials and methods. Moreover, Hogg (1980) found that the four phases, if they can be distinguished at all, may sometimes occur almost simultaneously. She also found that political factors can delay or accelerate them disproportionately with respect to other processes. For the purposes of providing shelter and reconstructing homes, it may in the end be better to follow the more straightforward set of phases proposed by UNDRO (1984): predisaster, the immediate relief period (impact to day 5), the rehabilitation period (day 5 to 3 months), and the reconstruction period (3 months onward).

The reconstruction process
Reconstruction is a predictable process which usually occurs on the same site in a definable manner and over an estimable timespan (Fig. 6.16). While it is going on, settlements will be in a state of flux in which there is intense competition for prime locations. There will be competition for resources, in which the winning parties will be those whose economic status has been affected least by the disaster. Banks, which have control over the supply of investment funds, and builders, whose trade will be booming, are usually favoured, as are businesses that form part of a large chain (which will have absorbed much of their losses). Such enterprises generally acquire the most advantageous sites and reconstruct quickest. The less financially secure businesses may be shaken out to the periphery of the disaster area. Moreover, reconstruction commonly requires 2–3 times more area, as it is likely to take place to a lower density than that of

Figure 6.16 Demolition at Sant'Angelo dei Lombardi, four years after the 1980 southern Italian earthquake. Few settlements are completely relocated after disaster and many are rebuilt *in situ* in similar architectural style to that which was destroyed. Demolition should thus be used with care, as building materials can often be recycled into the reconstruction process, land ownership patterns must be preserved and it may be possible to restore much of the urban fabric.

a devastated city core. Hence, during reconstruction the relationship between city and region will change in terms both of planning and economics.

Prior economic relationships based on location do not necessarily hold true after the disaster and equity among the victims tends to be eclipsed by the effect of market forces, in which those sectors of the economy that have traditionally been powerful rapidly regain control. Wealthier citizens tend to be the first to rebuild their homes, often with developmental improvements and modifications, as these people have access to capital and credit facilities. The poorer the individual, the more dislocation he will suffer and the longer it will take him to be resettled permanently. Such individuals can gain some protection from market forces by forming minority groups on ethnic, national or religious lines, if such are appropriate. The combined bargaining or purchasing power may help ease their difficulties, but in general residents of low socio-economic status must wait longest to reconstruct their homes and find employment.

The destruction of housing may pose a problem of unpaid mortgages. If the balance is not waived, victims may be too indebted to buy a new home, but if it is waived, then home-buyers are given an unfair advantage over renters. One alternative may be to institute insurance and require mortgagees to purchase it and thus protect their loans. Studies of loan-granting institutions in California, however, show that they seem to be much more worried about the effects on loan repayments of factors such as divorce than about the possibility that natural disasters will destroy the collateral (Palm et al. 1990).

The demographics of communities which have suffered major disasters indicate that two opposing tendencies may be present. Governments may encourage permanent emigration after disaster has struck, in order to relieve pre-existing problems such as homelessness and unemployment. This can be achieved by prevailing on other nations to relax immigration restrictions and work permit requirements for the relatives of people who immigrated at an earlier phase. This can be regarded as a form of international aid. On the other hand, once the reconstruction process begins, migrants may come back from abroad to participate. On aggregate, in the present century 10 per cent of population losses caused by disasters have been replaced within three years and 50 per cent within seven. Moreover, the boom period of post-disaster reconstruction can act as a magnet for migrants. Thus Skopje in the former Jugoslavia doubled in population rapidly after the 1963 earthquake (Greene 1987), as did Popayan in Colombia after the 1981 tremors (Davis 1987).

Rather than the unavailability of capital and labour, uncertainty tends to delay recovery. This can mean uncertainty about planning provisions and the availability of financial resources. Nevertheless, the spirit of the people may have a great deal to do with how quickly they reconstruct. National or regional pride may be capable of forcing the pace.

Practical reconstruction planning
The quality of the post-disaster environment depends upon several factors, among them the size of the area affected, the degree of continuing disruption, the effectiveness of relief efforts, the quality of communication, social attitudes in the disaster area and with regard to it and the level of social support. In large measure it also depends on the efforts made by government to plan for and facilitate reconstruction. Note, however, that ambitious planning usually fails!

Disaster plans should address the problem of reconstruction. It is the prime opportunity to introduce measures of hazard mitigation and vulnerability reduction. The plan should include means of facilitating the compulsory purchase of land in high-risk areas. For example, urban development could be removed in this manner from the banks of a river and an urban floodway park substituted under municipal jurisdiction. The disaster plan should also consider the possibility of permitting rapid change in the local building codes in order to accommodate information about hazards that has been gained in the disaster. In this respect, decision-making may also need to be streamlined in the field of planning, in order not to miss opportunities.

Ideally, cities should attempt to prepare a reconstruction plan before disaster strikes. The more readily the realities of destruction and damage are faced, the more abundant the opportunities to reduce suffering and benefit from creativity in the reconstruction process. However, faced with many other problems to resolve, and minimal funding for such tasks, few planning offices carry out studies of disaster risk and of the feasibility of reconstructing damage that has yet to happen, however inevitable the impact is. Public and political opinion, moreover, is not behind such tasks.

There are several ways in which post-disaster reconstruction can be financed, all of which have attendant advantages and drawbacks. First, aid can simply be granted to eligible survivors, though this will not stimulate the local economy and may encourage dependence on relief. Secondly, materials can be distributed free to self-help groups. This option encourages local initiative but may still result in aid-dependency. Thirdly, houses and other buildings can be sold to the people who need them. This is seldom an appropriate option in areas were personal resources have been depleted by the disaster. Fourthly, donated materials can be sold at subsidized prices and the revenues thus generated used to buy more supplies. These four strategies can be applied to the more substantial forms of temporary shelter as well as to permanent reconstruction.

Alternatively, loan programmes can be instituted. They can be classified into four types. The first, normal **long-term loans**, encourages a return to predisaster monetary conditions, but such opportunities are seldom accessible to victims who are not credit-worthy. A second possibility is that victims who cannot raise a deposit on their own may be able to do so if given **matching loans** by government or charities. A third alternative is that the national government exchequer offer **loan guarantees** in order that they be made accessible to low-income groups. Finally, **revolving loans** can enable repayments to be recycled in the disaster area as new loans. This, in effect, results in the creation of a non-profit credit union by an initial injection of capital.

Having decided on the method of aid, governments can avail themselves of several strategies in order to facilitate post-disaster shelter and housing. Before the event, tents, bedding and rebuilding materials can be stockpiled. This, of course, requires adequate foresight and investment (in both the materials and the means of storing them). Then, after the event, the prices of building materials can be subsidized or controlled in order to prevent speculation or other forms of profiteering. This method tends to disrupt local market economies and is difficult to enforce if the government is weak or disorganized. Alternatively, groups can be encouraged to pool their resources and purchase the necessary shelters or materials collectively in order to use their increased economic power to obtain a better price.

Where permanent housing does not take too long to plan and construct, governments can use collective resources to accelerate the process as a means of bypassing the temporary housing stage. Rather than rebuilding for the people (thus damaging the local economy), it may be expedient to construct a "model house", into which new safety provisions have been incorporated, to serve as an example to local builders. Alternatively, "core housing" can be supplied, in the form of a simple, mass-produced frame or solid core which survivors or local builders can adapt to particular needs.

Capacity to implement plans depends on the differentiation of social and political power and the availability of resources (Mileti 1980).

Reconstruction in the Third World
Third World disaster zones are often neglected areas. In the worst cases, health services may be inadequate and geared to the demands of the wealthy. The rate of illiteracy may be high, transport and communications poor and the distribution system for goods and commodities inadequate. The productivity of agriculture may be low and vulnerable to decimation, and pastures may be stocked with too many free-range animals. Natural resources such as forests may be depleted and women may work hard as well as rear children. The search for a solution to such problems forces people to inhabit marginal and hazardous lands. In fact, areas that are susceptible to disasters have been on the edge of the core–periphery relationship for centuries. They usually lack political force or lobbying power, and religion tends to play a dominant ideological rôle in their culture. They have attracted only a small share of developmental aid resources, while their own potential for improvement remains untapped, thus trapping them in an endless poverty cycle.

Disaster may be the opportunity to loosen the grip that poverty has on such societies (Fernandez 1979). Natural catastrophe tends to focus attention (however briefly) on the area and to produce and disseminate new information about it. Linkages may be fostered with the international aid community and lead to more frequent transactions with the outside world and with central government. The stimulus given by events and the examples set by aid workers may foster local leadership, while the plight of the area may cause the local community to align itself with groups seeking similar goals elsewhere. In such areas the need for assistance simply does not abate with the end of a disaster aftermath (if indeed it ever really ends at all). Hence, provisional facilities such as clinics and shelter may still function years later. Disasters may cause a process of selectivity in subsistence economies in which, sadly, the death and destruction caused to humans and their livestock reduce the excessive pressure on the land which is one of the causes of chronic poverty.

Despite these observations, which are as tentative as they are hypothetical, there have been few long-term studies of recovery from disaster in the Third World, and relatively little is known about the general relationship between disasters and development. What is known is, unfortunately, not encouraging. For example, after the 1976 earthquake in Guatemala there was some progress towards safer reconstruction. The number of adobe buildings, which constitute a particular source of seismic vulnerability, was drastically reduced, although those that remained were not significantly strengthened. Despite the best efforts of relief agencies, even very modest anti-seismic building methods did not diffuse well and the prevailing materials and construction techniques were those which individuals preferred, regardless of the risk.

References

Adams, M., L. Glaze, M. Sheridan 1991. Monitoring Colima volcano, Mexico, using satellite data. *Bulletin of Volcanology* **53**,

Alexander, D. E. 1980. The Florence floods: what the papers said. *Environmental Management* **4**, 27–34.

Alexander, D. E. 1989. Preserving the identity of small settlements during post-disaster reconstruction in Italy. *Disasters* **13**, 228–36.

Alexander, D. E. 1991. Natural disasters: a framework for research and teaching. *Disasters* **15**, 209–26.

Ali, A., D. A. Quadir, O. K. Huh 1989. Study of flood hydrology in Bangladesh with NOAA satellite AVHRR data. *International Journal of Remote Sensing* **10**, 1,873–91.

Anderson, W. A. 1969. Social structure and the rôle of the military in natural disaster. *Sociology and Social Research* **53**, 242–52.

Anderson, W. A. 1970. Military organizations in natural disaster: established and emergent norms. *American Behavioural Scientist* **13**, 415–22.

Aysan, Y. 1987. Homeless in 42 m². *Open House International* **12**, 21–6.

Baker, E. L. 1991. Hurricane evacuation behaviour. *International Journal of Mass Emergencies and Disasters* **9**, 287–310.

Bardo, J. W. 1978. Organizational response to disaster: a typology of adaptation and change. *Mass Emergencies* 3, 87–104.

Bates, F. W., W. T. Farrell, J. K. Glittenberg 1979. Some changes in housing characteristics in Guatemala following the February 1976 earthquake and their implications for earthquake vulnerability. *Mass Emergencies* **4**, 121–33.

Belardo, S., K. R. Karwan, W. A. Wallace 1984. Managing the response to disasters using microcomputers. *Interfaces* **14**, 29–39.

Belardo, S., H. L. Pazer, W. A. Wallace, W. D. Danko 1983. Simulation of a crisis management information network: a serendipitous evaluation. *Decision Sciences* **14**, 588–606.

Belardo, S. & W. A. Wallace 1989. "Gaming" as a means for evaluating decision support systems for emergency management response. In *Simulation in emergency management and technology*, J. Sullivan & R. Newkirk (eds), 113–17. La Jolla, California: Society for Computer Simulation.

Berke, P. & N. Stubbs 1989. Automated decision support systems for hurricane mitigation planning. *Simulation* **53**, 101–9.

Bianchi, R., R. Casacchia, A. Coradini, A. M. Duncan, J. E. Guest, A. Kahle, P. Lanciano, D. C. Pieri, M. Poscolieri 1990. Remote sensing of Italian volcanoes. *EOS: American Geophysical Union, Transactions* **71**, 1,789–91.

Blikra, L. H. 1990. Geological mapping of rapid mass movement deposits as an aid to land-use planning. *Engineering Geology* **29**, 365–76.

Blythe, K. & G. P. Nash 1980. Aerial infrared photography for floodplain investigations. *Institution of Water Engineers and Scientists, Journal* **34**, 425–34.

Booth, B. 1977. Mapping volcanic risk. *New Scientist* **75**, 743–5.

Borchardt, G. 1991. Preparation and use of earthquake planning scenarios. *California Geology* **44**, 195–203.

Brown, A. G., K. J. Gregory, E. J. Milton 1987. The use of Landsat Multi Spectral Scanner data for the analysis and management of flooding on the River Severn, England. *Environmental Management* **11**, 695–701.

Brunsden, D., J. C. Doornkamp, P. G. Fookes, D. K. C. Jones, J. H. M. Kelly 1975. Large-scale geomorphological mapping and highway engineering design. *Quarterly Journal of Engineering Geology* **8**, 227–53.

Burkhart, F. N. 1987. Experts and the press under stress: disaster journalism gets mixed reviews. *International Journal of Mass Emergencies and Disasters* **5**, 357–67.

Burkhart, F. N. 1991. Journalists as bureaucrats: perceptions of "social responsibility" media rôles in local emergency planning. *International Journal of Mass Emergencies and Disasters* **9**, 75–87.

Cain, J. M. & M. T. Beatty, 1968. The use of soil maps in the delineation of flood plains. *Water Resources Research* **4**, 173–82.

Carroll, J. M. (ed.) 1983. *Computer simulation in emergency planning*. La Jolla, California: Society for Computer Simulation.

Cavanagh, J. & F. Johnston 1976. Earthquakes and prefabs. *Ecologist* **6**, 104–6.

Christensen, L. & E. Ruch 1980. The effect of social influence on response to hurricane warnings. *Disasters* **4**, 205–11.

Comfort, L. K. 1991. Designing an interactive, intelligent, spatial information system for international disaster assistance. *International Journal of Mass Emergencies and Disasters* **9**, 339–53.

Congressional Research Service 1984. *Information technology for emergency management*. Washington, DC: Library of Congress, US Government Printing Office.

Cooke, R. U. 1984. *Geomorphological hazards in Los Angeles*. London: Allen & Unwin.

Curran, P. 1985. *Principles of remote sensing*. Harlow, England: Longman.

Cutter, S. L. 1991. Fleeing from harm: international trends in evacuations. *International Journal of Mass Emergencies and Disasters* **9**, 267–85.

Daines, G. E. 1991. Planning, training and exercising. In *Emergency management: principles and practice for local government*, T. E. Drabek & G. J. Hoetmer (eds). Washington, DC: International City Management Association.

Davis, I. 1977. Emergency shelter. *Disasters* **1**, 23–40.

References

Davis, I. 1978. *Shelter after disaster*. Headington, Oxford: Oxford Polytechnic Press.

Davis, I. 1987. Shelter and housing in disaster prone areas. *Open House International* **12**, 66–71.

Doornkamp, J. C., D. Brunsden, D. K. C. Jones, R. U. Cooke, P. R. Bush 1979. Rapid geomorphological assessments for engineering. *Quarterly Journal of Engineering Geology* **12**, 189–204.

Drabek, T. E. 1985. Managing the emergency response. *Public Administration Review* **45** (special issue), 85–92.

Drabek, T. E. 1986. *Human system response to disaster: an inventory of sociological findings*. New York: Springer.

Drabek, T. E. et al. 1969. Social processes in disaster: family evacuation. *Social Problems* **16**, 336–49.

Drabek, T. E. & G. J. Hoetmer (eds) 1991. *Emergency management: principles and practice for local government*. Washington, DC: International City Management Association.

Dynes, R. R. & E. L. Quarantelli 1975. *Organizational responses to major community crises*. Columbus, Ohio: Disaster Research Center.

Earthquakes and Volcanoes 1989. Vol. 21, no. 4.

Ehlers, M., G. Edwards, Y. Bedard, 1989. Integration of remote sensing with geographic information systems: a necessary evolution. *Photogrammetric Engineering and Remote Sensing* **55**, 1,619–28.

Fahmi, K. J. & J. N. Alabbasi 1989. Seismic intensity zoning and earthquake risk mapping in Iraq. *Natural Hazards* **1**, 331–40.

Fernandez, A. 1979. Relationship between disaster assistance and long-term development. *Disasters* **3**, 32–6.

Fischhoff, B., C. Hohenemser, R. E. Kasperson, R. W. Kates 1978. Handling hazards. *Environment* **20**, 16–20, 32–7.

Foster, H. D. 1980. *Disaster planning: the preservation of life and property*. New York: Springer.

Francis, P. W. & G. L. Wells 1988. Landsat Thematic Mapper observations of debris avalanche deposits in the central Andes. *Bulletin of Volcanology* **50**, 258

Frank, T. D. 1984. The effect of change in vegetation cover and erosion patterns on albedo and texture of Landsat images in a semi-arid environment. *Annals of the Association of American Geographers* **74**, 397–407.

Greeman, A. 1975. Oxfam building polyurethane foam houses for refugees. *New Scientist* **68**, 530.

Greene, M. R. 1987. Skopje, Yugoslavia: seismic concerns and land use issues during the first twenty years of reconstruction following a devastating earthquake. *Earthquake Spectra* **3**, 103–17.

Guha-Sapir, D. & M. F. Lechat 1986. Informational systems and needs assessment in natural disasters: an approach for better disaster relief management. *Disasters* **10**, 232–7.

Gupta, R. P. & B. C. Joshi 1990. Landslide hazard zoning using the GIS approach: a case study from the Ramganga catchment, Himalayas. *Engineering Geology* **28**, 119–32.

Hall, D. K. & J. Martinec 1985. *Remote sensing of ice and snow*. London: Chapman & Hall.

Handmer, J. (ed.) 1987. *Flood hazard management: British and international perspectives*. Norwich, England: Geobooks.

Harker, G. R. & J. W. Rouse 1977. Flood plain delineation using multi-spectral data analysis. *Photogrammetric Engineering and Remote Sensing* **43**, 81–7.

Heffron, E. F. 1977. Interagency relationships and conflict in natural disaster: the Wilkes-Barre experience. *Mass Emergencies* **2**, 111–19.

Herman, R. E. 1982. *Disaster planning for local government*. New York: Universe Books.

Hogg, S. J. 1980. Reconstruction following seismic disaster in Venzone, Friuli. *Disasters* **4**, 173–85.

Howard, J. & R. Mister 1979. Lessons learnt by Oxfam from their experience of shelter provision 1970–1978. *Disasters* **3**, 136–44.

IDI 1981. *The physical and social consequences of a major Thames flood*. London: International Disaster Institute.

Kates, R. W. & D. Pijawka 1977. From rubble to monument: the pace of reconstruction. In *Disaster and reconstruction*, J. E. Haas, R. W. Kates, M. J. Bowden (eds), 1–23. Cambridge, Mass.: MIT Press.

Kertesz, A. 1979. Representing the morphology of slopes in engineering geomorphological maps with special reference to slope instability. *Quarterly Journal of Engineering Geology* **12**, 235–41.

Kienle, J., K. G. Dean, H. Garbeil 1990. Satellite surveillance of volcanic ash plumes, applications to aircraft safety. *EOS: American Geophysical Union, Transactions* **71**, 266.

Kreimer, A. 1979. Emergency, temporary and permanent housing after disasters in developing countries. *Ekistics* **46**, 361–5.

Kreps, G. A. 1983. The organization of disaster response: core concepts and processes. *International Journal of Mass Emergencies and Disasters* **1**, 439–66.

Kreps, G. A. 1990. The Federal emergency management system in the United States: past and present. *International Journal of Mass Emergencies and Disasters* **8**, 275–300.

Lang, T. E. & J. D. Dent 1980. Scale modelling of snow-avalanche impact on structures. *Journal of Glaciology* **26**, 189–96.

Latina, C. 1987. The long-term performance of prefabricated housing after Italian earthquakes. *Open House International* **12**, 27–38.

Lindell, M. K. & R. W. Perry 1991. Understanding evacuation research. *International Journal of Mass Emergencies and Disasters* **9**, 133–6.

McLuckie, B. J. 1973. *The warning system: a social science perspective*. Washington, DC: National Weather Service, NOAA.

Malingreau, J. P. & X. Kasawanda 1986. Monitoring volcanic eruptions in Indonesia using weather satellite data: the Colo eruption of July 28 1983. *Journal of Volcanology and Geothermal Research* **27**, 179–94.

Maruyasa, T. et al. 1964. Statistical analysis of landslides and related phenomena on aerial photographs. *Journal of the Japan Society of Photogrammetry* **1** (special vol.), 93–100.

Mathewson, C. C. & R. G. Font 1974. Geologic environment: forgotten aspect in the land use planning process. In *Geologic mapping for environmental purposes*, H. F. Ferguson (ed.), 23–8. Boulder, Colorado: Geological Society of America.

Matson, M. 1984. The 1982 El Chichón volcano eruptions: a satellite perspective. *Journal of Volcanology and Geothermal Research* **23**, 1–10.

May, P. J. 1985. *Recovering from catastrophes: federal disaster relief policy and politics*. Westport, Connecticut: Greenwood Press.

Mileti, D. S. 1980. Human adjustment to the risk of environmental extremes. *Sociology and Social Research* **64**, 328–47.

Mileti, D. S. & J. H. Sorensen 1988. Planning and implementing warning systems. In *Mental health response to mass emergencies*, M. Lystad (ed.), 331–45. New York: Brunner-Mazel.

Mitchell, J. K., N. Devine, K. Jagger 1989. A contextual model of natural hazard. *Geographical Review* **79**, 391–409.

Mouginis-Mark, P. J., D. C. Pieri, P.W. Francis, L. Wilson, S. Self, W. I. Rose, C. A. Wood 1989. Remote sensing of volcanoes and volcanic terrain. *EOS: American Geophysical Union, Transactions* **70**, 1,567–75.

Mouginis-Mark, P. J., S. Rowland, P. W. Francis, T. Friedman, H. Garbeil 1991. Analysis of active volcanoes from the Earth Observatory System. *Remote Sensing of the Environment* **35**, 1–12.

Muñoz, A. V. 1989. Assessment of earthquake hazard in Panama based on seismotectonic regionalization. *Natural Hazards* **2**, 115–32.

Nakano, T. & I. Matsuda 1984. Earthquake damage, damage prediction and countermeasures in Tokyo, Japan. *Ekistics* **51**, 415–20.

Olson, R. S. & D. C. Nilson 1982. Public policy analysis and hazards research: natural complements. *Social Science Journal* **19**, 89–103.

PAHO 1981. *Emergency health management after natural disaster*. Washington, DC: Pan American Health Organization.

Palm, R. I., M. E. Hodgson, R. D. Blanchard, D. I. Lyons, 1990. *Earthquake insurance in California: environmental policy and individual decision-making*. Boulder, Colorado: Westview Press.

Pantelic, J. 1991. The link between reconstruction and development. In *Managing natural disasters and the environment*. A. Kreimer & M. Munasinghe (eds), 90–4. Washington, DC: Environment Department, World Bank.

Parker, D. 1981. The value of hazard zone mapping: Water Authority Section 24(5) surveys in England and Wales. *Disasters* **5**, 120–4.

Penn, S. 1984. Colour-enhanced infra-red photography of landslips. *Quarterly Journal of Engineering Geology* **17**, ii–v.

Perry Jr, J. B. & D. P. Meredith 1978. *Collective behaviour response to social stress*. St Paul, Minnesota: West.

Perry, R. W., M. K. Lindell, M. R. Greene 1981. *Evacuation planning in emergency management*. Lexington, Mass.: Lexington Books.

Petak, W. J. 1985. Emergency management: a challenge for public administration. *Public Administration Review* **45**, 3–6.

Phillips, B. D. 1986. The media in disaster threat situations: some possible relationships between mass media reporting and voluntarism. *International Journal of Mass Emergencies and Disasters* **4**, 7–26.

Quarantelli, E. L. 1981a. Disaster planning: small and large, past, present and future. *American Red Cross EFO Division disaster conference proceedings*. Alexandria, Virginia: Eastern Field Office, American Red Cross.

Quarantelli, E. L. 1981b. The command post point of view in local mass communications systems. *Communications: International Journal of Communication Research* **7**, 57–73.

Rasid, H. & M. A. H. Pramanik 1990. Visual interpretation of satellite imagery for monitoring floods in Bangladesh. *Environmental Management* **14**, 815–21.

Rothery, D. A., P. W. Francis, C. A. Wood 1988. Volcano monitoring using short wavelength infrared data from satellites. *Journal of Geophysical Research* **93**, 7,993–8,008.

Rubin, C. B. 1991. Recovery from disaster. In *Emergency management: principles and practice for local government*, T. E. Drabek & G. J. Hoetmer (eds). Washington, DC: International City Management Association.

Rubin, C. B. & D. G. Barbee 1985. Disaster recovery and hazard mitigation: bridging the intergovernmental gap. *Public Administration Review* **45**, 57–63.

Sauchyn, D. J. & N. R. Trench 1978. Landsat applied to landslide mapping. *Photogrammetric Engineering and Remote Sensing* **44**, 735–41.

Scanlon, J., S. Alldred, A. Farrell, A. Prawzick 1985. Coping with the media in disasters: some predictable problems. *Public Administration Review* **45** (special issue), 123–33.

Scanlon, T. J., R. Luukko, G. Morton 1978. Media coverage of crisis: better than reported, worse than necessary. *Journalism Quarterly* **55**, 68–72.

Sims, J. H. & D. D. Baumann 1983. Educational programmes and human response to natural hazards. *Environment and Behaviour* **15**, 165–89.

Sorenson, J. H. 1991. When shall we leave? Factors affecting the timing of evacuation departures. *International Journal of Mass Emergencies and Disasters* **9**, 153–65.

Stallings, R. A. 1991. Ending evacuations. *International Journal of Mass Emergencies and Disasters* **9**, 183–200.

Steinbrugge, K. V. 1982. *Earthquakes, volcanoes and tsunamis: an anatomy of hazards*. New York: Skandia America Corporation.

Stephenson, R. S. 1981. *Understanding earthquake relief: guidelines for private agencies and commercial organizations*. London: International Disaster Institute.

Sullivan, J. & R. T. Newkirk (eds) 1989. *Simulation in emergency management and technology*. La Jolla, California: Society for Computer Simulation.

Swanson, D. A., T. J. Casadevall, D. Dzurisin, R. T. Holcomb, C. G. Newhall, S. D. Malone, C. S. Weaver 1985. Forecasts and predictions of eruptive activity at Mount St Helens, Washington (USA), 1975–1984. *Journal of Geodynamics* **3**, 397–423.

Teng, W. L. 1990. AVHRR monitoring of US crops during the 1988 drought. *Photogrammetric Engineering and Remote Sensing* **56**, 1,143–6.

Toigo, J. W. 1989. *Disaster recovery planning, managing risk and catastrophe in information systems*. Englewood Cliffs, New Jersey: Prentice-Hall.

Toulmin, L. M. 1987. Disaster preparedness and regional training on nine Caribbean islands: a long-term evaluation. *Disasters* **11**, 221–34.

Turner, J. S. & D. K. Lilly 1963. The carbonated water tornado vortex. *Journal of Atmospheric Science* **20**, 468–71.

UNDRO 1979. *Disaster prevention and mitigation: a compendium of current knowledge*. Vol. 10, *Public information aspects*. Geneva: Office of the United Nations Disaster Relief Co-ordinator.

UNDRO 1982. *Shelter after disaster: guidelines for assistance*. Geneva: Office of the United Nations Disaster Relief Co-ordinator.

UNDRO 1984. *Disaster prevention and mitigation: a compendium of current knowledge*. Vol. 11, *Preparedness aspects*. Geneva: Office of the United Nations Disaster Relief Co-ordinator.

UNDRO/UNESCO 1985. *Volcanic emergency management*. New York: United Nations Press.

Varnes, D. J. et al. 1984. *Landslide hazard zonation: review of principles and practice*. Paris: UNESCO Press.

Walker, A. S. & C. J. Robinove 1981. Annotated bibliography of remote sensing methods for monitoring desertification. *US Geological Survey Circular* 851, 1–25.

Walsh, S. J., D. R. Butler, D. G. Brown, L. Bian 1990. Cartographic modelling of snow avalanche path location within Glacier National Park, Montana. *Photogrammetric Engineering and Remote Sensing* **56**, 615–21.

Walter, L. S. 1990. The uses of satellite technology in disastermanagement. *Disasters* **14**, 20–35.

Weather 1988. The storm of 15-16 October 1987. *Weather* **43**, 65–142.

Weinroth, J. 1989. A model for the management of building evacuation. *Simulation* **53**, 111–19.

White, G. F. 1975. *Flood hazard in the United States: a research assessment.* Report PB-262-023. Boulder, Colorado: Institute of Behavioural Sciences, University of Colorado.

White, G. F. & J. E. Haas 1975. *Assessment of research on natural hazards.* Cambridge, Mass.: MIT Press.

Whitlow, R. 1986. Mapping erosion risk in Zimbabwe: a methodology for rapid survey using aerial photographs. *Applied Geography* **6**, 149–62.

Yoo, K. H. & M. Molnau 1987. Upland erosion simulation for agricultural watersheds. *Water Resources Bulletin* **23**, 819–27.

Select bibliography

Alexander, D. E. 1991. Information technology in real time for monitoring and managing natural disasters. *Progress in Physical Geography* **15**, 238–60.

Anderson, M. B. 1991. Which costs more: prevention or recovery? In *Managing natural disasters and the environment*, A. Kreimer & M. Munasinghe (eds), 17–27. Washington, DC: Environment Department, World Bank.

Anderson, W. A. 1969. Disaster warning and communication processes in two communities. *Journal of Communication* **19**, 92–104.

Appropriate Technology, 1990. Issue on disaster management. *Appropriate Technology* **17**.

Arnold, C. 1984. Planning against earthquakes in the United States and Japan. *Earthquake Spectra* **1**, 75–88.

Baker, E. J. 1979. Predicting response to hurricane warnings: a reanalysis of data from four studies. *Mass Emergencies* **4**, 9–24.

Barrett, E. C., K. A. Brown, A. Micallef (eds) 1991. *Remote sensing for hazard monitoring and assessment: marine and coastal applications in the Mediterranean region.* New York: Gordon & Breach.

Bates, F. L., C. D. Killian, W. G. Peacock 1984. Recovery, change and development: a longitudinal study of the 1976 Guatemalan earthquake. *Ekistics* **51**, 439–45.

Belardo, S., A. Howell, R. Ryan, W. A. Wallace 1983. A microcomputer-based emergency response system. *Disasters* **7**, 215–20.

Belcher, J. C. & F. L. Bates 1983. Aftermath of natural disasters: coping through residential mobility. *Disasters* **7**, 118–28.

Bolin, R. C. & P. A. Bolton 1983. Recovery in Nicaragua and the USA. *International Journal of Mass Emergencies and Disasters* **1**, 125–44.

Bolin, R. C. & L. Stanford 1991. Shelter, housing and recovery: a comparison of US disasters. *Disasters* **15**, 24–34.

Burkhardt, F. N. 1991. *Media, emergency warnings and citizen response.* Boulder, Colorado: Westview Press.

Caldwell, N., A. Clark, D. Clayton, K. Malhotra, D. Reiner 1979. An analysis of Indian press coverage of the Andhra Pradesh cyclone disaster of 19 November 1977. *Disasters* **3**, 154–68.

Carter, D., G. W. Heath, G. Hovmork, H. Sax 1989. Space applications for disaster mitigation and management. *Acta Astronautica* **19**, 229–49.

Clary, B. B. 1985. The evolution and structure of natural hazard policies. *Public Administration Review* **45** (special issue), 20–9.

Comfort, L. K. (ed.) 1988. *Managing disaster: strategies and policy perspectives.* Durham, North Carolina: Duke University Press.

Cotecchia, V. 1978. Systematic reconnaissance mapping and registration of slope movements. *International Association for Engineering Geology, Bulletin* **17**, 5–37.

Davis, I. (ed.) 1981. *Disasters and the small dwelling.* Oxford: Pergamon.

Davis, M. & S. T. Seitz 1982. Disasters and governments. *Journal of Conflict Resolution* **26**, 547–68.

Dishaw, H. E. 1967. Massive landslides. *Photogrammetric Engineering* **32**, 603–9.

Drabek, T. E. 1990. *Emergency management: strategies for maintaining organizational integrity.* New York: Springer.

Drennon, C. B. & W. G. Schleining 1975. Landslide hazard mapping on a shoestring. *American Society of Civil Engineers, Proceedings; Journal of the Surveying and Mapping Division* **101**(SU1), 107–14.

D'Souza, F. 1986. Recovery following the Gediz earthquake: a study of four villages of western Turkey. *Disasters* **10**, 35–52.

El Bareidi, M. 1982. *Model rules for disaster relief operations–B.* New York: UNITAR.

Geipel, R. 1982. *Disaster and reconstruction: the Friuli (Italy) earthquakes of 1976,* trans. P. Wagner. London: Allen & Unwin.

Geipel, R. 1990. *The long-term consequences of disasters: the reconstruction of Friuli, Italy, in its international context, 1976–88.* Heidelberg: Springer.

Gillespie, D. F. & C. L. Streeter 1987. Conceptualizing and measuring disaster preparedness. *International Journal of Mass Emergencies and Disasters* **5**, 155–76.

Goltz, J. D. 1984. Are the news media responsible for the disaster myths? A content analysis of emergency response imagery. *International Journal of Mass Emergencies and Disasters* **2**, 345–68.

Gordenker, L. & T. G. Weiss 1989. Humanitarian emergencies and military help: some conceptual observations. *Disasters* **13**, 119–34.

Green, A., G. Whitehouse, D. Outhet 1983. Causes of flood streamlines observed on Landsat images and their use as indicators of floodways. *International Journal of Remote Sensing* **4**, 5–16.

Gruntfest, E. & C. Huber 1989. Status report on flood warning systems in the United States. *Environmental Management* **13**, 279–86.

Haas, J. E., R. W. Kates, M. J. Bowden (eds) 1977. *Reconstruction following disaster.* Cambridge, Mass.: MIT Press.

Hall, D. K., A. T. C. Chang, J. L. Foster 1986. Detection of the depth-hoar layer in the snow-pack of the Arctic Coastal Plain of Alaska, USA, using satellite data. *Journal of Glaciology* **32**, 87–94.

Handmer, J. 1988. The performance of the Sydney Flood Warning System, August 1986. *Disasters* **12**, 37–49.

Healy, R. J. 1969. *Emergency and disaster planning.* New York: John Wiley.

Hiroi, O., S. Mikami, K. Miyata 1985. A study of mass media reporting in emergencies. *International Journal of Mass Emergencies and Disasters* **3**, 21–50.

Howard, J. A., E. C. Barrett, J. U. Hielkema 1978. The application of satellite remote sensing to monitoring of agricultural disasters. *Disasters* **2**, 231–40.

Ives, J. D. 1981. Mapping of mountain hazards. *Impact of Science on Society* **32**, 79–88.

Jackson, T. J., J. C. Ritchie, J. White, L. LeShack 1988. Airborne laser profile data for measuring ephemeral gully erosion. *Photogrammetric Engineering and Remote Sensing* **54**, 1,181–6.

Journal of Communication 1987. Communicating hazards: special issue. *Journal of Communication* **37**, 10–131.

Kanakubo, T. & S. Tanioka 1980. Natural hazard mapping. *GeoJournal* **4**, 333–40.

Kartez, J. D. & M. K. Lindell 1987. Planning for uncertainty: the case of local disaster planning. *American Planning Association, Journal* **53**, 487–98.

Kent, R. C. 1987. *Anatomy of disaster relief: the international network in action.* London: Francis Pinter.

Kirschenbaum, A. 1992. Warning and evacuation during a mass disaster. *International Journal of Mass Emergencies and Disasters* **10**, 1–114.

Kockelman, W. J. & E. E. Brabb 1979. Examples of seismic zonation in the San Francisco Bay region. In *Progress on seismic zonation in the San Francisco Bay region*, E. E. Brabb (ed.), 73–84. US Geological Survey Circular 807, 73–84.

Kreimer, A. 1984. Housing reconstruction after major disasters as a vehicle for change. *Ekistics* **51**, 470–5.

Kunreuther, H. 1978. *An interactive modelling system for disaster policy analysis.* Boulder, Colorado: Natural Hazards Research and Applications Information Center.

Lang, K. & G. E. Lang 1976. Planning for emergency operations. *Mass Emergencies* **1**, 107–17.

Leenaers, H. & J. P. Okx 1989. The use of digital elevation models for flood hazard mapping. *Earth Surface Processes and Landforms* **14**, 631–40.

Lewis, J. 1979. The Tamil Nadu cyclone, November 1977: a comparison of newspaper reports. *Disasters* **3**, 123–5.

Lewis, J. 1984. Vulnerability to a cyclone: damage distribution in Sri Lanka. *Ekistics* **51**, 421–31.

McKay, J. M. 1983. Newspaper reporting of bushfire disaster in southeastern Australia: Ash Wednesday 1983. *Disasters* **7**, 283–90.

McKay, J. M. & B. Finlayson 1982. Observations on mass media reporting and individual motivation to obtain a flood inundation map: River Torrens, Adelaide, South Australia. *Applied Geography* **2**, 143–53.

McLoughlin, D. 1985. A framework for integrated emergency management. *Public Administration Review* **45** (special issue), 165–72.

Marston, S. A. (ed.) 1986. *Terminal disasters: computer applications in emergency management.* Boulder, Colorado: Natural Hazards Research and Applications Information Center.

Norton, R. 1980. Disasters and settlements. *Disasters* **4**, 339–47.

Oliver-Smith, A. 1991. Successes and failures in post-disaster resettlement. *Disasters* **15**, 12–23.

Perry, R. W. 1985. *Comprehensive emergency management.* Greenwich, Connecticut: JAI Press.

Perry, R. W. & A. H. Mushkatel 1984. *Disaster management: warning response and community relocation.* Westport, Connecticut: Quorum.

Powell, R. F., L. D. James, D. E. Jones Jr 1980. Approximate method for quick flood plain mapping. *American Society of Civil Engineers, Proceedings; Journal of the Water Resources Planning and Management Division* **106**(WRI), 103–22.

Public Management 1989. The value of emergency preparedness (special issue). *Public Management* **71**, 2–27.

Quarantelli, E. L. 1982. General and particular observations on sheltering and housing in American Disasters. *Disasters* **6**, 277–81.

Rango, A. & V. V. Salomonson 1974. Regional flood mapping from space. *Water Resources Research* **10**, 473–84.

Seydlitz, R., J. W. Spencer, S. Laska, E. Triche 1991. The effects of newspaper reports on the public's response to a natural hazard event. *International Journal of Mass Emergencies and Disasters* **9**, 5–29.

Sheffi, Y., H. Mahmassani, W. B. Powell 1982. A transportation network evacuation model. *Transportation Research* **16A**, 209–18.

Sinha, A. K. & S. U. Avrani 1984. The disaster warning process: a study of the 1981 Gujarat cyclone. *Disasters* **8**, 67–73.

Snarr, D. N. & E. L. Brown 1978. Post-disaster housing in Honduras after Hurricane Fifi: an assessment of some objectives. *Mass Emergencies* **3**, 329–50.

Snarr, D. N. & E. L. Brown 1982. Attrition and housing improvements: a study of post-disaster housing after three years. *Disasters* **6**, 125–31.

Sorenson, J. H. & P. J. Gersmehl 1980. Volcanic hazard warning system: persistence and transferability. *Environmental Management* **4**, 125–36.

Spirgi, E. H. 1979. *Disaster management: comprehensive guidelines for disaster relief*. Bern: H. Huber.

Stephens, L. H. & S. J. Green (eds) 1979. *Disaster assistance: appraisal, reform and new approaches*. New York: New York University Press.

Stratton, R. M. 1989. *Disaster relief: the politics of intergovernmental relations*. Lanham, Maryland: University Press of America.

Sylves, R. T. & W. L. Waugh, Jr (eds) 1990. *Cities and disaster: North American studies in emergency management*. Springfield, Illinois: Charles C. Thomas.

UNDRO 1976. *Guidelines for disaster prevention and management*. Vol. 3, *Management of settlements*. Geneva: Office of the United Nations Disaster Relief Co-ordinator.

UNDRO 1978. *Disaster prevention and mitigation: a compendium of current knowledge*. Vol. 5, *Land use aspects*. Geneva: Office of the United Nations Disaster Relief Co-ordinator.

US National Research Council 1980. *Disasters and the mass media*. Washington, DC: National Academy Press.

Walker, J. A., G. E. Ruberg, J. J. O'Dell 1989. Simulation for emergency management: taking advantage of automation in emergency preparedness. *Simulation* **53**, 95–100.

White, R. M. 1972. National Hurricane Warning Programme. *Bulletin of the American Meteorological Society* **53**, 631–3.

Wieczorek, G. F. 1984. Preparing a detailed landslide-inventory map for hazard evaluation and reduction. *Association of Engineering Geologists, Bulletin* **21**, 337–42.

Wolensky, R. P. & E. J. Miller 1983. The politics of disaster recovery. In *The sociological galaxy: sociology towards the year 2000*, C. E. Babbitt (ed.), 259–70. Harrisburg, Pennsylvania: Beacon's Press & the Pennsylvania Sociological Society.

Wolman, M. G. 1971. Evaluating alternative techniques of floodplain mapping. *Water Resources Research* **7**, 1,383–92.

Youd, T. L. & D. M. Perkins 1987. Mapping of liquefaction severity index. *Journal of Geotechnical Engineering* **113**, 1,374–92.

Zelinsky, W., L. A. Kosinski 1991. *The emergency evacuation of cities*. Savage, Maryland: Rowman & Littlefield.

Ziony, J. I. & J. C. Tinsley 1983. Mapping the earthquake hazards of the Los Angeles region. *Earthquake Information Bulletin* **15**, 134–41.

CHAPTER SEVEN

Medical emergencies

This chapter is devoted to the medical and sanitary consequences of disaster, including the direct effects of death and injury and the indirect effects associated with disruption of public hygiene and consequent risks of disease transmission. **Disaster epidemiology** is the study of the occurrence and rates of death, injury and disease in populations affected by natural or man-made disaster (Lechat 1976; see below). The emphasis may be either academic or practical, and the corpus of information is derived as much from field experience of disasters as from medical and statistical theory. Rates of death are known as **mortality**, while rates of physical injury and disease are known as **morbidity**. Morbidity that involves cuts, burns. concussion, broken limbs or other physical injury is known as **trauma**.

Major health needs after disaster include: the recovery and disposal of the dead; search, rescue and care of the critically injured; monitoring and control of the progress of communicable diseases; feeding, housing and caring for displaced populations; and, as a corollary, accounting for population movements in order to assess the number of people requiring care and attention. Steps must also be taken immediately after the impact to determine the pattern of functionality of hospitals and medical services in the disaster area (Beinin 1985). In this context, Figure 7.1 shows the structure of operational units in a typical emergency medical plan. But experience of medical planning in the United States suggests that it is often carried out in isolation from disaster preparations made by other bodies, such as police and firemen. If this occurs, expertise, information and resources are unlikely to be shared adequately (Tierney 1985). Hence the medical component of disaster preparedness should be integrated into a general, community-wide plan.

In many parts of the world there is a major difference in medical capabilities between neighbouring countries. Hence, small international

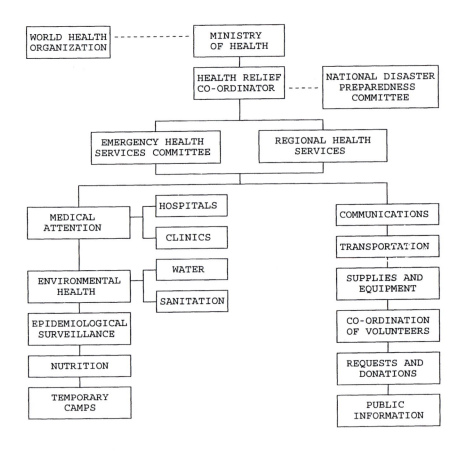

Figure 7.1 The organization of medical relief after a natural disaster (modified from PAHO 1981: 13).

journeys of medical equipment and personnel can sometimes contribute greatly to making the response adequate. However, past experience shows, paradoxically, that field hospitals (for example, military units) which arrive at the scene of the disaster 48 hours or more after the event generally provide routine and back-up medical care, not primary assistance (Coultrip 1974).

The phases of a medical emergency
The phases of the medical emergency caused by a natural disaster closely follow those of the more general civil and logistical crisis. As discussed in earlier chapters, the **impact phase** will vary in duration with the nature of the geophysical agent which causes the disaster. In sudden or profound impacts, some medical facilities may be rendered unusable (it seems that

hospitals have an unfortunate tendency to collapse catastrophically in major earthquakes – see Fig. 7.2, Koegler & Hicks 1972).

The **emergency-isolation phase** will be characterized by community self-help as the only form of relief. Existing manpower and supplies will be utilized in the medical operation until organized medical teams, and crates of supplies, appear from outside the disaster area (in this context it should be noted that sanitation workers, nurses and medical orderlies with good basic training may be needed in much greater quantities than doctors; see Quarantelli 1983). Shortages will begin to occur, principally of the following items: anaesthetics, analgesics and sedatives (to kill pain), antibiotics (to prevent the spread of infection), antiseptics, sera (to

Figure 7.2 Hospitals are focal points in the immediate aftermath of disaster. However, they are also sources of vulnerability. The maternity wing of this hospital at Sant'Angelo dei Lombardi collapsed during the 23 November 1980 earthquake, killing doctors, nurses, mothers and babies.

prevent tetanus developing in wounds), splints, sterile dressings, syringes and needles, water sterilization preparations and X-ray films.

The **rehabilitation phase** will involve several typical activities, such as encouraging medical officials to return to hospitals (perhaps by ensuring the safety of their families and looking after their personal needs), rehousing damaged clinics or parts of hospitals under temporary shelter and sorting out newly arrived medical supplies. Wrong and irrelevant supplies pose a problem of the time and effort needed to cope with them. Unfortunately, it is not uncommon for the principal multinational drug companies (or even foreign governments) to use natural disasters as an opportunity to dump used, time-expired and surplus medical supplies under the name of "aid". In this respect, rarely will more than 10 tonnes of medical supplies be needed after an individual disaster, yet there have been cases of more than twelve times as much material being imported into the disaster area. In such cases, needs should be assessed and donation co-ordinated better, and donors should exercise proper restraint.

Death and injury

Usually, only earthquakes and violent flooding appear to have the capacity to cause death tolls over 100,000, although volcanic eruptions could potentially do so. Whatever the total number of casualties, in sudden impact disasters only 5–10 per cent are likely to require prolonged hospital care, and the pattern will probably be well established within a week of the event.

In synthesis, Figure 7.3 shows a likely temporal pattern of medical consequences in a sudden impact disaster. Most of the deaths and injuries will occur within a matter of hours of time zero, with only a tiny fraction taking place later (deaths as a result of injuries sustained, injury during aftershocks or secondary disasters, etc.). Disease outbreaks, if they occur, will take place at the end of the incubation period for the disease in question (here given as four days), and their impact will depend on the level of sanitary control. Obviously, the type of medical emergencies, and the number and pattern of casualties, will differ with respect to the various disaster-producing agents. For example, flash floods and storm surges may cause many deaths and leave few people injured (French et al. 1983). Perhaps the most complete data on deaths and injuries in natural disasters are available for earthquakes.

Earthquake casualties
Coburn et al. (1989) analyzed mortality in earthquakes and found it to be heavily concentrated in the largest events and in particular places. In the first 90 years of the twentieth century there were over 1,000 earthquakes

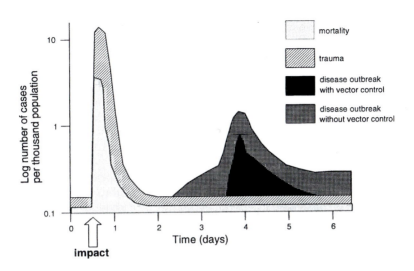

Figure 7.3 Temporal patterns of mortality in a sudden-impact natural disaster.

of magnitude >7.0. At least 650 were lethal, and a total of 1,300,000 people died. The five largest of these events were responsible for over half of these deaths, and the twenty largest for almost 80 per cent of them. Only 3 per cent involved death tolls of more than 10,000. Nine countries account for 80 per cent of recorded deaths in earthquakes and almost 50 per cent have occurred in China, where population totals and densities are very high, building stock tends to be weak and earthquakes are frequent and strong.

In major earthquakes, deaths are likely to exceed 10 per cent of the population of certain settlements near to the epicentre, while a few of the worst-affected places may have mortality rates of up to 85 per cent. Overall, it has been suggested (PAHO 1981) that injuries will probably be three times as numerous as deaths, although a strict definition for the term "injured" has never been established (my own research suggests that earthquakes of moderate magnitude – roughly between 5 and 7 – are most likely to produce three injuries to every one death; Alexander 1985). The number and proportion of both injuries and deaths fall off with distance from the epicentre until only isolated examples of each can be found. The complexity of the urban environment (in terms of, for example, housing type, state of maintenance and density of buildings) and the lack of uniformity in geological factors mean that the pattern of deaths and injuries will almost always be complex.

Figure 7.4, which is based on Wallace's spatial model of disaster (Fig. 1.5), gives four hypotheses concerning the possible distribution of casualties (Alexander 1989). In the concentric model, deaths exceed

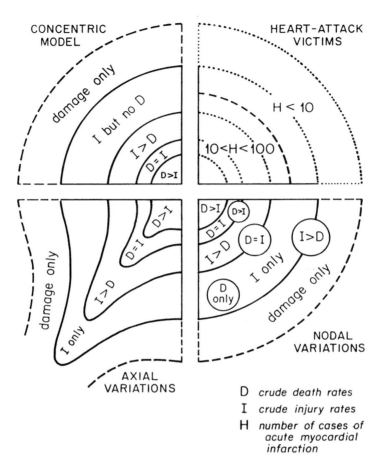

Figure 7.4 Four hypotheses of the spatial distribution of casualties in earthquake disasters.

injuries in the epicentral area, where building collapse is concentrated. In a narrow band further out from the epicentre, deaths and injuries occur in approximately equal proportions, beyond which injuries predominate (and building collapse is more sporadic). Moving out towards the periphery of the affected area, damage becomes progressively lighter and there are first no deaths and then no injuries either. In earthquakes that involve impressive strong motions but little structural collapse, most of the mortality may result from heart attacks (acute myocardial infarction). The pattern is likely to be less detailed with, in a populous area, tens of victims close to the epicentre and a handful at greater distances (Katsouyani et al. 1986).

The first two models shown in Figure 7.4 assume that conditions are isotropic, but it may be fairer to expect axial or nodal variations. For example, the fault line may extend isoseisms and hence create an axis along which structural failures, and hence casualties, are distributed (Berberian 1978). Furthermore, the concentration of population into towns and cities, and the corresponding agglomeration of vulnerable buildings, may lead to nodes at which particular patterns of casualty are concentrated – anomalously with respect to the overall concentric pattern. Some of these models have been tested for the 1978 Tabas-e-Golshan (Iran) and 1980 Irpinia (southern Italy) earthquakes and found to have a certain amount of validity (Slosek 1986).

As mentioned in Chapter 2, several factors influence the death and injury toll exacted by earthquakes (Lomnitz 1970). To begin with, the type, density and state of maintenance of houses is important, including whether they are aseismic or anti-seismic, and whether they contain many or few occupants. As Figure 7.5 shows, different relationships between seismic intensity and mortality are likely for different construction types. In addition, research in Japan suggests that failure to suppress fires caused by earthquakes may increase the eventual death toll up to tenfold (Scawthorn et al. 1981).

Secondly, the time of day governs aggregate patterns of human activity and hence is broadly related to states of readiness or vulnerability. Vulnerable activities which tend to be regulated by particular times of day include the journeys to and from work, the evening promenade, congregating in church or sitting in a cinema, and sitting around the family dinner table (all of which constitute active behaviour), and sleeping at night or during the heat of the afternoon (which constitutes passive behaviour). Hence, a different pattern of death and injury can result if people are at work or in school, rather than at home and asleep. Emergency

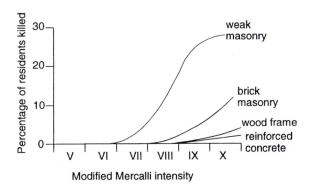

Figure 7.5 Seismic intensity, building type and mortality in earthquakes (Coburn et al. 1989: 115).

medical planners and hospitals should take these factors into account when designing disaster plans.

Thirdly, the most serious loss of life and injury occur where population densities are high; and hence fully occupied high-rise buildings constitute particular centres of vulnerability, as do older, more densely occupied historic settlements. And lastly, the effectiveness of immediate search and rescue can have a considerable influence on the overall death toll.

The risk of death in earthquake is highly variable. In Turkey, the ratio of deaths to buildings and structures destroyed in earthquakes averages 8.5:100. But in Lice during the 1975 earthquake the ratio was 132:100 and 2,385 people were killed (1 in 6.8 of the population). As the earthquake occurred at midday on the eve of Ramadan the houses that collapsed were full of women preparing the traditional feast, and these were the people who died (Davis 1978). The extremely low death:injury ratio (1:100) in the 1989 Loma Prieta (California) earthquake resulted from several particular factors. Deaths were mainly caused by the collapse of individual sections of road bridge, while injuries were much more widespread. Despite significant sources of danger, much had been done to abate seismic risk in the area (Bolin 1990).

When buildings collapse their inhabitants may be crushed by falling beams, or suffer injuries to the skull or thorax from falling objects (parts of the building or room furnishings). Multiple fractures and serious spinal injuries appear to be most closely associated with steel, concrete, stone and brick buildings, rather than adobe or wooden ones, which tend to generate simpler injuries. About half the people who suffer serious cranial injuries die within 24 hours, and open wounds or internal haemorrhaging may rapidly prove fatal unless treated (Jones et al. 1990). The relationship between time and survival rates is shown schematically in Figure 7.6.

De Bruycker et al. (1985) found that being trapped under the rubble increases the victim's chance of being injured fivefold. Trapped victims may be prevented from breathing by extreme pressure on the chest or by suffocation: both masonry and concrete structures can generate lethal quantities of dust when they collapse. Only 2–6 hours after an earthquake less than half of the people trapped under rubble are likely to be still alive. Those who are not killed by their injuries may die of shock or hypothermia after about 48 hours (the number of "miracle" salvations achieved by rescue teams days after the impact is invariably small). Even uninjured victims trapped under collapsed buildings may die fairly quickly of exposure or dehydration. Injured victims need to be kept warm and stable, which cannot be done if they are trapped beneath rubble. Moreover, people who are old, frail or unwell are unlikely to survive in such conditions as long as healthy young adults (Olson & Olson 1987).

In the 1976 Guatemalan earthquake women, small children and the

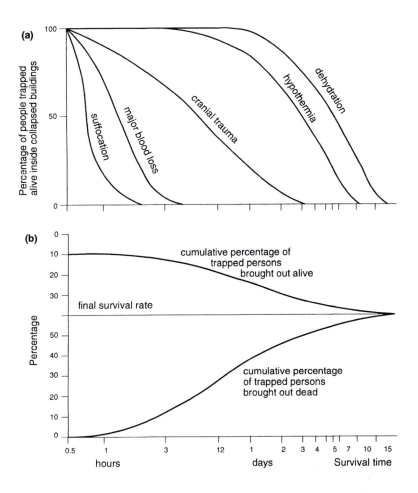

(a)

Percentage of people trapped alive inside collapsed buildings

100

50

0

suffocation

major blood loss

cranial trauma

hypothermia

dehydration

(b)

0

Percentage

10

20

30

40

50

40

30

20

10

0

cumulative percentage of trapped persons brought out alive

final survival rate

cumulative percentage of trapped persons brought out dead

0.5   1   3   12   1   2   3 4 5 7 10   15

hours   days   Survival time

Figure 7.6 Survival time for trapped earthquake victims (Coburn et al. 1989: 125).

elderly were more at risk than other groups (Glass et al. 1977). However, a study of 3,618 villagers affected by the 1980 southern Italian earthquake found no differences in mortality rates according to age or sex (De Bruycker et al. 1985). Despite this, it does appear that in sudden-impact disasters the very young, the elderly and seriously ill or handicapped people lack the perception and mobility to protect themselves as well as more favoured groups. In southern Italy in 1980 one third of victims trapped under rubble were eventually killed, whereas only 0.3 per cent of people who were not trapped died.

Little information is available on the distribution of injuries in earthquakes. In the 1968 Khorasan earthquake in Iran, only 3.3 per cent of 11,000 injured people required prolonged inpatient care (Rennie 1970). And in the 1964 Alaskan earthquake, which occurred in a vast but relatively unpopulated area, 115 people were killed but only 50 were injured enough to require prolonged hospital treatment. The expected pattern is broadly as follows (PAHO 1981). There will be a mass of injured patients having minor cuts and bruises which can be treated rapidly without admission to hospitals. Care must be taken when examining them to ensure that they are not suffering from serious shock or complications, or from hidden internal injuries. A smaller group will have simple fractures and larger contusions, while the smallest group will have serious multiple injuries requiring surgery and, perhaps, intensive care. The need for expert diagnosis on the development of complications may mean that this last group of patients must be transferred to hospitals with special facilities (cardiac or burns units, for instance). The same may be true when intensive care units in an epicentral disaster zone are overloaded by the number of casualties brought in from the immediate catchment area.

Earthquake scenarios for urban industrial areas predict a break-down of serious injuries into about 50 per cent simple lacerations, 25 per cent contusions, 20 per cent fractures (Italy), or 27 per cent surgical cases, 23 per cent orthopaedic cases, 15 per cent cardiac cases, 10 per cent neurosurgical, 10 per cent shock, 6 per cent severe burns and 5 per cent smoke inhalation and toxins (southern California). Crush injuries and fractures of the clavicle have predominated in some events, but not in others. Paraplegia cases numbered 1.5 per thousand in Guatemala, 1976, while heart attack (acute miocardial infarction) victims increased by 50 per cent in the Greek earthquake of 1981 (Trichopoulos et al. 1983). Reported ratios of serious to slight injuries vary from 1:9 to 1:30.

Other disasters

Tsunamis
The likelihood of death by drowning or through the impact of debris or structural collapse increases nonlinearly with the run-up height and recurrence interval of tsunamis (Fig. 7.7). Besides the rather obvious facts that most casualties occur in populous areas, at the coast and close to the point of genesis of the waves (where run-up heights are greatest) few data exist which might help elucidate the pattern of casualties.

Volcanic eruptions
Autopsies performed on some of the 57 victims of the 1980 Mount St Helens eruption indicate several causes of death: blast and projectile injuries, second- and third-degree burns, asphyxia by ingesting ash and mucus, and internal burns (Eisele et al. 1981). People caught within volcanic blast or pyroclastic flow zones tend to be consumed if temperatures are very high (human hair singes at 120°C) and dehydrated, tanned and mummified if temperatures are not excessively high (Baxter

1990). Table 7.1 summarizes the impacts of eruption upon health and some means of preventing them (of which evacuation is the simplest and most effective).

As the table shows, because of the toxic or irritating gases and ash which they produce, volcanic eruptions can exacerbate respiratory diseases such as asthma, bronchitis, emphysema, pneumoconiosis, pneumonitis and pulmonary vascular diseases. However, impacts will vary according to the resistance of individual victims. Table 7.2 gives toxicity information and threshold concentrations for the principal volcanic fluids (mainly gases). In addition, ash particles smaller than 10 μm in diameter are often produced in explosive eruptions and can be respired: for example, over 90 per cent of the ash particles produced in the May 1980 Mount St Helens eruption could be respired (Baxter et al. 1986).

There is little relationship between the frequency of volcanic eruptions and death tolls, as human vulnerability characteristics (for example, proximity and population density) are the main determinants of casualty

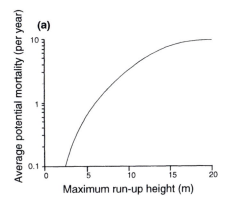

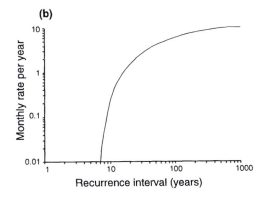

Figure 7.7 Expected mortality rates in Hawaii as a function of tsunami run-up height and recurrence interval (Cox & Morgan 1984: 75–7).

471

Table 7.1 The impact of volcanic eruptions on human health and some preventative measures (Baxter et al 1986: 85).

Eruptive event	Consequences	Health impact	Preventative measures
Explosions	lateral blast, rock fragments, air shock waves	trauma; skin burns, lacerations from broken windows	evacuation, minimization of exposure to flying glass
Hot ash release	nuées ardentes ash flows and falls, lightning	skin and lung burns, asphyxiation, electrocution	evacuation
Melting ice, snow or rain accompanying eruption	mudflows, floods	engulfing, drowning	evacuation, diversion barriers
Lava	lava flow, forest fires	engulfing and burns (rare)	evacuation, diversion barriers
Gas emissions: CO, CO_2, HF, H_S, SO_2	pooling in low-lying areas and inhalation	asphyxiation, constriction of the airways	evacuation, breathing apparatus for geologists
Radon	radiation exposure	lung cancer	evacuation
Earthquakes	building damage	trauma	evacuation, anti-seismic construction

levels. Nevertheless, starvation and disease are no longer likely to be as important as they were in the past, while pyroclastic flows and lahars will continue to be major sources of mortality and morbidity in eruptions (Fig. 7.8a). Figure 7.8b shows that historically the populous lands of the Indonesian archipelago, where explosive eruptions are frequent, have had the lion's share of the global death toll.

Floods
PAHO (1981) argued that a death: injury ratio of 1:6 is likely in floods. Cochrane (1975) found a ratio of one casualty per two houses destroyed in floods and hurricanes, and a ratio of one serious injury to every four light ones. The main source of death in floods is drowning. In developed countries particular risks seem to be associated with unwillingness to abandon vehicles or buildings when it is necessary to seek safety (though it is not clear what effect this has on the death statistics). The disruptions caused by flooding appear to be the source of many medical problems, among them respiratory disease, stress-related ailments, lymphoma, leukemia and miscarriage. As usual, the elderly are particularly at risk. Rather, particular health hazards result from the release of dangerous chemicals into raging floodwaters: these include ammonia, sulphuric acid and pesticides, all of which can cause poisoning.

Table 7.2 Toxicity and health effects of volcanic gases (after Blong 1984, Table 3.14).

Volcanic fluid	Formula	Description	Toxicity threshold mg/m^3	ppm	Human health impact upon contact
Ammonia	NH$_3$	pungent, colourless gas	79	100	skin, eye, nose and throat irritation
Carbon dioxide	CO$_2$	odourless, colourless gas	9,000	5,000	asphyxiation in very high doses
Carbon monoxide	CO	odourless, colourless gas	115	100	blood poisoning: lethal in high concentrations
Fluorine	F$_2$	pale yellow gas	0.2	0.1	caustic irritation; bone degeneration
Hydrochloric acid	HCl	colourless gas or fuming liquid	7	5	Irritation of eyes and respiratory tract
Hydrofluoric acid	HF	colourless gas or fuming liquid	2	3	Skin corrosion and mucous membrane irritation
Hydrogen sulphide	H$_2$S	pungent, colourless and flammable gas	28	20	Irritation and asphyxiation
Sulphur dioxide	SO$_2$	pungent, colourless gas or liquid	13	5	Inflammation of skin, eyes, nose and throat
Sulphuric acid	H$_2$SO$_4$	colourless oil liquid	1	—	Burns, dermatitis, respiratory tract inflammation

Hurricanes

As stated in Chapter 4, most deaths in hurricanes occur by drowning. Thus, in the 1970 Bangladesh cyclone, 14.2 per cent of the population (240,000 people) died, but mortality varied from 4.7 per cent inland to 46.3 per cent on the coast and virtually 100 per cent in some fishing villages. Middle-aged men showed the highest survival rate, while women, especially those over 60 years old, were more likely to be killed or seriously injured. Hospital admissions may also include crush symptoms and lacerations. In fact, severe abrasions to the arms, chest and thighs have been noted in victims and survivors of cyclonic storm surges who have clung desperately to trees to avoid being swept away, and this has been dubbed the "cyclone syndrome". Moreover, in the Andhra Pradesh (India) cyclone of 1970 most of the 1,519 victims in Guntur village died when wind and rain caused houses to collapse (Sommer & Mosley 1972).

(a) Cause of death in volcanic eruptions, 1600–1987

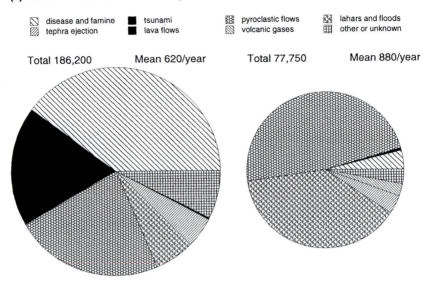

◻ disease and famine	■ tsunami	▨ pyroclastic flows	▨ lahars and floods
▨ tephra ejection	■ lava flows	▨ volcanic gases	▦ other or unknown

Total 186,200   Mean 620/year   Total 77,750   Mean 880/year

(b) Death toll in volcanic eruptions by region, 1600–1987

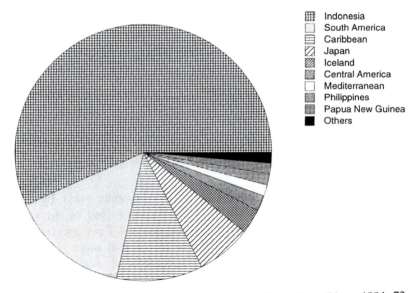

▦	Indonesia
◻	South America
▤	Caribbean
▨	Japan
▨	Iceland
▨	Central America
◻	Mediterranean
▤	Philippines
▥	Papua New Guinea
■	Others

Figure 7.8 Deaths in volcanic eruptions, 1600–1980s (data from Blong 1984: 72, and Tilling 1989: 260).

Tornadoes
The main form of fatal or non-fatal injury in tornadoes is serious damage to the head (cerebro-cranial trauma), followed by crush wounds to the chest or other parts of the trunk. Fractures are the most common non-

fatal injury, followed by lacerations and other soft-tissue damage, but injuries are often multiple and wounds may be heavily contaminated. As a result of wound infection post-treatment sepsis can affect up to one quarter of patients. Finally, both psychopathology and stress (e.g. fear of the weather) seem to be common consequences among the survivors of tornado disasters.

In the tornadoes of 28 March 1984 in North and South Carolina, a survey of 955 casualties (the majority of those killed and injured) revealed 6 per cent fatalities, 27 per cent hospital cases and 67 per cent who were treated and released (Glass et al. 1980). Analysis of 2,575 tornadoes occurring in the United States over the period 1952–73 showed that 497 of them caused 3,125 deaths. Almost half of these tornadoes caused only one fatality, while a mere 26 of them caused 1,180 deaths (Fujita 1973). Thus, as with earthquakes, a few events of large magnitude cause disproportionately large numbers of casualties. The pattern of tornado casualties in the United States is highly variable from year to year, as data for the decade 1970–80 show (see Fig. 7.9).

Tornadoes with longer tracks tend to cause more deaths (not surprisingly!), and it appears that deaths are usually confined to the track

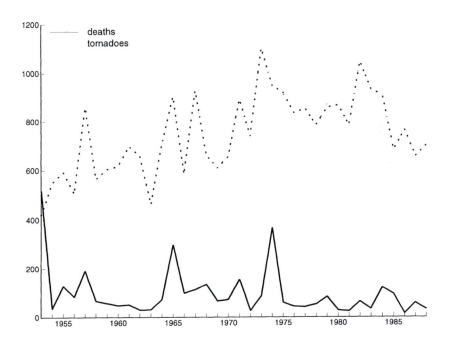

Figure 7.9 Number of tornadoes and associated deaths in the USA, 1953–88 (data from Office of Emergency Preparedness 1972, Vol. 1: 36, and Fujita 1987).

itself, while serious injuries predominate on its margins. As with other forms of sudden-impact disaster, the over-sixties are more at risk than other age groups, and women above the age of 40 seem to experience proportionately more fatalities (for reasons that are by no means clear). Unless they evacuate promptly, the occupants of mobile homes run a considerably higher risk of death in tornadoes than occupants of permanent buildings, and activities such as driving are also hazardous. Glass et al. (1980) estimated the relative risk of death in a tornado as 3 per 1,000 for the occupants of houses, 23 per 1,000 for people in cars and 85 per 1,000 for people in mobile homes (cf. Eidson et al. 1990). Finally, it appears that there is a roughly one-to-one relationship between casualty totals and the number of houses destroyed in tornadoes (Cochrane 1975).

Environmental fires

In many cases, fires kill people by asphyxiation or poisoning through the inhalation of smoke or toxic gases. In other instances, burns may be the principal injury. The major pathophysiological impacts of external burns (i.e. excluding burns acquired by inhaling hot gases) are on the pulmonary and cardiovascular system. When burns cover at least 40 per cent of body area, the lungs cannot function adequately, yet in many cases the toxic effect of smoke inhalation may mean that they are called upon to work harder than usual. If more than half of total body area receives major burns, a reduction of about 70 per cent in the output of the heart may occur. Moreover, the damage caused to blood vessels may lead to severe side-effects, such as thrombosis. Fluids, serum and large amounts of protein are lost through the damaged tissues, causing severe imbalance to the electrolytes, water content, metabolism and serum proteins of the body. In addition, nitrogen imbalance can result in weakness and weight loss. In all instances, the vital functions of the body must be stabilized through intensive care.

Heatwaves

High atmospheric temperatures can cause heatstroke, heat exhaustion, heat syncope or heat cramps, as well as sunburn, where appropriate. The most serious, and often fatal, condition is heatstroke, which is diagnosed when rectal temperature is at least 40.6°C. The patient may become delirious, stupefied or comatose, and the average chance of death is about one in seven. Diagnosis rates are low compared to incidence, as death is often attributed to stroke *per se* and as the traces of heatstroke disappear as the body gradually cools.

Although human beings can adapt to hot conditions and thus develop a certain resistance to heatstroke, the condition is most common in cities, where ambient temperatures may be higher than the surrounding countryside, and where summer wind velocities may be decreased by tall buildings. Infants under the age of one are particularly vulnerable, but the old are much more at risk: they may already suffer from a fragile health status, or may be taking medicines which predispose them to heatstroke (mainly drugs that inhibit perspiration and hence thermo-regulation).

Severe cold

At the other end of the temperature scale, severe cold may encourage death by stroke, ischemic heart disease or pneumonia (Killian & Graf-Baumann 1981). At body temperatures of 95°C or less, hypothermia sets in, irritating the myocardium and causing ventricular fibrillation (Collins 1983). In water (which conducts heat away much better than does air), an unclothed man will become helpless from hypothermia in less than half an hour if the temperature is as low as 5°C. The effects of exposure are combatted by increased rate of metabolism, which requires remarkably large bodily reserves (1,700 kcal are needed to counteract the effect of wet clothing at low temperatures). Hence, the exhausted survivors of floods may be unable to muster the necessary bodily resistance and their condition will deteriorate progressively. In this respect, children have relatively large surface areas in relation to weight and metabolism which increases their vulnerability to exposure (especially if they are suffering from protein energy malnutrition; see Ch. 8), while tissue insulation is low in babies, which increases theirs too.

Primary hypothermia is caused by the direct effects of cold, while the secondary kind results from damage to the central nervous system, the metabolic system or the heart (and is often the more serious of the two; Anderson & Rochard 1979). As with heatstroke, so with hypothermia, treatment involves restoring the correct "core" temperature of the body while, in serious cases, ensuring that bodily electrolytes and other fluids are not dangerously unbalanced and that the heart functions correctly. Intensive care facilities may thus be needed. The curve of mortality with age is almost identical to that for heatstroke, such that once again babies and old people are the main risk groups.

Having reviewed the causes of mortality and morbidity in natural disasters, we will now consider how mass casualty situations are managed.

Triage

The principle of first come first served is inadequate during mass emergencies, when there will be too few doctors to see all patients. **Triage** consists of rapidly classifying patients on the basis of what benefit can be expected from immediate medical attention, not with regard to the seriousness of their injuries. The highest priority is given to patients when some simple medical care can dramatically improve their immediate or long-term prognosis. The lowest priority is given to moribund patients who require considerable attention for questionable benefit, and to the lightly injured who can wait for assistance without coming to further harm (Baker 1980).

Triage is carried out in two stages: at the scene of the disaster, or wherever patients are recovered, their injuries are classified to determine who should receive priority transport to hospital. Then, at the ambulance bay of the hospital they receive triage again to determine who should be given priority care. Figure 7.10 shows how the two points of triage relate to transportation and care facilities. Special tags have been designed to standardize the process and render the triage officer's instructions clear and unambiguous (Fig. 7.11), and these are attached to the patient when he or she is classified. In the 1980 El Asnam earthquake in Algeria, four categories of triage were used: I – vital functions affected; II – serious injuries needing operation but able to wait up to 12 hours after receiving first aid; III – moribund cases; and IV – slightly injured, who with proper supervision could afford to wait for treatment (Baker et al. 1974).

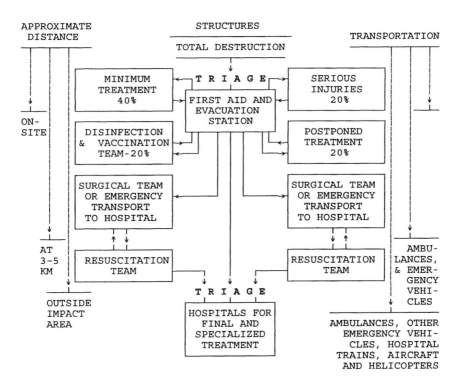

Figure 7.10 Organization of medical facilities and transport after disaster (after Manni 1982: 198).

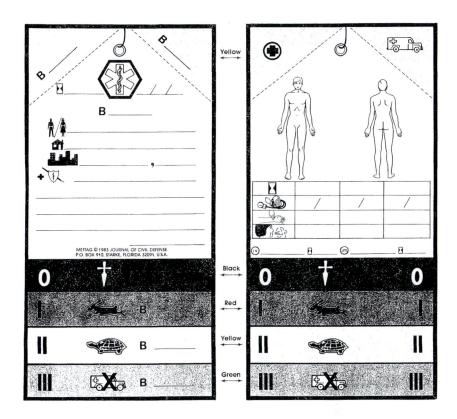

Figure 7.11 The "Mettag" triage tag (reproduced with permission from Mettag, Inc.).

Guzmán et al. (1989) were concerned about the subjectivity of assessments of a victim's state of health made using the triage tag. They proposed an index for the assessment of trauma, based on the status of the patient's respiratory, cardiovascular and central nervous systems, the type of injury (nature of the lesion) and the part of the body affected. For each factor a rating of 0 (minimal), 1 (moderate) or 2 (severe) is assigned and the result summed. The patient's condition is then classified according to his index value: ambulatory (A) for values 0–1, moderately serious (MS) for 2–4, critical but recoverable (CR) for 5–7, and critical but unlikely to recover (CRU) for values of 8 or more. A new triage card has been devised to reflect this index system.

A more detailed quantitative assessment of the severity of injury can be made using anatomical or physiological scales. The former include the **abbreviated injury scale** (AIS) and **injury severity score** (ISS), which are both measures of tissue damage. To obtain an AIS rating the body is divided into seven regions: external, head and face, neck, thorax, abdomen and pelvic contents, spine, and extremities and pelvic bones. Injuries are classified according to location and severity (from minor to critical and maximum). The derivative ISS measure is the sum of squares of the highest AIS score in each of the three most heavily injured areas. It can be compared with as measure known as LD_{50}, the injury severity "dose" which is lethal for 50 per cent of the patients under study.

One problem with anatomical measures of injury severity is that the patient's condition may change over time. Physiological measures are more flexible in this respect and can thus reflect the quality of treatment and care that the patient receives. One such scale is the **trauma score**, which is based on seven circulatory, respiratory and neurological variables, all of which can be reassessed at any particular moment in time. There is also a simplified measure called the **RPV assessment method**, in which the index is based on respiration and pulse rates and best verbal response. The last of these is used to determine whether the patient is wakeful or comatose. Besides quickness, the advantage of RPV is that the user requires no medical training.

Most demands for hospital assistance come within the first 24 hours after a sudden-impact disaster. New demand for hospital beds tends to fall dramatically after 3–5 days and to return to predisaster levels within 10 days, but patients may appear in two waves (Fig. 7.12). The first will be composed of those injured survivors brought directly to the hospital from its catchment area, or from that of nearby medical centres which have been damaged or overwhelmed with patients. The second consists of referrals from other medical centres, who may arrive several days after the disaster has struck.

Operational decisions on how, when and where to direct medical resources will inevitably have to be taken in the absence of a full knowledge of the spatial extent of the disaster and the number of casualties. But in most cases hospitals that have not been put out of action by the disaster will be able to cope if triage is strict and administration is efficient, given that it is unlikely that more than 10 per cent of the catchment population will require any form of treatment, and the medical resources will rapidly be augmented by reserves from outside the disaster area. The problem of supplies is usually one of distribution, rather than absolute availability, and it can be solved by planning post-disaster delivery patterns before the event.

Death and injury in American disasters
The United States is one of the few countries for which the epidemiology of natural disasters has been assessed in a systematic manner, although the number of casualties is low by the standards of countries such as Perú

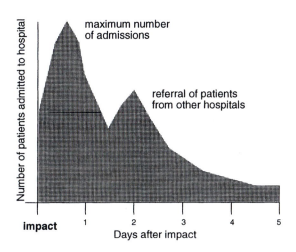

Figure 7.12 Admission of patients to a hospital following a sudden-impact natural disaster (modified from Seaman et al. 1984: 24, De Ville De Goyet et al. 1976: 95).

and Bangladesh. The largest death toll is caused by lightning strikes (in isolated incidents), tornadoes (in concentrated impacts) and urban snow (over widely dispersed areas), all of which involve death tolls of more than 5 per 10 million of the population per year. The largest totals of injuries occur in hurricanes (120/10 million/yr) and tornadoes (90/10 million/yr). It is very difficult to establish adequate norms or averages for mortality and morbidity caused in the USA by tsunamis, floods and earthquakes.

Essentially, the number of traumatic injuries exceeds the number of deaths in cyclones, hurricanes and tornadoes. Instead, deaths exceed traumatic injuries in avalanches, earthquakes, floods, landslides, tsunamis and volcanic eruptions. Little or no death or traumatic injury is caused by drought, hailstorms and windstorms. In so far as trends are discernable, death tolls seem to be increasing in avalanche, earthquake and windstorm disasters; they are stable in landslides, urban snow events and volcanic eruptions; and they appear to be decreasing in floods, hurricanes, lightning strikes and tornadoes (White & Haas 1975).

Although these trends are reasonable estimates for the United States, the picture appears different elsewhere and at the world scale. The number of deaths in floods, for example, is probably on the increase. Although major tornado disasters are rare outside North America, earthquakes and volcanic eruptions take a much greater toll of human life in parts of Asia and South America. But as the deaths and injuries caused by natural disasters are not internationally notifiable, a precise picture of their global distribution in space and time is exceedingly difficult to obtain.

Death and traumatic injury are only half of the medical picture of disasters: there is also likely to be some disruption of public hygiene and disease control, leading to an increased risk of disease transmission during the chaotic aftermath period. As the following section will show, disasters are not automatically – perhaps not even frequently – associated with epidemics, and hence the risk may be more theoretical than real.

Disease rates

During the twentieth century, enormous changes have been wrought in world patterns of disease. However, the impact of disease control has been relatively small in developing countries, where communicable diseases are still usually the leading cause of mortality and morbidity. The world's poor run excessive risk of contracting diarrhoea and dysentery, measles, whooping cough, diphtheria, respiratory diseases, skin ailments, intestinal parasites, tuberculosis, malaria and meningitis. Poor hygiene may lead to, for example, louse-borne typhus or relapsing fever where humans are heavily infested with the insect carriers. Furthermore, in developing countries disease may be held at bay only by rather precarious prevention programmes. When these are interrupted, however briefly, the pathogens may resume their growth.

The World Health Organization has estimated that two thirds of the population of the Third World lack access to uncontaminated drinking water (WHO 1986). Natural disasters thus increase a risk of disease transmission that is already high. Nowhere is this more so than in the fast-growing "primate" cities of the Third World, where sanitary conditions are normally very precarious: for example, the water and sewage systems of Cairo were designed for a population of 2 million yet are currently being used by 11 million (Davis 1987). In fact, provision of safe water supplies should be a prime consideration after disasters, especially for large urban populations who are dependent on a single source. Areas where water is obtained from many unconnected sources tend to be less at risk. Even so, in a 1973 survey of water pumps in use in the Punjab Province of Pakistan, it was found that 85 per cent showed some degree of faecal contamination. Under these circumstances, the impact of disasters must be kept in its proper perspective: contamination after a flood would lose its significance (and the pre-existing bacterial content might itself be attributed wrongly to the flood).

Although not a particularly common consequence, the rate of disease transmission may increase after a natural disaster as a result of several factors (Blake 1989). On occasion, it may stem directly from the physical event. For example, sewer and water mains may rupture together and, in the mixing that follows, there may be faecal contamination of the drinking water supply. Alternatively, routine programmes of disease control may be disrupted by the disaster. In this context, the highest risk is that of endemic diseases (those which were prevalent in the disaster area before the impact; for example, cholera). Where there is much homelessness and migration of survivors, overcrowding and poor sanitation in refugee camps may lead to a form of social promiscuity in which diseases will transmit with relative ease among a population weakened by hunger and exhaustion. But in general, disease epidemics will only occur if the population is not immune (and if it includes carriers), if susceptibility to disease increases (e.g. through malnutrition), or if the transmission rates increase (e.g. through overcrowding).

Several particular sources of disease transmission can be mentioned in the context of natural disasters. Flies may breed on refuse, dead bodies or carcasses and spread enteroviruses, *Shigella* and conjunctivitis. Increase in the breeding sites for mosquitoes may result in a rise in the incidence of malaria, dengue, yellow fever or encephalitis, if any of these diseases is locally endemic. Floodwaters may spread the micro-organisms that cause leptospirosis and typhoid fever, although this is more likely to result from disruption and contamination of water supplies than directly from the floods (PAHO 1982a).

Post-disaster disease rates and health service responses

Despite the apparent risk (see Table 7.3), communicable disease has proved to be much less of a problem after disaster than was commonly supposed (Table 7.4). The risk of significant disease rates in rural areas usually tends to be minimal, but gastroenteritis is a notable hazard in camps overcrowded with survivors. Urban water-borne diseases are generally avoided by prompt public hygiene measures, such as hyper-chlorination of the mains water supply. Outbreaks of typhoid or hepatitis are usually no more than the normal pattern of endemic occurrence, while meningitis tends to be rare, although these diseases may be thrown into high relief by the disaster. As many diseases have incubation times which are more or less long (see Table 7.5), the discovery of cases before a length of time equivalent to the incubation period has passed since the disaster means that infection must have antedated it.

Several common responses to the health problem posed by disaster are inappropriate. For example, as corpses tend not to pose a health hazard, hasty liming, communal interment or mass cremation do no good medically and tend to demoralize the bereaved relatives. Moreover, contrary to popular belief, spraying people with disinfectant and bulldozing rubble will not have any effect on disease rates. Although

Table 7.3 Theoretical risk of communicable disease, by disaster type (Blake 1989: 10)

Disaster	Person-to person*	Water-borne†	Food-borne‡	Vector-borne§
Cold spell	low	low	low	low
Earthquake	medium	medium	medium	low
Famine	high	medium	medium	medium
Fire	low	low	low	low
Flood	medium	high	medium	high
Heat wave	low	low	low	low
Hurricane	medium	high	medium	high
Tornado	low	low	low	low
Volcanic eruption	medium	medium	medium	low

* Shigellosis, streptococcal skin infections, scabies, infectious hepatitis, pertussis, measles, diphtheria, influenza, tuberculosis, other respiratory infections, giardiasis, AIDS, meningococcal meningitis, venereal diseases, pneumonic plague.
† Typhoid and paratyphoid fevers, cholera, "sewage poisoning", leptospirosis, infectious hepatitis, shigellosis, campylobacteriosis, Norwalk agent, salmonellosis, *Escherica (E.) coli* (enteroinvasive and enteropathogenic), amoebiasis, giardiasis, cryptosporidiosis.
‡ Typhoid and paratyphoid fevers, cholera, "food poisoning", infectious hepatitis, shigellosis, campylobacteriosis, salmonellosis, *E. coli* (enterohemorrhagic, enterotoxigenic, entroinvasive and enteropathogenic), amoebiasis, giardiasis, cryptosporidiosis.
§ Louse-borne typhus, plague, relapsing fever, malaria, viral encephalitides.

requiring medical workers to undergo routine prophylaxis is a good idea, campaigns of indiscriminate vaccination inevitably fail and waste precious resources. In fact, according to UNDRO (1982), there are eight possible reasons why mass vaccination is usually not warranted after disaster:

(a) If the disease is not present in the area, there is little point in vaccinating against it.
(b) Water-borne diseases such as cholera and typhoid are best combatted by ensuring that drinking water is sterilized against them or comes from uncontaminated sources.
(c) Mass vaccination cannot protect rapidly enough to be of much use after disasters, as immunity does not develop immediately and a follow-up vaccination may be required after 7–28 days.
(d) Even when properly administered, typhoid vaccines are only 70–80 per cent effective and cholera vaccines only 50 per cent effective.
(e) Vaccination may give people a false sense of security, so that they no longer take other precautions against diseases.
(f) Vaccines often have side-effects, some of which may be severe.
(g) Mass vaccination wastes scarce manpower, time, financial resources, vaccines and logistical capabilities.
(h) In a chaotic disaster aftermath it is virtually impossible to keep account of all people who have been vaccinated, all who have not yet been inoculated and all who need a follow-up vaccination. The only

Table 7.4 Outbreaks of communicable disease attributable to natural disasters, as detected in post-event studies by the US Federal Centers for Disease Control, 1970–85 (Blake 1989: 9)*

Year	Country (state)	Disaster	Outbreaks
1970	Peru	earthquake	none
	USA (Texas)	hurricane	none
1971	Truk (Micronesia)	hurricane	balantidiasis
1972	USA (South Dakota)	flood	none
	USA (Pennsylvania)	flood	none
	Nicaragua	earthquake	none
1973	Pakistan	flood	none
1974	Sahel (West Africa)	famine	none
1976	Guatemala	earthquake	none
1978	Zaire	famine	none
	USA (Texas, Oklahoma)	tornado	none
	Trinidad	volcanic eruption	none
	Dominica	hurricane	none
	Marshall Islands	flood	respiratory
1980	Marshall Islands	hurricane	none
	Mauritius	hurricane	typhoid fever
	USA (Washington)	volcanic eruption	none
	USA (various states)	heatwave	none
	USA (Texas)	hurricane	none
1982	Chad	famine	none
	USA (Illinois)	tornado	none
1983	Bolivia	flood	none
1984	Mauritania	famine	none
	Mozambique	famine	none
	Bolivia	famine	none
1985	Puerto Rico	flood	none
	Colombia	volcanic eruption	none†

* Excludes snow disasters.
† Wound infections occurred, but not epidemics.

Table 7.5 Periods of incubation and communicability of some common infectious diseases (after Benenson 1980)

Disease	Period of incubation (days)	Period of cumminicability
Bacillary dysentery	1–7	≤28 days
Blennorrhea	5–12	10 months
Botulism	½–1½	—
Brucellosis	5–21	—
Diphtheria	2–5	≤28 days
Infectious parotitis	12–26	≤9 days
Leptospirosis	4–19	—
Meningococcal meningitis	2–10	rapid
Poliomyelitis	3–21	≤42 days
Scarlet fever	1–3	10–21 days
Tetanus	4–21	—
Tuberculosis	28–84	some weeks
Typhoid fever	7–21	variable
Varicella	14–21	≤27 days
Viral hepatitis A	15–50	30–50 days
Viral hepatitis B	45–160	100–160 days
Whooping cough	7–21	≤21 days

way that this can be done is to ban the movement of people, which may have very serious consequences for relief efforts.

The most justifiable exception to these arguments may be a carefully administered programme to vaccinate children or other vulnerable people, preferably those who form well-defined groups whose progress can be monitored.

Disaster epidemiology
Although no strict quantitative definitions exist, we can regard an **outbreak** as the simultaneous and presumably related occurrence of several cases of the same condition, in excess of the usual (seasonal) figures, or **background levels**. An **epidemic**, in so far as there is a definition, is any uncontrolled outbreak of a communicable (i.e. infectious or contagious) disease, affecting several or more people. **Pandemics** are usually international or very wide-travelling, and involve simultaneous epidemics of the same condition at several or many widely dispersed locations.

In planning to contain an epidemic the following should be considered: the incidence, prevalence and case fatality rates (with daily updates); incubation period for the disease in question; the source of infections, reservoir and vector; rapidity of spread and potential to affect hitherto untouched areas; effectiveness of specific control measures and of surveillance; and the criteria used to define when the epidemic is over (PAHO 1982b).

Epidemiology can be defined as "the study of the distribution and determinants of health-related events in human populations" (Gregg 1989: 3). Disaster epidemiology involves the study of independent variables, such as the agent (the disaster itself), the environment (which it affects) and the host (affected people), and dependent variables or outcomes (the disease or injury condition). Six types of study can be used. First, **case studies** involve detailed description and analysis of the condition of an individual who has the given disease or injury. This must be compared with the general picture. Secondly, **descriptive studies** require analysis of the whole affected population. Thirdly, **ecological studies** involve comparing differences in rates of the disease or injury among populations. Care must be taken to avoid the so-called **ecological fallacy**, the assumption that relationships observed at the population level are also true at the individual level.

Fourthly, in **case-control studies**, a group of subjects who have the disease or condition is compared with a group which does not. Factors that can be studied include the degree of exposure to the disease and the special characteristics of the host population. Fifthly, **cohort studies** require definition and comparison of the varying degrees of exposure to the hazard among groups within the host population. They thus tend to be long in duration, and indeed a historical cohort study may be needed, in which degrees of exposure are analyzed from the more distant past. Sixthly, **intervention study** involves planned experimentation with exposure to the hazard or use of naturally occurring experimental conditions. For example, the 1981 earthquake in Greece provided a sort of "natural laboratory" for the controlled study of the incidence of acute miocardial infarction (fatal heart attack; Trichopoulos et al. 1983).

Controls are necessary in most epidemiological studies, as people who have the disease or condition being studied are likely to have had a different degree of exposure to risk from those who are free of it. This can only be ascertained by establishing a "baseline" group whose degree of contact with the risk is known.

Epidemiological rates
Mortality, morbidity and the occurrence of disease should be expressed as rates, in which the numerator is the measured quantity and the denominator is some average, constant or total figure (Rose & Barker 1979). Many epidemiological rates are composed of *the frequency of an observed state or event divided by the total number of people to whom this state or event might occur* (the population at risk). "Population" always means *all* persons to be measured, regardless of whether they have contracted the given condition or not. Hence, "mortality" is the frequency of death and "morbidity" the frequency of illness, injury or disability. For comparative purposes, these rates are standardized by expressing them per 1,000 or 100,000 of the population.

Epidemiological rates are expressed either statically as prevalence or dynamically as incidence. The **prevalence rate** is the proportion of a defined group having a given condition at one point in time: if a long period of time is required to measure how many people have the given condition, the value should be termed a **period prevalence rate**. The **incidence rate**, on the other hand, is represented by the proportion of a defined group developing a condition within a stated period. One problem with this statistic is that the rate of discovery or notification of cases may go up as a result of better surveillance associated with compilation of the measure. Figures rise also if patients who are cured become reinfected and therefore appear twice in the compilation of incidence rates for a given period. Measured incidence gives **crude incidence rates**, which must be standardized if populations are to be compared. Apart from the use of standard population sizes varying from 100 to 10,000,000, the age-group distribution of the population may also be standardized, and the categories used are commonly 0–4, 5–14 and ≥15 years, which correspond broadly to infants, children and adults. This is necessary because behavioural and immunological characteristics differ between each group, and hence so do the medical risks. Finally, among other measures there is the **case fatality rate**, which is the number of deaths, divided by the total number of cases, including all mortality and morbidity values. This serves to illustrate the proportion of cases which prove fatal.

Disease control is best achieved by calculating epidemiological rates and using them in a surveillance system.

Epidemiological surveillance
The difficulty of stopping disease transmission by indiscriminate pro-phylaxis indicates that the best means is to track the progress of a selection of relevant conditions in order to know when or where the incidence rates rise, and hence where to concentrate medical resources in preventing an epidemic (Parrish et al. 1964). Thus, surveillance of a disaster area is carried out in order to increase the available information on disease, injury or death rates so as to prevent or control outbreaks. It is carried out primarily by using standard statistical collection procedures, unofficial community sources and reports by relief workers and government organizations. It should be noted, however, that better surveillance tends to increase the number of diagnoses of diseases, and hence incidence rates obtained immediately after disasters should be treated with some caution (PAHO 1982c).

Technically, the objective is one of early detection to facilitate rapid response, while operationally, the avoidance of wasteful "blanket" protection measures enables medical resources to be redirected to more productive uses. As one important task of epidemiological surveillance teams is to investigate rumours of disease outbreaks, the social objective is one of making people feel safe by discounting and preventing alarmist rumours. In the aftermath of the 1976 Guatemala earthquake 30 rumours were investigated and disproved, mainly in remote rural areas, but also in Guatemala City (De Ville De Goyet et al. 1976).

Existing surveillance will probably cover diseases that are locally endemic, amenable to control, of public health importance and internationally notifiable. Under international agreements supervised by the World Health Organization, information is collected from around the world on the incidence and prevalence of certain communicable diseases and controllable illnesses. This can be complemented by new surveil-lance, consisting of more focused, symptom-based monitoring of diseases and conditions that are directly related to the disaster and which, again, are amenable to control.

In practical terms, surveillance should encompass diseases which break out at normal times, as well as those which spread because of the disaster. It should not neglect diseases (such as cholera, malaria and typhoid fever) which, though they may not usually be present in the disaster area, are of importance to world health programmes. Surveillance depends on systematic reporting by telephone, telegraph, telex or fax of a variety of medical conditions. These include both bacteriologically confirmed cases of diseases and patients with suspected clinical syndrome (i.e. apparent symptoms). Symptoms should be reported by broad, general categories, such as diarrhoea, cough, diarrhoea with blood or mucus, fever without diarrhoea, etc. Thus, any reports or rumours of abnormal incidence of disease should rapidly be investigated in the field using a multidisciplinary team of researchers and specialist physicians.

Setting up an epidemiological surveillance system involves a series of practical steps. First, a limited group of diseases and conditions is identified for monitoring, and appropriate symptomatic indicators are selected (see Table 7.6). In refugee camps (see Ch. 8) surveillance can be linked to problems of nutrition, in which case **anthropometry** (e.g. weight-for-height, girth-by-height measures, etc.) will be used to

Table 7.6 Epidemiological surveillance list

Famine not expected	Developing famine situations
Diagnosed diseases:	*Major causes of mortality:*
Meningococcal meningitis morbillus (measles) pertussis (whooping cough) typhoid fever viral hepatitis	diarrhoea and cholera dysentery malaria malnutrition meningococcal meningitis morbillus (measles) pneumonia
Partially classified symptoms:	*Epidemic diseases requiring immediate action:*
diarrhoea with fever diarrhoea without fever fever with cough fever without cough or diarrhoea	cholera malaria meningococcal meningitis morbillus (measles) poliomyelitis typhus
Partially related to the crisis:	
general surgery trauma	
May be unrelated to the general crisis situation:	*Special nutritional deficiencies:*
freezing or hypothermia general medicine obstetrics psychological disturbance	beriberi microcytic anaemia (iron deficiency) macrocytic anaemia (folate and B12 deficiency) pellagra scurvy xerophthalmia

- Sum the total number of cases for each category, Σ
- Sum the grand total number of cases, Σ

determine whether patients are receiving sufficient food. Secondly, the limits of the disaster area and the location of its medical centres, hospitals and clinics are established. Next, rapid statistical sampling methods are used at medical centres to ascertain the prevalence rates of the conditions to be monitored. Presumed or confirmed cases are identified on the basis of particular symptoms or complaints, and priorities are established for field investigation and rapid screening of reports of suspected outbreaks. Local epidemiologists can be used here.

A reporting system is set up, including monitoring of the numbers of patients discharged from hospitals, as well as those who are admitted. Data are interpreted at the national level: they must be compared with predisaster rates in the disaster area and adjusted for seasonal fluctuations. Finally, as the medical emergency created by the disaster draws to a close, data should be integrated into a national epidemiological system and made part of a plan for continued surveillance, perhaps as part of a national epidemiological observatory.

References

Alexander, D. E. 1985. Death and injury in earthquakes. *Disasters* 9, 57–60.

Alexander, D. E. 1989. Spatial aspects of earthquake epidemiology. *Proceedings of the International Workshop on Earthquake Injury Epidemiology for Mitigation and Response* (July), 82–94. Baltimore: The Johns Hopkins University Press.

Anderson, T. W. & C. Rochard 1979. Cold snaps, snowfall and sudden death from ischemic heart disease. *Canadian Medical Association, Journal* 121, 1,580–3.

Baker, F. J. 1980. The management of mass casualty disaster. In *Priorities in multiple trauma*, H. W. Meislin (ed.), 149–57. Germantown, Maryland, & London: Aspen Publications.

Baker, S., B. O'Neill, W. Haddon et al. 1974. The injury severity score: a method for describing patients with multiple injuries and evaluating emergency care. *Journal of Trauma* 14, 187–96.

Baxter, P. J. 1990. Medical effects of volcanic eruptions: I. Main causes of death and injury. *Bulletin of Volcanology* 52, 532–44.

Baxter, P. J., R. S. Bernstein, A. S. Buist 1986. Preventive health measures in volcanic eruptions. *American Journal of Public Health* 76 (suppl.), 84–90.

Beinin, L. 1985. *Medical consequences of natural disasters*. New York: Springer.

Benenson, A. S. (ed.) 1980. *Control of communicable diseases in man*, 12th edn. Washington, DC: American Public Health Association.

Berberian, M. 1978. Tabas-e-Golshan (Iran) catastrophic earthquake of 16 September 1978: a preliminary field report. *Disasters* 2, 207–19.

Blake, P. A. 1989. Communicable disease control. In *The public health consequences of disasters*, M. B. Gregg (ed.), 5–10. Atlanta, Georgia: Federal Centers for Disease Control.

Blong, R. J. 1984. *Volcanic hazards: a sourcebook on the effects of eruptions*. Orlando, Florida: Academic Press.

References

Bolin, R. (ed.) 1990. *The Loma Prieta earthquake: studies of short-term impacts.* Boulder, Colorado: Institute of Behavioral Sciences, University of Colorado.

Coburn, A. W., A. Pomonis, S. Sakai 1989. Assessing strategies to reduce fatalities in earthquakes. *International Workshop on Earthquake Injury Epidemiology for Mitigation and Response*, 107–32. Baltimore: The Johns Hopkins University Press.

Cochrane, H. C. 1975. *Natural hazards and their distributive effects.* Boulder, Colorado: Institute of Behavioural Sciences.

Collins, K. J. 1983. *Hypothermia: the facts.* New York: Oxford University Press.

Coultrip, R. L. 1974. Medical aspects of US disaster relief operations in Nicaragua. *Military Medicine* **139**, 879–83.

Cox, D. C. & J. Morgan 1984. *Local tsunamis in Hawaii: implications for warning.* Honolulu, Hawaii: Hawaii Institute of Geophysics.

Davis, I. 1978. *Shelter after disaster.* Oxford: Oxford Polytechnic Press.

Davis, I. 1987. Safe shelter within unsafe cities. *Habitat International* **5**, 5–15.

De Bruycker, M., D. Greco, M. F. Lechat 1985. The 1980 earthquake in southern Italy: mortality and morbidity. *International Journal of Epidemiology* **14**, 113–17.

De Ville De Goyet, C., E. Del Cid, A. Romero, E. Jeannee, M. Lechat 1976. Earthquake in Guatemala: epidemiological evaluation of the relief effort. *Bulletin of the Pan American Health Organization* **10**, 95–109.

Eidson, M. et al. 1990. Risk factors for tornado injuries. *International Journal of Epidemiology* **19**, 1,051–6.

Eisele, J. W., R. L. O'Halloran, D. T. Reay et al. 1981. Deaths during the May 19, 1980, eruption of Mount St Helens. *New England Journal of Medicine* **305**, 931–6.

French, J., R. Ing, S. Von Allmen, R. Wood 1983. Mortality from flash floods: a review of National Weather Service Reports 1969–82. *Public Health Reports* **98**, 584–8.

Fujita, T. T. 1973. Tornadoes around the world. *Weatherwise* 26, 56–62, 78–82.

Fujita, T. T. 1987. *US tornadoes, part 1: 70-year statistics.* Chicago: Department of Geophysical Sciences, University of Chicago.

Glass, R. I., J. J. Urrutia, S. Sibony, H. Smith, B. Garcia, L. Rizzo 1977. Earthquake injuries related to housing in a Guatemalan village. *Science* **197**, 638–43.

Glass, R. I. et al. 1980. Injuries from the Wichita Falls tornado: implications for prevention. *Science* **207**, 734–8.

Gregg, M. B. (ed.) 1989. *The public health consequences of disasters 1989.* Atlanta, Georgia: Federal Centers for Disease Control.

Guzmán, N., M. X. Paz, N. R. Moreno, F. Niño 1989. Design and validation of a practical instrument for trauma assessment. *Disasters* **13**, 154–65.

Jones, N. P., F. Krimgold, E. K. Noji, G. S. Smith 1990. Considerations in the epidemiology of earthquake injuries. *Earthquake Spectra* **6**, 507–28.

Katsouyani, K., M. Kogevinas, D. Trichopoulos 1986. Earthquake-related stress and cardiac mortality. *International Journal of Epidemiology* **15**, 326–30.

Killian, H. & T. Graf-Baumann 1981. *Cold and frost injuries: rewarming damages: biological, angiological, and clinical aspects.* New York: Springer.

Koegler, R. R. & S. M. Hicks 1972. The destruction of a medical centre by earthquake: initial effects on patients and staff. *Western Journal of Medicine (California Medicine)* **116**, 63–7.

Lechat, M. F. 1976. The epidemiology of disasters. *Royal Society of Medicine, Proceedings* **69**, 421–6.

491

Lomnitz, C. 1970. Casualties and behaviour of populations during earthquakes. *Bulletin of the Seismological Society of America* **60**, 1,309–13.

Manni, C. 1982. Italian earthquake. In *Mass casualties: a lessons learned approach*, R. A. Cowley, S. Edelstein, M. Silverstein (eds), 195–201. Washington, DC: US Department of Transportation.

Office of Emergency Preparedness 1972. *Disaster preparedness*, Vol. 1. Washington, DC: Executive Office of the President, US Government Printing Office.

Olson, R. S. & R. A. Olson 1987. Urban heavy rescue. *Earthquake Spectra* **3**, 645–58.

PAHO 1981. *Emergency health management after natural disaster*. Washington, DC: Pan American Health Organization.

PAHO 1982a. *Environmental health management after natural disasters*. Washington, DC: Pan American Health Organization.

PAHO 1982b. *Emergency vector control after natural disaster*. Washington, DC: Pan American Health Organization.

PAHO 1982c. *Epidemiological surveillance after natural disaster*. Washington, DC: Pan American Health Organization.

Parrish, H. M., S. A. Baker, F. M. Bishop 1964. Epidemiology in public health planning for natural disasters. *Public Health Reports* **79**, 863–7.

Quarantelli, E. L. 1983. *Delivery of emergency medical services in disasters: assumptions and realities*. New York: Irvington.

Rennie, D. 1970. After the earthquake. *Lancet* **2**, 704.

Scawthorn, C. Y. Yamada, H. Iemura 1981. A model for post-earthquake urban fire hazard. *Disasters* **5**, 125–31.

Slosek, J. 1986. The spatial distribution of morbidity and mortality caused by earthquakes. M.S. thesis, University of Massachusetts, Amherst.

Sommer, A. & W. H. Mosley 1972. East Bengal cyclone of November 1970: epidemiological approach to disaster assessment. *Lancet* **I**, 1,029–36.

Tierney, K. J. 1985. Emergency medical preparedness and response in disasters: the need for inter-organizational co-ordination. *Public Administration Review* **45**, 77–84.

Tilling, R. I. 1989. Volcanic hazards and their mitigation: progress and problems. *Reviews of Geophysics* **27**, 237–69.

Trichopoulos, D. et al. 1983. Psychological stress and fatal heart attack: the Athens (1981) earthquake natural experiment. *Lancet* **I**, 441–4.

UNDRO 1982. *Disaster prevention and mitigation: a compendium of current knowledge*. Vol. 8, *Sanitation aspects*. Geneva: Office of the United Nations Disaster Relief Co-ordinator.

White, G. F. & J. E. Haas 1975. *Assessment of research on natural hazards*. Cambridge, Mass.: MIT Press.

World Health Organization 1986. Communicable diseases after natural disasters. *Weekly Epidemiological Record* **11**, 79–81.

Select bibliography

Abrams, T. 1990. The feasibility of prehospital medical response teams for foreign disaster assistance. *Prehospital and Disaster Medicine* **5**, 241–6.

Alter, A. J. 1970. Environmental health experiences in disaster. *American Journal of Public Health* **60**, 475–80.

Select bibliography

Baskett, P. & R. Weller (eds) 1987. *Medicine for disasters*. Bristol, England: John Wright.

Beinin, L. 1981. An examination of health data following two major earthquakes in Russia. *Disasters* **5**, 142–6.

De Boer, J. & T. W. Baillie 1980. *Disasters: medical organization*. Oxford: Pergamon.

De Bruycker, M., D. Greco, I. Annino et al. 1983. The 1980 earthquake in southern Italy: rescue of trapped victims and mortality. *World Health Organization, Bulletin* **61**, 1,021–5.

De Ville De Goyet, C. & M. F. Lechat 1976. Health aspects in natural disasters. *Tropical Doctor* **6**, 152–7.

Duffy, J. C. (ed.) 1990. *Health and medical aspects of disaster preparedness*. New York: Plenum.

Durkin, M. E., C. C. Thiel, Jr, J. E. Schneider, T. De Vriend 1991. Injuries and emergency medical response in the Loma Prieta earthquake. *Seismological Society of America, Bulletin* **81**, 2,143–66.

Gómez, N. G. & N. De Sarmiento 1987. Post-disaster behaviour of some communicable diseases. *Disasters* **11**, 235–40.

Gueri, M. & H. Alzate 1984. The Popayan earthquake: a preliminary report on its effects on health. *Disasters* **8**, 18–20.

Heckman, J. D. (ed.) 1992. *Emergency care and transportation of the sick and injured*, 5th edn. Chicago: American Academy of Orthopaedic Surgeons.

Kravis, T. C. & C. G. Warner (eds) 1987. *Emergency medicine: a comprehensive review*, 2nd edn. Rockville, Maryland: Aspen Systems Corporation.

Logue, J. N., H. Hansen, E. Struening 1981. Some indications of the long-term health effects of a natural disaster. *Public Health Reports* **96**, 67–79.

Logue, J. N., M. E. Melick, H. Hansen 1981. Research issues and directions in the epidemiology of health effects of disasters. *Epidemiological Reviews* **3**, 140–62.

Mackenzie, E. J. 1984. Injury severity scales: overview and directions for future research. *American Journal of Emergency Medicine* **2**, 537–49.

Noji, E. K. 1991. The medical consequences of earthquakes: co-ordinating the medical and rescue response. *Disaster Management* **4**, 3–11.

Noji, E. K. & K. T. Sivertson 1987. Injury prevention in natural disasters: a theoretical framework. *Disasters* **11**, 290–6.

PAHO 1990. *International health relief assistance: a guide for effective aid*. Washington, DC: Pan American Health Organization.

Palmer, S. R. 1989. Epidemiology in search of infectious diseases: methods in outbreak investigation. *Journal of Epidemiology and Community Health* **34**, 311–14.

Reis, N. D. & E. Dolev (eds) 1989. *Manual of disaster medicine: civilian and military*. New York: Springer.

Rose, G. & D. J. P. Barker 1979. *Epidemiology for the uninitiated*. London: British Medical Association.

Seaman, J. 1980. The effects of disasters on health: a summary. *Disasters* **4**, 14–18.

Seaman, J., S. Leivesley, C. Hogg 1984. *Epidemiology of natural disasters*. Basel: S. Karger.

Shears, P. 1991. Epidemiology and infection in famine and disasters. *Epidemiology and Infection* **107**, 241–51.

Smith, M. 1991. Water and sanitation for disasters. *Tropical Doctor* **21** (suppl. 1), 30–7.

Toole, M. J. 1992. Communicable disease epidemiology following disasters. *Annals of Emergency Medicine* **21**, 418–20.

World Health Organization 1989. *Coping with natural disasters: the rôle of local health personnel and the community*. Albany, New York: World Health Organization.

CHAPTER EIGHT

The Third World

Disasters in developing countries

Diagnosis

Fifty to sixty developing countries are extremely susceptible to natural catastrophe. It is easy to show that citizens of the poor nations run the highest risk of death in natural disasters (see also Fig. 8.1):

	High income countries	Middle income countries	Low income countries
Mortality per 1000 km^2	1	8	4,857
Mortality per 1000 population	19	28	69,116
Mortality per event	125	500	33,003,925

The disaster areas of developing countries often show the same depressing list of characteristics: productivity is low, infrastructure is inadequate, available resources are under-exploited, political and economic marginality have prevailed for centuries, and religion is the main source of ideology (Jeffery 1982). Moreover, almost half of the world's population of 5,000 million has an average per capita income of no more than $270. Disaster mitigation and recovery represent a crippling and often impossible burden to such people: for example, of the 53,500 families who lost their homes in the 1986 San Salvador earthquake, about 40,000 could not afford to reconstruct them (Davis 1987).

For the sake of analysis, one can make a distinction between overcrowding of dangerous urban sites and marginalization in rural environments. These two problems are not, however, mutually exclusive: for example, the earthquake of February 1976 in Guatemala affected rural and urban populations about equally. Sixty-four per cent of the national population were impacted (a total of 3.4 million people), and 1.2 million people were left homeless as a result of damage or destruction received

495

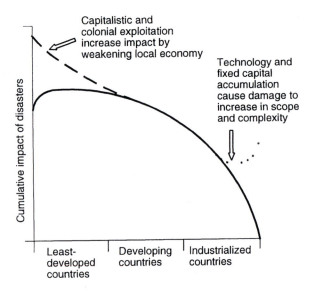

Figure 8.1 Level of economic development and severity of natural disaster (modified after O'Keefe, in Torry 1979: 531).

by 222,000 dwellings. One-third of the homeless were slum dwellers from Guatemala City, while the remainder were rural dwellers or inhabitants of small towns and villages distributed throughout the disaster area. In all cases, the largest impact of the disaster was on low-income housing (Kreimer 1978).

The urban vulnerability problem is a very severe one (Havelick 1986). In many developing countries demographic growth rates are about 2–3 per cent and the population doubles in 20–25 years; yet urban growth rates of 4–7 per cent cause the population of many major cities to double in only 12–15 years. Worse still, rural–urban migration often causes the number of inhabitants in slum and squatter settlements to double every 5–7 years, giving densities of up to 150,000 persons per km², which may be ten times the level in old, established quarters and orders of magnitude greater than the density of rich areas (Davis 1978). In some of the Latin American metropolises three-quarters of the population are living in makeshift accommodation, which tends to be extremely vulnerable: for example, the vast majority of the 59,000 houses destroyed in Guatemala City in the 1976 earthquake where occupied by the urban poor. But overpopulation is a symptom of the Third World's difficulties much more than it is a cause: in the absence of state aid, large families are the only guarantee of social security (Fig. 8.2).

The rural problem is one of vulnerability stemming from the progressive neglect, repression or deprivation of the poor. We can term it **marginalization**, which Blaikie (1985: 125) defined as follows:

Figure 8.2 Family group from a rural village on the Fijian island of Viti Levu. In Third World countries large and extended families are a form of insurance against many vicissitudes, including disasters. To some extent they make up for the lack of a social security system. On the other hand, however, overpopulation can severely tax the carrying capacity of marginal land, especially when this is stressed by the impact of natural calamity.

> Used in the context of peasants in lesser developed countries, marginalization has tended to imply the process by which they lose the ability to control their own lives (where they live and derive their income, what crops or stock they produce, how hard and when they work).

Overpopulation increases vulnerability as much in the rural case as in the urban one. Although densities are much more sparse than urban agglomerations, where the opportunities for agriculture are severely limited to small areas, rural populations may exceed 1,000 persons per km^2, as in some of the volcanic hazard zones of Indonesia and the Philippines. Yet an area can be considered overpopulated at much smaller densities if its "carrying capacity" is slight. Thus, in the hill tracts of Nepal, which are subject to flash floods, landslides and accelerated erosion, rural population densities have in places reached 15 people per cultivated hectare. Holdings have become smaller and more fragmented, while individuals have unequal access to land and capital. Blaikie (1985: 133)

497

argued that many such mountainous areas now suffer from "a common syndrome of environmental deterioration, demographic pressure, political and economic subordination and the partial preservation of an ancient or feudal social and economic structure".

In both rural and urban contexts, overpopulation and marginalization are often greatly exacerbated by repressive patterns of land ownership. For example, in northeast Brazil drought is the result of irregular and scarce rainfall. But the disaster that it provokes stems from the destruction that lack of rain causes to peasant agriculture, which is made all the more vulnerable by a grossly inequitable land tenure system. In many cases, the only solution to the problem of vulnerability of the poor is land reform, which is unlikely to occur without a successful revolution.

In conclusion, Susman, O'Keefe & Wisner (1983: 279) stated the political diagnosis of marginalization in unequivocal terms:

> It is believed that the international division of labour among rich and poor countries, and market forces within the poor underdeveloped capitalist economies of the Third World, cause the poorest of the poor to live in the most dangerous places. This is a matter of complex social allocation – both of the social surplus product (national income) and of locations within the space-economy. The issues of income distribution are entangled with the issues of "spatial social justice".

Poverty, vulnerability and militarization

In 1977 the global expenditure on armaments was $350,000 million and since then it has more than doubled. In the Third World expenditure on military hardware and forces cannot be dissociated from development, which it inhibits. As it is also difficult to dissociate the effect of natural disasters from problems of development, disaster impacts must, at the very least, be linked indirectly to military impacts.

Militarization leads to the concentration of wealth and reduction of investment in development projects. The increasing poverty that results can lead to political destabilization, which may lead to repression. The effects of this may mingle with those of natural disaster, leading to a series of human consequences that includes mortality, famine, refugee migrations and the destruction of home and livelihood (Turton 1991). The vicious circle is known as the **poverty–repression–militarization cycle** (Fig. 8.3). It appears that the only way out of this situation is disarmament, which may need to be accomplished as a unilateral policy, in order to divert some of the expenditure into peaceful development and disaster mitigation.

Disaster relief and the Third World

In 1979 a us National Academy of Science committee concluded that the purpose of international aid is to complement rather than supplant what a

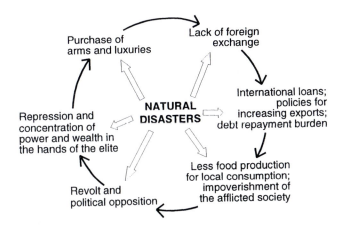

Figure 8.3 The poverty–repression–militarization cycle (after COPAT 1981).

developing country stricken by disaster can do to help itself. It also stated that such aid should help fulfil the overall societal goals of the country in question, not merely help individual victims survive the disaster (US National Research Council 1979).

These precepts, however, highlight a series of difficulties commonly experienced when aid is supplied by industrialized countries to developing nations. For instance, experts from developed countries often have difficulty in distinguishing disaster-related needs from the chronic problems that are characteristic of the world's poorest nations. As a result, deficiencies in infrastructure (for example, in sanitation and water supplies) may falsely be attributed to the disaster. Thus relief requirements may be overstated with respect to needs created strictly by the catastrophe itself. Although, as we shall see below, it does not usually make sense to dissociate development from disaster, generally it indicates better planning if the two problems are linked explicitly rather than by default.

When planning is inadequate or insensitive, relief aid may attempt to impose standards that are wholly inappropriate to a developing area. Thus, many forms of technology suited to industrialized countries will not work well in developing societies. Techniques and technical standards that assume the existence of an extensive service network cannot be transferred easily to countries which lack such resources (for example, workshops for maintaining vehicles or machinery). Likewise, sophisticated information systems may not work in areas where local people and officials are unaccustomed to them, or, if they work in a purely technical sense, they may not be understood. Lastly, management systems imported from developed countries may not be appropriate or may not function (for example, procedures for formulating and enforcing building codes).

In synthesis, aid from the developed to the developing world tends to suffer from three deficiencies. First, it is often supplied without adequate analysis of need and cost-effectiveness. Secondly, it is often given on the basis of political and military alliance, thus mirroring the flow of development loans and arms supplies (the Middle Eastern oil-producing countries tend to support the Muslim world, while the United States uses aid to reinforce its strategic interests). Hence, at best such aid can only be considered as "benign self interest". Thirdly, recent history confirms that much aid constitutes the unloading of surplus products, regardless of their utility (El-Khawas 1976; see Ch. 6).

Whatever the eventual balance of suffering and destruction, disasters with abrupt impacts and high death tolls (such as major earthquakes) elicit a much quicker and more fullsome donor response than creeping disasters such as drought. Large campaigns immediately after dramatic events tend to elicit a greater rate of donation than slow, steady projects such as irrigation schemes or agricultural extension. Thus, Wijkman & Timberlake (1984: 128), who stated that "the major disaster problems are essentially unsolved development problems", could be interpreted as meaning both that development would reduce the risks and that it has been neglected under the current donor system.

The Office of the United Nations Disaster Relief Co-ordinator (UNDRO)

One major initiative in favour of the Third World was the establishment on 14 December 1971 of the Office of the United Nations Disaster Relief Co-ordinator (UNDRO), which is based in Geneva, although it has a Liaison Office at the UN headquarters in New York. The head of the organization is the Disaster Relief Co-ordinator, who is an Under-Secretary General of the UN and is responsible directly to the Secretary General. Although it has reciprocal agreements with other parts of the organization, UNDRO is a separate entity within the UN Secretariat and is responsible for organizing and mobilizing international relief efforts after disaster has struck, as well as encouraging governments to mitigate, prevent and prepare for future disasters. UNDRO consists of a Relief Co-ordination and Preparedness Branch and a Prevention and Support Branch. In developing countries there are Resident Co-ordinators, who are associated with the UN Development Programme, UNDP.

The mandate of UNDRO charges it to mobilize and co-ordinate relief efforts in order to supply rapid and effective aid to countries stricken by disaster, most of which are in the Third World. Members of the UNDRO staff are often sent to disaster areas around the world in order to help governments estimate damage and relief needs and to co-ordinate relief efforts. During 1983–4, for example, UNDRO participated in 43 emergencies and in consultation with the government of each affected country launched 26 relief appeals. The organization seeks to remove obstacles that impede the rapid delivery of disaster relief to countries in need. It asks donor countries and recipients of aid to reduce or streamline bureaucratic procedures for the transfer of goods or personnel involved in disaster aid. Where necessary, it may also negotiate a reduction in tariffs or the cost of transporting relief supplies.

UNDRO has specialized in the development and application of methods for analyzing vulnerability to natural disasters and in the promotion of legislation, land-use planning and other less expensive methods of combating disaster impacts. In fact, its staff members and consultants are often asked by national governments to give advice on how to develop appropriate structures for disaster relief, prevention and mitigation. If the recommended measures are beyond the means of the government in question, UNDRO may either fund the project itself or help find a suitable donor. But UNDRO is not lavishly endowed and hence cannot finance major relief efforts. It does, however, have a modest trust fund from which small disbursements are made each year for emergency relief used to meet some of the immediate post-disaster needs of the countries it helps.

While the United Nations' organizations are a valuable and necessary part of disaster aid, they have not succeeded entirely in preventing recurrent failure of international relief efforts. In the first place, UNDRO cannot work in a particular country without the express permission of the host government; and it cannot work well without the support of that government. Moreover, although experience has taught valuable lessons to UNDRO and the other relief agencies, characteristic mistakes are still made, as the following example shows.

Deficiencies in aid: the volcanic gas disaster in Cameroon, 1986
In September 1986 Lake Nyos, in the interior of the West African state of Cameroon, emitted lethal quantities of volcanic gases (see Ch. 2). Within 48 hours, 1,887 people and about 10,000 head of livestock had been killed; 2,913 people (including 975 children of school age) had fled the area and international aid mechanisms had been set in motion (Kling et al. 1987). The calamity is worthy of analysis principally for what it tells us about the nature and quality of assistance to the Third World and about the relationship – or the lack of one! – between disasters and development.

International interest in the catastrophe was stimulated principally by the high death toll and unusual nature of the agent that caused it, although such eruptions are not entirely uncommon in Cameroon (in fact, Lake Monoun had killed 37 people in a similar event only two years previously; Sigurdsson et al. 1987). The world response led to the importation of 800 tonnes of goods and victuals into the area and to the construction of a new airstrip. Many foodstuffs that were donated appear to have perished in storage, while some other items, such as the gas masks sent by France, proved useless.

Aid was a very transient phenomenon. It arrived suddenly in huge quantities and then it ceased, leaving a problem of how to put the donated commodities to good use. Yet the Cameroon Government seems to have had surprisingly little to say about the quantity and quality of what was supplied by the international relief community. Unnecessary aid seems to have stimulated corruption among a small minority of those to whom it was entrusted, but a situation of glut would have depressed the market value of the items in question and hence reduced the opportunities to speculate in them. In any case, superfluous relief goods (tents, blankets, etc.) created a problem of storage. At the same time a large influx of foreign visitors – mainly aid workers, scientists and journalists – led to temporary difficulties in managing them, thus reducing the time available for local officials to concentrate on problems actually caused by the disaster.

Essentially, the wide range of goods supplied from abroad did not do much to ameliorate the problems created by the disaster. Free distribution of food and shelter may have retarded the pace of rehabilitation, while there was a clear lack of aid of the kind that would have stimulated a return to normality by encouraging long-term development and economic activity. The airstrip proved to be the only item of infrastructure that was built. Meanwhile, in rainy weather the relief vehicle convoys actually contributed to the impassability of the region's sparse and atrocious roads.

The real needs of the survivors were substitute housing, roads, schools and livestock. As they had little choice but to continue living in the disaster area, they would also have benefited from continuous monitoring of the hazard in order to detect future risks. As the aftermath of the Lake Nyos catastrophe wore on, there was little indication that these needs would be met effectively: six weeks after the event there were signs neither that the authorities were planning for resettlement nor that they were preparing for reconstruction. Opportunities for the effective use of what was provided were restricted, as officials were, understandably, rather ill-equipped to cope with sudden, massive influxes of relief. Apparently, the aid did not include sufficient of the required organizational and administrative help.

In summary, we may regard the Cameroon volcanic gas disaster and its aftermath as a good illustration of how insensitive the international relief system can be. It also shows how lack of attention to survivors' long-term needs can perpetuate the difficulties of a marginalized area in a poor developing country. The "window of opportunity" offered by disaster, in which the attention of national officials and donor countries is focused on the affected area, undeniably came to nothing. For this one can blame both the international relief community, which failed to provide leadership and the right sort of aid, and national administrators, who failed to use the opportunity to request what was really needed.

Famine and starvation

Food shortage and its consequences are fundamental and recurrent problems in many developing countries. But though famine and starvation are catastrophic, they do not fall within the classification of "natural disasters" adopted here. They are, however, frequently caused by drought, flood, desertification and other phenomena considered in the first half of this work. They will thus be considered as consequences of natural disaster, bearing in mind, of course, that just as often they result from civil or international war, repression, tyranny or economic blockade. In fact, the socio-economic, political and military causes so frequently blend with the effects of natural disaster that it would be counterproductive to attempt to separate them. Accordingly, the following sections examine the nature and causes of famine, the social and economic underpinnings of food shortage, and the strategies used to combat it, with particular emphasis on food aid.

The individual impact of famine

For the individual, famine usually results in **protein-energy malnutrition**, or PEM. This can lead progressively to weakness, excessive weight loss, diarrhoea, anorexia, immobility and finally death. But undernutrition can lead to failure to function and is often as important as outright starvation. PEM is usually first to manifest itself in small children and can take any of several forms (De Ville De Goyet et al. 1978): nutritional marasmus involves severe wasting away of fat and muscle and results from prolonged starvation; kwashiorkor is characterized by oedema, involving skin and hair changes; and marasmic kwashiorkor combines wasting and oedema. The accompanying conditions include anaemia and deficiencies of vitamin A (xerophthalmia and blindness), vitamin B1 (beriberi), niacin (pellagra), vitamin C (scurvy) and vitamin D (rickets). In addition, intestinal worms are commonly present during malnutrition, especially ascaris (roundworms) and hookworms.

Starvation is related to a person's daily protein-energy intake. Hence, the World Health Organization (WHO Working Group 1986) defines a **survival ration** as 1,500 kcal/day, a **maintenance ration** as 1,800 kcal/day, and a (modest) **recovery ration** as 2,000 kcal/day (see Fig. 8.4). About

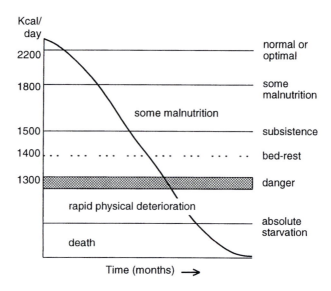

Figure 8.4 Human calorie intake and survival (Aall 1979: 432).

1,400 kcal/day constitutes a **bed-rest survival ration**, which is conducive to apathy and extreme vulnerability to disease. It is, in fact, no more than a **short-term survival ration**: people suffering from exposure or exhaustion would need extra food in order to be able to recover. In reality, immunity against disease and the strength to keep up some semblance of normal activity cannot usually be maintained with an intake less than 2,200–2,500 kcal/day and heavy manual work can require up to 3,500 kcal/day (De Ville De Goyet et al. 1978).

As Table 8.1 shows, separate norms have been established for babies, infants and children. Moreover, breastfeeding mothers could not be expected to keep children alive on only 1,400 kcal/day: women who are pregnant require at least 350 kcal more energy and 15 g more protein than those who are not, while lactation needs 550 kcal and 20 g, respectively, above normal requirements. These are all, of course, minimum figures: by way of comparison, an average adult male from a developed country eats about 3,000 kcal per day, 95 per cent of which is converted to body heat.

The indicators and causes of famine
The indicators of famine or potential famine include infant mortality rates, proportions of underweight and undernourished people within the community, per capita food production and problems of food distribu-

Table 8.1 The minimum amount of energy required by average human beings in various circumstances (after De Ville De Goyet et al. 1978: 10).

Group	Height	*Daily intake per person:*						
		Emergency subsistence (few weeks)		Temporary maintenance (some months)		Recommended intake (generally)		
	cm	MJ*	kcal$_{th}$	MJ	kcal$_{th}$	MJ	kcal$_{th}$	
0–1 yrs	<75	3.4	810	3.4	810	3.4	810	
1–3 yrs	75–95	4.6	1,100	5.4	1,290	5.7	1,360	
4–6 yrs	96–116	5.4	1,290	6.7	1,600	7.7	1,840	
7–9 yrs	117–136	6.3	1,510	7.5	1,790	9.2	2,200	
10+ yrs male	>136	7.1	1,700	8.4	2,010	11.7	2,800	
						12.6	3,000	
10+ yrs female	>136	6.3	1,500	7.5	1,790	9.2	2,200	
						10.3	2,460	
Pregnant women	>136	8.0	1,910	9.2	2,200	10.7	2,560†	
Lactating women	>136	8.0	1,910	9.2	2,200	11.5	2,750	
Average per day per person		≈6.3	1,500	≈7.5	1,790	9.2	2,200	

* 1 MJ (megajoule) ≡ 239 kcal$_{th}$, 1,000 kcal$_{th}$ ≡ 4.184 MJ.
† Second half of pregnancy.

tion. But when analyzing the phenomenon of starvation, one must first understand the situation that prevails in the absence of famine – i.e. the background rates of infant mortality and malnutrition.

First, it is as well to remember that infant mortality is *normally* high in the Third World: for instance, in the 34 lowest income countries it averages 114 per 1,000 live births, and in the Indian subcontinent it varies between 100 and 150 per 1,000. It is highest of all among the African countries, reaching 206 in Sierra Leone, 201 in Burkina Faso and 197 in the Gambia (Whitehead 1989). By contrast, in Western Europe it varies between 9 and 14 and in Japan it is about 7.1. As one might expect, the risks begin during pregnancy. Whereas healthy, slim women in European countries may have an average of 13 kg of body fat, the body fat of Gambian women often falls to 8 kg during the hungry season. If pregnancy coincides with the seasonal food shortage, a Gambian expectant mother will lose 5 kg of fat, at a time when she should instead be accumulating it. Under these circumstances up to 30 per cent of babies are born weighing less than 2.5 kg, and many will not survive (Whitehead 1989).

Secondly, subclinical malnutrition is widespread all the time in many poorer countries (Green 1986). In the Third World as a whole, 2 per cent of children are severely underweight (<70 per cent of normal weight-for-height) while 20 per cent more are moderately underweight (70–90 per cent of normal weight-for-height). The World Health Organization (WHO) estimates that 100 million children in the world currently have moderate or severe protein-energy malnutrition (Torun & Viteri 1984). And children are particularly affected by famine, in which serious undernutrition (as represented by <80 per cent of normal weight-for-height) may reach 6–50 per cent of the population (Table 8.2; see, e.g. De Waal 1989a, Galvin 1988, Scrimshaw 1987).

Thirdly, there is the question of food production, which should be considered in the light of the population problem. The calculated population doubling rates are 436 years in Western Europe and 3,465 years in Sweden, but 36 years in Asia and only 23 years in Africa. In the last of these continents, which contains 22 of the world's poorest countries, population will rise from 384 million to 645 million by AD 2000 and 1,270 million by AD 2025. This will inevitably lead to more widespread malnutrition, as it is totally out of step with both actual and potential rates of agricultural improvement. In India, which is the world's second most populous country, since 1976 the "Green Revolution" has so increased cereal production that severe malnutrition (e.g. the proportion of children 40 per cent underweight) has largely disappeared. But this is mainly the result of specialized programmes targeted at particular worst cases; general malnutrition still persists, and disasters can easily exacerbate it.

Since 1960 per capita food production has declined by about 20 per cent in 37 of the 39 African countries south of the Sahara. Per capita cereal production in black Africa has fallen 60 per cent, and the margin between production in years when the harvest does not fail and those when it does has narrowed considerably, while a wage-earning economy has not developed to the extent that sufficient people can buy the food that they cannot produce. It is thought that more than a million people may thus have died during the Sahelian famine of 1984–5 (De Waal

Table 8.2 Prevalence of undernutrition in some populations affected by famine (Toole & Foster 1989: 82).

Population*	Year	Percentage of acute undernutrition†
Kampuchean refugees	1979	10.0–18.0
Ogaden refugees in Somalia	1980	28.0–39.0
Mozambique	1983	12.0–28.0
Mauritania	1983	8.2–17.2
Niger	1984	11.5–12.2
Burkina Faso	1985	6.0–10.0
Tigrean refugees in Sudan	1985	13.8–50.0
Eastern Sudan (not refugees)	1985	17.0
Ethiopia (Korem camp)	1985	70.0
Ethiopia (northern Shoa)	1985	26.0

* Based on surveys of children less than 5 years old.
† Defined as weight-for-height <80% of median World Health Organization reference population.

1989b). Hence, it can fairly be said that the balance between environment and development has broken down (Timberlake 1984).

Such disequilibrium is easily seen in agronomic factors. The list is a depressing one and includes the cultivation of marginal or unstable lands, dependence on a single crop or kind of livestock, seasonality in the availability of food, poor methods of food production, weak transportation and distribution systems, inaccessibility of the local area, and limited forecasting and response capacity on the part of government. Globally, rising population pressure and density in marginal lands is increasing the risk of famine. Overgrazing is becoming more widespread, soil erosion is on the increase, desertification is encroaching on once productive ranges, pastures and cultivated fields, and the carrying capacity of land is thus more easily exceeded, with consequent environmental degradation that is hard to counteract or reverse.

Yet, despite the food production crisis, absolute lack of food is relatively rare in these times of rapid intercontinental transport and communications. If it occurs, it tends to be restricted to small, isolated rural areas and to bear a strong relationship to social, political or military turmoil (Wisner et al. 1982). Sadly, there have been cases in which governments or international relief agencies have caused the problem by distributing free food and thus contributing to the failure of normal coping mechanisms, which usually depend on the sale of reserve food supplies (Lofchie 1975). Moreover, in the event that crops fail, it is clearly dangerous for the afflicted population to depend too closely on imported foodstuffs. Delays in the arrival of food aid may leave the population for whom it is destined to starve in the meantime. There is also a risk that the foods supplied will be unpalatable to a population accustomed to a very different diet, and hence the recipients may find the donated food very difficult to assimilate.

Lack of food can lead to starvation, but the severity of famine (in terms of its speed of onset and the ensuing mortality) stems mostly from the failure of the market and the normal system of exchange and distribution. Unemployment, hyperinflation or loss of purchasing power may eventually mean that the poor and disadvantaged are unable to buy the food they need for survival. Famine then results from the collapse of the local market system and may increase if those who have food hoard it in order to drive up the price (Seaman & Holt 1980). Thus, starvation can occur where there *is* enough food to feed the population (Scrimshaw 1987). The reason for this is that the effects of the food shortage depend on the state of the commercial market, on normal systems of food purchase and distribution (including reciprocal arrangements, loans between kin and among wider social groups), on the degree to which alternative employment is available and on the extent to which grain, livestock or cash reserves have been accumulated.

Another vital factor is whether a subsistence economy is responsible for all or part of the food supply to the community and whether, if it fails, it can be substituted by a cash economy in order to buy in food supplies. Thus, where socio-economic adaptation to conditions of recurrent crop failure is extensive, famine may cause widespread hardship and child mortality, but little social disruption (of the kind caused by unplanned mass migration) and few adult deaths. On the other hand, where mechanisms evolved to cope with recurrent shortfalls in food production have consistently failed, subsistence economies have been forced into adopting a sort of cash system with taxation to pay for their future needs.

A grim illustration of this situation is given by the 1972–3 northern Ethiopian famine, in the Province of Wollo (Seaman et al. 1977). Lack of food was a highly localized phenomenon that resulted mainly from reduction in the availability of wage labour needed to supplement the failure of subsistence farming. Crop failure led to a huge rise in the market price of foodgrains (as a result of hoarding), massive outmigration of people who had no cash to buy food and starvation of those who were unable to find work in cities and cash-crop areas. In this way an unanticipated crop failure can trigger a famine if the distribution system cannot cope with the shortage or if the failure causes a sudden rise in prices in an area where people have few or no reserves of cash.

In synthesis, famine tends to be selective in its impact. Hence, the countries with the lowest per capita incomes are also those of the least calorie supply (Fig. 8.5) and the greatest risk of starvation. In such nations, the groups which run the greatest risk are infants, small children, old people, pregnant and lactating women; undernourished and chronically ill people with low bodily reserve; landless, unemployed and underemployed families; ethnic, religious and political minorities; and refugees and displaced or dispossessed persons. For each of these groups, starvation has several sets of causes, including environmental factors such as the occurrence of drought, flood, desertification or other natural disaster. Political instability is also important, as it often betokens a failure of the socio-economic mechanisms that keep people fed. The collateral factors include widespread, endemic poverty and the habitual presence of seasonal malnutrition.

Strategies to combat famine and malnutrition
When faced with cases of starvation and malnutrition, relief planners must consider first how to ensure optimal health wherever possible and, secondly, what is the absolute minimum diet necessary for the human body to function: in cases of widespread malnutrition and food shortage they will mainly have to work within the latter constraint.

UNICEF, the United Nations Children's Fund has proposed a concerted effort to halve the child mortality rate by combating disease and

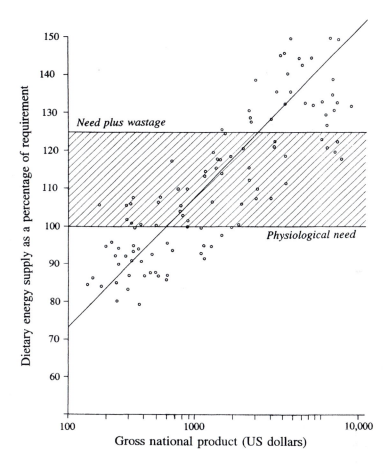

Figure 8.5 Dietary energy supply and gross national product (Whitehead 1989: 85).

malnutrition with four simple and inexpensive techniques. The first is **oral rehydration treatment** (ORT), using a combination of salt and sugar, which could prevent the deaths of 3 million children each year from diarrhoeal dehydration. Secondly, regular monitoring of the growth and nutritional status of children would enable mothers and health workers to share explicit information on health practices. Thirdly, breastfeeding should be promoted to combat the malnutrition which has often resulted from dependence on substitute foods of dubious nutritional value or unreliable supply. Fourthly, children should be immunized against diphtheria, whooping cough, tetanus, measles, poliomyelitis and tuberculosis. The programme of universal child immunization adopted by 74 national governments under UNICEF auspices could save the lives of 5 million children each year. These strategies need to be adopted at village level, but can be implemented by local volunteers with limited medical training and using relatively small medical resources.

Infant malnutrition and disease should be tackled by careful treatment feeding to restore the child's diet. Drugs should be used very sparingly (procaine penicillin being the best antibiotic), though immunization against morbillus (measles) is often necessary, as seriously under-nourished children are often greatly at risk of death from this disease. But as the root of the problem is the **nutrition–infection complex** (De Ville De Goyet et al. 1978), aid must tackle problems of sanitation, prevention of disease and infection, and supply of adequate nutrition together if conditions are to be improved. Hence, getting food to people in severe nutritional hardship is vital, but so are nutritional education, water supply and sanitation, child welfare and disease control.

Apart from the logistical, nutritional and alimentary problems of supplying food aid to communities in need, there is also a problem of recognizing famine situations as they evolve. Repeated anthropometric surveys and comparisons are needed. Anthropometric data must be related to other factors, such as the nutritional status of the population. The norms for survival and for basic and supplementary rations must be established (see Fig. 8.6), and sufficient food aid supplied, in time to save the recipients of the aid from dying of starvation. Experience shows that this is not always as easy and straightforward as it seems. For example,

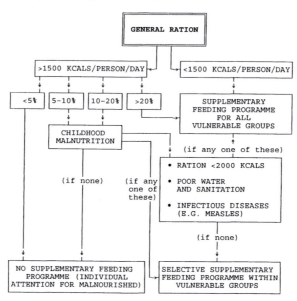

Figure 8.6 The indications for implementation of supplementary feeding programmes in famine relief (Toole & Foster 1989: 84).

510

Table 8.3 lists some of the rations supplied by the relief agencies to counteract famine during an Ethiopian drought. Although the basic foodstuffs were fairly similar (and presumably palatable to the recipients), the calorific value per person-day of the rations varied by as much as 87 per cent (compare figures given in Table 8.3 with the minimum energy requirements listed in Table 8.1).

Aid programmes
Aid for famine situations is dogged by the problem that there has been excessive preoccupation with short-term crises instead of long-term development. In this book I have drawn attention to the fact that it is

Table 8.3 Examples of some general rations distributed in Somalia* and Ethiopia† after drought disaster (Young 1987: 104; Tresalti et al. 1985).

Distributing organization	Ration size	kcal/person/day
*UN High Commission for Refugees	*Per person per month:* 6 kg maize, 2 kg rice, 2 kg flour, 500 g oil, 500 g sugar, 250 g tea, 250 g meat, 50 g powdered milk	1,392
†Red Cross	*Per person per month:* <6 yrs: 3 kg corn soy milk 7–15: 7.5 kg wheat grain Over 15: 15 kg wheat grain	380 875 1,750
†Co-ordinating body of various Christian charities	*Per family of five per month:* 45 kg bulgur wheat 45 kg dried skimmed milk 3.6 kg oil	1,325
†World Vision	*Per person per month:* 15 kg wheat flour or grain 1 litre oil	1,780
†International Committee of Red Cross Societies	*Per person per month:* 12 kg wheat flour 2 kg beans 1.2 kg oil	2,026
†Ethiopian Red Cross	*Per person per month:* 15 kg wheat flour 1 kg sugar 1 litre butter oil	1,793
†Save the Children Federation (USA)	*Per person per month:* 15 kg wheat flour 4 kg soya fortified sorghum grits 2 kg dried skimmed milk 1 litre oil	2,484
†CARE (Sudan)	*Per person every ten days:* 5 kg wheat flour or grain 0.6 kg beans 0.3 kg oil	2,230

easier to encourage donors to give money for well-publicized emergencies than for chronic long-term situations. Thus, popular support may be greatest for governments that tackle crises at the expense of more persistent but less spectacular problems. Nevertheless, many government aid programmes aim to generate new wealth among the poorer sectors of the community by instituting projects that tend to be large-scale and involve investment in major works, such as dams or irrigation schemes. Socio-economic and health benefits are difficult to demonstrate in the short and medium term, while such programmes can wax and wane with the prevailing political milieu of the country in question.

Aid from charities tends to be smaller in scale: in fact, according to Cuny (1983) imported relief seldom exceeds 40 per cent of all aid supplied. But charities can first bring interim help where there is a crucial need and, secondly, stimulate interest in and acceptance of health-related programmes of a more permanent nature. Being less dependent on central governments, they are more resilient in times of trouble. Moreover, low-key aid and low-technology methods can often succeed in times of political turmoil or general shortage, especially as charities are not usually dependent on central government to sustain them.

All things considered, the international agencies and charities need to demonstrate results, achievements and a high ratio of funds spent in the field to those collected from donors. In the past this has impelled many of them to concentrate on disaster relief rather than development assistance, as the latter tends to have higher overhead costs and longer realization times. Yet, however much it is needed, development cannot take place if nutritional levels are insufficient for people to be able to work.

Food aid

In simple terms, 1 tonne of dry, processed food occupies about 2 m^3. Dry rations for 10,000 people for a week weigh about 36 tonnes and constitute three 12-tonne truckloads. In addition, about half a tonne of food can be carried in one four-wheel vehicle, and this represents the daily ration of about 20 families, or supplementary food for 250 children (Knott 1987). However, considerable planning will be required to ensure that food and distribution teams reach their destinations: truck breakdowns, bad weather or poor roads can all lead to delays, and for every ten vehicles at least one is likely to be out of action at any given time, however reliable they are.

This is the logistical side of one of the world's great commodities: food aid. In its early years, the World Food Programme, which was set up by the Food and Agriculture Organization in 1963, earmarked between 30 and 85 per cent of its resources for the victims of natural disaster. But on a world scale, only 10 per cent of food aid is used for relieving disasters and feeding refugees: 20 per cent is distributed free to the poor and 70 per cent is given or sold to Third World governments, who may resell it to increase their revenues (Warnock 1987).

Fluctuations in the availability of surpluses and the market price of such commodities exert a very strong influence on the amount of food aid supplied. Moreover, political considerations often override need: in 1974, for example, more than half of US food aid went to countries in which American strategic interests were paramount. In 1991 the debilitating effect of cheap food surpluses caused many developing nations to threaten to impede the formulation of a new General Agreement on Tariffs and Trade (GATT) unless their own markets could be freed of the depressing effect resulting from commodities donated by the industrialized countries.

Often, food aid will go to urban populations (and especially urban elites, if these are in need) rather than rural subsistence farmers. For the latter, there may be no possibility of spreading the burden of crop failure among the general population. The gross domestic product of peasants tends to be less than half (or even less than one third) that of the national average and well below levels in urban areas. For example, in the mid-1970s the people of the Andes were able to obtain only three quarters of the protein and five sixths of calories they required. But despite such clear demonstrations of need, on average only 10–15 per cent of total development capital in the world's poorest countries has been invested in rural projects.

Food relief can create budgetary dependence and a balance of payments crisis, especially when the aid flow ceases and must be substituted by indigenous efforts (often by purchasing food abroad). This is a very real problem, although it can be tackled by giving financial support to the national exchequer. Lastly food relief is considered inefficient, especially in comparison with the flexibility of cash aid. In fact, both may be ineffective if not used intelligently: food aid tends to limit diseconomies at times of chronic need and cannot always be supplanted by increasing the buying power of the national government.

Such problems have raised both criticisms and defences of food aid. These were summarized by Green (1986: 298), who noted that by lowering prices aid may discourage local food production. On the other hand, this may help stop inflation and make food accessible to people who have very low wages. Many of these are likely to be rural people, who will probably not be reached by aid that is concentrated on urban recipients. Yet, by combating malnutrition, aid can preserve the ability to produce and it can be targeted specifically at rural residents. And in more general terms, aid can be beneficial if it stabilizes the food economy, though care should be taken to ensure that growers (rather than wholesalers) receive a fair price and Third World food exporters are not ruined. However, food aid may undermine government programmes of agricultural development, though it can be tied to programmes of rural progress.

In order to enhance the benefits and minimize the drawbacks of food aid a reliable strategy is needed (Eldridge 1989). First, food should be purchased regionally, not imported over vast distances. This not only cuts down transportation time, but also stimulates regional economies and makes use of the fact that food shortages are seldom absolute, but are usually the result of market and distribution imbalances. Secondly, food must be obtained rapidly, if necessary by diverting it from less important destinations: hence the need for flexibility. Thirdly, strategists must be sensitive to local markets. After the 1976 Guatemala earthquake, for example, imported food surpluses did enormous economic damage to local farming, which lost its market. Lastly, wherever possible, local expertise should be substituted for foreign advisors, in order to avoid "aid dependency" and to utilize local knowledge and connections better.

Concerning the management of food aid, there is a tendency to send surplus products to Third World countries, and this involves the following problems. North American and European wheat surpluses will be of no value unless the recipients have some means of turning them into foods such as bread. Fishmeal (surplus from Norway), wheat-soya blend (US surplus product) and dried skim milk (surplus from the developed world's cheese production) will not be useful unless they are a diet that is acceptable to the recipient populations. Hence, donors should interest themselves in the utilization of donated foods, which are often simply dumped at the frontier and go bad in the sun or rain.

Aid should firstly be targeted towards people who are truly in need and secondly delivered in a form that is culturally, environmentally and clinically capable of being used. Problems will arise if developed countries simply consider the Third World as a convenient dumping ground for food surpluses, or if countries which receive aid become dependent on it. Such aid may inhibit the diversification and development of local agriculture.

In summary, food aid should aim to tackle famine problems but not inhibit local initiatives. There is no single formula to solve this problem.

An example of how famine can be artificial
The following example, which is admittedly one of the worst cases of its kind, illustrates how a famine disaster can be as much the artefact of the perception of those responsible for relief operations as it is the result of an original disaster.

In April 1978 approximately 200,000 refugees from Burma arrived in Bangladesh and had to be cared for by the Bangladesh Government and participating UN organizations (FAO, UNDRO, UNHCR, UNICEF, WPF and WHO). About 10,000 died, 7,000 of whom were children. Why? The food, technological expertise and knowledge of nutritional needs were available, but the relief effort was bungled (Aall 1979).

As Figure 8.4 shows, in order to survive a person must receive at least 1,500 kcal/day. All values given in the figure were reduced by 100 kcal/day to allow for the small body size of the Burmese who arrived in Bangladesh. Yet these refugees actually received mixed cereal and rice dry rations worth only 1,300 kcal/day. The emergency persisted month after month and malnutrition built up. But why was such a low basic ration supplied? In the first place, the Bangladeshi Government could not justify offering more than it did to its own destitute peoples, especially during a time of persistently large shortfall in the needs for rice. The aim appears also to have been one of "inducing" the refugees to return to Burma. In the second place, the international agencies failed to recognize the preconditions for a serious famine. The refugees had few resources of food stocks, cattle which could be slaughtered, or jewellery that could be sold, and what they had was used up within a few weeks.

After two months, the Food and Agriculture Organization decided to provide a supplementary ration, but it was already too late. It was proposed to supply an average of 1,000 kcal/day per person, but in fact only 160 kcal/day was achieved. At that moment in time it proved too difficult to distribute a supply of food equivalent to 500 kcal twice a day to 100,000 refugees. Supplementary feeding had failed to achieve its aims and it was too late to alter the basic dry ration.

Essentially, the international agencies and the Bangladesh Ministry of Health took too long to accept the situation as it actually was. Investigations were carried out in the refugee camps in order to determine the state of sanitation and care. Sample surveys were conducted to determine whether mortality rates were high. They were, moreover, in an area where the "background" mortality involves rates of 3.8 per 10,000 of the population per week (i.e. 2 per cent per year).

First, mortality among the refugees increased steadily. Then the food supply partially broke down, meaning that rations were reduced for a few weeks. There was a measles epidemic, and average mortality increased to 850 per cent of the expected level, and in some camps to 2,000 per cent. This meant at least 60 deaths per week where 4 were expected. If such circumstances had prevailed for a year, up to a third of the population of camps would have died. In one camp of 20,000 refugees, 1,800 malnourished children needed supplementary "treatment feeding" on a round-the-clock basis. Yet at this time the UN High Commission for Refugees was still looking for "other reasons" why the mortality should have been high (Aall 1979).

Eventually, the FAO requested a supplementary ration, to be provided over a two month period. It also promoted a pattern of food aid that would strengthen vulnerable groups (children, pregnant women and lactating mothers), who were to receive **treatment feeding** of 3–5 meals per day. Patients who were suffering from diseases caused primarily by malnutrition (such as kwashiorkor) were to receive intensive care.

Hence, measures were taken after eight months of fluctuating but extremely high mortality. This fell in response to supplementary feeding programmes, when these were implemented, thus clearly showing the correlation between lack of food distribution and high death rates.

This disastrous failure of relief is, one hopes, unlikely to be repeated in the future, as the agencies concerned have learned from their mistakes. The problems were principally ones of perception, involving a lack of understanding of the situation until it became too late to solve it. There was a lack of agreed norms for adequate nutrition levels, which should have been determined and publicized with clarity and firmness. There was also a lack of awareness of the logistical problems of implementing *ad hoc* and stop-gap feeding programmes for large populations in isolated, underdeveloped areas.

The plight of the Burmese refugees in Bangladesh was neither the first nor the last disaster of its kind. It did, however, reinforce the need for norms concerning the determination of what constitutes adequate full and supporting dry rations, the nature and timing of supplementary feeding programmes for vulnerable groups and the implementation of treatment programmes for severely malnourished children. We now consider the question of refugees in more general terms.

Refugees

The United Nations High Commission for Refugees (UNHCR) estimated in 1980 that 5–6 million refugees had crossed international boundaries, while a similar number were displaced within their own country (and hence are not technically considered to be refugees). If one includes all people who seek sanctuary from warfare, persecution or social or natural catastrophe, the figure might unofficially be placed in the range 50–60 million. Conservative estimates suggest that 6–10 million people fled Vietnam during the war of 1965–73, while by 1982 Pakistan had received 2.7 million refugees of the conflict in Afghanistan. These figures prompt one to ask whether world interest in the problem is sufficient to enable such large number of people to be fed and housed.

The refugee question is both an aid problem and a political one, although public opinion tends to regard it as merely a question of relief, rather than calling for an examination of why so many refugees were generated. The main pressure points on a world scale relate primarily to war, and include Sub-Saharan Africa and Southeast Asia. The impact of natural disasters merges with that of armed conflict, all the more so given recent trends towards the politicization of aid and the struggle for control of both devastated areas and relief supplies to their populations. Drought and war may combine to destroy livestock, crops and livelihoods, propelling survivors into desperate unplanned mass migrations.

Who and what is a refugee?
According to the 1951 United Nations Convention on the Status of Refugees, they are defined in official terms as

> Any person who owing to well-founded fear of being persecuted for reasons of race, religion, nationality, membership of a particular social group or political opinion, is outside the country of his nationality and is unable, or owing to fear is unwilling, to avail himself of the protection of that country; or who, not having a nationality and being outside the country of his former habitual residence, is unable, or having such fear is unwilling, to return to it. (See D'Souza 1981a).

In this respect, it is notable that the world's most influential nation, the United States of America, has ratified the 1967 UN Protocol on Refugees, but not the 1951 convention as cited above. Moreover, the latter agreement is full of legal procedural problems and loopholes by which a country may evade its responsibility to care for refugees.

Despite the official wording, defining and classifying refugees in practical terms poses severe problems. The researcher Julius Holt has argued that some sort of balance is probably struck between the tendency of journalists to go for the highest possible numbers and the tendency of governments to minimize the official estimates (Holt 1981a). Trends in the number of refugees are hard to identify, as, like the level of conflict around the world, international interest in the problem tends to wax and wane. But several factors may be contributing to increases. First, the rapid rise in population in many Third World countries causes pressure on land and on towns (and therefore on governments) and increases the vulnerability of certain groups, particularly the landless. Secondly, there are signs of increasing marginalization of groups of people in Third World economies. Finally, economic recession and stringency in developed countries is reducing the aid given to resettle refugees.

International assistance and its problems
Refugee situations tend to require a very different kind of disaster relief to those impacts which do not displace populations. A refugee is at the absolute bottom of the social scale, cut off from normal means of gaining a living and often from the support of the family. Often, he or she is constrained to seek sanctuary in an unfamiliar culture. Moreover, refugees on the move may be exhausted by exertion, exposure or disease

517

and may lack access to local markets. They therefore require specialized assistance (D'Souza 1981a).

UNHCR was set up in 1951 in order to achieve any of the following three aims: voluntary repatriation of international refugees, or their resettlement in their country of arrival, or their resettlement in a third country of asylum (usually a Western developed nation). In recent years there has been a tendency for developed countries to limit their reception of refugees and to offer substitute aid to facilitate resettlement in Third World countries (D'Souza 1981b).

This strategy has encountered four main problems. First, the definition, and the protection that it implies, exclude all people who have been displaced within their native country. Secondly, the victims of civil or international war are not specifically protected under the convention (although they are covered under a more limited convention adopted in Africa by the Organization of African Unity). Thirdly, it may be difficult for refugees to produce documentary evidence that establishes beyond doubt that their fear of persecution is justified, and hence avoid being classified simply as economic migrants in search of better prospects. For example, in many cases religious persecution can easily be confused with economic deprivation. This fundamentally affects their rights to protection under the 1951 UN convention. Finally, UNHCR cannot in any way enforce refugees' rights of asylum. Thus, although it would be illegal under the 1951 convention forcibly to repatriate a refugee to the country from which he or she has fled, the convention does not make it illegal for the host country *not* to offer aid. Such responsibilities are defined by the UN's 1948 Declaration on Human Rights, but this carries no legal force and is not binding on member nations.

Relief operations and their problems

Once a refugee problem has emerged and the host country makes a formal request for aid, the international community begins to mount a relief operation. UNDRO and UNHCR may be involved, as will be a series of non-governmental relief organizations (NGOs). The following is a list of the most well-known examples of the latter: some have formal ties with the UN or with national governments and most are based in the Western developed countries:

ARC	American Refugee Committee
CARE	Co-operative American Relief
CHARITAS	Catholic Relief Organization
FAO	Food and Agriculture Organization (UN affiliate)
HTA	Help the Aged
IRC	International Rescue Committee
MSF	Médicine Sans Frontières

OXFAM	Oxford Famine Relief Organization
PAHO	Pan-American Health Organization (WHO affiliate)
SCF/UK	Save the Children Fund (UK)
SCF/US	Save the Children Federation (USA)
UNDRO	Office of the UN Disaster Relief Co-ordinator
UNHCR	United Nations High Commission for Refugees
UNICEF	United Nations Children's Fund
WCC	World Council of Churches
WHO	World Health Organization
WFP	World Food Program
WV	World Vision
YMCA	Young Men's Christian Association

In addition, an International Committee and a League represent 132 national Red Cross and Red Crescent Societies, with more than 210 million members worldwide.

Many of the agencies maintain staff permanently in the field, and together with the host government these are often the people who are best placed to assess the need for aid. In other cases the host government will communicate that need through its embassies in the donor countries. Information must be gathered and verified concerning the gravity of the situation, the priorities among needs, the cost of required aid and who can supply which type of aid, and by what means. Then the NGOs mount a formal relief appeal to governments or private donors for funds.

Experience over the last two decades indicates that there are various common problems with the supply of aid to refugees (US National Research Council 1979). First, it can take a long time to detect and supply needs. Stocks of food can take 4–6 months to order, purchase, ship, unload and distribute, whereas starvation is often an immediate risk. To compound the problem, field assessment is not always systematic and on occasion may be subjective. The professionalism of voluntary agencies varies widely, although the major charities have built up considerable volumes of field expertise. Nevertheless, large sums of money may be committed and spent on the basis of very superficial assessments of need.

On occasion, insufficient attention may be given to the indicators of a developing refugee crisis, such that no attempt is made to solve the problem before it becomes chronic. But in any case, the international donor system tends not to allow substantial allocations of funds to be made until the refugee problem has reached dramatic proportions. Meanwhile, host countries may develop unrealistic expectations of the aid to be provided. On the other hand, superfluous aid may be given and efforts may be duplicated, as refugee situations are often very imprecise, and the level of communication between field representatives and central institutions, or between different organizations participating in the relief effort, may be insufficient.

As mentioned in the discussion of UNDRO, though they are usually the main source of relief, United Nations agencies can only operate in a particular country with the explicit permission of the host government, and their mandates are often narrowly defined, enabling them to do certain tasks but not others. The more flexible and experienced of the NGOs can sometimes provide aid quicker than more powerful UN agencies if these are held up by the host government's bureaucracy. But there have been instances of private relief organizations falling out with the UN agencies, and of mutual misunderstanding of each other's rôles. No rôles can involve policing, yet in recent years aid has often been used as a political weapon, either through being expropriated (or destroyed) in the field, or by being withheld at source. Funds, moreover, have sometimes been curtailed for political reasons.

In synthesis, many problems of supplying aid to refugees are common to disaster relief in general. Any failure to optimize relief may in the end increase the mortality tax on the displaced population.

Refugee camps
One aspect of the relief and management of refugee situations that is often lamentably misunderstood is the design of camps and the provision of sanitation, shelter, clinics and feeding arrangements. According to one expert in the field (Cuny 1977), the total costs of designing and constructing a refugee camp properly and efficiently are less than the extra operating costs of a substandard one. The more haphazard the layout of a camp, the more difficult and expensive it is to run. Conversely, a well designed and run camp enables refugees to recover more quickly from the rigours of displacement and migration: the incidence of disease is reduced, a feeling of security is promoted and less administrative effort is required. Layout, in fact, depends critically on the objectives of refugee management. Thus, camps require different plans if refugees are to be integrated with the surrounding community or isolated as a self-sufficient social unit.

Cuny (1977) identified three stages in the occupancy of a newly developed refugee camp. First, the refugees are assigned shelter, which they occupy with minimal social participation. Secondly, they reorganize themselves to create or re-establish social groupings. Eventually, in cases where it is feasible, permanent rebuilding replaces temporary shelter and lasting facilities are installed.

Refugee populations that are temporarily resettled do not consist simply of an uprooted village or community and cannot be treated as such (Holt 1981a). Commonly, villages, neighbourhoods and even nuclear families disperse during flight. The camp can be regarded as a collection of people from a badly damaged society, and it may be rendered unstable by the periodic influx of new groups of arrivals. There may also be different ethnic, religious or caste groups which in their original setting would be controlled by local authorities or by a balance of power and interests.

But there are several reasons why refugee camps do not usually turn out to be battlegrounds. The camp will be foreign ground to most refugees, who do not automatically acquire rights within it. The refugees will be dependent to a greater or lesser degree on outside officials, who can exert some control over them. Moreover, the usual social patterns do not necessarily disappear entirely, although social rank may dwindle away with the wealth that was lost in the disaster which led to the migration. Finally, given half a chance, refugees in the camp will quickly evolve social and communal patterns of survival and organization.

The basic elements of a refugee camp are housing, sanitation, water supply points, clinics and feeding centres (Holt 1981b). Outside, there may be some *ad hoc* opportunities for agriculture or grazing, or for labouring employment. Housing may be organized in segregated groups relating to pre-existing social constraints, religion or caste, and to ensure the stability of the camp it is best to preserve the social order as completely as possible. Buildings should be spaced out in order to reduce the risk of fires. Water supply and drainage are critical factors. Poor drainage can lead to the spread of disease and insanitary conditions. Water needs break down to about 15–20 litres per person per day for general use, 20–30 litres for feeding and 40–60 litres for hospitals and clinics (McAdam 1987). Access to the camp is critical in terms of the need to receive supplies as easily as possible; and access should be ensured within the camp to latrines, water supply points and waste disposal areas, otherwise these will not function (Cuny 1977).

The absence of any of the facilities discussed above is dangerous: inadequate waste disposal increases the risks of fire and pollution, while absence of sanitation will increase the insect population and the rates of disease transmission. Overcrowding of camps can lead to shortages of food and medicines, can increase the risk of fire breaking out and can seriously compromise sanitation facilities. Such a state of social promiscuity can also increase the rates of disease transmission.

The diseases most commonly found in refugee camps are diarrhoea, dysentery, measles, whooping cough, malaria, tuberculosis and scabies (Lusty 1979). Relief workers who fail to immunize themselves may fall victim to any of these. As illustrated in the section on famine and starvation (above), infant and child mortality in refugee camps has sometimes been remarkably high when feeding programmes have been absent or inefficient. During the 1980s, death rates in some northeast African camps reached 30–50/10,000/day, and if peak values such as these were to be sustained for a year the entire refugee population would be wiped out. Among Ugandan refugees in 1981, infant mortality reached 600/1,000 live births, while after the Bangladesh floods of 1974 children in the 1–4 years age group died at a rate of 86.5/1,000 over the period of the emergency. Both infants and small children thus tend to sustain the highest death rates in famine, but for all refugee populations there is a clear relationship between undernutrition and mortality.

Health and nutrition in refugee populations
The health status of a refugee depends on the following (Seaman 1981): his or her health status upon arrival in the camp; whether enough food is supplied by the relief agencies, or whether food can be generated to a limited extent outside the camp; whether he or she is capable of eating unfamiliar food, given that relief supplies may be very different to whatever constitutes local diet; and, for babies, whether milk supplies are available from nursing mothers or preprepared sources.

In refugee camps the breeding conditions for diseases can be controlled by installing proper sanitation, water supply and waste disposal facilities, building and staffing clinics, isolating the sick and lowering the density of living conditions for the general population. They may be one place where it makes sense to institute a programme of general vaccination, especially among vulnerable groups such as children. Usually, initial vaccinations must be followed up by a second innoculation, and hence the population must be impressed with this need and adequate records must be kept of who has been vaccinated (this is not easy among large populations which probably lack all forms of identification papers).

Nutrition is often a particularly pressing problem. In camps which contain thousands, or perhaps tens of thousands, of refugees, the only way to ascertain their general nutritional and health status is by objective sample survey. Nutritional studies involve **anthropometry**, or the relationship between the measurable characteristics of the human body and its calorie intake (WHO Working Group 1986). Thus, inadequate diet may lead to wasting (emaciation), and in children to restriction of growth (stunting). Methods of measuring these symptoms must be simple and robust, and include tabulating mid-upper arm circumference, girth or body weight against height (*Disasters* 1981). Height or weight can also be tabulated against a child's age (for example, 105 cm is the approximate average height of a 5 year-old), but in certain cultures age may not be known. In any instance, the relationships involved are nonlinear as, under normal circumstances, growth slows down in the mid-teens. Each measurement must be compared against tables of anthropometric relationships which are judged to be normal in a reference population of well-fed individuals with the same ethnic characteristics. The cut-off point, below which the patient is taken to be malnourished, is arbitrary, but is usually taken to be 75, 80 or 85 per cent of the mean values for the reference population (or is based on standard deviations from these means).

Anthropometry is used in the initial diagnosis of protein-energy malnutrition (PEM) and its most common manifestation in young children, the disease marasmus. Hence, measuring efforts are often concentrated on children under 6 years old or less than 115 cm tall (Chen et al. 1980). In northeast African refugee camps in 1985, up to 70 per cent of such children were suffering from PEM, with the attendant horrors of vitamin A-related blindness and scurvy resulting from vitamin C deficiency. Through the lack of immunity which commonly goes with PEM, diseases such as morbillus (measles) can be fatal: in some of the worst African emergencies one third to one half of all children contracting measles died. Thus, relatively common complaints, such as diarrhoea and pneumonia have proved fatal among PEM sufferers, often in a majority of cases, and simple prophylaxis, such as measles vaccination and revaccination, can save many lives.

The provision of food to refugees can take any of several forms. Where facilities exist for the recipients to prepare their own meals, the ingredients can be distributed, bartered or sold to them. If this is not possible, meals can be prepared in a central kitchen for direct distribution, providing that such facilities are available, of course. "Preventative" feeding involves giving out a supplementary ration to individuals or groups judged to run a special risk of malnutrition, while "curative" or "therapeutic" feeding signifies giving a special diet to patients who need to be restored to health. In view of the considerations on refugee health discussed above, at least 1,900 kcal (containing sufficient proteins, minerals and vitamins) should be supplied to each person, regardless of age.

Hence, health care, food supply and the maintenance of hygiene are three factors that are most important to the good management of refugee camps.

The impact of floods upon developing countries

So far this chapter has considered three of the most important problems of developing countries in which natural disasters play a rôle: marginalization, famine and refugees. This survey concludes by looking at the most lethal – and arguably the most important – natural hazard which the Third World faces, namely floods, and at Bangladesh, the country worst affected by flooding. Accordingly, the present section will discuss the impact on developing countries of floods and the storm surges caused by tropical cyclones (hurricanes). It thus extends to the Third World the treatment of these topics given in Chapter 3.

Floods are the most frequent disaster (40 per cent of all cases) to occur in the Third World, especially if the storm surges associated with tropical cyclones are considered. About 15 per cent of the world's population is at risk from the high winds, intense rainfall and elevated water levels associated with cyclone landfall, and most are located in developing countries. Impacts are often profound: death tolls in excess of 200,000 have been recorded when cyclones have made landfall at the Asian river deltas; and during the year 1985–6 coastal and river flooding affected 1.6 million people in six countries (US National Research Council 1987).

The damage to crops, infrastructure and housing, and the negative impacts on health and sanitation caused by floods and cyclones are particularly severe in the populous floodplains and coasts of many Third World states. For example, one in 20 people in India is vulnerable to flooding: in the summer of 1978, 66,000 villages were flooded, 2,000 people were drowned, crops were lost from 3.5 million ha and 40,000 cattle were killed. And the impacts on smaller countries are often no less profound. Dispersed island nations and countries with long coastlines are perhaps not likely to suffer damage to a large part of their national territory. But small states, that consist of single islands (as are common, for example, in the Caribbean) or are located around major river floodplains, may be particularly vulnerable to the economic devastation that floods cause. Certain exceptions tend to confirm these rules. For example, in French Polynesia cyclones of the 1982–3 season involved only low probability that any individual living in this scattered island group would suffer damage. But excessive centralism in the political administration of these islands led to a low level of disaster preparedness and mitigation, amounting to basic negligence in the face of high vulnerability (Dupon 1984). Similar problems have been noted in the Hawaiian Islands, even though, like French Polynesia, they are to be considered among the most developed oceanic islands (Mitchell 1985).

The effects of floods on developing nations can be summarized according to three categories. First, the consequences for human health include death, physical injury, disease transmission, malnutrition, shock, degeneration of morale and loss of motivation. Secondly, the consequences for agriculture include loss of crops, food stocks, seeds, stored products and agricultural wages, damage to farmed land, death or dispersion of livestock and rising commodity prices. Thirdly, the impact on settlement and the economy comprises damage to housing, buildings and infrastructure, loss of household effects, depreciation of property, reduced output, loss of business income and rising prices. Though these are also the consequences of floods in industrialized nations, in the Third World lack of resouces creates some distinctive and particularly serious outcomes.

The significance and duration of impact will, of course, depend on the level of resources available in the affected country to facilitate recovery, as much as on the scope of devastation. In this respect, it may be better not to think of damage in terms of the absolute value of resources destroyed, but as a percentage of GNP or GDP. In any event, there will be a level of destruction of basic resources (principally food stocks) against which it is not feasible to hold and manage reserves. At this point, the level and efficiency of international aid will determine the outcome of attempts at recovery.

The following sections will consider in detail the effect of floods and cyclones on human health, agriculture and housing, which are among the most important factors to be influenced by such disasters.

Medical, public health and nutritional consequences of floods

The impact of floods upon human health and nutrition is often particularly serious in Third World countries. It can be classified according to the type of effect and its relation to the timescales created by the emergency and its aftermath. Death is obviously the most serious health effect. Characteristically, it occurs by drowning, due to impact (falling trees, collapsing buildings, etc.), or due to exposure and isolation. Injury, in the form of physical trauma and associated wound infection, has similar causes. Other health effects include respiratory diseases due to exposure, enteric diseases due to inadequate sanitation and water supply, diseases associated with ecological changes (in the habitat of rodents, insects, etc.), and malnutrition with associated high levels of infant mortality (when food sources are destroyed).

Levels of immediate mortality will depend on the rapidity of impact and effectiveness of prior evacuation programmes, if these exist (De Ville De Goyet & Lechat 1976). One extreme is represented, for example, by the very high death tolls in storm surge floods in Bangladesh (see next section), where there has generally been no warning, no evacuation and a near-instantaneous impact. In contrast, recent cyclone impacts in the Caribbean have involved only a handful of deaths, thanks to adequate forewarning and rigorous evacuation. As in other types of disaster, old people, infants, the handicapped and the sick are very susceptible to both death and injury, as a result of their reduced awareness and low capabilities for self-preservation (Glass et al. 1977).

Physical trauma is primarily a consequence of flash floods and floods associated with cyclones. Fast-flowing water may carry debris, buildings subjected to high hydrostatic or hydrodynamic pressure may burst or collapse, high winds may blow over trees and structures and carry debris at high velocity, and any of these factors may cause injury to people who have not been evacuated. Data from Andhra Pradesh (India) in 1977 indicate that most physical trauma involved minor cuts and bruises: in 1979 in the Dominican Republic wound infection followed such injuries, as a result of poor sanitation and health care (both examples were cyclone disasters).

The November 1977 cyclone in Andhra Pradesh led to widespread incidence of shock and exposure, which may be compounded by high levels of stress among the affected population. If these problems are accompanied by lack of hygiene, sanitation and health care, the survivors will tend to manifest low levels of resistance to disease. Cold conditions, for example, can lead to notable increases in respiratory diseases, pneumonia, angina and lumbosacral neuralgia, according to evidence from Perú and Bolivia (Gueri et al. 1986, Telleria 1986).

Clearly, floods exert a profoundly negative effect upon water supply, sewage disposal and general sanitation in developing countries. Wells and rivers may receive significant faecal contamination; hence the morbidity profile of the affected population tends to change as the flood waters ebb and diseases multiply and incubate. The changes in incidence rates will depend on several factors: the severity of disruption to public hygiene, the effectiveness with which sanitation and water supply problems are tackled, the degree of overcrowding among survivors, the level and degree of organization of health care and epidemiological surveillance and the profile of diseases that are endemic to the affected area (Bissell 1983, Hederra 1987).

Leptospirosis broke out in the wake of the 1975 floods in Recife, Brazil, and yellow fever in 1966 in Argentina. Floods have also caused the typhoid incidence rate to increase by 50 per cent and that of scarlet fever to double (Beinin 1979). More dramatic rates of increase in gastrointestinal complaints have been observed repeatedly after floods and cyclones in Latin America and the Indian subcontinent, always as a result of microbiological pollution of faecal origin. In Bolivia, for instance, flooding related to El Niño in 1983 led to 70 per cent increases in the incidence rate of salmonellosis diseases (to 60 per 1,000), and the under-fives were the principal group to be affected (Telleria 1986).

Floods may disrupt the habitat of fauna to the extent that medical problems arise, or are worsened, in human populations. For example, African and Latin American epidemiological data suggest increased incidence of snake bite. Small mammal ecosystems may be disrupted and, although some rodents may drown, others will escape and perhaps cause increases in metaxemic diseases in humans. The vectors and reservoirs of diseases such as haemorrhagic fever may thus be altered by flooding. Leptospirosis, which has achieved worldwide distribution, is carried by rodents, dogs, pigs and cattle and can be transmitted in contaminated water during floods. In tropical countries, the more humid conditions associated with persistent or slowly subsiding floodwaters may stimulate the breeding of anthropophilic household mosquitoes and lead to malarial infestations (Moreira Cedeño 1986). The El Niño floods of 1983 caused such outbreaks in rural Ecuador (Hederra 1987), while in Haiti flooding caused by a cyclone in 1963 resulted in a thirteen-fold increase in malaria, especially among babies under the age of one.

Analyses of agricultural data indicate that the widespread destruction of crops, food stocks and food distribution systems that major floods and cyclones cause may lead to food shortages and malnutrition, especially among infants and in the poorest, most isolated sectors of the community. After Hurricanes David and Frederick in Dominica in 1979, protein-energy malnutrition was widespread, especially among the under-fives, as evidenced by weight-for-age data (Wit & Gooder 1981). Stunting, emaciation, scabies, diarrhoea, dehydration and low resistance to disease could all result from this, but the distribution of adequate rations tends to reduce the problem considerably. Bolivian data indicate that after floods the nutritional status of survivors varies directly with the quality of health care and volume of assistance received locally.

In some cases, the countries most affected by flood impacts also have high prevalence rates of endemic disease, poor sanitation and health care, repeated shortages of food and overcrowding among the most vulnerable groups. For instance, flooding in 1980 in Pakistan led 60–75 per cent of the survivors to fall sick with diseases such as cholera, malaria, dysentery and pneumonia. About 5 per cent of the average family budget was spent on treating ailments provoked or exacerbated by the floods (Sikander 1983). In the Department of Piura (Perú), floods related to El Niño in 1983 led to extremely high infant morbidity and mortality. The prevalence rates per thousand for new-born babies were, respectively, 400 and 18 for respiratory diseases, and 300 and 21 for gastrointestinal diseases. These figures represent increases of 200–270 per cent over pre-flood figures (Gueri et al. 1986).

Experience indicates that most medical and sanitation problems related to flooding can be foreseen, if only there are sufficient resources for study and mitigation (Seaman et al. 1984). For example, it is unlikely that a disease which is not endemic to the affected area will reach epidemic proportions unless, that is, it happens to be strongly prevalent in relief workers from outside the area who have prolonged contact with survivors. The primary requirement, as always, is for data and information: on population characteristics, migration patterns, housing conditions, the prevailing pathology, basic sanitation conditions, disease vectors and available resources. In the case of cyclones, the initial aftermath may call for sanitation workers and paramedical personnel to a greater extent than doctors, and also for well co-ordinated relief initiatives. Hence it may be necessary to rescue survivors, bury or cremate the dead, treat cases of shock or exposure, immunize groups of people (if appropriate), provide drinking water, clear and clean up wells and stop faecal contamination. Particular attention may need to be paid to food, water and medical supplies, housing and sanitation in small, isolated villages. Epidemiological surveillance and control may later become vital factors if the morbidity profile of survivor groups fails to stabilize.

In summary, flood and cyclone disasters tend to differ from earthquake and landslide catastrophes in the sense that disease transmission is a much more probable consequence when large populations in developing countries are involved.

Consequences of floods for agriculture

Major flooding, especially that associated with cyclones, can lead to serious destruction in the agricultural sector of the economy, which may be the principal source of domestic and foreign earnings. High winds, torrential rains and the storm surge associated with cyclones that make landfall in coastal agricultural areas may have a variety of effects on plants (Hammerton et al. 1984; see Fig. 8.7), including defoliation and stripping, snapping of limbs, trunks and stems, bending and whipping back of stems, and lifting and exposure of roots, especially where soils and root systems are shallow. Heavy rain exerts a beating action that may break off leaves and shred crops, or spatter them with mud. The effect on vegetable crops can be severe. Falling branches and trees may damage or smother understoreys of lower-growing crops, while flooding that lasts for more than one or two days may kill crops and perhaps cover their remains with deposited sediment.

Terraces, dykes, channels and irrigation systems may be washed out by overflowing rivers or by a coastal storm surge. Paddy systems may be destroyed and soil damaged by rill, gully or landslide erosion. Positive effects in the form of soil enrichment by sedimentation are, of course, possible, but they usually only produce a net benefit if planned for in advanced – that is, if the local agricultural system is adapted to their occurrence. Otherwise, they tend to be outweighed by the destruction caused by such flooding. Moreover, land flooded by seawater may become excessively saline and hence toxic to crops. Finally, although livestock may be crushed or hit by falling trees or debris from buildings, most animals are lost by drowning. If the animals are not killed, herds or flocks may be dispersed during the impact phase.

In essence, the effect of floods on agriculture may be summarized as: crop and livestock loss; seed, root stock or breeding stock loss; income and employment loss; and loss of production potential. Although fruits, tubers and root crops can often partially be salvaged after a cyclone, persistent and widespread flooding tends to cause the total loss of vegetables and pulses. In traditional economies of either a cash or subsistence kind, the loss of crops may lead to significant reductions in a household's resource base for a long period. Moreover, the resources of

Figure 8.7 In 1985 Hurricanes Eric and Nigel passed through the Fijian Islands and caused extensive damage to pine forests. At this site on the island of Viti Levu, high winds uprooted trees and torrential rains caused extensive landsliding.

individual farmers may have to be devoted to procuring food and repairing houses, rather than to purchasing agricultural items or paying labourers to work in the fields. Growers' associations may lack adequate cash balances to carry them through to the return of profitability.

Hence, although land reclamation or structural reorganization of the agricultural sector may be vital responses to flood damage, cash aid is often the principal requirement at all levels of the economy. In many cases, small country economies that are dependent on agriculture for subsistence and exports cannot survive the devastation caused by hurricane-force winds and associated flooding without considerable aid. However, not all the effects of monetary aid are positive. Chung (1987) noted that in Fiji after cyclones they could be subtly debilitating:

As more relief is provided, dependency increases and the efforts and commitments to rehabilitation decrease. The direct result is the increased vulnerability of the community because traditional coping mechanisms are often conveniently ignored. Relief measures, instead of complementing rehabilitation by enabling farmers to get on with the task of recovery, often undermine recovery by competing for the scarcest of resources, that is, money.

In any case, whatever the seriousness of their plight, farmers will not usually be able to wait for assistance more than one or two months after the impact, so agricultural reconstruction will have to be planned and financed rapidly if it is to succeed.

The drawing up of agricultural rehabilitation plans, and the forwarding of requests for international aid to put them into effect, must be preceded by a field survey of losses. Some problems impede such endeavours (Hammerton et al. 1984). First, roads, telephone lines and vehicles may have been damaged, thus hampering movement and communication. Secondly, it is often difficult to estimate crop losses if planted areas are fragmentary. Thirdly, visual estimates of losses, assumptions about fruiting potential and estimates of plant physiological damage may be incorrect. For example, physiological damage to sugar cane in Fiji after Hurricanes Eric and Nigel in 1985 was not detected until harvests were reduced by a value of $7.5 million (Chung 1987).

Agricultural rehabilitation may conveniently be divided into short- and long-term phases. In the short term, given adequate seed distribution, quick-maturing crops may be cultivated in order to achieve local self-sufficiency in food and supplement relief supplies. If the political will and bureaucratic organization are sufficient, the long term may involve improvements in rural infrastructure and agricultural support systems. More fundamentally, much thought should be given to the dominant cultivars. "Hurricane-resistant" crops may be those which possess the following attributes: a short production cycle, low growth stature, amenability to simple storage or to being kept in the ground, high protein and energy concentrations, and acceptability to both the producer and the consumer. In general, the agricultural destruction caused by the disaster may present an opportunity to replant greater areas of cash and subsistence crops, replace less economical crops and introduce high-yield, low-vulnerability crop strains. Moreover, floods and hurricanes usually exhibit high seasonality, and the main cycles of cultivation can perhaps be timed not to coincide with them.

Despite these prospects, there is a caveat emptor to agricultural reconstruction. In many developing nations droughts and food shortages are as frequent as floods and hurricanes (Brammer 1987). Floodplain soils tend to be eminently cultivable when not under water. To the poor villager, the food production involved in risking crop loss by tempest or inundation may represent the key to survival and improved standards of living. This makes it difficult to ascertain the net costs of restricting floodplain use.

Impact of floods on housing

Major shanty towns that are susceptible to floods exist in many of the Third World's principal metropolises, including Calcutta and Delhi (India), Dhaka (Bangladesh), Guayaquil (Ecuador), Lagos (Nigeria),

Manila (Philippines), Monrovia (Liberia), Port Moresby (Papua New Guinea), Recife (Brazil) and San Juan (Puerto Rico). In addition, flash floods are a major problem for the poor in cities such as Rio De Janeiro (Brazil) and Port-au-Prince (Haiti).

The effect of floods and hurricanes on housing may be profound. Although few studies are available that give details, certain generalizations have been obtained, and by way of example four of them are reported here (Hughes 1982). First, while all buildings decay, the rate varies with the materials used in their construction and the degree of periodic maintenance applied to them. Frequent flooding can accelerate the decay process by giving rise to salt efflorescence, wet rot or other damaging phenomena. Secondly, a wooden building will float upright because its centre of buoyancy is above its centre of gravity. Thirdly, the weight of concrete below water is reduced by half, and hence concrete strip foundations are more easily displaced. Finally, air pockets trapped in any cavity of a building may lead to devastating bursts if compressed. Careful study of building performance in floods, and of acceptable local architectural traditions, could lead to the correction of most deficiencies such as these.

Evidence of storm surge impacts in the Dominican Republic (Jeffery 1982) and in Orissa, India (Mishra & Prakash 1982), indicates that the reconstruction of traditional dwellings may be rapid and may require relatively little government assistance if local resources (such as wood) are available. In part, this is because little differentiation of labour is involved in building in very traditional societies. It has been argued, however, that many inhabitants of traditional vernacular housing would much rather move into more technologically advanced buildings which are less cheap and easy to reconstruct, but which involve greater levels of protection, status and comfort. In any case, in countries such as Pakistan, flood survivors of all social ranks commonly have to spend perhaps 10 per cent of family income on housing reconstruction (Sikander 1983). Furthermore, the massive need for housing generated by such disasters means that government or relief agency assistance may be limited to furnishing materials and expertise, or to selective rebuilding.

Certain housing situations in developing countries pose particularly intractable problems. The *barrios* of Latin America are precarious settlements commonly located on steep hillsides or in narrow ravines (i.e. on land of low value). They are often afflicted by landslides and flooding, yet the dwellings are unlikely to have any form of resistance to disaster impact, and the population densities in such settlements are usually higher than those of ordinary cities and towns. Analogous conditions exist in the poorer riverine, coastal or estuarine settlements of many developing countries. The essence of the problem is poverty and the unplanned migration of poor or dispossessed people (Jeffery 1982).

Conclusion

As stated at the beginning of this chapter, a population that suffers restriction or reduction in the control of its basic resources of livelihood can be considered as marginalized. There are several ways in which floods may worsen the process of marginalization among the poor of the Third World. Absence of funds, or of their adequate local distribution, may inhibit the reconstruction of agriculture, housing or infrastructure, hence leading to grinding poverty. The devastation caused by inundation or high winds may be used by governments as an opportunity to reorganize and redevelop areas in a way that excludes or oppresses the poor (in this context, given the concentration of power in very few hands in certain developing countries, state development may mean little more than institutionalized private capitalism). And the damage to health care and sanitation facilities, or the simple lack of them, may lead to debilitation of the affected population and thus retard economic recovery even further. Yet, as the analyses reported above suggest, each of these problems is solvable if sufficient aid and satisfactory mitigation planning are applied to the disaster area. But, as always, the world's most vulnerable nations tend to be the most populous and least wealthy.

The case of Bangladesh

In order to gain an understanding of the complexities of multiple hazards, and their relation to poverty and economic development, we focus on Bangladesh, as in a world context it is probably the nation most affected by natural catastrophe, especially floods. From 1960–81 it suffered 63 disasters, with the loss of 655,000 lives. Of these events, 37 were tropical cyclones, which killed 386,200 people (Murty et al. 1986). Whatever their origin, natural catastrophes are repetitive in Bangladesh: for example, less than a year after the devastating 1988 floods, on 26 April 1989, a tornado struck the area of Manikganj, a few tens of kilometers west of the capital, Dhaka, and caused 800–1000 deaths; serious floods occurred in 1990 and then in May 1991 sea surges associated with a series of cyclones killed about 140,000 inhabitants on the Bay of Bengal coasts.

Extreme poverty and demographic pressure exacerbate the suffering caused by such events and increase vulnerability to future disasters. As in the poorer parts of Africa, famine is an ever-present menace (it is estimated that 30,000 people died in the food shortages that followed the 1974 flood). Moreover, floods, droughts and cyclonic sea surges undermine efforts to develop Bangladesh: hence, in the year 1988–9, 45 per cent of the country's development budget was used to pay for their impacts.

The following account draws considerable inspiration from the seminal works of Brammer (1987, 1990a & b) and Khalil (1990). A comprehensive geological account of the Brahmaputra river system can be found in a paper by Coleman (1969) and a critique of current plans for training the rivers of Bangladesh appears in an article by Boyce (1990). Remote sensing of floods has been described by Ali et al. (1987, 1989) and Rasid & Pramanik (1990).

Physical geography of Bangladesh
Bangladesh (Fig. 8.8) occupies 144,836 km^2 in the northwest part of southern Asia and consists of a fertile deltic floodplain surrounded by uplands which in the Chittagong Hill Tracts reach a maximum of 1,230 m above sea level. Of the national territory, floodplains occupy about 80 per cent, hills cover 12 per cent and terraces comprise the remaining 8 per cent. Most of the plains are at risk from flooding during the monsoon season.

The climate of Bangladesh is warm and humid, with annual average temperatures in the range 24.4–26.7°C, but subject to fluctuations. Annual rainfall is greater than 5,000 mm in the northeast: in the northwest and southeast of the country it reaches 2,500 mm, while in the centre and west it is only 1,250 mm. There are three seasons, which are separated by periods of transition: winter (November–February), summer

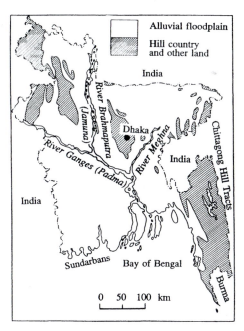

Figure 8.8 Bangladesh and its rivers.

(March–May) and monsoon (June–October). Rainfall totals in excess of 200 mm occur for five successive months in the west and 7 months in the northeast, while in the extreme northeast 5 months have more than 500 mm. However, the date at which the rainy season begins varies considerably from year to year, especially in the west. Concerning the regional water balance, the pre-*kharif* period is that in which soil moisture resulting from rainfall exceeds half potential evapotranspiration (PET). The length of this period varies from more than 60 days in the southwest to less than 20 days in the northeast, but the total varies from year to year. The *kharif* period is that in which soil moisture exceeds 100 per cent of PET, and varies from fewer than 180 days in the west to more than 280 days in the northeast. The drought season is known as *rabi*.

Human geography and hazard vulnerability

Culturally, Bangladesh comprises the eastern part of Bengal. It was ruled by the Indian Mughal dynasties from 1608–1757 and then by the British until 1947, while after the Partition it became East Pakistan, under the jurisdiction of West Pakistan. The War of Liberation lasted for 9 months during 1971, cost 3 million lives and created 10 million refugees, but it brought independence for the first time in many centuries. Since then, martial law has frequently been imposed in order to quell political violence stimulated by (among other factors) electoral malpractice; but the 1991 elections appear to have been legitimate.

The population of Bangladesh, which is predominantly Muslim, is currently about 117 million, with a density of 800–850 persons per km^2, one of the highest national figures in the world. Conservative estimates suggest that the total number of inhabitants will rise to 140 million by the year 2000 and 250 million by AD 2050, which makes the country a veritable "demographic bomb". Although the government is trying to reduce the average birth rate from its current 2.6 per cent to 1.8 per cent, the present rate of urban growth (8 per cent) is nearly twice the Third World average of 4.5 per cent.

The capital, Dhaka, was founded in 1608 by the Mughal emperors of Delhi. It currently occupies 414 km^2 but, according to its urban plan, is destined to expand to 777 km^2. The population is about 5–6 million and from 1961 to 1974 it grew by 212 per cent. Given that there is no sign of any slowing in its growth, by 2000 Dhaka will probably have 10 million inhabitants. At present 85 per cent of the population of the city lives below the poverty line: half of these people are unable to satisfy more than 80 per cent of their basic nutritional needs. Many of them are squatters, about one third of whom migrated to the city because poor economic conditions had made them landless or destitute, and one quarter of whom were driven there by the impact of floods, hurricanes or famines. Dhaka suffers from lack of quantity, quality and equity in all its

534

urban services. For example, to reach a minimum acceptable educational standard it needs at least 1,800 schools and 1,100 institutes. Bangladesh, in fact, is among the world's five poorest nations, with an annual per capita income of only US$170 and an illiteracy rate of 76 per cent.

Despite the furious pace of urban growth, 85 per cent of the population lives in the country's 68,000 rural villages (Fig. 8.9). After the administrative reforms of the 1980s the nation was divided into four divisions, 64 districts and 460 *upazilas*. The latter are associations of municipalities which are empowered to devise autonomous local development plans. They therefore have considerable potential as a base for civil defence planning, including flood protection measures.

Agriculture is clearly the dominant economic activity of Bangladesh. It provides 46 per cent of GDP, 73 per cent of jobs and more than 80 per cent of export earnings (industry contributes only 9.5 per cent to GNP). A total of 12.4 million hectares of land is cultivated, 80 per cent in rice, 6 per cent in jute and 4 per cent in wheat. A law of 1987 facilitated the redistribution of land owned by the government to those who worked it, but in overall terms, land reform has succeeded in shifting neither the social nor the economic balance of the country, nor has it reduced the vulnerability to natural catastrophe of farmers.

The rivers
Bangladesh is little more than a great river delta or alluvial floodplain. It contains about 250 perennial rivers, of which 56 originate outside the country – in Tibet, India, Bhutan and Nepal. Only 7.5 per cent of the total catchment area falls inside Bangladesh and 90 per cent of water

Figure 8.9 Bengali fishing village on the banks of the River Ganges. The houses are located barely one metre above the mean water level and flooding is an annual occurrence.

discharge (1,360,000 million m^3 per annum) originates outside. From this it is clear that hydrological regulation in Bangladesh is a truly international problem. The principal rivers (see Fig. 8.8) are:

(a) the Ganges, whose lower course is known as the Padma: basin area 1,113,700 km^2, length 2,478 km, average maximum discharge 299,000 cumecs.
(b) the Brahmaputra, whose lower reaches are known as the Jamuna: basin area 934,990 km^2, length 2,900 km, average maximum discharge 317,000 cumecs.
(c) the Meghna, an easterly affluent of the Padma–Jamuna complex: length 800 km, half of which lies in Bangladesh.

The drainage basins of these rivers occupy about 2 million km^2, while the area of delta is 60,000 km^2. In front of the latter is the world's largest sedimentary structure, the Bengal Fan, which extends 3,000 km to the abyssal plain off the Sri Lankan coast and is 1,500–2,000 m thick. During the construction of the delta, the rivers have changed course on many occasions. For example, the present-day Tista derived from the ancient Atrai in the late eighteenth century, while the lower Ganges (Padma) emerged in the early nineteenth century from the Arian Khan.

The annual sediment discharge of all rivers in Bangladesh is about 2,500 million tonnes, which is equivalent to 1,000 million m^3. It is not known how much of this material reaches the sea, but it is clear that continuous erosion and deposition make the Ganges–Brahmaputra delta an extremely mobile landform. In fact, lateral shifts of more than 800 m per year have been observed in the course of the Brahmaputra River. Its bedforms are among the largest and most mobile in the world: point bars more than a kilometre long have migrated hundreds of metres downstream in a single day during flood stage. Another cause of river channel changes in Bangladesh is seismicity. Many cases of liquefaction occurred in the saturated sands of the delta during the catastrophic earthquake of 1757, which also caused the unification of the Padma and Jamuna Rivers, that until then had flowed separately into the Bay of Bengal. In addition, the 1950 Assam earthquake altered the hydrography of the Brahmaputra such that bed levels were raised in its upper reaches, channel capacity was correspondingly reduced and overbank flooding thus became more common.

Relationship between floodplains, floods and agriculture
The cultivation of crops in Bangladesh tends to be finely attuned to differences in climate and soils, such that damage results mainly from floods that are unexpected or particularly intense, and not from normal seasonal flooding. Three main varieties of rice are grown (Fig. 8.10): *aus*

Figure 8.10 Rice paddies and nucleated villages in rural Bangladesh. Settlement is often located on slight rises in elevation, such as those related to former point bars, or between abandoned meander loops.

and *aman* are mainly cultivated during the wet season (when 80 per cent of the total annual precipitation falls), while *boro* is principally raised by irrigating paddies during the dry season. Floods may uproot crops or damage the plants, alluvium may bury them and river banks may erode into cultivated land. The most damaging floods occur in June and August–September, i.e. early or late in the *kharif*. Although there may be substantial losses, *aman* crops tend paradoxically to be large in flood years, in part because areas which escape severe inundation are stimulated both biologically and economically to increase production.

Bangladeshi floodplains are divided into six physiographical categories (Brammer 1990a):

(a) **Piedmont plains** comprise the steepest parts of the floodplains and are rapidly inundated during the rainy season. Their soils are sandy or clayey, with greater proportions of clay in the basin floors.

(b) **Active floodplains** show irregular topography as a result of rapid changes in the position of channels during flood stages. Their soils are sandy and silty, and rapid overbank spillage of rivers has a strong impact during the wet season.

(c) **River meander floodplains** have a topography that consists of crests and basins, with an altitudinal difference of no more than 5 m. The soils of the basins are sandy, while those of the crests are more sandy–silty. Except where rivers burst their banks, floods here result mainly from the direct effect of rainfall, and the depth of flooding varies in relation to local topography.

(d) **Large floodplain basins** are clayey and have river banks 2–5 m above their bottoms. Flooding mainly results from rainfall, but in some areas is also caused by rapid overbank spillage from rivers. During the dry season the centres of the basins do not dry out entirely.

(e) **Estuarine floodplains** have a height difference of no more than 2 m. Flooding here is caused by river spillage, intense rainfall or the surge associated with cyclones arriving from the Bay of Bengal. During the dry season the soils of these areas become saline.

(f) **Coastal floodplains** are almost without height differences and become flooded during high tides or cyclones via the many channels and estuaries which they contain. The tidal range in Bangladesh is 5.6 m. In the southwest of the area mangrove coasts can be found (the Sunderbans).

The difference in altitude between floodplain crests and bottoms is generally 1 m on coastal floodplains, 1–2 m on estuarine floodplains and 2–5 m on the three types of river floodplain.

Six altitudinal zones have been defined on the basis of varied flooding propensity, and each has a characteristic potential for cultivation (Brammer 1990a: 15):

(a) **High land** exists above normal flood levels. Where the terrain is permeable, cereals, pulses, jute, fruit, vegetables and spices can be cultivated, whereas on impermeable land various kinds of rice are cultivated.

(b) **Medium-high land** is flooded to about 90 cm during the monsoon season. Its propensity for cultivation is that of high land.

(c) **Medium-low land** is flooded to 90–180 cm during the monsoon season. Rice is cultivated here under irrigation; jute and senape are cultivated on a local scale.

(d) **Low land** is normally flooded 180–300 cm during the monsoon. Aman rice, irrigated boro rice and pulses are cultivated locally.

(e) **Very low land** is normally flooded more than 3 m during the monsoon. Boro rice is cultivated here.

(f) **Bottom land** is located between the other categories of terrain; this land does not entirely desiccate during the dry season. Boro rice is cultivated here.

Causes and consequences of flood disasters

Floods in Bangladesh consist of those that are **seasonal** (i.e. normal for the season in which they occur), and those that are **contingent** (i.e. unexpected and generally damaging). The second category comprises four

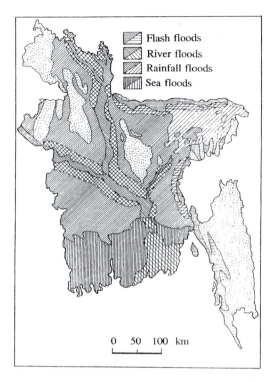

Flash floods
River floods
Rainfall floods
Sea floods

0   50   100   km

Figure 8.11   Distribution of types of flooding in Bangladesh (data from Brammer 1990a: 15–17).

types (Brammer 1990a; see Fig. 8.11). The first type is **flash floods**, which are caused by heavy monsoon or pre-monsoon rains that fall on the mountains and hills adjacent to the floodplains. They carry a heavy sediment load and hence distribute sediment irregularly over the land and in channels, thus raising the beds of rivers. Flash floods occurring in April and May can damage boro rice paddies located on basin floors.

River floods occur most easily around present-day river channels in response to particular rainfall events. When the banks are breached (as often happens on the right bank of the Brahmaputra) these floods deposit large quantities of sediment on the adjacent plains. They occur between May and September mainly as a result of heavy regional storms or the melting of the Himalayan snowpacks. They can, however, be aggravated by intense local rainstorms. Floods of this type occurring between July and the beginning of August can help the growth of Aman rice, but those occurring in June tend to damage immature crops, while those taking place at the end of August or in September damage Aman rice in the flowering part of its cycle.

Rainfall floods are an effect of localized precipitation. During the monsoon season they occur rapidly on low-lying areas and have a damaging impact on the production of jute and aus rice. Their effect on Aman rice is the same as that of river floods. Lastly, **cyclonic (sea) flooding** occurs when cyclones make landfall in Bangladesh from the Bay of Bengal. The friction of wind on the surface of the sea causes a storm surge to move inland.

Soils develop rapidly in Bangladesh, and sediments of varying age contribute to a wide variety of pedological catenas. The impact of floods on soil fertility is not an effect of sedimentary deposition (Fig. 8.12). In the first place, most floodplains tend to be inundated by rainwater rather than river water. Their soils tend to be acid (pH 4.5–5.5), whereas those of river floodplains are alkaline (pH 7.0–8.4), above all after the calcareous floods of the Ganges. In any case, recent flood sediments contain little organic material and little of the phosphorus and nitrogen which plants require as nutrients. Instead, the increase in fertility caused by flooding is derived partly from algae (including those that fix nitrogen) that grow on plants and on the surfaces of soil particles. It also results from decomposition of vegetal material killed by the flooding, and from oscillation between reducing and oxidizing conditions in the soil as a result of the periodic flooding, which helps break down minerals and release nutrients.

Problems of soil fertility are sometimes aggravated by schemes designed to combat the effect of floods. These projects tend to use three strategies: building dykes or river levées with drainage of water from the floodplains by tidal action or by pumping; building banks and dykes without drainage of water from the protected basins; or building banks and dykes with sluices which permit controlled flooding of the protected basins. The first of these methods precludes the natural refertilization of the basins by periodic inundation. Thus the soil, especially that of slightly higher land, loses its fertility rapidly. In each case it is necessary to design ways of recycling organic matter to soils impoverished by cultivation or flood regulation.

It has been suggested that deforestation in Assam and Nepal has increased by 60 per cent the sediment load of the international rivers that reach Bangladesh (soil loss in the Himalayas averages about 60 cm/1,000 years). This may have caused river bed levels to rise about 5–7 m, thus decreasing bankfull stage and increasing their flooding propensity. In fact, a study of the 125 km of the Ganges located between the Indian border of Bangladesh and the confluence with the Brahmaputra suggests that mean bed levels have risen in upper reaches and fallen in the lower ones (Bangladesh Research Bureau 1989).

This view has not gone unchallenged. For example, Ives & Messerli (1989) argued that there is no direct evidence that Himalayan deforestation has resulted in increased discharges of water and sediment along the

Figure 8.12 Sand deposits on agricultural land after the 1988 Bangladesh floods. The fertility of this land has been diminished, rather than increased, by the overwash.

major rivers that drain them. Moreover, the rising bed level and increased tendency to flood of the lower Ganges may be a consequence of the Farakka Barrage (in western India), which has made the downstream flow regime more extreme. But whatever its consequences, deforestation is an established fact in the subcontinent: over the period 1961–81 forest cover in Nepal was reduced from 82,000 km² to 31,000 km², while in recent decades that of Bangladesh has declined from up to 23,000 km² to little more than 12,000 km². Moreover, for the period 1976–85 annual deforestation rates exceeded 2,400 km² collectively in Bangladesh, Bhutan, India and Nepal. There is little doubt that the runoff coefficient increases on deforested lands, as do the slope and discharge of channels that drain them.

Tropical cyclones and floods
Particular attention deserves to be given to the fourth cause of flooding in Bangladesh, the landfall and sea surge of tropical cyclones. The first written account of a cyclone disaster is dated 1584. Though historical records are incomplete, in present times an average of 16 cyclones occur in the Bay of Bengal each year, with particular concentration in two periods: April–May and October–December. As it is situated almost at sea level, the Brahmaputra–Ganges delta is particularly vulnerable to sea surges associated with the passage of cyclones. In fact, the entire coastal plain of Bangladesh is at risk, and especially the unstable sand banks known as *chars*, on which population densities have become high during the twentieth century.

The cyclone of 12–13 November 1970 caused a storm surge 7 m high. On the coastal plains at least 224,000 people (16.5 per cent of the local population) died, 85 per cent of families were left without shelter and the total of homeless was 600,000. Two thirds of fishing activities were destroyed, along with 125,000 animals and 127,000 ploughs. The World Bank supplied US$185 million to finance a reconstruction plan.

The shortage of land in Bangladesh was such that survivors reoccupied the disaster zone within two months of the event. After the catastrophe the Bangladeshi Government initiated measures of structural defence without paying sufficient attention to safeguarding agriculture, to the need for alarm systems and to settlement patterns. As structural measures are unable to reduce the risks to zero, the construction of embankments and dykes may in the end increase, rather than decrease, the risk of flood damage. In fact, despite some notable advances in community protection planning, the destruction was remarkably similar in the cyclone of 29 April 1991, which involved a storm surge 6–9 m high.

Whereas maximum wind speeds reached 185 km/hr in the 1970 cyclone, they exceeded 240 km/hr in the 1991 event, which had a central pressure of only 938 mb. The latter killed an estimated 150,000–200,000 coastal villagers, directly affected 15 million residents of 74 *upazilas* and did huge amounts of damage to the port of Chittagong. Some 281,000 tonnes of rice and other crops were destroyed and nearly half a million head of livestock perished. The cost of damage was estimated at US$1,385 million. Relief workers have argued that more than 2,000 multipurpose cyclone shelters are required in the area affected by the 1991 cyclone.

Impact of the "greenhouse effect"

Apart from the usual effect of floods, the so-called "greenhouse effect" could cause even greater disasters in Bangladesh. The possible impacts include: rise in sea level, together with permanent inundation of coastal zones and floodplains; rise in the mean atmospheric temperature; intensification of climate (rains, floods and droughts); and migration inland of the salt water front. The level of the sea, in fact, already rises an average of 60 cm during the period of monsoon winds, which blow from the southwest and the Bay of Bengal, impeding rapid drainage of the rivers of Bangladesh.

But the "greenhouse effect" is unlikely to have a simple impact on Bangladesh. In the first place, the rise in sea level that it is likely to provoke will be slow compared to other environmental changes in the area. Moreover, because the physical geography and geomorphology of the region are very complex, the rise in sea level could have different effects from place to place. Finally, from one year to another the local climatic, hydrological, agricultural and geomorphological variations tend to be larger than the forecast local impact of the "greenhouse effect". The demographic and environmental dynamism of Bangladesh could influence strongly the degree to which marine transgression occurs.

The alternation of erosion and deposition of sediment on river banks causes dynamic morphological changes which are difficult to predict without the aid of detailed monitoring. Strong earthquakes in the surrounding mountainous areas may increase the solid load of rivers, which could accelerate their already rapid rates of lateral migration. In addition, crustal movements caused by active tectonics could raise up new areas of land (as occurred in 1897), while the Sylhet Basin in northeast Bangladesh is currently undergoing subsidence as a result of consolidation among the sediments of which it is formed. Subsidence may explain why the delta of the Meghna and Jamuna Rivers has grown remarkably little during the last 200 years, despite enormous recent increases in the solid load of the rivers that rise outside the national territory (Milliman et al. 1989).

Although the true impact is difficult to predict, it is likely that the "greenhouse effect" will cause the salt-water front to move upstream, although it will be difficult to distinguish this process from the changes caused by the growth of irrigation, which will deplete the fresh water of the low floodplains. Finally, in coastal areas the deposition of fluvial sediment and similar phenomena may in the end counteract the incursions of the sea, such that the "greenhouse effect" is limited there. The principal impact will probably occur upstream of the coastal plains – given that the water level of rivers will be increased – such that they will have a greater propensity to break their banks. However, an accurate assessment of the impact of the "greenhouse effect" will require continued and detailed monitoring of a wide range of environmental variables.

The floods of September 1988

In an average year riverine and rainwater floods cover about 20 per cent of the land area of Bangladesh, while in an exceptional year 30–40 per cent is inundated. The incidence of floods, and the damage which they cause is, however, irregular (Fig. 8.13). In September 1988 58 per cent of the nation was flooded, which is significantly more than in previous great floods, such as those of 1954, 1955, 1974 and 1987.

Most of the annual rains fall in the period between June and the middle of October, while the thaw of snows in the Himalayas takes place from May to June, causing an increase in rainfall. Normally, the River Meghna floods between May and June, the Brahmaputra between June and

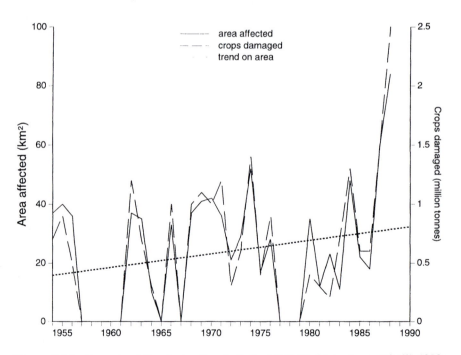

Figure 8.13 Impact of floods in Bangladesh, 1954–88 (data from Khalil 1990: 386).

August and the Ganges between July and September. But in 1988 the time difference between flood peaks was reduced to 3 days, and hence the rivers burst their banks for 13–15 days consecutively. From 20 August to 1 September most of the Brahmaputra Basin, and on occasion also the Meghna and Ganges Basins, was covered by cumulo-nimbus clouds, indicating heavy rains. After 2–3 days of continuous rain a moderate flood was predicted, but after 12–13 days the situation had become very serious indeed (Rasid & Pramanik 1990). As the Moon was at perigee, the flood tide of 29 August was high and hence the mouths of the great rivers were obstructed by seawater.

Table 8.4 compares the impacts of the 1987 and 1988 floods. Although water levels were very similar, the 1987 floods were caused largely by rain falling on Bangladesh itself, while the 1988 inundation resulted from rainfall across much of the subcontinent and hence from rising river stages. In one of the more usual floods the impact would occur over 3–4 days. However, in 1988, 122,000 km^2 of land were inundated in less than

Table 8.4 Comparison of the impacts of the 1987 and 1988 floods in Bangladesh (Brammer 1990a: 18).

	1987	1988
Area flooded (km^2)	57,000	84,059
Peak flood level at Serajganj (Brahmaputra) (m)	14.57	15.12
Peak flood level at Hardinge Bridge (Ganges) (m)	14.80	14.87
Lag time of flood peak between Hardinge Bridge and Serajgang (days)	34	3
Deaths	1,657	2,379 (1,797)*
Rice production lost (million tonnes)	3.5	2.0
Crop damage (million ha)	†	7.16
Head of cattle lost	64,700	172,000
Poultry lost	206,000	410,000
Houses destroyed or damaged (millions)	2.5	7.2
Hospitals flooded	†	45
Clinics flooded	†	1,400
Schools flooded	†	8,481
Industrial units damaged	†	14,000
Trunk roads damaged (km)	1,523	2,935
Rural roads damaged (km)	15,107	65,892
Road bridges damaged	1,102	898
Railway lines damaged (km)	†	638
Railway bridges damaged	†	34
Levées damaged (km)	1,279	1,990
Drainage canals damaged	222	283
Overal cost (million US$)	1,211.7	1,138.1

* Also given as 1,410 drownings and 387 deaths caused by diarrhoea.
† Data not available.

48 hours. Flooding in two thirds of Dhaka left 30 million people homeless, while a further 20 million were seriously affected. The Dhaka–Narayanganj–Demra embankment, which was constructed in the 1960s, threatened to collapse when it came under the duress of the floodwaters, thus putting at risk the 500,000 inhabitants of the land southeast and west of the capital. This illustrates the risk of placing too much reliance on purely structural methods of flood defence.

Floods tend to affect the various social classes in different ways, according to differences in their incomes, possessions and savings (Zaman 1991). They tend in particular to accentuate the process of impoverishment of the most vulnerable groups. For example, surveys were conducted on the nutritional status of babies aged 6 to 35 months, who survived the 1988 floods. They indicated that the most undernourished babies came from families that had no land to cultivate and were therefore unable to store large quantities of rice at home. Poverty was also strongly linked to disease. During the 1988 floods drinking water samples were obtained at 28 locations in Dhaka, and coliform bacteria were found in concentrations of 10^3–10^6/100 ml of water. Concentrations as high as these could be correlated directly with the incidence of diarrhoeal disease, especially in the poorer quarters of the city.

Flood defence measures

Without substantial international agreements, Bangladesh will have little opportunity to manage its hydrological resources and their associated risks. Until very recently only structural measures of flood defence were instigated. The Bangladesh Water Development Board has inaugurated eight large dam projects. Dredging is used selectively to remove fluvial sediments from congested water courses, but as the sediment transport and deposition rates are so high it is not considered capable of reducing the overall flood risk significantly.

Most structural measures proposed in order to reduce flooding in Bangladesh are controversial. Upstream flood control basins and methods of drawing down or diverting water reserves would require engineering works that are too large and costly to be practicable, as well as international co-operation on a scale of which there is currently no sign. It has been proposed that the World Bank and donor communities spend $6,000–10,000 million on a major levée system for the Ganges–Brahmaputra system (Boyce 1990). However, many risks would be created by raising a major, mobile-bed watercourse above the level of the surrounding land, as the Chinese have found to their cost in attempts to contain the Hwang Ho and Yangtse Rivers on their eastern floodplains.

Civil defence has been organized in Bangladesh only very recently, with the convening of a National Committee for the Prevention of Disasters, including two subcommittees of experts. Among the non-structural measures under consideration are the construction of a helicopter landing pad in every *upazila*, arrangements for storing food, local flood refuge sites and the reorganization of villages in more nucleated form on more elevated sites. One major priority is the motorization of rural transport, which may be achieved largely by adapting irrigation pumps for use as outboard motors on light river craft (an example of the application of intermediate technology).

It is deemed possible to predict floods in Bangladesh 2–3 weeks in advance of their occurrence if adequate attention is given to remote sensing of the condition of the Himalayan snowpacks, to meteorological satellite data and to monitoring of water and sediment discharges and climatic factors. Remote sensing, in fact, is now widely used in Bangladesh to map recent changes in the position of river courses and in the extent of tropical forests and mangrove swamps, to map flooded and floodable areas and to monitor meteorological and glaciological phenomena (Ali et al. 1987, 1989).

Bangladesh is one of the nations which has shown the greatest need for shelter after civil conflict and natural disaster. The international relief agencies have applied technology from the developed world to this problem. However, in general, both advanced and appropriate technology solutions have met with little success among the homeless. In 1972–3, after the war of liberation, the international agencies constructed 450,000 shelters, but were unable to adapt local building techniques sufficiently to solve the housing problem and reduce the vulnerability of the homeless. On the other hand, using traditional methods and materials, the local populations were able to build more than a million homes, without particular help from the international agencies, thus proving that indigenous solutions are possible (Davis 1978).

Obstacles to natural disaster mitigation in Bangladesh
Bangladesh has taken the initiative to create a scientific structure capable of predicting floods, using environmental monitoring and measuring techniques of all kinds, together with specially developed mathematical models. But the situation is often very complex and dynamic, involving the alternation of rain falling on Bangladesh and on neighbouring nations, the effect of rainfall on the melting of Himalayan snows and other forms of environmental feedback. Hence, although flood forecasting is now a national reality, not all situations are amenable to prediction.

For a nation as poor as Bangladesh, the principal obstacles to increased crop yields (and to the introduction of the "Green Revolution" of high-yielding varieties of rice and grain) are the uncertainty of risks and the cost of storing food stocks in preparation for shortages. On the basis of this, some economists have suggested that risk management, rather than engineering construction, should constitute the basis of flood mitigation strategies.

The government of Bangladesh argues that New Delhi has not yet responded positively to its initiatives in the field of international river management. The proposed measures tend to be blocked by the long debate over the respective merits of regionalism versus bilateralism. But multilateral river management requires co-operative efforts to collect physical, hydrological, ecological, economic, topographic and demographic data. A 1977 agreement established the Indo-Bangladesh Joint River Commission for the administration of the Ganges, and for data collection relative to that river. Bilateral agreements on flood defence have also been finalized between Bangladesh and Nepal, Bhutan and China, but they lack the funds and the political will to bring them to fruition.

Predictions concerning the future state of Bangladesh are not reassuring (Alam 1990, Ali 1990). The number of families was 14.8 million in 1984 and will be 23.2 million in the year 2000. The cultivable land per head of the population was 0.12 ha in 1984 (average farm size was 0.91 ha, subdivided into about 10 parcels), but if the greenhouse effect is coupled with the expected growth in population there could be only 0.065 ha per capita by 2010. The current rice deficit is 13 per cent, requiring an importation of 2 million tonnes per year. Yet there is little prospect of bringing more land into cultivation. By 2000 population growth will cause 130,000 ha of agricultural land to be urbanized. Moreover, some forecasters argue that the greenhouse effect will submerge 30 per cent of the nation under an average of 3 m of water. If these predictions are correct, food production in Bangladesh will fall by 30–40 per cent, with disastrous consequences for the fast-growing population.

Even though foodstocks must of necessity be obtained from abroad, imported victuals have tended to have a negative effect on the local economy. In the 1970s a high rate of importation of rice depressed the market, ruining those farmers who were not able to sell their produce on the black market in India. Moreover, food distribution has resulted in substantial foreign interference in the administrative functions of Bangladesh, without encouraging indigenous participation in the civil service (even though foreigners are unlikely to acquire the local knowledge that indigenous people possess). For these reasons it is now considered better to stimulate home production than to saturate the market with imported foodgrains.

In synthesis and despite recent events, as Figure 8.13 shows, there is no conclusive evidence that flooding is on the increase and food production on the decline in Bangladesh, despite the gloomy prognostications described above. Undoubtably, the past 40 years have been marked by increasing fluctuations in the occurrence of floods and the productivity of agriculture, but without a longer run of data it is impossible to say whether these represent a trend or part of a larger cycle.

References

Aall, C. 1979. Disastrous international relief failure: a report on Burmese refugees in Bangladesh from May to December 1978. *Disasters* **3**, 429–34.

Alam, M. 1990. Geotectonics and subsidence of the Ganges–Brahmaputra delta of Bangladesh and accompanying drainage, sedimentation and salinity problems. In *Sea level rise and coastal subsidence: problems and strategies*, J. D. Milliman & S. Sabhasri (eds). New York: John Wiley.

Ali, A. 1990. Impacts of greenhouse effects on floods in Bangladesh: some deductive physical reasonings. *Bangladesh Quest* **2**, 26–8.

Ali, A., D. A. Quadir, O. K. Huh 1987. Agricultural, hydrologic and oceanographic studies in Bangladesh with NOAA AVHRR data. *International Journal of Remote Sensing* **8**, 917–25.

Ali, A., D. A. Quadir, O. K. Huh 1989. Study of flood hydrology in Bangladesh with NOAA satellite AVHRR data. *International Journal of Remote Sensing* **10**, 1,873–91.

Bangladesh Research Bureau 1989. Bangladesh flood: regional and global perspectives. Conference abstracts.

Beinin, L. 1979. Sanitary consequences of inundations. *Disasters* **3**, 213–16.

Bissell, R. A. 1983. Delayed impact infectious disease after a natural disaster. *Journal of Emergency Medicine* **1**, 59–66.

Blaikie, P. 1985. *The political economy of soil erosion in developing countries*. Harlow, England: Longman.

Boyce, J. K. 1990. Birth of a megaproject: political economy of flood control in Bangladesh. *Environmental Management* **14**, 419–28.

Brammer, H. 1987. Drought in Bangladesh: lessons for planners and administrators. *Disasters* **11**, 21–9.

Brammer, H. 1990a. Floods in Bangladesh, I. Geographical background to the 1987 and 1988 floods. *Geographical Journal* **156**, 12–22.

Brammer, H. 1990b. Floods in Bangladesh, II. Flood mitigation and environmental aspects. *Geographical Journal* **156**, 158–65.

Chen, L. C., A. K. M. Chowdhury, S. L. Hoffman 1980. Anthropometric assessment of energy-protein malnutrition and subsequent risk of mortality among pre-school age children. *American Journal of Clinical Nutrition* **33**, 1,836–45.

Chung, J. 1987. Fiji, land of tropical cyclones and hurricanes: a case study of agricultural rehabilitation. *Disasters* **11**, 40–8.

Coleman, J. M. 1969. Brahmaputra River: channel processes and sedimentation. *Sedimentary Geology* **3**, 129–239.

COPAT 1981. *Bombs for breakfast*. London: Committee on Poverty and the Arms Trade.

Cuny, F. C. 1977. Refugee camps and camp planning: the state of the art. *Disasters* **1**, 125–44.

Cuny, F. C. 1983. *Disasters and development*. New York: Oxford University Press.

Davis, I. 1978. *Shelter after disaster*. Headington, Oxford: Oxford Polytechnic Press.

Davis, I. 1987. Safe shelter within unsafe cities. *Open House International* **12**, 5–15.

De Ville De Goyet, C. & M. F. Lechat 1976. Health aspects in natural disasters. *Tropical Doctor* **6**, 152–7.

De Ville De Goyet, C., J. Seaman, U. Geijer 1978. *The management of nutritional emergencies in large populations*. Geneva: World Health Organization.

De Waal, A. 1989a. Famine mortality: a case study of Darfur, Sudan, 1984–1985. *Population Studies* **43**, 5–24.

De Waal, A. 1989b. *Famine that kills*. Oxford: Oxford University Press.

Disasters 1981. Anthropometric tables. (Special issue on refugees.) *Disasters* **5**, 306–8.

D'Souza, F. 1981a. Who is a refugee? Definitions and assistance. *Disasters* **5**, 173–5.

D'Souza, F. 1981b. *The refugee dilemma: international recognition and acceptance*, 2nd edn. London: Minority Rights Group.

Dupon, J.-F. 1984. Where the exception confirms the rule: the cyclones of 1982–1983 in French Polynesia. *Disasters* **8**, 34–47.

Eldridge, C. 1989. Thought for food: suggestions for a systematized approach to emergency food distribution operations. *Disasters* **13**, 135–53.

El-Khawas, M. 1976. A reassessment of international relief programmes. In *The politics of natural disaster: the case of the Sahel drought*, M. H. Glantz (ed.), 77–100. New York: Praeger.

Galvin, K. A. 1988. Nutritional status as an indicator of impending food stress. *Disasters* **12**, 147–56.

Glass, R. I., J. J. Urrutia, S. Sibony, H. Smith, B. Garcia, L. Rizzo 1977. Earthquake injuries related to housing in a Guatemalan village. *Science* **197**, 638–43.

Green, R. H. 1986. Hunger, poverty and food aid in Sub-Saharan Africa: retrospect and potential. *Disasters* **10**, 288–302.

Gueri, M., C. Gonzáles, V. Morin 1986. The effect of the floods caused by "El Niño" on health. *Disasters* **10**, 118–24.

Hammerton, J. L., C. George, R. Pilgrim 1984. Hurricanes and agriculture: losses and remedial actions. *Disasters* **8**, 279–86.

Havelick, S. W. 1986. Third World cities at risk: building for calamity. *Environment* **28**, 6–11, 41–5.

Hederra, R. 1987. Environmental sanitation and water supply during floods in Ecuador (1982–1983). *Disasters* **11**, 297–309.

Holt, J. 1981a. Refugee resettlement: economic, political and social viability. *Disasters* **5**, 209–13.

Holt, J. 1981b. Camps as communities. [Special issue on refugees.] *Disasters* **5**, 176–9.

Hughes, R. 1982. The effects of flooding upon buildings in developing countries. *Disasters* **6**, 183–94.

Ives, J. D. & B. Messerli 1989. *The Himalayan dilemma: reconciling development and conservation*. London: Routledge.

Jeffery, S. E. 1982. The creation of vulnerability to natural disaster: case studies from the Dominican Republic. *Disasters* **6**, 38–43.

Khalil, G. 1990. Floods in Bangladesh: a question of disciplining the rivers. *Natural Hazards* **3**, 379–402.

References

Kling, G. W., M. A. Clark, H. R. Compton, J. D. Devine, W. C. Evans, A. M. Humphrey, E. J. Koenigsberg, J. P. Lockwood, M. L. Tuttle, G. N. Wagner 1987. The Lake Nyos gas disaster in Cameroon, West Africa. *Science* **236**, 169–75.

Knott, R. 1987. The logistics of bulk relief supplies. *Disasters* **11**, 113–16.

Kreimer, A. 1978. Post-disaster reconstruction planning: the cases of Nicaragua and Guatemala. *Mass Emergencies* **3**, 23–40.

Lofchie, M. F. 1975. Political and economic origins of African hunger. *Journal of Modern African Studies* **13**, 551–67.

Lusty, T. 1979. Notes on health care in refugee camps. *Disasters* **3**, 352–4.

McAdam, R. 1987. Engineering management in refugee camps. *Disasters* **11**, 110–12.

Milliman, J. D., J. M. Broadus, F. Gable 1989. Environmental and economic implications of rising sea level and subsiding deltas: the Nile and Bengal examples. *Ambio* **18**, 340–5.

Mishra, D. K. & H. R. Prakash 1982. An evaluation of the Andhra Pradesh cyclone shelters programme: guidelines for the Orissa programme. *Disasters* **6**, 250–5.

Mitchell, J. K. 1985. Prospects for improved hurricane protection on oceanic islands: Hawaii after Hurricane Iwa. *Disasters* **9**, 286–94.

Moreira Cedeño, J. E. 1986. Rainfall and flooding in the Guayas river basin and its effects on the incidence of malaria 1982–1985. *Disasters* **10**, 107–11.

Murty, T. S., R. A. Flather, R. F. Henry 1986. The storm surge problem in the Bay of Bengal. *Progress in Oceanography* **16**, 195–233.

Rasid, H. & M. A. H. Pramanik 1990. Visual interpretation of satellite imagery for monitoring floods in Bangladesh. *Environmental Management* **14**, 815–21.

Scrimshaw, N. S. 1987. The phenomenon of famine. *Annual Reviews of Nutrition* **7**, 1–21.

Seaman, J. 1981. Principles of health care. [Special issue on refugees.] *Disasters* **5**, 196–204.

Seaman, J. & J. Holt 1980. Markets and famines in the Third World. *Disasters* **4**, 289–99.

Seaman, J., J. Holt, J. Rivers 1977. The effects of drought on an Ethiopian province. *International Journal of Epidemiology* **7**, 31–40.

Seaman, J., S. Lievesley, C. Hogg 1984. *Epidemiology of natural disasters*. Basel: Karger.

Sigurdsson, H., J. D. Devine, F. M. Tchoua, T. S. Presser, M. K. Pringle, W. C. Evans 1987. Origin of the lethal gas burst from Lake Monoun, Cameroun. *Journal of Volcanology and Geothermal Research* **31**, 1–16.

Sikander, A. S. 1983. Floods and families in Pakistan, a survey. *Disasters* **7**, 101–6.

Susman, P., P. O'Keefe, B. Wisner 1983. Global disasters: a radical interpretation. In *Interpretations of calamity*, K. Hewitt (ed.), 263–83. Boston: Allen & Unwin.

Telleria, A. V. 1986. Health consequences of floods in Bolivia in 1982. *Disasters* **10**, 88–106.

Timberlake, L. 1984. *Africa in crisis: the causes, the cures of environmental bankruptcy*. Washington, DC: Earthscan.

Toole, M. J. & S. Foster 1989. Famine. In *The public health consequences of disasters*, M. B. Gregg (ed.), 79–89. Atlanta, Georgia: Federal Centers for Disease Control.

Torry, W. I. 1979. Anthropological studies in past environments: past trends and new horizons. *Current Anthropology* **20**, 517–40.

Torun, B. & F. E. Viteri 1984. Protein-energy malnutrition. In *Tropical and geographic medicine*, K. S. Warren & A. A. F. Mahmoud (eds). New York: McGraw-Hill.

Tresalti, E., F. Abdulle, H. Ismail 1985. Nutritional problems of refugees: three years' experience in the Somali camps. In *Emergency and disaster medicine*, C. Manni & S. I. Magalini (eds), 196–204. New York: Springer.

Turton, D. 1991. Warfare, vulnerability and survival: a case from southwestern Ethiopia. *Disasters* **15**, 254–64.

US National Research Council 1979. *Assessing international disaster needs*. Washington, DC: National Academy Press.

US National Research Council 1987. *Confronting natural disasters: an International Decade for Natural Hazard Reduction*. Washington, DC: National Academy Press.

Warnock, J. 1987. *The politics of hunger*. London: Methuen.

Whitehead, R. 1989. Famine. In *The fragile environment*, L. Friday & R. Laskey (eds), 82–106. Cambridge: Cambridge University Press.

WHO Working Group 1986. Use and interpretation of anthropometric indicators of nutritional status. *Bulletin of the World Health Organization* **64**, 929–41.

Wijkman, A. & L. Timberlake 1984. *Natural disasters: acts of God or acts of man?* Washington, DC: Earthscan.

Wisner, B., D. Weiner, P. O'Keefe 1982. Hunger: a polemical review. *Antipode* **14**, 1–16.

Wit, J. M. & P. Gooder 1981. Nutritional statuts of hospitalized pre-school children in Dominica, before and after Hurricane David. *Disasters* **5**, 93–7.

Young, H. 1987. Selective feeding programmes in Ethiopia and East Sudan, 1985-1986. *Disasters* **11**, 102–9.

Zaman, M. Q. 1991. The displaced poor and resettlement policies in Bangladesh. *Disasters* **15**, 117–25.

Select bibliography

Aga Khan, S., H. B. Talal et al. 1986. *Refugees: the dynamics of displacement*. London: Zed Books.

Anderson, M. B. & P. J. Woodrow 1991. Reducing vulnerability to drought and famine: developmental approaches to relief. *Disasters* **15**, 43–54.

Anderson-Burley, L. 1973. Disaster relief administration in the Third World. *International Development Review* **15**, 8–12.

Chowdhury, A. M. R. 1988. The 1987 flood in Bangladesh: an estimate of damage in twelve villages. *Disasters* **12**, 294–300.

Currey, B. 1992. Is famine a discrete event? *Disasters* **16**, 138–44.

Curtis, D., B. Hubbard, A. Shepherd 1988. *Preventing famine: policies and prospects for Africa*. London & New York: Routledge.

Cutler, P. 1984. Famine forecasting: prices and peasant behaviour in Northern Ethiopia. *Disasters* **8**, 48–56.

Dick, B. & S. Simmonds 1983. Refugee health care: similar but different? *Disasters* **7**(4), 291–303.

Dietz, V. J. 1990. Health assessment of the 1985 flood disaster in Puerto Rico. *Disasters* **14**, 164–70.

Disasters, 1981. Special issue on refugees. *Disasters* **5**.

Select bibliography

Dudasik, S. 1982. Unanticipated repercussions of international disaster relief. *Disasters* **6**, 31–7.

Duffield, M. 1990. From emergency to social security in Sudan, Part I: The problem. *Disasters* **14**, 187–203.

Duffield, M. 1990. From emergency to social security in Sudan–Part II: The donor response. *Disasters* **14**, 322–34.

Feldman, S. & F. McCarthy 1983. Disaster response in Bangladesh. *International Journal of Mass Emergencies and Disasters* **1**, 105–24.

Godfrey, N. 1986. Supplementary feeding in refugee populations: comprehensive or selective feeding programmes? *Health Policy and Planning* **1**, 283–98.

Green, S. 1977. *International disaster relief: towards a responsive system.* New York: Pergamon/UNITAR.

Hagman, G. 1985. *Prevention better than cure: human and environmental disasters in the Third World.* Geneva: Swedish Red Cross and League of Red Cross and Red Crescent Societies.

Harrell-Bond, B. E. 1986. *Imposing aid: emergency assistance to refugees.* Oxford: Oxford University Press.

Havelick, S. W. 1986. Third World cities at risk: building for calamity. *Environment* **28**, 6–11, 41–5.

Islam, M. A. 1974. Tropical cyclones: coastal Bangladesh. In *Natural hazards: local, national, global*, G. F. White (ed.), 19–25. New York: Oxford University Press.

Kattelmann, R. 1990. Conflicts and cooperation over the floods in the Himalaya-Ganges region. *Water International* **15**, 189–94.

Loescher, G. I. & L. Monahan (eds) 1990. *Refugees and international relations.* Oxford: Oxford University Press.

Long, F. 1978. The impact of natural disasters on Third World agriculture: an exploratory survey of the need for some new dimensions in development planning. *American Journal of Economics and Sociology* **37**, 149–63.

Mercer, A. 1992. Mortality and morbidity in refugee camps in eastern Sudan, 1985–90. *Disasters* **16**; 28–42.

Montgomery, R. 1985. The Bangladesh floods of 1984 in historical context. *Disasters* **9**, 163–72.

Rahmoto, D. 1988. Peasant survival strategies in Ethiopia. *Disasters* **12**, 326–44.

Rasid, H. & B. K. Paul 1987. Flood problems in Bangladesh: is there an indigenous solution? *Environmental Management* **11**, 155–73.

Rogge, J. R. (ed.) 1987. *Refugees: a Third World dilemma.* Totowa, New Jersey: Rowman & Littlefield.

Sewell, W. R. D. & H. D. Foster 1976. Environmental risk: management strategies in the Developing World. *Environmental Management* **1**, 49–59.

Shahabuddin, Q. & S. Mestelman 1986. Uncertainty and disaster avoidance behaviour in peasant farming: evidence from Bangladesh. *Journal of Development Studies* **22**, 740–52.

Simmonds, S., P. Vaughan, S. W. Gunn (eds) 1983. *Refugee community health care.* Oxford: Oxford Medical Publications, Oxford University Press.

Toole, M. J. & R. J. Waldman 1990. Prevention of excess mortality in refugee and displaced populations in developing countries. *Journal of the American Medical Association* **263**, 296–302.

Torry, W. I. 1980. Urban earthquake hazard in developing countries: squatter settlements and the outlook for Turkey. *Urban Ecology* **4**, 317–27.

CHAPTER NINE

Disasters and socio-economic systems

A continuing theme of this book is that natural hazards and disasters always involve a combination of physical impact and human vulnerability and response. In the past much emphasis has been laid on understanding the physical aspects of earthquakes, volcanic eruptions, floods, landslides, and so on (Scheidegger 1975), and there is no doubt that this is a necessary prelude to good hazard management. But many social scientists would argue that the human aspects of natural catastrophe are more important than the physical ones as a result of the finely developed ability of human beings to put themselves at risk (Hewitt 1983). Accordingly, "last but not least" we consider the effects of risk, hazard and disaster on societies and economies. Aspects of this were, of course, reviewed in previous sections: for example, though warning and evacuation were considered in Chapter 6 under the heading of logistics, their sociological interpretations were also dealt with. The present chapter focuses on individual and collective reaction to hazard and disaster. It ends with an evaluation of the impact of natural catastrophe on human history.

The sociology of disasters

Besides the damage and physical injuries which they cause, natural disasters also have a profound impact on survivors, who can be considered sociologically, as communities, and psychologically, as individuals. In both cases there is a wealth of definable regularities in human perception and behaviour. To begin with, the very fact of human settlement in high risk zones implies social attitudes connected with "bounded rationality" (Burton et al. 1978) Often, the net social benefit to be derived from living in such areas has not been calculated, and hence "optimizing man" must be replaced with a "satisficer", whose limited perception of risks or alternatives results in continued occupation of land that has the potential to be devastated by geophysical forces.

The response to disaster by the social system will depend on its inherent characteristics and dynamics. But it will also depend on the characteristics of the disaster: its cause, predictability, controllability, frequency of occurrence, speed of onset, duration, scope of impact and destructive potential.

Despite the disruption and chaos that tend to occur in the immediate aftermath of natural catastrophe, anti-social forms of behaviour, panic and apathy are uncommon reactions. Individual actions are likely to be rational and socially oriented, although potentially unco-ordinated. People who end up as leaders in disaster usually have a well-defined rôle, which they play with the benefit of prior experience, appropriate skills and a certain sense of detachment from the proceedings (Fritz 1957). If such people (emergency managers, medical and emergency personnel, volunteers, and so on) suffer conflicts of loyalty between the care of their immediate family and the imperative nature of their work, the problem can be reduced by planning to take care of their next of kin.

Disasters unify societies (Fritz 1961). During the aftermath, survivors tend not to flee the impact area, but instead large numbers of people and huge quantities of supplies (which are often useless and unsolicited) tend to arrive in the area in what has been termed a **convergence reaction**. The participants in this process have been classified into five groups: people returning to the area who normally reside there; anxious relatives; volunteers and relief workers; curious sightseers; and people seeking to profit by the disaster (Fritz & Mathewson 1957).

Community functions and disasters

A community can be defined as a geographical grouping of people into interacting social units organized to provide the basic social functions of daily life (Dynes 1970). As such an organism will not be designed to cope with the impact of disaster, it is likely to be disrupted by such an event, but as pre-existing social functions break down, new ones better adapted to current conditions take their place: hence the paradox that disasters create both confusion and social integration. At heart, the community will uphold a core of imperative values, including the care of survivors and living victims, and the restoration and maintenance of services, public order and general morale.

Human beings associate with one another in social and economic terms for a series of reasons, which constitute four normal community functions (Wenger 1978): **production, distribution and consumption; socialization** and social participation; **social control** (laws, societies, by-laws, rules and regulations); and **mutual support** (groups, families, etc.). Generally, there is little awareness of any overall system of priorities among these functions, and the importance of any one of them relative to the others will vary with the type of society and nature of its individual participants. "Society" in this context must be defined broadly to include various scales of social grouping, with respect to different age-groups and social ranks, and different degrees of association, from co-workers and local communities to major ethnic and national groups.

The normal characteristics of a society will profoundly affect its reaction to disasters and its ability to cope with their impact. Two opposing tendencies are particularly important, namely the degree of conflict and the level of integration inherent in the society (Quarantelli & Dynes 1970). Generally, the larger the social grouping under consideration, the less close-knit it is, which involves a lowering of the intensity of interaction between social groups. In addition, the degree to which the community has experienced disasters on previous occasions and developed the capability to manage crises will affect its ability to cope with present and future impacts and will govern the level of resources it sets aside for the next extreme event. Community reaction to natural catastrophe may vary from total shock and the breakdown of order (which is extremely rare) to routine behaviour in cases where the disaster is expected and fairly commonplace.

A social system is in **crisis** (Dynes & Aguirre 1979) when its traditional institutions and structure have been significantly destroyed, neutralized or discarded by its members, and replaced by something more appropriate (such as a military junta in place of an elected government). We can also define a less extreme situation, the **community emergency**, in which the traditional structure of society adapts to meet new demands or the strain placed upon it by a new and unexpected situation (Demerath 1957). In either circumstance, the nature of normal community functions, and the main priorities of society, are altered for the duration of the crisis and, in the case of an enduring need for greater public safety, permanently.

All four of the principal community functions undergo changes during the immediate aftermath of a natural disaster (Wenger 1978). The production–distribution–consumption function alters drastically. Social welfare temporarily replaces the profit-based market system for the distribution of basic necessities such as food, clothing and shelter. Production and distribution units that have been damaged or destroyed, or that produce goods and services which are not relevant to particular needs created by the disaster, are shut down until the normal workings of society have been restored.

Social participation is strengthened, in that the disaster creates its own tasks, and formal barriers between groups are temporarily removed. During the immediate aftermath, disasters tend to act as social levellers, in that all classes and groups of people are threatened or affected, and all are thus induced to participate in the relief effort.

Social control undergoes a change in priorities. During the emergency phase following a disaster, many rules and regulations that are normally essential to the functioning of society may become irrelevant and will temporarily be suspended. Emphasis will be placed instead on security and the proper control of the convergence reaction. Hence, the regulation of movement in and around the disaster area, and the control of access to particular parts of it where damage is severe or hazards linger, become important. In this respect, self-seeking behaviour, such as looting, is usually rather less important than may be supposed (unless social tensions that antedate the disaster remain high after it). Although suicides and murders are thrown into sharp relief, there is seldom evidence that natural disasters actually increase the rates, even if the causes are ostensibly related to the losses sustained during such impacts (Alexander 1982).

Finally, mutual support receives high priority after disaster, and both survivors and relief workers often manifest considerable altruistic concern for the welfare of others and for that of the community in general.

The question of society in disaster prompts one to return to the classification of organizations as emerging, expanding, adapting and disbanding (see Ch. 6). One form of emergent organization is the pressure group or public interest group. Studies reveal that these tend to have distinct characteristics, i.e. they may be limited to fewer than 100 members with an active core of half a dozen or so, they have a simple, non-hierarchical structure and they are not aligned with established political parties (Stallings & Quarantelli 1985). They can emerge before, during or after disasters and are a natural outcome of normal social processes. Their purpose is not necessarily to oppose authority and their existence is not necessarily a negative factor, but even if it were they cannot be eliminated by planning measures. To survive, such organizations must first crystallize from the perceived common needs of their adherents. They must then strive for recognition and become institutionalized in the structure of power and socio-political relations pertaining to the disaster area (Ross 1980). In the long term they must become integrated into wider concerns than those of the disaster that spawned them.

Community conflict and disaster

There are six possible reasons why the level of conflict in communities tends to be reduced during the immediate aftermath of a natural disaster (Quarantelli & Dynes 1976). First, the disaster constitutes an exogenous threat and hence tends not to amplify existing community cleavage. Secondly, the threat tends to lack ambiguity, as the disaster agent can usually be identified and perceived clearly. Thirdly, a high value is given to community interests and a low value to personal ones and self-seeking behaviour. Fourthly, the immediate and imperative need to solve problems related to the disaster tends to distract the attention of the community from other grievances. Fifthly, the power of the impact reduces the tendency to recall past conflicts and anticipate future ones. It strengthens community identification. And finally, in the short term, disasters tend to affect both the rich and the poor, the powerful and the disadvantaged. They thus act as social levellers, although it must be stressed that their long-term effect tends to be precisely the opposite.

Hence, differences within the community tend to be put aside and conflicts held in abeyance during the emergency phase (Dynes & Quarantelli 1971). But various factors may cause the conflicts to be intensified afterwards (Quarantelli & Dynes 1976; Drabek 1986: 229). These include reduction in social control and participation, fragmentation of traditional community groupings, weakening of the system of rights and obligations and racial or ideological discrimination in the distribution of relief. Lack of leadership may lead to disorientation and institutional paralysis, especially where political differences arise between officials at different levels of the administrative hierarchy. Such conditions offer much scope to vested interests, and frequently lead to struggles between rival groups for the control of relief supplies. Finally, there is the plight of the poor, who will probably have lost proportionately more in the disaster than the rich.

Research after the May 1980 eruption of Mount St Helens indicated that there were fewer interpersonal conflicts (fights, domestic disputes, and so on) during the immediate aftermath, but such problems were much amplified after one year, when delayed stress reactions had come to the fore (Blong 1984: 140). Enduring community conflict can impose severe delays on reconstruction, as in the case of the December 1988 Armenian earthquake (see Ch. 2), where there are unmistakable signs that conflict between Azerbaijani and Armenian ethnic groups imposed a virtual moratorium on recovery in some areas that were damaged by the tremors.

There are two other potential negative effects of disasters on communities in the long term (Barton 1970). First, loss of interest in the disaster area on the part of the authorities may result in decline in the quality and comprehensiveness of social and medical services. Secondly, lack of funds for the reconstruction and regeneration of local production may have economic effects such as loss of livelihood, unemployment, decreased productivity and, eventually, the outmigration of disadvantaged workers and personnel. At this point, community conflict may resurface, as these effects tend to accentuate pre-existing cleavages, injustices and inequalities in the social system.

One phenomenon that may act in favour of social cohesion is kinship. The family unit tends to gain in strength and solidarity during disasters and to strengthen its links with friends and relatives. However, this tendency is not universal and is most likely to occur when recovery from physical damage is rapid, the rôle of the family is central to recovery processes and the local community is a stable one (Drabek 1986: 275). One illustration of the central rôle of the family is that after sudden impact disaster, search and rescue is carried out first by the survivors, who tend to give priority to helping friends and relatives, particularly close family. At this point people who are socially isolated may easily be neglected.

But familial relationships may undergo a more than superficial change during disaster. For example, in Cyprus it was found that disaster led to hastier and less premeditated marriages (Loizos 1977). Moreover, research in many settings (e.g. Williams & Parkes 1975, Placanica 1985) suggests that disasters involving high mortality lead to correspondingly high birth rates.

Panic

Panic is one extreme phenomenon associated with disasters that can be viewed in both sociological and psychological terms (Johnson 1987). The word has, of course, also been applied to economic behaviour connected with the fall of stock market prices, but this section is concerned with more spontaneous forms of reaction.

Panic is difficult to define because it tends to merge into other less extreme forms of behaviour (Quarantelli 1954). Hence it constitutes one end of a continuum from fear of a highly specific threat to anxiety about a vague one. However, it can be regarded as an instinctive asocial reaction to some threat that is perceived to be tangible and immediate. Panic is not an anti-social reaction, as it does not encompass organized behaviour. Instead, it causes the individual to undergo a spontaneous, uncontrolled withdrawal from the social structure, although one that focuses on what will happen next. People who panic see the potential threat to their lives as immediate and their survival as dependent on reaction that is too rapid for logical weighing of alternatives. Hence it is a non-rational, rather than an irrational, reaction. It does not allow the possibility of group action, and no measures are taken to combat the hazard except self-removal.

Four conditions may lead to the development of panic. First, there must be a crisis, involving the interruption of normal events and the abrupt manifestation of an immediate, unforeseen threat to human life. Secondly, the individual may feel that he or she is unable to escape from an impending threat, and that the likely consequence is entrapment leading to annihilation. A real physical impediment is not necessary for panic to be stimulated; it is sufficient for the feeling of entrapment to develop. However, the supposed blockage of the escape route must relate to the immediate physical danger involved in being trapped, or the subject may begin to reason rather than panic. Thirdly, when an individual feels that he is unable to prevent the worst consequences of an immediate threat to his safety, he may feel helpless and isolated. Other conditions which contribute to the development of panic include previous experience of danger, which may leave a person highly sensitized to its recurrence. In addition, the presence of a crowd may be a contributing factor, as panic is definitely contagious. The degree to which the cultural matrix of society contributes in times of crisis to the generation of panic is practically unknown (Alexander 1990).

Panic depends on *contextual conditions* (such as lack of social ties between panic participants, added to a high likelihood that a crisis will develop with some potential for panic) and *immediate conditions* (e.g. a sense of isolation, powerlessness or entrapment; Quarantelli 1977).

Apart from hysteria, which represents a complete withdrawal of an individual's perceptive functions, the most likely consequence of panic is flight. This is not random behaviour, but is guided by the perceived locations of danger and safety, as people always tend to flee *from* something (Quarantelli 1976). Perception is concentrated on the way out of danger. The route to safety is chosen on the basis of the physical layout of terrain or buildings, habitual patterns of behaviour (such as always using the kitchen door rather than the front hall door), and what other people do. Panic may lead to the most rational course of action under the circumstances (i.e. reaching safety as quickly as possible) or it may lead the subject into danger. Movement (such as running, driving or crawling) will tend to be directed away from perilous objects, as far as the subject is able to perceive these, although there is no guarantee that the route towards presumed safety will be free of hazard (Quarantelli 1977).

In summary, panic is irrational, asocial, impulsive, non-functional, maladaptive and often inappropriate action. But compared with other reactions to disaster it is a relatively uncommon phenomenon. However, it has only been studied seriously with regard to building fires, crowd behaviour, military action, structural collapses and, to a lesser extent, earthquakes. Thus, little is known about its rôle and potential in most natural disaster situations.

Disaster psychology

According to the psychologist A. J. W. Taylor, humanity faces three forms of adversity in the modern age: **psychotism**, or mental ill health, **somatoticism**, or physical ill health and **socioticism**, or the breakdown of social and environmental supports (Taylor 1984: 446). While they may principally be responsible for the last of these, disasters can of course profoundly affect the first two.

Attitudes to disaster
Personal values tend to alter in response to the overwhelming experience
of surviving the sudden impact of a natural disaster. Part of this change
will be transient, in that the individual will revert to more workaday
attitudes once the crisis has passed and the emergency is over. But
disasters also act as landmarks in human consciousness, tremendous
experiences that punctuate the flow of life. The catastrophe will act as a
yardstick against which to measure all subsequent experience and as a
point of reference in time, in that when other significant experiences
occur there will be an inveterate tendency to place them on a timescale
relative to the date of the disaster.

During the aftermath phase, high value tends to be given to community
interests and low value to personal ones, and generally there is greater
consensus on what is good, just and correct. The altruism and increased
social support that tend to proliferate in the wake of disaster enable many
individuals to conquer their feelings of abandonment, to reaffirm their
existences in the face of tragedy and to benefit from goodwill. This brief
period has been referred to as **therapeutic community** (Gist & Lubin
1989: 71), or **post-disaster eutopia** (Wolfenstein 1957: 193), and it can
help mitigate or reduce the stress that recovery and reconstruction place
on the survivors.

But there is often a pervasive need to apportion blame for the carnage
and destruction caused by disasters (Bucher 1957, Drabek & Quarantelli
1969). It is perhaps easier to do this legitimately when dealing with the
effects of dam bursts, fires and collisions than with the impact of
earthquake, volcanic eruption or flood. However, there may be a
powerful (and often justified) tendency to blame the authorities for lack
of preparation and speculative builders for the poor quality of damaged
housing. The desire to apportion blame is one of the most pervasive
aspects of disaster aftermaths. It tends to become focused on specific
individuals or groups who, rightly or wrongly, are judged to have caused
at least part of the misery and destruction. This condition stems from
feelings of anger, which are in turn a natural reaction that signifies a need
to fight against adversity. Anger is very difficult to cope with unless it is
focused – hence the desire for a scapegoat.

Fatalism and activism are two opposing traits in human attitude to
emergency situations. The former may include supplication and prayer,
but is most likely to manifest itself in the form of passive behaviour.
Activism can often be channelled into useful relief efforts, although these
require organization and prior preparedness. On occasion, however, it
can take the form of activities that have no direct productive value during
the emergency, such as political agitation. Alternatively, the continuum
between passive and active approaches to disaster can be subjected to a
more sophisticated classification. Thus, Janis & Mann (1977) argued that

five psychological conditions represent different individual patterns of coping with disaster. First, if the risks are not deemed serious enough to take protective action, the result is termed **unconflicted inertia**. Secondly, if the risks are not judged serious when the easiest protective action is chosen, the result is termed **unconflicted change** to a new course of action. Thirdly, if it is not considered reasonable to hope for a better protective strategy, the result is termed **defensive avoidance**. Fourthly, if there is not expected to be time to search for a new means of avoiding danger, the result is called **hypervigilance**. Finally, all reactions not covered in the last four cases are termed **vigilance**.

Although much research has been done on the impact of natural and technological disaster on human psychology, there are remarkably few consensuses, which suggests that the problem is a complex one. The following sections will deal with the definition of victims, the types of psychological impact that they are likely to suffer (including formal models such as Wallace's "disaster syndrome"), and the causes and consequences of mental health problems in disaster. Finally, the rôle of psychiatric help will be evaluated.

The psychological victims of natural disaster

It is worth noting at the outset that mental health specialists use the terms "clinical" and "trauma" in a rather different way to physicians (Erickson 1976, Newman 1976). The former term refers to accurate diagnosis of profound conditions or states, while the latter describes the negative impact of events on a person's psychological or emotional stability (cf. Ch. 7).

The emotional casualties of disaster can be divided into primary and secondary victims. The former experience directly the losses caused by the impact, and the latter witness the destructiveness or suffer bereavement but are not directly involved. The degree of psychological impairment tends to be lesser in secondary than in primary victims, as does the degree to which the patient utilizes counselling services, if these are available. It should be noted here that in general mental health counselling is not easily accepted by disaster victims, who may see it as carrying a degree of social stigma.

At a more detailed level, Dudasik (1980) classified the survivors of disaster into four categories according to decreasing levels of personal involvement. **Event victims** are those whose potential for psychological disorder stems from direct involvement with the disaster as primary victims. **Context victims** may be psychologically overwhelmed by the rigours of the aftermath, but do not experience the impact itself. **Peripheral victims** may suffer, for example, through kinship ties with victims who have died, been injured or suffered losses. They may themselves suffer grief about the present situation and anxiety about the future. Finally, **entry victims** may react psychologically to the death and destruction which they witness upon entering the disaster area during the aftermath.

As Table 9.1 shows, psychological reactions to disaster may be specific to the age of the victim. In this respect, children and the elderly are often particularly vulnerable. The former, if they are unable to understand and rationalize the event, may suffer from phobias, sleep disturbances, loss of interest in schoolwork and aggressive or undisciplined behaviour (the category most at risk may be 8–12 year olds). The latter, especially if they live alone and lack adequate support systems, may suffer from depression and a sense of hopelessness. In any event, both groups are likely to benefit from any specialized help that can be given.

The psychological consequences of disaster
The impact of natural disasters on the mental health of survivors can be divided into three categories: mental illness (severe **psychopathology**; Perry 1979), mental health problems and problems associated with daily

Table 9.1 Age-related mental health problems encountered in survivors of disaster. These reactions to stress may manifest themselves immediately or days or weeks after the impact (after Lystad 1987: 5)

Preschool	Latency age	Preadolescent and adolescent	Adult	Senior citizen
Confusion	Confusion	Aggressive behaviour	Anger	Accelerated physical decline
Crying	Depression	Changes in peer group friends	Loss of appetite	Agitation or anger
Fear of abandonment or of strangers	Fears about own safety	Confusion	Loss of interest in everyday activities	Apathy
Immobility	Fighting	Headaches or other physical complaints	Psycho-somatic problems such as stomach ulcers or heart trouble	Confusion
Irritability	Headaches or other physical complaints	Poor performance		Depression, withdrawal
Loss of bowel or bladder control	Inability to concentrate	Withdrawal or self-isolation	Sleep problems	Disorien-tation
Thumb-sucking	Poor performance		Withdrawal, suspicion or irritability	Increase in number of somatic complaints
	Withdrawal from peers			Irritability or suspicion
				Memory loss

563

life. Emotional disturbance and negative forms of stress may be experienced to a greater or lesser degree of seriousness. In most cases, severe disturbance is likely to affect only a minority of victims (perhaps 10 per cent), many of whom will recover rapidly, but lighter forms of emotional problems are often remarkably persistent during disaster aftermaths. The proportion of survivors affected varies among different events from none to about three quarters. The longevity of symptoms also seems to vary markedly.

One of the first attempts to classify the psychological effects of disaster was made after the 1963 Skopje (former Yugoslavia) earthquake, when it was estimated that 75 per cent of survivors were suffering from mild disturbances and 10 per cent manifested severe psychopathological reactions (in the light of subsequent disasters these appear very high proportions). Psychiatrists found examples of the following states: mild stupor, escape reactions, childishness and increased suggestibility, depression, hysterical amnesia, passivity and psychosomatic disturbance (Arvidson 1969). In Managua, Nicaragua, after the 1972 earthquake, large increases were observed in neurosis, psychosis, mental retardation, personality disorder and brain disorders (cerebral organic syndrome).

In more general terms, serious mental problems resulting from disaster may be classified for diagnosis into acute stress, acute psychosis, mental retardation and the effect of head injuries. Each involves a different set of characteristic symptoms. Thus, the victims of acute stress tend to suffer loss of memory, cognition or orientation. They may find difficulty in making a decision and suffer from a certain flatness of emotions. These are symptoms of **post-traumatic stress disorder** (Nolen-Hoekeema & Morrow 1991). Acutely psychotic patients also tend to manifest subdued behaviour, including depression and apathy. But they may undergo phases of bizarre thinking or energetic and potentially anti-social behaviour. The effects of head injuries can sometimes be similar to those of psychosis (see below), but proper diagnosis should reveal that the underlying mental disorganization is lacking. In contrast, mentally retarded individuals may manifest infantile speech and attitudes and react to the disaster by disorganized or disoriented behaviour. Such problems can usually be alleviated by providing adequate support and guidance.

Stress tends to be the most common and widespread of these states (Glass 1970). Warheit (1985) defined it as "an altered state of an organism produced by agents in the psychological, social, cultural, or physical environments [which] . . . produces deleterious physical or mental health effects for certain individuals". He defined the following common sources of stress: an individual's biological constitution or psychological characteristics, his or her cultural matrix, interpersonal relationships, the social structure and the geophysical environment (Warheit 1979). In addition, the most pressing problems of the disaster aftermath (such as lack of adequate and easily accessible medical care) may have particularly high potential for creating stress among survivors.

The "disaster syndrome"

The tornado which devastated part of the city of Worcester, Massachusetts, on 9 June 1953 took 94 lives. It was also the first natural disaster to have its aftermath investigated by researchers from the US National Academy of Sciences. One of these, A. F. C. Wallace, proposed a general model for human psychological reaction to unexpected sudden impact disasters (Wallace 1956). His **disaster syndrome** constitutes a *psychologically determined defensive reaction pattern*. It applies to victims who have not been severely injured in physical terms and consists of four stages. In stage 1 the victim is dazed, stunned, apathetic, passive, aimless or immobile. This state varies in duration from minutes to hours or days, according to the character of the victim and the seriousness of "wound shock" received from physical injuries.

In stage 2 the subject manifests extreme suggestibility, altruism, gratitude for help and anxiousness to know whether familiar people and places have survived the disaster. The impact of personal losses is minimized and the victim is concerned principally for the welfare of family or community. This stage may last for some days. Stage 3 involves a mildly euphoric state of identification with the damaged community and enthusiastic participation in repair and rehabilitation initiatives. It resembles a revival of neighbourhood spirit and may last for weeks. Finally, the euphoria dissipates in stage 4 and more usual ambivalent attitudes return. Criticisms and complaints are expressed and there is an awareness of the long-term problems which the disaster has caused. This stage can usually be detected in subjects several weeks after the impact.

In symbolic terms the syndrome represents the "destruction of the whole world". The victim's first response is one of withdrawing perceptual contact with such a grim possibility. He or she may be found amid the destruction staring and unwilling to leave the scene of the disaster. The remainder of the syndrome consists of restoring by degrees the forms of behavioural organization which the subject had used before the disaster. This can begin in the form of bodily contact with relief workers or medical personnel, and continue with personal participation in the relief effort.

The syndrome occurs when the impact is sudden and unexpected, when it destroys much of the immediate physical and cultural environment and kills or injures many people (or threatens to do so), and especially when the individual victim is not trained to cope with its consequences. The more sudden, unexpected or catastrophic the impact, and the less trained the victims, the more profound the syndrome.

Field investigations suggest that incidence of the syndrome is unrelated to cultural or ethnic factors. In the Worcester tornado of 1953 about one third of uninjured or slightly injured survivors manifested it to some degree, and the incidence of stage 1 declined with time after the impact. In other cases researchers have found only one sixth, or even smaller proportions, of the survivors to be affected to any degree (Dynes & Quarantelli 1976: 235). It appears that people escape the syndrome in several ways; for example, if they become hysterical or are severely injured, or if they have predefined rôles to play, which absorb their attention. Thus, firemen and doctors tend to acquire some psychological immunity against disaster impacts.

One other characteristic of disasters that deserves comparison with the disaster syndrome is the so-called "death imprint" or "survivor syndrome". If the former represents the symbolic destruction and reconstruction of the victim's world, the latter is a sort of "redestruction". The tendency to relive the event has been observed to be so strong in some victims that it constitutes a semi-permanent source of inner terror. A shadow is cast over human relationships, borne of despair, deep depression and feelings of utter hopelessness. At worst, the victim enters a state of "psychological numbing", in which the rational distinction between life and death becomes blurred (Lifton & Olson 1976). Titchener & Kapp (1976) described this state as a "temporary collapse of the ego", which in the case of the 1973 Buffalo Creek flood in Pennsylvania took from six months to two years to be rectified.

The causes of psychological impairment in disasters
The extent of mental health problems in disaster appears to depend on the following variables: type, duration and recurrence interval of disaster, degree of personal loss, and potential for controlling the outcome (Berren et al. 1980). The level of psychological impact depends on the suddenness and unexpectedness of the impact, the intensity and persistence of threat experienced, how dramatic the events witnessed are (especially if major loss of life is witnessed), the degree of personal and community upheaval and the degree of long-term exposure to the visual signs of the catastrophe (Green 1982).

It is not easy to ascertain which characteristics of disasters are most likely to cause mental health problems. Some of the most dramatic events have not provoked great or long-lasting psychological problems. But, essentially, the mental health impacts seem to be greatest where victims have long been aware of the danger, and the impact is very sudden, intense and unexpected. Conversely, warning and preparation tend to allow the kind of socialization that reduces the risk of psychological disturbance. Indeed, high awareness of persistent or recurrent threats can lead to the development of "disaster subcultures", in which the risk becomes part of everyday experience and thinking. The subcultures can

566

either help or hinder the provision of emergency relief: they are very diverse phenomena and vary in character from the mystical to the technological (Wenger 1972). Such collective states of mind also tend to dilute the mental health impact.

The character, severity and duration of psychological problems which survivors experience may be intimately linked to the nature of their losses. Long-term homelessness, dislocation, unemployment, conflict or confusion are all likely to, at the very least, lower morale. Bereavement (Fig. 9.1) and injury both create particular patterns of psychological damage. For example, depression, dysphoria, sleeplessness and appetite disturbances are likely to manifest themselves in some combination in more than two thirds of people who have been bereaved of a spouse or someone of similar importance to them. Depression and grief are the most common negative consequences, although there may be a risk of suicide, alcoholism or excessive recourse to sedatives. Parents who have been bereaved of children suffer particularly, especially from grief, while loss of spouse seems to affect older widows and widowers less than younger ones. In this context, loneliness as a result of bereavement can be divided into personal and social components, the former resulting

Figure 9.1 Graveyard for some of the 264 victims of the 1985 mudflow in the Stava Valley of the Italian Dolomites. Bereavement and associated feelings of loss are highly significant components of the psychological impact of many disasters.

from loss of companionship and the latter from underinvolvement in community. Both are a function of low morale and pessimism of the bereaved person (Bahr & Harvey 1979).

The psychological impact of injury appears to be greatest when the appearance of the victim is strongly altered (as by burns), when his or her functioning is impaired (as by spinal injuries) or when serious head injuries have been inflicted. The last of these can result in coma followed by post-traumatic amnesia, the respective lengths of which form part of the measure of injury severity. Upon regaining consciousness, patients with head injuries often suffer from disinhibition, including symptoms of irritability, talkativeness and childishness. Children with head injuries appear particularly susceptible to psychiatric disorders of these and other kinds.

The perception of almost all aspects of disaster varies among those who experience it (see next section). While this may complicate, or at least individualize, the psychological impact, there are some notable regularities. First, uncertainty tends to increase stress levels among those who are subject to it. Secondly, demoralization may ensue if the disaster is perceived to be unmanageable. Thirdly, perception responds to experience: for example, people who have not lived through a particular form of disaster tend to underestimate the hazard. However, it is less easy to generalize about the rôle of experience among people who are not facing the threat of disaster for the first time. In some instances, it may be assumed that the worst form of impact will not recur, while in others the onset of a high-risk period may lead to acute anxiety.

In psychological terms, losses can be collective as well as individual, witness the destruction of community, which has been termed **collective trauma**. In this respect, the seriousness of physical impacts is important. Psychological disorders tend to be magnified if a large proportion of a community's resources are destroyed by the disaster, if the local environment is in some way profoundly altered after the event or if it proves impossible for victims to return to some semblance of the predisaster reality (for example, if permanent relocation is necessary). Other factors that stimulate mental problems include mutation of the social fabric and inability to control developing situations. However, the social fabric of a community is often more likely to be strengthened than weakened by disaster, and hence its effect on mental health is therapeutic.

Relief workers: a special category of psychological victim

Volunteer relief workers may find that their perception is severely distorted by the enormity of the disaster and the strain of working with it (Hartmann & Allison 1981, Duckworth 1986). Some may experience the so-called **Magna Mater complex**, in which the subject tries to assume all the problems of the moment as his own, which leads rapidly to tiredness and demoralization. Others may instead experience feelings of omnipotence, the so-called **Jehovah complex**, feeling that they are able to solve all existing problems, which is, of course, never the case.

Taylor (1984) found that professional workers stand up to stress better than volunteers in disaster situations. However, if they are in the front line of the fight against the disaster, they too may be particularly susceptible to stress disorders. One source of high stress levels can be found in working relations, including the effect of personal injury or similar impacts upon fellow team members, or simply the sight of much death, injury and destruction. Concurrently, committing errors or being unable to solve practical problems can lead to a sense of failure.

Occupational pressures also create stress. Thus, hazardous working conditions, pressing deadlines, long hours, excessive amounts of work and physical or emotional demands on the worker can all take their toll on his or her emotional resources. The final culprit is organizational pressures, including inadequate chain of command (see Ch. 6), conflict between or within organizations and problems of defining rôles. In almost all cases, the stress reaction will be transient, but it may be severe. Physical reactions include lower back pain, nausea, faintness and dizziness, headaches, muscle cramps and upset stomach. Cognitive problems, which result from the fact that comprehension is usually reduced under stress, include memory problems, loss of objectivity, slowness in thinking and loss of concentration. Psychological difficulties include fear, guilt, anger and feelings of isolation.

Often, the sources of stress in disaster relief workers cannot be tackled, especially if they result from the physical impact itself. Resistance to stress can be gained by training, experience and practice gained before the emergency. On the job, problems can be alleviated by adequate briefing of personnel, rotation of workers among low- and high-stress positions, substitution of workers whose effectiveness is seen to diminish and provision of support where necessary. After the event, tensions can be liberated, and valuable experience shared, by involving workers in a debriefing session in which the progress of the emergency is discussed.

Psychiatric help in disasters

Given the generalizations described above, it should be possible to predict the likely psychological impact of a major natural disaster and thus the level of psychiatric help required by survivors (Krell 1978). To summarize the preceding discussion, there is unlikely to be a high incidence of acute psychopathology, although depression will be widespread. For most of those victims who are affected by it, depression will set in directly after the event, but for a minority its onset will be delayed (see the section on Wallace's "disaster syndrome"). Depressive symptoms

569

will be highest among the bereaved, severely injured (but lucid) patients, people suffering from mental problems before the disaster, survivors whose socio-economic status is low, individuals facing multiple sources of stress and those whose sources of social support are very restricted. Shock, numbness, anxiety, guilt, somatic complaints (e.g. loss of appetite or sleep), and even physical deterioration and increased mortality may accompany depression.

Essentially, the mental health victims of disaster create two kinds of practical problem, neither of which is necessarily exclusive of the other: some require guidance and support, while others indulge in anti-social behaviour that, for the common good, must be restricted. The latter category poses an imperative for psychiatrists working in the field, who must make their diagnoses with great rapidity. People who cannot accept rules or schedules, who are vociferous troublemakers and who indulge in petty theft may be reacting to psychopathological disorder, but without the benefit of time it is often extremely difficult for the psychiatrist to ascertain the emotional root cause of such behaviour.

The psychological consequences of disaster may persist or worsen if they are not tackled promptly. Both the causes and the consequences of the problem need to be treated. Stability needs to be restored to the socio-economic fabric of the community, and sources of mutual support need to be identified and encouraged. Groups that are particularly vulnerable to mental health problems should be identified and given psychiatric assistance. Finally, cultural norms need to be understood and respected, especially when psychiatrists are not members of the local cultural group.

Perception of natural hazards and disasters

Whatever the physical reality, disasters must also be viewed in terms of how they are perceived and estimated. Though it is in some respects a subfield of psychology, hazard perception has been studied mainly by geographers (Whyte 1986). In this context, **perception** can be defined as the individual organization of stimuli for the purposes of cognition and recognition. The following regularities have commonly been observed in the human perception of natural hazards (see Burton et al. 1978: 106–7):

(a) It is difficult to estimate the magnitude and frequency of many geophysical events, and hence, under normal circumstances, people find it difficult to identify specific threats to their safety. They tend to adjust their knowledge of hazards by revising their ideas on the basis of any new information received. But if the fundamental understanding is wrong, then they will have more confidence in their own ideas than these merit. Perception of natural hazards and disasters is therefore a very imprecise entity.

(b) People who have had frequent or recent experience of disaster tend to be more knowledgeable about the issue and sensitive to it. The severity of risks undergone and the length of time that the attendant consequences have been endured are important in this context. Thus, at Shrewsbury in England, Parker & Harding (1979) found that awareness of flood hazards depended on the frequency and recency of impacts. However, they also found that complacency followed the construction of a dam which was (wrongly) held to solve the flood problem once and for all.

(c) Severe or intense events can elicit responses that are highly influenced by the personalities of the respondent.

(d) Hazards that are likely to exert a strong influence on the life of an individual tend to stimulate the person in question to become more knowledgeable and wise about the risks involved.

(e) The degree of adjustment to risks which people manifest is often linked to factors of personality, such as inherent fatalism or ability to face up to danger.

(f) Wealthier people perceive hazards with greater accuracy and make more and better adjustments than do poorer folk.

(g) Social pressures and mores can either encourage or retard the adoption of adjustments to hazard in quite a different manner to individual predilections.

(h) Local officials and politicians usually manifest a degree of awareness of hazards which is closely related to what they perceive the public expects of them (Drabek 1986: 339).

(i) People in many Third World countries perceive the rôle of God in disasters to be much more immediate, simple and direct than do people in industrialized nations, who tend to see Him as "benign but removed" (Baumann & Sims 1974).

Leaving aside the question of divine intervention, Mileti (1980) noted that perception can be divided into that pertaining to the likelihood of damage and that relating to the rôle of mitigation. Preston et al. (1983) described the psychological adjustments necessary to reduce "cognitive dissonance", which they defined as the psychological discomfort that arises when two conflicting beliefs are held simultaneously, as when a person perceives his environment to be hazardous, but continues to live in it. They also noted that people tend to overestimate the impact of disastrous or sensational hazards and to underestimate that of pervasive hazards which claim only small numbers of victims each time they strike. But Kastenbaum's so-called "Law of Inverse Magnitude" states that death, destruction or loss must increase ("by an undetermined but powerful constant"; Turner 1976) if the same level of psychological impact is to be maintained while physical or emotional distance from the disaster is increased (Kastenbaum 1974). However, according to Anderson (1967: 304), the more chronic and well-known the threat, the more integrated it will be with the local culture, the more uniform will be the reaction to it and the less shifting will be the focus of concern.

Slovic et al. (1974) hypothesized that a rational individual will appraise the likelihood of disaster, examine the range of alternative mitigating actions, evaluate the consequences of particular alternatives and choose one or several actions. But it seems that people find it difficult to evaluate more than a few options at once, and that they do so sequentially rather than concurrently. According to Slovic et al. (1977) people are resistant to change and hence may distort new evidence on risk to their own ingrained perceptions. It seems that, regardless of intelligence, relatively few people are equipped to have an accurate perception of risk levels or to make reliable decisions about them. Turner (1979) argued that people tend to discount, overlook or ignore the risk of natural disaster for various reasons: they may feel themselves to be invulnerable, they may make incorrect assumptions about the hazard, they may have faith in precautions that are unlikely to function, or they may be unable to understand or assimilate the necessary information.

There have been few fundamental studies that have linked hazard perception to behaviour *and* the psychological state of the individual. In this respect, Kilpatrick (1957) found that people tend to adopt a dominant percept, usually a familiar one, in the light of which to view events. If they cannot do so, they mainfest considerable degrees of suggestibility, though their actions remain logical and appropriate in terms of how they perceive events. Though active responses constitute the best way of changing one's perception, in a crisis many individuals tend to withdraw themselves from events and to resort to familiar behaviour and percepts, however inappropriate these may be. The stress caused by ambiguous, inconclusive or inadequate perceptual cues may make the subjects depressed, though the resolution of such conflicts may be greeted with elation.

In a slightly different context, Burton, Kates & White (1978: 102) argued that accuracy of hazard perception is tempered by both the extent to which the resources of the threatened place are needed and the social problems of the population at risk. They noted considerable variation in the extent to which natural hazard risks are perceived. Japanese research and my own studies both suggest that hazard perception is greater and more accurate among the middle-aged than among teenagers (Shimada 1972, Alexander 1990). Generally, it appears that older, more experienced people have better perceptions than younger ones, and country dwellers have a better idea of the risks than townsfolk, but such regularities are easily altered by particular sets of circumstances.

Spatial mobility in rapidly changing societies (such as the USA) means that increasing numbers of people come to live in unfamiliar surround-ings, where they have severely limited perceptions of risk and experience of impact (White & Haas 1975). Thus, in the United States, people who move onto floodplains are often either too poor to be seriously concerned about flooding, or rich enough to be preoccupied with other socio-economic questions. The former group tends to look upon floods as an unavoidable affliction, while the latter is frequently very vocal and influential in encouraging structural solutions to the flood problem, whether or not these eventually reduce the risk (James 1973).

Although, as noted, familiarity with hazards does tend to sharpen perception, some people fall foul of the so-called "gambler's fallacy", which postulates that if a particular event occurs in one year it is less likely to recur soon afterwards. While this may be true for earthquakes that discharge the strain which has accumulated on faults over time, it is untrue for hurricanes, floods and many other hazards whose future occurrence is unconstrained by the past.

In general, the negative characteristics of hazard perception are more striking than the positive ones. In Western societies, which tend to be overconditioned by the electronic media, hazards may become exalted as a media spectacle. Many people have weak or inaccurate images of the effects of disasters (Kilpatrick 1957), and the common consensus appears to be that such eventualities always happen to someone else. Where natural hazard impacts are rare but potentially very serious, it is likely that a majority of residents will deny that the risk exists or make a show of ignoring it. In some cases, this may mask considerable personal insecurity. In cases where impacts are more frequent but their effects appear hard to modify, the prevailing attitude may be one of fatalism and passivity. In yet other cases, the thresholds of both *awareness* and *action* have been crossed and attempts are made to combat hazards and prevent losses. In extreme circumstances the last of these cases may involve mass migration or relocation.

In summary, the level of perception will depend on the ability to estimate risk and perceive its causes, the level of past experience with hazards, the propensity to deny that a risk exists, the level of access to appropriate information and the size of the unit analyzed. It also depends on the importance of the hazard to the local community (White 1974). Much work remains to be done in the field of hazard perception, especially with regard to comparison between different cultures. More-over, as Whyte (1986: 259) pointed out, few data exist that indicate to what extent people regard hazards as natural or man-made, even in the United States, where most of the perception studies have been carried out.

Natural hazard perception is intimately related to the question of risk, especially as the latter can often be divided into its objective and subjective forms – i.e. measured or calculated risk and perceived risk. Hence, the next section deals with the emerging discipline currently striving to quantify the risks of human life or make them explicit, and thus to reduce them.

Risk assessment and management

The natural hazards that prevail in the Third World are mostly still the ancient ones that afflict agriculture and settlement: drought, flood and earthquake (Sewell & Foster 1976). The developed world also faces a range of more complex hazards, which, however, usually have a less profound impact upon society, as more resources are available to mitigate them and cushion their effects. The nature of risk thus varies with level of economic development.

Societies are not and cannot be risk free. But they have evolved in a manner that allows them to operate within specific levels of tolerance of natural and man-made hazards, hence most human activity involves both risk and benefit. The limits of tolerable risk in societies are defined by both law and common practice. There will always be potential for these limits to be exceeded, and a wide and expanding spectrum of hazards capable of causing destruction exists. On the one hand, human behaviour is increasing the scope for negative impacts by becoming more diverse, but on the other, research and public education are broadening the range of possible mitigating adjustments.

There are large gaps in society's understanding of risk and its management. Risk–benefit analysis is not used as widely as it could be. Assumptions and uncertainties go unrecognized or unconsidered, such that decisions are not made under ideal circumstances and risks are assumed without being understood. Irrational attitudes often prevail, such that people are willing to assume certain major risks but very unwilling to take other smaller ones. Hence, the concept of safety tends to be defined vaguely and inconsistently by society (Foster 1980).

Nevertheless, there are some regularities. For example, Starr (1969) defined four laws of the acceptability of risk. First, it is proportional to the cube of real or imagined benefits associated with the risk. Secondly, the public will accept risks derived from voluntary activities which are about one thousand times greater than those which it would tolerate from involuntary activities which would generate comparable benefits. Hence, the tolerance of risks created by certain hazardous sports is thought to be three orders of magnitude greater than that pertaining to earthquakes or dam failure. Thirdly, the acceptable level of risk is inversely proportional to the number of individuals exposed to it. Fourthly, the level of risk tolerated for voluntarily accepted hazards is similar to that resulting from disease. Therefore, the general level of risk caused by disease forms a "psychological yardstick" for evaluating the acceptability of other risks (see Fig. 9.2). Starr's "laws" are debatable and not necessarily universally valid. However, the natural disease rate may indeed represent the maximum level of risk that Western society will tolerate (unless one includes the nuclear threat).

The natural hazard risk in a highly developed society is approximately equal to 1/100,000,000,000 fatalities per person-hour of exposure. This represents a minimum background level of risk: smoking, for example, involves risks that are 50,000 times greater, while commercial aviation, using motor vehicles and falling ill each involve risks that are 100,000 times greater. There are, of course, spatial variations that increase the natural hazard risk rate sufficiently to make it locally worth mitigating, and this analysis is definitely not valid for Third World countries, where risk levels of all kinds are generally higher.

Risk analysis

Risk to society is defined as the size of a societal hazard multiplied by its probability of occurrence (Okrent 1980). It can be quantified as the chance of a given number of people being killed in a particular impact or accident per unit of time (e.g. per year). For example, it has been calculated that the benefits of putting the best available hazard mitigation solution into operation in California outweigh the costs that would thus be incurred by 6.2 to 1. In the case of all hazards that significantly affect this state, the benefits of mitigation outweigh the costs in monetary terms,

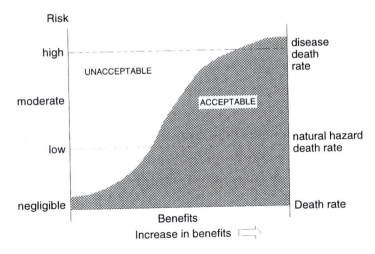

Figure 9.2 Risk-benefit pattern from involuntary exposure to hazard (after Starr 1969: 1,234). Copyright 1969 by the AAAS.

with ratios that vary from 1.3:1 to 137:1 (the cost of a comprehensive statewide hazard mitigation programme would, however, exceed $6,000 million). But not all risks and benefits can be measured, and many are ignored when people make decisions. There are many intangible factors (such as the value of a human life), and hence risks are often difficult to express in numerical terms. Yet this is possible, providing sufficient precise information on them is available. As society's resources are limited, the greatest hazards in a particular location should receive the highest proportion of expenditure on risk reduction.

The phenomenon of risk can be divided into two forms: **objective risk** is that which can be calculated from statistical data on past events, while **perceived risk** is the assessment of hazard made subjectively by individuals. The latter is apt to differ from the former in proportion to the significance given by the public to particular events. Perceived risk is determined by any of the following means (Burton & Pushchak 1984). First, the probability of an event of given magnitude may be calculated by assuming that it is going to be very similar to a better known one that has happened in the past. Similarly, reference may be made to those events that can be recalled, and those that are remembered most easily are regarded as the likely standard for future occurrences. Alternatively, perceptions may be anchored to an original experience, and not substantially altered by whatever has happened since. Together with objective knowledge, these judgements form the basis of risk as it is perceived by lay people.

Risk analysis consists of identifying the risk, measuring or estimating it and evaluating it (Covello & Mumpower 1985). The process involves both the science of measurement and the art of judgement in order to determine the acceptability of particular risks. Terms borrowed from the literature on technological hazards include "release" (the rate at which the hazard strikes), "exposure" (the vulnerability of populations per unit time) and "dose rate" (impact per person).

Once it has been estimated, the occurrence of risk over time can be expressed as a simple probability equation:

$$R = \sum P(E).C.t$$

where
R = risk level per unit time
E = events, expressed in terms of probability, P
C = consequences of the events
t = some unit of time

Risk as perceived can instead be described by the following relationship:

$$R = C^p.Ca.P(E)$$

where
Ca = causes
C = consequences raised to a power, p, as a result of fears generated by perception

In terms of risk analysis, the distinction between "optimizers" and "satisficers" (see Ch. 1) can be refined further. According to Burton et al. (1978: 49), there are several ways in which the degree of risk incurred by particular actions can be assessed. Hence, a person who considers all known possible outcomes and rationally chooses the most lucrative uses **expected utility methods**. One who has a (perhaps deliberately) inaccurate view of known outcomes uses **subjective expected utility methods**. And an individual who for personal reasons chooses a compromise between utility and risk that in objective terms is less than optimal uses **bounded rational methods**.

Much of the complexity of risk analysis stems from the interconnection between risk and external influences. It cannot, for example, be separated adequately from cultural factors, which largely determine what sorts and levels of risk people are willing to take or tolerate. Thus value judgements are often paramount. Moreover, politics can be instrumental in determining attitudes to some risks, yet there are insufficient links between scientific risk analysis and political decision-making (O'Riordan 1982).

Risk analysis is a valid technique only if it fulfils a series of appropriate criteria (see Fischoff et al. 1982). Thus it should state the probability as well as the size of impact to be expected, and the latter should be expressed in relation to the size of the population at risk. It should state clearly all the conditions and assumptions on which it is based and, if these are not constant, the outcome should be allowed to vary in proportion to the degree of uncertainty. Furthermore, the confidence limits of predictions should be given, and the means of deriving them explained. Multiple hazards should be evaluated, or partial risk distinguished from total risk, and exposure should be divided into voluntary and involuntary types. Finally, risk analysis should be distinguished from risk policy: the former explains the situation, while the latter offers a prescription for changing it.

Inconsistencies in risk perception and management

Although in theory there is a point at which the cost of risk reduction is balanced by the savings in risks reduced, society tends to set arbitrary tolerance levels on the basis of its perception of risks and priorities for their management (see Fig. 9.3). Sadly, one is forced to conclude that to reduce risks rationally is not one of society's principal aims.

The explicit consideration of societal risk calls into question the definition of a natural hazard or disaster. Consider, as an example, the disaster which occurred on 31 July 1976 in Big Thompson Canyon, Colorado. The canyon was struck by a flash flood resulting from a very intense thunderstorm. It has been estimated that about 4,000 people were

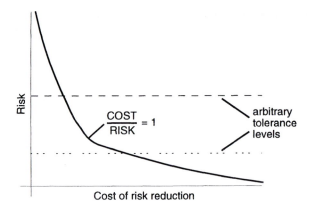

Figure 9.3 Relationship between risk and cost of risk reduction (Fischhoff et al. 1978: 34).

in the canyon at the time, many of whom were holiday-makers. Of these, 139 died and four were recorded as missing, while $41 million of damages were caused. There had been a total lack of prior preparation and mitigation: possible strategies not used include flash-flood monitoring, warning and evacuation systems, information dissemination and restrictions on the use and occupance of buildings in the canyon. Virtually every fatality could have been avoided by using these methods, but efforts after the disaster to impose restrictions on land use and occupance in the canyon were fought vigorously by developers. Clearly, different protagonists required different levels of safety, and a consensus had not been achieved.

Another illustration of societal risk is given by the dam safety problem in California (Okrent 1980). Starting from the hypothesis that 250,000 people could be killed by the failure of a single dam, the risk of such a failure has been quantified as 1 in 5,000 dams per year; but the probability of any individual dam failing in any given year may be as high as 1 in 100. The near-collapse of the Van Norman Dam in California, as a result of soil liquefaction during the 1972 San Fernando Valley earthquake, was only recognized as a hazard at the site after the dam had been constructed. Had dam failure been complete and the reservoir full, the number of fatalities could have exceeded 50,000. In response to this event, the State of California passed a dam-safety law, which specifies that each dam under its jurisdiction must be reviewed and found "safe". But the law does not oblige the state to publicize the risks that inevitably remain when a dam is declared safe, and these risks are not reduced by such a finding. Moreover, privately administered dams are not subject to inspection under the terms of the law, and hence the risks that they pose are largely unknown.

In 1976 the US Geological Survey advised the Governor of California that the collapse of aseismic buildings would cause thousands of deaths during a major West Coast earthquake, and that such an impact was highly likely. But four years later the City of Los Angeles had imposed neither prohibitions on the use of aseismic buildings nor had it formulated evacuation plans down stream of vulnerable dams. The fact that the state chose not to devote significant financial resources to these problems indicates that they have been given low priority and the risks have at least partially been discounted.

In practical terms, the sums dispersed on saving lives vary greatly, and with little regard to the actual probability of death in any particular hazard (Foster 1980). In recent decades, the French have spent $30,000 per life to prevent deaths in road accidents and $1 million per life saved in aviation accidents. The British have spent $10,000 to preserve the life of an agricultural worker, and $20 million to ensure the safety of each high-rise apartment dweller. In the USA the expenditure on avoiding deaths in radiation emissions from nuclear power plants has reached $1,000 per person-rem, or $5 million per premature death avoided. Yet deaths caused by the pollution emanating from coal-fired power plants are valued at only $30,000, because the majority of victims are reckoned to be chronically ill old people, who have already ceased their economically productive lives (yet the victims of radiation will not necessarily be young!). These discrepancies indicate that the criteria for expenditure are independent of the true relative risk of death: and in this respect little attempt has been made to put natural hazard risk into a context of other societal risks (see, however, Petak & Atkisson 1982).

Social surveys demonstrate that a person's understanding of risk is more or less directly proportional to his or her experience of it (Van Ardsol et al. 1964). For example, despite annual flooding of Darlington, Wisconsin, people continue to live there. The high level of hazard awareness in the area suggests that residents have calculated intuitively that the potential benefits of residing there outweigh the potential risks of loss. In other cases, however, there may be a widespread desire to avoid the loss but, through poverty or other constraints, an inability to do so.

Problems of rational hazard management

Essentially, society pursues more than one goal: employment, food, water supply, energy production and transportation all have to compete with the need for public safety. In principle, a thing is safe if its risks are judged to be acceptable (Okrent 1980). But there may be little consensus on what is acceptable, especially as most risks cannot be reduced to zero, at least not without intolerable expense. Effective risk management encounters a series of obstacles. To begin with, risks and benefits are not evenly distributed across society, and in certain cases one group may run the risk and another reap the benefits. The problem is often one of poverty, in that disadvantaged groups are often the ones that bear the greatest risks in all fields, natural disasters included.

Another problem is that the intensity of risk does not increase evenly with degree or length of exposure to a hazard. This makes it difficult to determine a "safe dose". Furthermore, over time, risks, and the responses to them, do not necessarily remain fixed. Society has the potential both to put itself at risk and to withdraw from risks that it does not wish to entertain. Mitigation and risk reduction can be applied but can also lapse. Hence, risk levels are somewhat dynamic. On the other hand, social, economic and locational inertia reduce society's capacity to respond to risk. At the same time, the complexity of risk makes it difficult to manage. Multiple hazards may threaten an area, but time, money and levels of interest may be sufficient to enable only one or two risks to be reduced effectively.

The dichotomy between actual and perceived risk weakens attempts to reduce risk and promotes instead a conflict of objectives. Concurrently, not enough may be known for cause to be linked to effect, yet caution is difficult to quantify in respect of risk reduction strategies, while the strategies themselves may result in losses for some groups (for example, lost production may be an unavoidable consequence of evacuation drills for the staff of a factory). Finally, there are likely to be institutional weaknesses. On occasion, the same agency is responsible for both promoting and controlling the risk-taking activity. Alternatively, the responsibility for risk reduction may be divided among several agencies, such that it is hard to apply safety measures effectively.

Strategies for risk management

Several methods can be employed to reduce societal risk (Martin & Lafond 1988). In this, it is perhaps unwise to distinguish too clearly between the threat of natural hazards and that pertaining to other types of impact, as the net effect is the same.

Risk aversion (Smith 1979) involves a decision to achieve the maximum possible risk reduction, regardless of the costs involved. This method does not allow one risk to be compared with another in order to apportion scarce resources; neither does it allow the balancing of costs and benefits. It appears to be acceptable as a strategy only with respect to certain advanced technologies (air travel and nuclear hazards) and virulent diseases. Hence, it is not viable in terms of comprehensive planning (Foster 1980).

Risk balancing helps determine what risks are acceptable and thus helps set goals for community mortality, morbidity and economic loss. This approach assumes that some positive level of risk is socially acceptable. The level must be determined by comparing the risk in question with similar hazards elsewhere, with other risks and with certain yardsticks for acceptability. Public and private resources are allocated to increase the overall safety level to the point at which it is publicly acceptable. Risk balancing can be achieved by the **revealed preference method**, in which the standards of risk tolerance which society currently applies to itself are examined.

As it is usually impossible to create absolute safety, the process of risk reduction is a complex one that involves balancing many alternatives and making many value-laden judgements. The following range of options can be used, or to some extent combined with each other. A generally acceptable and rigorous approach to risk management would incorporate most of these strategies to some degree.

(a) **Non-intervention** is a strategy that simply leaves it to market forces to determine whether an activity is acceptable or not. High-risk activities will have unacceptable costs associated with them.

(b) **Professional standards** can be employed by relying on technical standards to determine the appropriate level of risk.

(c) **Procedural approaches** involve designing regulations to prevent or limit activities that appear to be too risky.

(d) **Comparative approaches** can be employed to reveal societal preferences in the face of a varied catalogue of risks. These include comparison with natural background or pre-existing levels of risk, alternative ways of dealing with the risk, unrelated risks, or benefits.

(e) **Cost-benefit analyses** or cost-effectiveness criteria can be utilized to provide a rational economic criterion for risk reduction expenditures. Hazards that involve particularly expensive impacts must be mitigated. Unfortunately, however, benefits are notoriously difficult to measure.

(f) **Decision analyses** can be used to make the process of risk reduction explicit and hence understand better the criteria by which it is better achieved.

(g) The method of **expressed preferences** relies on public perception of risks in order to obtain a consensus on how resources should be devoted to their reduction.

Using cost effectiveness to determine the acceptable level of risk involves maximizing risk reduction achieved for each unit of expenditure. When additional expenditure results in insignificant increases in risk reduction, then nothing further will be done. To use this method, all hazards must be evaluated in terms both of their relative severity and of the relative cost of reducing their risks. However, some hazards can be reduced significantly at minimal cost; for example, by prohibiting access to dangerous areas.

In cost–benefit balancing the losses tolerated increase in proportion to the size of benefits resulting from the activity or risk-taking venture. The greatest levels of tolerance of a given risk pertain to cases in which the activity is important and cannot easily be modified or relocated out of harm's way.

Unfortunately, decisions about risk often produce inequality in society, in which people who do not have the economic or political power to avoid the risk are forced to tolerate it, whether they would prefer to or not (Smith 1990). Several alternative ways of tackling this problem have been outlined by Kasperson & Kasperson (1983). First, risks can be adjusted so that they maximize the summed welfare of all members of the community (the **utility condition**). Secondly, risks can be allocated according to people's ability to bear them (the **ability condition**). Thirdly, imbalance of risks can be compensated proportionately (the **compensation condition**). Finally, the allocation of risks is just only if it has the voluntary consent of those who bear the risk (the **consent condition**).

Wrong priorities in risk management effectively mean that people are being killed whose deaths can and should be prevented (this is, however, a relatively common feature throughout history). Although there are no signs that the public perception of risk will lead to a much more rational state of management, most governments contain at least some members who recognize the need to confront risks judiciously and rationally. Risk analysis teaches that there is usually an optimum balance between risk reduction measures and threats to public safety, and that beyond this the benefits of new mitigation measures are unobtainable or not worth achieving. As the resources of society are scarce, a national strategy of risk reduction is needed in order to apportion them efficiently. However, the particular interests and agendas of risk mitigators (architects, engineers, public officials, etc.) can make it difficult or impossible to determine risks objectively. For example, risks are easily overdramatized when there is money to be made out of mitigating them!

Besides the risks posed by natural hazards to individual and collective safety there is also a risk of material loss and costly damage. This prompts an examination of disasters in the light of the theory and practice of economics.

The economics of disaster

As stated in Chapter 1, natural disasters tax the global economy by at least $50,000 million a year, of which two thirds are accounted for by damages and losses, and the balance represents the costs of prevention and mitigation. Of the $30,000 million in damages, perhaps $12,000–18,000 million per year occurs in the form of large, catastrophic losses, such that the average loss from a major international disaster is about $350–500 million (Berz 1992). On average floods and drought cost only about one tenth of 1 per cent of the GNP of industrialized countries but twenty times as much (about 2 per cent) of the GNP of some badly afflicted Third World nations. Worldwide, growth rates have averaged about 3 per cent during most of the twentieth century, although this figure masks substantial national, regional and temporal variations. Thus it can be seen that in the long term natural disasters have little economic effect on industrialized countries, but (where they are recurrent) they can drastically slow the economic growth of developing nations.

A special case is represented by the territories of the former USSR, in which direct losses caused by natural disasters cost 3,500–4,000 million roubles per year, or about 0.4 per cent of GNP. The figure doubles if non-agricultural indirect losses are included and quintuples (to 2 per cent of GNP) if the impact on agriculture is added. These figures do not include medical or social insurance costs, which, however, are only moderate, in view of the low standard of care offered. Military spending in the former Soviet Union absorbed between 4 and 50 times more of GNP than did natural disasters. But in comparison with the USA, natural catastrophes in the USSR took 3.5–4 times more of national GNP, while military spending took only 2.5 times more (Porfiriev 1992). Thus, in the context of GNP the economic impact of natural disasters in the Commonwealth of Independent States resembles that of developing nations in some of the more febrile and hazardous areas of the world: economics, we might conclude, cannot be separated from politics.

Some economic effects of disaster
Economic considerations are important both before and after disaster has struck (Sorkin 1983). With regard to the former, both the state of vulnerability and degree to which this is generally known will have possible economic consequences. Thus, in the United States much attention has been given to assessing the negative impact of a future earthquake prediction (see Fig. 9.4; Cochrane 1984, Ellson et al. 1984). It is likely that once a damaging seismic event is forecast, the economy of the area involved will be affected by an "inverse multiplier". Property values will fall, consumers will save rather than spend, businesses will face reduced profits and unemployment will rise in a sort of chain-reaction. Those businesses that have substantial markets, with plant and capital invested outside the area, may move elsewhere, but those that have strong ties to the local area (and have taken precautions against impending damage) may elect to remain. Generally, smaller towns and those where there is only one industry or source of employment are likely to be the most vulnerable to the negative effects of earthquake predictions, while larger communities (such as major cities) are likely to

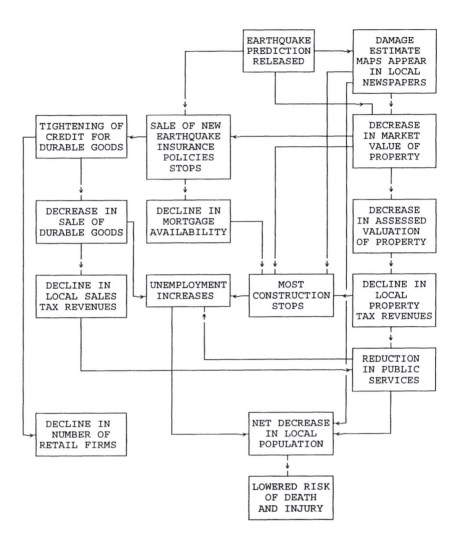

Figure 9.4 Likely consequences of a magnitude M ≥ 7 earthquake prediction (Rikitake 1982: 303). Reprinted by permission of Kluwer Academic Publishers.

find in their own diversity a capacity to resist decline, not least because substantial parts of them may be located outside the major seismic risk zone.

Once disaster has finally struck a community, the potential financial consequences include loss of tax revenues, loss of sales taxes through decline in or damage to local businesses, lowered debt ratio (borrowings in relation to taxable property) and reduced income from service charges. However, counter-measures can be taken. These include the transfer of general funds to a "disaster account", the completion of mutual aid agreements with neighbouring jurisdictions, self-insurance through pooling resources with other communities, borrowing based on tax anticipation promissory notes, the issuing of municipal bonds, participation in private or national insurance schemes and supplementary taxation.

Law suits are a particular problem of disaster aftermaths in some countries, notably the United States (see Kusler 1985). For instance, in the January 1982 storms in the bay region of San Francisco 6,500 homes and 1,000 businesses were damaged or destroyed at a cost of $280 million, of which $66 million of damage resulted from landslide impacts and the rest from flooding. But over the next four months lawsuits against various local authorities amounted to $18 million more than the total cost of damage.

In order to win a suit, the plaintiff must prove that whoever is being prosecuted (the local authority, for example) had a clear duty to protect against the hazard which caused the damage, that the damage occurred and that this was because of failure to carry out the duty. The defendant will win if any one of these conditions cannot be proved. Moreover, liability will probably not be assumed automatically, but may instead be substituted by the concept of "reasonable care" – i.e. the level of protection that could reasonably be expected under a possibly complex series of circumstances.

Disasters as a tax on society

Disaster damage costs are a form of tax which is distributed among the various sections of the community. People who are killed or who suffer serious physical injury pay the highest price, while those who suffer homelessness, loss or damage pay a lesser price. There is in fact a continuum from direct effects to indirect responses, where the costs of damage are apportioned among the community or nation at large (perhaps even the international community) in the form of voluntary donations and mandatory taxation (Fig. 9.5, Table 9.2). The arrangement of this taxation system may be planned or *ad hoc*, depending on whether the disaster was foreseen, and arrangements were made to apportion the costs, or whether it came as a surprise (in reality no natural disaster should take a government by surprise, but many administrations, when faced by pressing problems in other fields, may turn a blind eye to the risk). The apportionment of costs may also be explicit or concealed, depending on whether the general public is aware of who is paying for the disaster. When the costs of damage, mitigation and other forms of adjustment are concealed within general taxation it may be difficult to ascertain who bears the greatest financial burden.

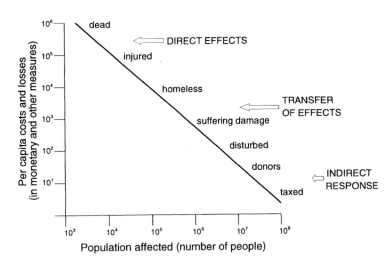

Figure 9.5 Distributive effects of natural hazards and disasters (Burton et al. 1978: 120).

Table 9.2 Distributive effects of natural hazards and disasters (after Burton et al. 1978: 120).

Small population affected, concentrated effects

- Death (mortality)
- Injury (morbidity):   trauma: severe (hospitalization)
                                      slight (out-patient treatment)
                          disease
                          starvation and malnutrition
                          psychological injury
- Bereavement
- Homelessness: permanent (reconstruction or migration)
                          temporary (repair)
- Unemployment
- Damage and destruction of assets and possessions
- Economic loss: loss of income or investment, indebtedness, bankruptcy
- Disruption of activities
- Voluntary donation
- Compulsory taxation

Large population affected, well distributed effects

This discussion of costs allocation leads to a consideration of questions of equity. Ethically, people who inhabit hazard zones should bear the cost of disasters and the responsibility for the risks which they run, if these are assumed voluntarily. Where losses are likely to be unavoidable, unforeseeable or induced by forms of public exploitation (such as the kind of speculation in construction that cannot be blamed on the eventual residents), then governments have a duty to indemnify their citizens (Burby et al. 1991). Two distinct systems of indemnification have evolved, although they are not necessarily mutually exclusive. First, governments may simply pay for disaster losses by offering survivors grants and loans, credit relief and other forms of subsidy. At its most unsophisticated, this has been nicknamed "forgiveness money", because it effectively forgives the victims for lack of a hazard mitigation strategy. Although quite complex reconstruction laws can be passed, which make structural mitigation mandatory, this method does not tend to encourage the adoption of a range of measures for reducing the impact of future hazards. Alternatively, insurance schemes can be initiated under the auspices of government bodies or private sector companies (see below). This is a means of committing present-day resources to hazard mitigation and the amelioration of losses in the future, but it demands that society have a sufficient surplus to be able to accumulate sufficient reserves.

Relative adequacy of adjustments to hazard: an economic approach
In the discussion of adjustment to hazard given in Chapter 1 the economic aspects were largely implicit. It is now time to consider them in their own right. Actual estimates of losses in disaster are always based on the fact that the geophysical event is complemented by a human response. Losses are a measure of human failure to adjust to environmental hazards. Natural events are so variable that a single level representative of the "threshold intensity for disasters" cannot be specified and, as human responses are equally variable, neither can a "threshold intensity of response".

The ratio of damage or loss to costs of adjustment varies from about 1:1 for droughts to 5–10:1 for floods and hurricanes (Burton et al. 1978: 77). Generally, costs rise and losses fall with more adequate adjustment (Russell 1970). In economic terms, the most justifiable expenditure on adjustment occurs where the two curves intersect (Fig. 9.6a). This represents A*, the minimum of the two sums, losses L, and costs C (Fig. 9.6b). In pure cost–benefit terms, no further expenditure is justified if rising marginal costs diverge from falling marginal losses, as the cost of each new adjustment will always be greater than the cost of losses (Fig. 9.6c). Hence A*, the most economical practicable adjustment, occurs at the intercept of the L + C curve (Fig. 9.6d).

An illustration of the importance of marginal costs is given by seismic retrofitting (see Fig. 9.7). In Mexico this accounts for 15–40 per cent of the total costs of new construction. However, there is a nonlinear relationship between investment in the seismic upgrading of buildings, size of earthquake and potential victims saved. For example, if loss of life is quantified in monetary terms, great savings can be achieved by strengthening the weakest 5 per cent of buildings, but the per capita savings are reduced by spending more or by attempting to ameliorate the effects of smaller earthquakes (Coburn et al. 1989).

In this situation the best course of action may be to spread the losses, to share them between the general public and those who have to suffer

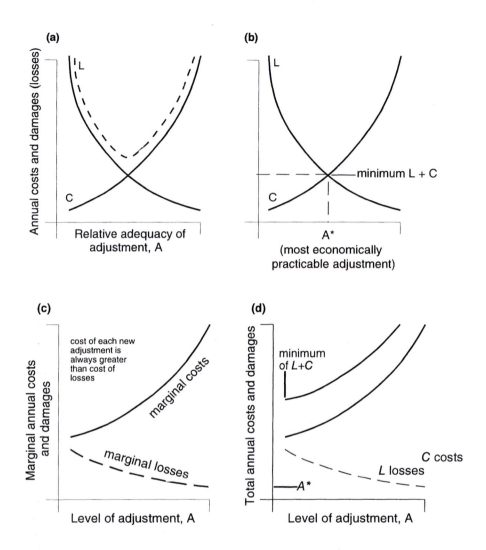

Figure 9.6 Costs and damages in relation to relative adequacy of adjustment to hazards (Russell 1970: 386, 389).

the disaster impact, or to share them between the greatest and least affected victims of the disaster (as in Fig. 8.1). This is accomplished by insurance, taxation, loans and grants (Burby et al. 1991). In essence, society transfers real income over time in order to smooth out the course of aggregate social welfare. In a similar vein, productivity must be averaged over the years to take account of periods in which production was reduced by the impact of disasters. In economic growth, in fact,

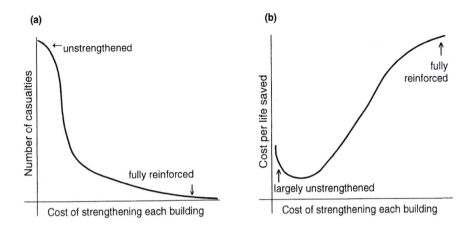

Figure 9.7 Economics of seismic retrofitting in terms of lives saved and relative costs (Coburn et al. 1989: 116).

setbacks are natural, and realism is the only economically viable way of facing up to disasters. Instead, people often put off their own need to face up to future losses, using fatalism, a naive belief in a hypothetical law of averages (in which recent disasters are judged not to recur until a decent interval of time has passed), or by underestimating the power of disaster impacts.

The nature of economic aid after disasters

Many losses sustained in disasters are compensated for by government-sponsored grants or loans. Aid of this kind frequently depends upon the recognition and classification of an impact zone as a "disaster area". Disbursements may be complemented by action taken to waive or suspend outstanding debts or the granting of loans and subsidies intended to help pay them off. Commentators have often termed this "forgiveness money", as it does nothing to encourage preparedness against disaster and mitigation of its risks. Indeed, such a strategy rewards the gambler and penalizes the prudent and virtuous citizen, who safeguards himself against the effects of natural hazards. Thus, relief money may actually encourage people to build or rebuild unsafe structures or structures in unsafe areas.

Insurance against disaster losses

In actuarial terms the risks posed by natural disasters are often difficult to assess and may be ambiguous. Thus, in the USA tidal surges caused by hurricanes were once the subject of extensive litigation, as standard home insurance covered damage caused by wind but not by water. The insurance costs of a major earthquake, for example, are rather poorly understood. Insurance companies, moreover, may adopt a strategy of discouraging the residents of hazardous areas from taking out private policies against natural hazard risk (or at least not encouraging them to do so). Under such circumstances, the best strategy may be for the government to supply insurance.

Research in the USA (Kunreuther 1974, Kunreuther et al. 1978) has shown that insurance companies do encourage individuals to take out policies if the risk is small. In such cases policies concerned with disaster losses are not costly and are encouraged by the insurance industry, on the principle that many premiums will pay for the losses sustained in few disasters. As insurance companies are concerned not to tarnish their public image, they will therefore insure property owners against earthquake damage. But flood insurance in the USA has been opposed by private insurance companies, as it is difficult to spread the premiums among policy holders and the damage is potentially very costly. Insurance against flood damage is not considered a realistic profit-making option, and hence the Federal Government took over the field after the Rapid City (South Dakota) flood of 1972 proved that a voluntary programme would not work. The US National Flood Insurance Act now enables the Federal Government to subsidize flood insurance costs by 90 per cent.

The insurance question actually consists of several alternative strategies (Kunreuther & Miller 1985). **Compulsory insurance** ensures that risk-takers set aside a portion of their income or benefits to indemnify against their expected future losses. This is potentially a fair option, but is not likely to be a popular one and may fail if participants cannot, or will not, pay the stipulated premiums. **Long-term insurance** (e.g. 50-year policies) will be necessary in the case of potentially high losses, as companies will need to build up capital investments against future disbursements. However, such a strategy would do little or nothing to discourage people from living in hazardous areas.

Separate coverage against natural disaster losses would ensure that people not running the risk would not be expected to indemnify against it. The best strategy here would involve a comprehensive disaster insurance package, and some sort of subsidy might be necessary as the value of property would be expected to fall in relation to the disclosure of risks and the probable cost of "economic" insurance rates charged on it. Accurate hazard zonation and recognition is essential to this strategy, and that is why American and German insurance companies have started to generate vast data banks on the incidence and location of natural disaster impacts (Rossi et al. 1983, Berz 1992). These companies are mostly involved in **reinsurance**, which is appropriate where individual private insurance companies are unable to bear the risks associated with particular hazard situations. Such institutions could be underwritten by a government pledge to protect them from losses which they could not sustain.

In New Zealand all buildings with fire insurance are assessed an extra 5 cents per $100 of insured value. Of the extra revenue, 90 per cent goes into an earthquake and war damage fund and the remainder is credited to the Extraordinary Disaster Fund, to be used for relief and rehabilitation in the event of tornadoes, landslides, volcanic eruptions or tsunamis (Falck 1991).

Insurance has also been studied from a sociological point of view. Mileti et al. (1984) found that the more earthquake insurance is purchased in a community, the fewer the anti-seismic measures and initiatives adopted. Studies reveal that people will not insure themselves against hazards unless the risk they run exceeds some arbitrarily set threshold, their assets seem deserving of protection and their income furnishes a necessary surplus. Moreover, people tend to be unwilling to acquire the necessary information to make such decisions rationally, unless prompted to do so by some particular event, such as a recent disaster (Kunreuther & Slovic 1982). Thus earthquake insurance (if it is available) is purchased mainly after a seismic disaster and much less in quiescent times. But field studies (Kunreuther et al. 1978) suggest that many of the most pessimistic householders will not purchase insurance, while some who feel that there is little chance of a natural hazard affecting them have watertight policies! In summary, the main reasons why people do not buy insurance when it is offered seem to be that they have a false sense of security or that they lack the information that would impel them to seek cover.

The economic impact of disasters in the USA
The shear size of the United States and the range of natural disasters that affects it make it a good illustration of the national economic impacts of disaster losses. It is also one country that has been particularly active in formulating policies to mitigate such impacts, though the results of this have been somewhat heterogeneous.

Over the period 1953–86, the United States Federal Government paid out disaster relief totals of more than $30 million in 35 different states. From 1 April 1974 to 30 June 1986 the president declared 348 major disasters under Public Law 93–288. Total Federal payments were $2,990 million, an average of $8.59 million per disaster. The 28 disasters in which more than $25 million was disbursed under the US Disaster Relief Fund accounted for 55.3% of funds used. The annual losses caused by natural hazards in the USA amount to more than $8,000 million and are expected to rise to $17,700 million by the next century.

By the year 2000 natural hazards in the USA may well become more costly than traffic accidents, health insurance premiums and comparable

591

risks (Petak & Atkisson 1982). Among the top nine hazards, only riverine flooding and expansive soil hazards are likely to decrease in per capita costs to the nation, and by less than 10 per cent. The others will probably double each decade, despite the key rôle of mitigation in reducing the rise in costs.

American Federal policy on disasters has been to treat them as a public responsibility with the costs of damage sustained by the few being borne by the many – i.e. taxpayers. The Small Business Act of 1952 instituted the granting of loans at low interest to owners of homes or businesses damaged by natural disaster. Until 1964 loans were offered at 3 per cent interest and were repayable over 20 years. After the Alaskan earthquake of that year, in which the Small Business Administration played a prominent rôle, the programme was generalized to include loans at 7–8 per cent interest, repayable over 30 years, to amortize loans for which the collateral had been destroyed in the disaster. Then, in the Southeast Hurricane Disaster Relief Act of 1965, up to $1,800 of each loan was "forgiven", or turned into an outright grant. This "forgiveness" ceiling was raised to $2,500 in the 1970 Disaster Relief Act. In 1972, floods at Rapid City, South Dakota, and the impact of Hurricane Agnes further liberalized policy; the size of grant was doubled and loans were dispensed at 1 per cent interest repayable over 30 years.

Essentially, this rising tide of "forgiveness" money meant that some survivors found themselves materially better off after the disaster than they had been before, while a few even managed to amass working capital as a result of government aid. But from 1974 onwards the picture changed. The period 1933–73 had seen the growth and entrenchment of the assumption that Federal action was necessary in order to combat natural hazards effectively. The Flood Control Act of 1936 set the pattern for almost four decades by encouraging reliance on structural methods, and discouraging or not giving equal weight to non-structural approaches. But the excessive cost of structural control and its lack of overall success caused the emphasis to change from the mid-1960s onwards, and a new strategy of more comprehensive policies began to emerge in the 1970s, including the principle that an individual should bear some responsibility for the risks he or she runs, and also for mitigating them.

In the USA there is now a trend towards compulsory insurance on any mortgaged or loan-supported property (or, in fact, on any collateral for loan repayments). This will increase the day-to-day cost to the individual, but reduce the collective burden when disaster strikes. Generally, however, the American public is not interested in sharing costs by purchasing natural hazard insurance, and it appears that a large percentage of the uninsured believe that they are under no risk. In California, earthquake insurance is available for a few dollars per $1,000 of property insured, with a 5 per cent deductible (both the premium and the deductible are halved in rates pertaining to the eastern seaboard of the USA), but fewer than 5 per cent of residents have actually purchased such insurance. Moreover, research by Palm et al. (1983) indicates that the banks and mortgage companies have limited perception of earthquake risks and, in any case, are much more concerned about factors such as divorce and separation as potential causes of loan repayment defaulting (cf. Anderson & Weinrobe 1986). The American public seems to regard natural hazard insurance as an investment from which a return is expected. People more readily buy insurance against high-probability low-loss events than against low-probability high-loss events.

Hitherto, current and recent economic trends in natural hazard impacts and mitigation have been examined. Economic history enables us to adopt a longer-term perspective: in fact, disasters can be looked upon as abrupt shocks to the economic system (Jones 1981). The density of human populations, their income level and social organization, the crops they grow and the animals they keep, all affect their vulnerability to impacts. Throughout history large and small disturbances have engendered a continuous series of adjustments in human activities. Indeed, some authors have seen the large disturbances caused by disasters as, in principle, no different from the workaday adjustments made in the economy to shifts in supply, demand and relative prices. In the next section, which deals with the historical dimension of disasters, economics remain a key issue, linked as they are in history to the very survival of cultures, states and populations.

An historical perspective

The historical record of natural disasters varies greatly from country to country. Iran, Italy and China, for instance, are extremely rich in long-term records of disasters and their impacts. In Egypt the "Nilometer" at Cairo gives a 3,500-year record of the flooding of the River Nile, which was so vital to agriculture (although one in which the records were sometimes altered to facilitate tax evasion!). On the other hand, in the Americas records of past disasters and their impacts are often extremely sketchy. Nevertheless, where records are sufficiently detailed and long-standing, we may ask several important questions. For instance, have natural disasters either directly or indirectly led to the destruction of human civilizations? What is the relationship between patterns of disaster and the path of economic development? Does the pattern of impact and effect vary from one part of the world to another? Does the severity of destruction of labour and capital vary also?

In order to answer these questions, it is useful to classify disasters in the following manner (Jones 1981): **geophysical hazards** include earthquakes, landslides, tsunamis and volcanic eruptions; **climatic hazards** comprise droughts, floods, hailstorms, hurricanes, torrential rains and windstorms; **biological hazards** involve crop disease outbreaks, epidemics, epizootics and locust invasions; and **social hazards** include the collapse of man-made structures, insurrection and repression, urban fires and warfare. In a book on natural disasters, *sic stantibus rebus*, we are primarily concerned with the first two categories of this list.

Geophysical disasters and history: volcanic eruptions

Major volcanic eruptions have sometimes been regarded as capable of eradicating the localized human cultures that they affect. The Greek island (or, more properly, islands) of Santorini (Thera) in the Aegean Sea lies 110 km north of Crete and 210 km southeast of Athens. This small and fragmented landmass comprises the remains of a volcano that erupted around 1500 BC in an immense cataclysm that formed a caldera 600–800 m deep and 83 km^2 in size. Tsunamis and earthquakes occurred, and 5 cm of tephra fell over eastern Crete and the islands of Karpathos and Rhodes. Archaeologists have speculated that the eruption caused the demise of the Bronze Age Minoan culture (3000–1470 BC), which was supplanted by the mainland culture of the Myceneans (see Antonopoulos 1992). But although the ashfall may have devastated agricultural production, and the tsunamis scuttled the Minoan fleet at anchor at Knossos, it took several decades before the Cretans were finally eclipsed by Micenæ. There is similar speculation regarding the demise of the Hindu state of Mataram after a volcanic eruption on Java in AD 1006, but again the evidence is inconclusive.

The Laki fissure eruption of 1783 in Iceland emitted 12 km^3 of lava, which flowed freely over several hundred km^2 (Thorarinsson 1970). Livestock fell victim to fluorine poisoning and bone diseases (osteo-fluorosis), accumulated ash in their digestive tracts or had their teeth abraded by sharp particles that had settled on the grazing lands. Yet even such a calamity as this did not spell the end of settlement on Iceland, even temporarily (Jackson 1982).

One can compare these circumstances with one of the greatest eruptions of modern times, that of Krakatoa (Java) in 1883 (Simkin & Fiske 1983; see Fig. 9.8). The statistics associated with this event are impressive: 17 km^3 of magma were erupted, and volcanic dust fell out of the atmosphere at 2,500 km from the volcano over about 827,000 km^2. During the paroxysmal phase of the eruption total darkness occurred 450 km from its centre and lasted 60 hours. A sound wave travelled 5,000 km and an atmospheric pressure wave 6,000 km, while tsunami waves 40 m high killed 36,000 inhabitants in 295 coastal fishing villages (Yokoyama 1987). Yet, despite the severity of the impact, the course of civilization remained unaltered.

Similarly, the largest eruption in recorded history is that of Tambora,

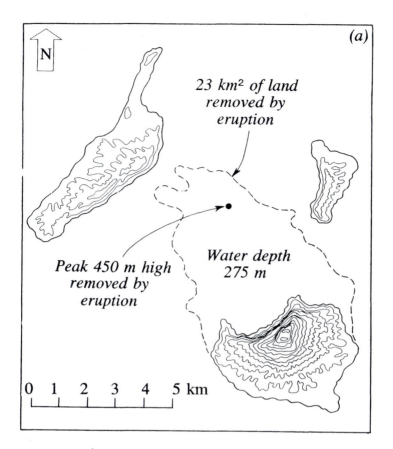

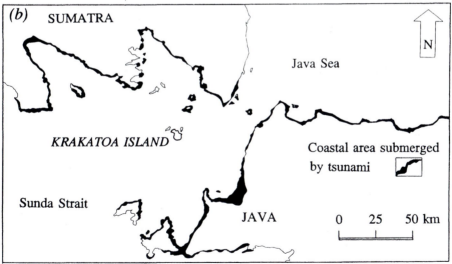

Figure 9.8 Distribution of effects of the 1883 eruption of Krakatoa (Decker & Decker 1981: 51).

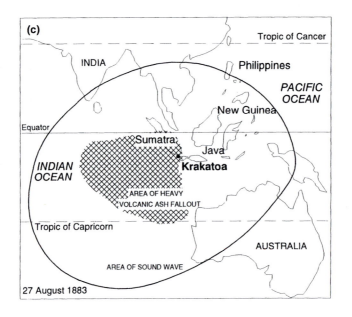

(c)

Tropic of Cancer

INDIA

Philippines

PACIFIC OCEAN

New Guinea

Equator

Sumatra

Java

Krakatoa

INDIAN OCEAN

AREA OF HEAVY VOLCANIC ASH FALLOUT

Tropic of Capricorn

AUSTRALIA

AREA OF SOUND WAVE

27 August 1883

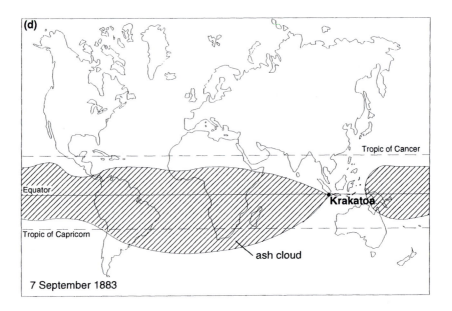

(d)

Tropic of Cancer

Equator

Krakatoa

Tropic of Capricorn

ash cloud

7 September 1883

on Sumbawa Island in Indonesia, which in 1815 erupted about 50 km^3 of magma, caused 90,000 deaths and affected climate worldwide (Stothers 1984). Like Krakatoa, its effect on human affairs was transient. But there is potential for eruptions more than ten times larger than this, as occurred in prehistoric times, and if such were to happen during the course of human history a major reorientation of society would probably be needed.

Geophysical disasters and history: earthquakes

Historically, 91 per cent of deaths caused by earthquakes have occurred in a belt of latitude from 25°–45°N, and 78 per cent in the band 30°–40°N. This tract contains Japan, central Asia, the Middle East and the Mediterranean lands (see Fig. 9.9). It is an area of high seismicity and ancient cultures, whose buildings and capital stock are in many cases in poor condition (see Ch. 2).

Yet though the impact of earthquakes upon particular cities could be great (Melville 1981), the death tolls in continental terms have been

Figure 9.9 This temple at Selinunte in Sicily was one of the largest and most robust of Magna Græcia. Since Classical times, however, it has been brought down by repeated earthquakes.

Figure 9.10 The destruction of Lisbon, Portugal (Davis 1978: 75) contemporary print of earthquake, tsunami and fire. Reprinted by permission of the New York Public Library, Astor, Lennox & Tilden Foundations.

slight. An estimated 830,000 people were killed in the Shensi earthquake in China in 1556, but this represents only 0.3 per cent of the Asian continental population of the time. Between 30,000 and 70,000 people died in the earthquake, fire and tsunamis that struck Lisbon, Portugal, in 1755 (Fig. 9.10), which was only 0.04 per cent of the contemporary European population. Yet the event prompted philosophers such as Voltaire to meditate deeply on the rôle of cataclysms in human social evolution. Hence, a strong imbalance can be observed between the impact of seismic disasters in Europe and Asia. In historical times Europeans have comprised about 21 per cent of the population of Eurasia, but have suffered only 0.7 per cent of the deaths in major earthquakes. Thus an Asian was 30 times more likely to die in a seismic disaster than a European (Jones 1981).

However, such observations are based on very imprecise data. In historical times it was rare that a tally of deaths and damages was compiled. Such information first began to appear at the time the early scientific societies were founded, and hence the 1692–93 earthquakes in eastern Sicily (Noto and Siracusa) were among the first to have their effects recorded and examined in the light of scientific curiosity (Hughes 1983). The six earthquakes that occurred in Calabria in 1783 had their impact carefully quantified by the Bourbon government's bureaucracy, the Papal Court and groups of scientific adventurers and travellers (Placanica 1985). But vast lands such as China felt no need to audit the toll of disasters, which in any case recurred with almost monotonous regularity.

598

Little evidence of "architectural Darwinism" can be found – in other words, there is little sign that the repeated collapse of buildings led to the "survival of the fittest" and hence improved structural practices. True, the palace of Knossos in Crete, built in about 1650 BC (and probably after a natural disaster) has a wooden ring beam that appears to have been designed to hold the walls together during earthquakes. In fact, Anatolia, Crete and northern Pakistan all show evidence dating back over 2,000 years of the use of timber bracing structures as an anti-seismic measure in areas that lack good supplies of wood. But these cases were the exception rather than the rule, as was the 1755 Lisbon earthquake, which led to more rational, fire- and earthquake-resistant reconstruction as part of a grand plan orchestrated by the Marquis of Pombal (Figs 9.10 & 9.11; França 1983).

Climatic disasters and history
In 610 seasons over the period AD 610 to 1619 one or more provinces of China received too little rain to enable crops to be grown effectively (Jones 1981). In one third of these cases extreme drought occurred. At the other extreme, from 206 BC to 1949 in China, 1,092 flood disasters were recorded. The breaching of the Hwang Ho (Yellow) River and Yangtse often led to the inundation of thousands of square kilometres of floodplain land. The former flows for 4,000 km of its 4,800 km length across loess deposits, which contribute a sediment load that may reach 40

Figure 9.11 Part of the Praça do Comércio in Lisbon, Portugal, as reconstructed by the Marquis de Pombal after the 1755 earthquake. Shear walls and fire-resistant partitions were incorporated into the principal buildings.

per cent of total discharge. The annual average area flooded is 8,300 km^2, and the deposition of silt on the bed of the river severely impedes drainage of the floodwaters. Even as late in history as 1880, up to 5 million people died in only three floods, if one includes both drownings and starvation through loss of crops. The debris flows of 1920 in Kansu Province, China, which also took place in loess deposits, killed about 100,000 people directly, while perhaps a similar number died from starvation or exposure to harsh winter conditions.

Floods and droughts in China tended to cause famines, of which 1,828 were recorded from 108 BC to AD 1911, amounting to almost one a year in at least one province. The causes of famine in China are best known for recent historical times, and studies of the period 1850–1932 suggest that drought was the main cause, followed by flooding (Wang & Zhao 1981). During this period starvation killed 5 per cent of the afflicted population, 13 per cent emigrated and 24 per cent barely survived.

Historical famines tended to result from absolute lack of food stocks rather than from the distribution and logistical problems that characterize their modern counterparts (Carlson 1982). Starvation often went hand in hand with disease, as in East Prussia during the period 1708–11, when 250,000 people died from these causes. Food shortage in France during 1692–94 may have led 2 million people to die of starvation, representing about 2 per cent of the continental population. Yet this toll pales into insignificance when compared with the Bengal famine of 1769–70, in which a third of the population (about 10 million people) died. The solutions to such problems could not be found without better social organization, improved communications and increased agricultural productivity. Until these were achieved, the crop failure tended to lead to a total scarcity of food, however much money was available in the economic system to purchase it.

Social historians consider that emigration to relieve the pressure on devastated land often took the form of rather aimless nomadism (Jones 1981). Its alternative was starvation, suicide or that true but neglected indicator of the strength of famine, cannibalism. The recovery from such disasters was usually sluggish: capital and working animals had been lost, skills had disappeared with the death or outmigration of craftsmen and manpower tended to be scarce. It might have taken 25 years to recover from a severe famine, at which time the next disaster would potentially be at hand. Economic historians estimate that the total failure of crops once a decade would require a saving of at least 10 per cent of gross output and 13 per cent of annual net production, which are very large surpluses even by today's standards.

Effects of disaster on capital accumulation

According to John Stuart Mill, disasters in Europe represented the equivalent of very fast consumption of goods which would in any case have been used up or worn out (Jones 1981). In the widest sense, these goods included cleared land, hedges, roadways, houses, farm buildings, livestock, implements, small-scale manufacturing plant and stocks of saleable items. Before the Industrial Revolution the local economies of Europe were cellular, in the sense of being near self-subsistence, and largely self-contained. War and pestilence tended to exert a profound impact upon labour, but destruction of capital goods by natural disasters was rare, and these remained protected. Moreover, there was an increasing tendency to conserve structures and capital so that rehabilitation after disaster could be accomplished rapidly.

Labour-saving devices were few and far between, but although the demand for labour remained high, human resources went unprotected, as improvements in political stability, environmental health and medicine were slow to occur or simply absent. In Asia, technological change did no more than keep pace with population growth, as labour and capital suffered equally from the profound, widespread and recurrent impacts of disasters.

Although technology developed at a pace that today would be regarded as very slow, it nevertheless permitted improvements in living standards to take place, and the expertise gained by experimentation and invention was not usually lost when disaster destroyed the capital goods themselves. However, advancements were often absorbed by the need to replace existing capital goods (which, if they were not destroyed by disaster, tended to wear out quickly), rather than to create new ones.

This situation is amply illustrated by the techniques used in mediaeval building. The poor built themselves weak, expendable shelter which could be replaced relatively easily when destroyed, while the rich, through ignorance of engineering principles, were forced to have their buildings overdesigned to withstand up to 12–20 times the maximum likely forces. The collapse of cathedrals (such as the loss of the nave at Utrecht during the storm of 1674) was costly and embarrassing, and required the long-term commitment of vast resources to rebuild. Hence, structures had to be designed on the basis of the available materials, as it was not yet possible to create new materials for particular structural tasks.

The problem of the survival of structures was particularly acute in Asia Minor. In Turkey in the 1500s, the Ottoman rulers invested enormous sums in standardizing the design of mosques, particularly after the devastating earthquake of 1509. Barrel vaults and round-headed arches were given cross-ties, and the stone triangles that linked them to domes and cupolas were made exaggeratedly thick, thus protecting structure and worshippers alike. In Istanbul, the slim minaret of the Mihrimah Sultan complex (built in 1564), which had survived the earthquake of 1766, collapsed in the tremors of 1894, thus disproving theories on the durability of elegance. On the other hand, in 1564 the architect Sinan built the Lala Mustafa Pasha mosque at Erzerum with a short, thick minaret in the hope that it would resist earth tremors. Little was done in Iran, however, where rural poverty led to the perpetuation of rammed earth construction that was – and still is – extremely vulnerable to seismic damage.

By the start of industrialization, the style of construction using a "stone front" and a "brick front" had spread across Europe and thus eliminated many of the weaker, more vulnerable structures. This was facilitated by the fact that interest rates had been low for centuries, such that the average annual growth rate in seventeenth-century England, for example, was six times the population growth rate. The mass production which began during the Industrial Revolution afforded a further measure of protection against both the physical and the economic impact of natural hazards. But China, by contrast, was dogged by very high interest rates for centuries, and when the time came industrialization was almost completely inhibited. Thus the vulnerability of Asia to disaster remained higher than that of Europe (Jones 1981).

References

Alexander, D. E. 1982. *The earthquake of 23rd November 1980 in Campania and Basilicata, southern Italy*. London: International Disaster Institute/Foxcombe Publications.

Alexander, D. E. 1990. Behaviour during earthquakes: a southern Italian example. *International Journal of Mass Emergencies and Disasters* **8**, 5–29.

Anderson, J. W. 1967. Cultural adaptation to threatened disaster. *Human Organization* **27**, 298–307.

Anderson, D. R. & M. Weinrobe 1986. Mortgage default risks and the 1971 San Fernando earthquake. *Journal of the American Real Estate and Urban Economics Association* **14**, 110–35.

Antonopoulos, J. 1992. The great Minoan eruption of Thera volcano and the ensuing tsunami in the Greek archipelago. *Natural Hazards* **5**, 153–68.

Arvidson, R. M. 1969. On some mental effects of earthquake. *American Psychologist* **24**, 605–6.

Bahr, H. M. & C. D. Harvey 1979. Correlates of loneliness among widows bereaved in a mining disaster. *Psychological Reports* **44**, 367–85.

Barton, A. M. 1970. *Communities in disaster: a sociological analysis of collective stress situations*. Garden City, New York: Anchor Books.

Baumann, D. D. & J. H. Sims 1974. Human response to the hurricane. In *Natural hazards: local, national, global*, G. F. White (ed.), 25–30. New York: Oxford University Press.

Berren, M. R., A. Beigel, S. Ghertner 1980. A typology for the classification of disasters: implications for intervention. *Community Mental Health Journal* **16**, 103–11.

Berz, G. 1992. Losses in the range of US$ 50 billion and 50,000 people killed: Munich Re's list of major natural disasters in 1990. *Natural Hazards* **5**, 95–102.

References

Blong, R. J. 1984. *Volcanic hazards: a sourcebook on the effects of eruptions.* Orlando, Florida: Academic Press.

Bucher, R. 1957. Blame and hostility in disaster. *American Journal of Sociology* **62**, 467–75.

Burby, R. J., B. A. Cigler, S. P. French, E. J. Kaiser, J. Kartez, D. Roenigk, D. Weist, D. Whittington 1991. *Sharing environmental risks: how to control governments' losses in natural disasters.* Boulder, Colorado: Westview Press.

Burton, I., R. W. Kates, G. F. White 1978. *The environment as hazard.* New York: Oxford University Press.

Burton, I. & R. Pushchak 1984. The status and prospects of risk assessment. *Geoforum* **15**, 463–76.

Carlson, D. G. 1982. *Famine in history, with a comparison of two modern Ethiopian disasters.* New York: Orbis Maryknoll.

Coburn, A. W., A. Pomonis, S. Sakai 1989. Assessing strategies to reduce fatalities in earthquakes. In *International workshop on earthquake injury epidemiology for mitigation and response*, 107–32. Baltimore: The Johns Hopkins University Press.

Cochrane, H. C. 1984. An economic evaluation of earthquake prediction under current and possible future conditions. In *Earthquake prediction*, T. Rikitake (ed.), 713–36. Tokyo: Terra Scientific Publishing & Paris: UNESCO Press.

Covello, V. T. & J. Mumpower 1985. Risk analysis and risk management: an historical perspective. *Risk Analysis* **5**, 103–20.

Davis, I. 1978. *Shelter after disaster.* Headington, Oxford: Oxford Polytechnic Press.

Decker, R. & B. Decker 1981. *Volcanoes.* San Francisco: W. H. Freeman.

Demerath, N. J. 1957. Some general propositions: an interpretative summary. *Human Organization* **16**, 28–9.

Drabek, T. E. 1986. *Human system responses in disaster: an inventory of sociological findings.* New York: Springer.

Drabek, T. E. & E. L. Quarantelli 1969. Blame in disaster: another look, another viewpoint. In *Dynamic social psychology*, D. Dean (ed.), 604–15. New York: Random House.

Duckworth, D. H. 1986. Psychological problems arising from disaster work. *Stress Medicine* **2**, 315–23.

Dudasik, S. W. 1980. Victimization in natural disaster. *Disasters* **4**, 329–38.

Dynes, R. R. 1970. *Organized behaviour in disaster.* Lexington, Mass.: Lexington Books.

Dynes, R. R. & B. E. Aguirre 1979. Organization adaptation to crisis: mechanisms of co-ordination and structural change. *Disasters* **3**, 71–4.

Dynes, R. R. & E. L. Quarantelli 1971. The absence of community conflict in the early phases of natural disasters. In *Conflict resolution: contributions of the behavioural sciences*, C. G. Smith (ed.), 200–4. South Bend, Indiana: University of Notre Dame Press.

Dynes, R. R. & E. L. Quarantelli 1976. Community conflict: its absence and its presence in natural disasters. *Mass Emergencies* **1**, 139–52.

Ellson, R. W., J. W. Milliman, R. B. Roberts 1984. Measuring the regional economic effects of earthquakes and earthquake predictions. *Journal of Regional Science* **24**, 559–79.

Erickson, K. T. 1976. Trauma at Buffalo Creek. *Society* **13**, 58–65.

Falck, L. R. 1991. Disaster insurance in New Zealand. In *Managing natural disasters and the environment*, A. Kreimer & M. Munasinghe (eds), 120–5. Washington, DC: Environment Department, World Bank.

Fischhoff, B., C. Hohenemser, R. E. Kasperson, R. W. Kates 1978. Handling hazards. *Environment* **20**, 16–37.

Fischhoff, B. et al. 1982. *Acceptable risk*. New York: Cambridge University Press.

Foster, H. D. 1980. *Disaster planning: the preservation of life and property*. New York: Springer.

França, J-A. 1983. *Lisboa pombalina e o illuminismo*. Lisbon: Bertrand.

Fritz, C. E. 1957. Disasters compared in six American communities. *Human Organization* **16**, 6–9.

Fritz, C. E. 1961. Disaster. In *Contemporary social problems*, R. K. Merton & R. A. Nisbet (eds), 651–94. New York: Harcourt, Brace, Jovanovich.

Fritz, C. E. & J. H. Mathewson 1957. *Convergence behavior in disasters*. Washington, DC: National Research Council, National Academy Press.

Gist, R. & B. Lubin (eds) 1989. *Psychosocial aspects of disaster*. New York: John Wiley.

Glass, A. J. 1970. The psychological aspects of emergency situations. In *Psychological aspects of stress*, H. S. Abram (ed.), 62–9. Springfield, Illinois: Charles C. Thomas.

Green, B. L. 1982. Assessing levels of psychological impairment following disaster: consideration of actual and methodological dimensions. *Journal of Nervous and Mental Disease* **170**, 544–52.

Hartmann, K. & J. Allison 1981. Expected psychological reactions to disaster in medical rescue teams. *Military Medicine* **146**, 323–7.

Hewitt, K. (ed.) 1983. *Interpretations of calamity*. Boston: Allen & Unwin.

Hughes, R. 1983. Historic disasters. *Disasters* **7**, 161–3.

Jackson, E. L. 1982. The Laki eruption of 1783: impacts on population and settlement in Iceland. *Geography* **67**, 42–50.

James, L. D. 1973. Surveys required to design non-structural measures. *Proceedings of the American Society of Civil Engineers, Journal of the Hydraulics Division* **99**(HY10), 1,823–6.

Janis, I. L. & L. Mann 1977. Emergency decision-making: a theoretical analysis of response to disaster warnings. *Journal of Human Stress* **3**, 35–48.

Johnson, N. R. 1987. Panic and the breakdown of social order: popular myth, social theory, empirical evidence. *Sociological Focus* **20**, 171–83.

Jones, E. L. 1981. *The European miracle: environments, economies and geopolitics in the history of Europe and Asia*. Cambridge: Cambridge University Press.

Kasperson, R. E. & J. X. Kasperson 1983. Determining the acceptability of risk: ethical and policy issues. In *Risk: symposium proceedings on the assessment and perception of risk to human health in Canada*, J. T. Rodgers, & D. V. Bates (eds), 135–56. Ottawa: Royal Society of Canada.

Kastenbaum, R. 1974. Disaster, death and human ecology. *Omega: Journal of Death and Dying* **5**, 65–72.

Kilpatrick, F. P. 1957. Problems of perception in extreme situations. *Human Organization* **16**, 20–2.

Krell, G. 1978. Managing the psychosocial factor in disaster programmes. *Health and Social Work* **3**, 139–54.

Kunreuther, H. 1974. Disaster insurance: a tool for hazard mitigation. *Journal of Risk and Insurance* **41**, 287–303.

Kunreuther, H., R. Ginsburg, L. Miller, P. Sagi, P. Slovic, B. Borkan, N. Katz 1978. *Disaster insurance protection: public policy lessons*. New York: John Wiley.

References

Kunreuther, H. & L. Miller 1985. Insurance versus disaster relief: an analysis of interactive modelling for disaster policy planning. *Public Administration Review* **45** (special issue), 147–54.

Kunreuther, H. & P. Slovic 1982. Decision making in hazard and resource management. In *Geography, resources and environment*, Vol. 2, R. W. Kates & I. Burton (eds), 153–87. Chicago: University of Chicago Press.

Kusler, J. A. 1985. Liability as a dilemma for local managers. *Public Administration Review* **45**, 118–22.

Lifton, R. J. & E. Olson 1976. The human meaning of total disaster: the Buffalo Creek experience. *Psychiatry* **39**, 1–18.

Loizos, P. 1977. A struggle for meaning: reactions to disaster among Cypriot refugees. *Disasters* **1**, 231–9.

Lystad, M. 1987. *Human problems in major disasters: a training curriculum for emergency medical personnel*. Rockville, Maryland: US Alcohol, Drug Abuse and Mental Health Administration.

Martin, L. R. G. & G. Lafond (eds) 1988. *Risk assessment and management: emergency planning perspectives*. Waterloo, Ontario: University of Waterloo Press.

Melville, C. P. 1981. Historical monuments and earthquakes in Tabriz. *Iran* **19**, 159–77.

Mileti, D. S. 1980. Human adjustment to the risk of environmental extremes. *Sociology and Social Research* **64**, 328–47.

Mileti, D. S., J. H. Sorensen, J. R. Hutton 1984. Social factors affecting the response of groups to earthquake prediction: implications for public policy. In *Earthquake prediction*, T. Rikitake (ed.), 649–58. Tokyo: Terra Scientific Publishing & Paris: UNESCO Press.

Newman, C. J. 1976. Children of disaster: clinical observations at Buffalo Creek. *American Journal of Psychiatry* **133**, 306–12.

Nolen-Hoekesema, S. & J. Morrow 1991. A prospective study of depression and post-traumatic stress symptoms after a natural disaster: the 1989 Loma Prieta earthquake. *Journal of Personality and Social Psychology* **3**, 115–24.

Okrent, D. 1980. Comment on societal risk. *Science* **208**, 372–5.

O'Riordan, T. 1982. *Environmentalism*, 2nd edn. London: Pion.

Palm, R. I. et al. 1983. *Home mortgage lenders, real property appraisers and earthquake hazards*. Boulder, Colorado: Natural Hazards Research and Applications Information Center.

Parker, D. J & D. M. Harding 1979. Natural hazard evaluation, perception and adjustment. *Geography* **64**, 307–16.

Perry, R. W. 1979. Detecting psychopathological reactions to natural disaster: a methodological note. *Social Behaviour and Personality* **7**, 173–7.

Petak, W. J. & A. A. Atkisson 1982. *Natural hazard risk assessment and public policy: anticipating the unexpected*. New York: Springer.

Placanica, A. 1985. *Il filosofo e la catastrofe: un terremoto del settecento*. Turin: Einaudi.

Porfiriev, B. N. 1992. The environmental dimension of national security: a test of systems analysis methods. *Environmental Management* **16**, 735–42.

Preston, V., S. M. Taylor, D. C. Hodge 1983. Adjustment to natural and technological hazards: a study of an urban residential community. *Environment and Behaviour* **15**, 143–64.

Quarantelli, E. L. 1954. The nature and conditions of panic. *American Journal of Sociology* **60**, 267–75.

Quarantelli, E. L. 1976. Human response in stress situations. In *Proceedings of the first conference and workshop on fire casualties*, B. M. Halpin (ed.). Laurel, Maryland: Applied Physics Laboratory, Johns Hopkins University.

Quarantelli, E. L. 1977. Panic behavior: some empirical observations. In *Human response to tall buildings*, D. J. Conway (ed.), 336–50. Stroudsburg, Pennsylvania: Dowden, Hutchinson & Ross.

Quarantelli, E. L. & R. R. Dynes 1970. Dissensus and consensus in community emergencies: patterns of looting and property norms. *Il Politico* **34**, 276–91.

Quarantelli, E. L. & R. R. Dynes 1976. Community conflict: its absence and its presence in natural disasters. *Mass Emergencies* **1**, 139–52.

Rikitake, T. 1982 *Earthquake forecasting and warning*. Dordrecht: D. Reidel.

Ross, G. A. 1980. The emergence of organization sets in three ecumenical disaster recovery organizations: empirical and theoretical exploration. *Human relations* **33**, 23–39.

Rossi, P. H., J. D. Wright, E. Weber-Burdin, J. Pereira 1983. *Victims of the environment: losses from natural hazards in the United States*. New York: Plenum Press.

Russell, C. S. 1970. Losses from natural hazards. *Land Economics* **46**, 383–93.

Scheidegger, A. E. 1975. *Physical aspects of natural catastrophes*. New York: Elsevier.

Sewell, W. R. D. & H. D. Foster 1976. Environmental risk: management strategies in the developing world. *Environmental Management* **1**, 49–59.

Shimada, K. 1972. Attitudes towards disaster defence organizations and volunteer activities in emergencies. In *Proceedings of the Japan–United States disaster research seminar: organizational and community responses to disasters*, 208–17. Columbus, Ohio: Disaster Research Center, Ohio State University.

Simkin, T. & R. S. Fiske 1983. *Krakatau, 1883: the volcanic eruption and its effects*. Washington, DC: Smithsonian Institution Press.

Slovic, P., B. Fischhoff, S. Lichtenstein, B. Korrigan, B. Combs 1977. Preference for insurance against probable small loss: implications for theory and practice of insurance. *Journal of Risk and Insurance* **44**, 237–58.

Slovic, P., H. Kunreuther, G. F. White 1974. Decision processes, rationality and adjustment to natural hazards. In *Natural hazards: local, national, global*, G. F. White (ed.), 187–205. New York: Oxford University Press.

Smith, P. J. 1990. Redefining decision: implications for managing risk and uncertainty. *Disasters* **14**, 230–40.

Smith, R. R. 1979. Mitigation, risk aversion and regional differentiation. *Journal of Regional Science* **19**, 31–45.

Sorkin, A. L. 1983. *Economic aspects of natural hazards*. Lexington, Mass.: Lexington Books.

Stallings, R. A. & E. L. Quarantelli 1985. Emergent citizen groups and emergency management. *Public Administration Review* **45**, 93–100.

Starr, C. 1969. Societal benefit versus technological risk. *Science* **165**, 1,232–8.

Stothers, R. B. 1984. The great eruption of Tambora and its aftermath. *Science* **224**, 1,191–8.

Taylor, A. J. W. 1984. Architecture and society: disaster structures and human stress. *Ekistics* **51**, 446–51.

Thorarinsson, S. 1970. The Lakigigar eruption of 1783. *Bulletin Volcanologique* **33**, 910–27.

Titchener, J. L. & F. T. Kapp 1976. Family and character change at Buffalo Creek. *American Journal of Psychiatry* **133**, 295–9.

Turner, B. A. 1976. The development of disasters: a sequence model for the analysis of the origin of disasters. *Sociological Review* **24**, 753–74.

Turner, B. A. 1979. The social aetiology of disasters. *Disasters* **3**, 53–9.

Van Ardsol, M. G., J. Sabagh, F. Alexander 1964. Reality and the perception of environmental hazards. *Journal of Health and Human Behaviour* **5**, 144–53.

Wallace, A. F. C. 1956. *Human behavior during extreme situations*. Washington, DC: National Academy of Sciences.

Wang, S.-W. & A.-C. Zhao 1981. Droughts and floods in China, 1440–1979. In *Climate and history: studies in past climates and their impact on man*, T. M. L. Wigley, M. J. Ingram, G. Farmer (eds), 271–88. Cambridge: Cambridge University Press.

Warheit, G. J. 1979. Life events, coping, stress and depressive symptomatology. *American Journal of Psychiatry* **136**, 502–7.

Warheit, G. J. 1985. A propositional paradigm for estimating the impact of disasters on mental health. *International Journal of Mass Emergencies and Disasters* **3**, 29–48.

Wenger, D. E. 1972. DRC studies of community functioning. In *Proceedings of the Japan–United States disaster research seminar: organizational and community response to disasters*. Columbus, Ohio: Disaster Research Center, Ohio State University.

Wenger, D. E. 1978. Community response to disaster: functional and structural alterations. In *Disasters: theory and research*, E. L. Quarantelli (ed.), 17–47. Beverley Hills, California: Sage.

White, G. F. (ed.) 1974. *Natural hazards: local, national, global*. New York: Oxford University Press.

White, G. F. & J. E. Haas 1975. *Assessment of research on natural hazards*. Cambridge, Mass.: MIT Press.

Whyte, A. V. T. 1986. From hazard perception to human ecology. In *Themes from the work of Gilbert F. White*. Vol. 2, *Geography, resources and environment*, R. W. Kates & I. Burton (eds), 240–71. Chicago: University of Chicago Press.

Williams, R. M. & C. M. Parkes 1975. Psychosocial effects of disaster: birth rate in Aberfan. *British Medical Journal* **II**(5966), 303–4.

Wolfenstein, M. 1957. *Disaster: a psychological essay*. Glencoe, Illinois: Free Press.

Yokoyama, I. 1987. A scenario of the 1883 Krakatau tsunami. *Journal of Volcanology and Geothermal Research* **25**, 157–65.

Select bibliography

Advances in Behaviour Research and Therapy 1991. Impact of natural disasters on children and families: proceedings of a symposium held in San Francisco, November. *Advances in Behaviour Research and Therapy* **13**.

Baker, G. W. & D. W. Chapman 1962. *Man and society in disaster*. New York: Basic Books.

Barkun, M. 1977. Disaster in history. *Mass Emergencies* **2**, 219–31.

Baumann, D. D. & J. H. Sims 1978. Flood insurance: some determinants of adoption. *Economic Geography* **54**, 189–96.

Blaustein, M. (ed.) 1991. Natural disasters and psychiatric response. *Psychiatric Annals* **21**, 516–65.

Bolin, R. C. 1982. *Long-term family recovery from disaster*. Boulder, Colorado: Natural Hazards Research and Applications Information Center.

Burton, I. 1972. Cultural and personality variables in the perception of natural hazards. In *Environment and the social sciences: perspectives and applications*, J. F. Wohlwill, D. H. Carson (eds), 184–95. New York: American Psychological Association.

Burton, I. & R. W. Kates 1964. The perception of natural hazards in resource management. *Natural Resources Journal* **3**, 412–41.

Claxton, R. H. (ed.) 1986. Investigating natural hazards in Latin American history. *Studies in the Social Sciences* **25**, 1–167.

Cohen, R. E. & F. L. Ahearn, Jr 1980. *Handbook for mental health care of disaster victims*. Baltimore: The Johns Hopkins University Press.

Covello, V. T. & J. Mumpower 1985. Risk analysis and risk management: an historical perspective. *Risk Analysis* **5**, 103–20.

Dacy D. C. & H. Kunreuther 1969. *The economics of natural disaster: implications for federal policy*. New York: The Free Press.

Douglas, M. & A. Wildavsky 1982. *Risk and culture: an essay on the selection of technical and environmental dangers*. Berkeley: University of California Press.

Dynes, R. R., B. De Marchi, C. Pelanda (eds) 1987. *Sociology of disasters: contribution of sociology to disaster research*. Milan: Franco Angeli.

Friesema, H. P. et al. 1979. *Aftermath: communities after natural disaster*. Beverly Hills, California: Sage.

Fritz, C. E. & H. B. Williams 1957. The human being in disasters: a research perspective. *Annals of the American Academy of Political and Social Science* **309**, 42–51.

Gillespie, D. F., R. W. Perry, D. S. Mileti 1974. Collective stress and community transformation. *Human Relations* **27**, 767–78.

Glenn, C. 1979. Natural disasters and human behaviour: explanation, research and models. *Psychology* **16**, 23–36.

Gleser, G. C., B. L. Green, C. N. Winget 1981. *Prolonged psychosocial effects of disaster: a study of Buffalo Creek*. New York: Academic Press.

Goltz, J. D., L. A. Russell, L. B. Bourque 1992. Initial behavioural response to a rapid onset disaster. *International Journal of Mass Emergencies and Disasters* **10**, 43–69.

Grosser, G. H., H. Wechsler, M. Greenblatt (eds) 1964. *The threat of impending disaster: contributions to the psychology of stress*. Cambridge, Mass.: MIT Press.

Gurung, S. M. 1989. Human perception of mountain hazards in the Kakani-Kathmandu area: experiences from the Middle Mountains of Nepal. *Mountain Research and Development* **9**, 353–64.

Handmer, J., E. C. Penning-Rowsell (eds) 1990. *Hazards and the communication of risk*. Aldershot, England: Gower.

Heathcote, R. L. 1990. *Managing the droughts: perception of resource management in the face of drought hazard in Australia*. Dordrecht: Kluwer.

Jackson, E. L. 1982. The Laki eruption of 1783: impacts on population and settlement in Iceland. *Geography* **67**, 42–50.

Johnson, B. B. & V. T. Covello (eds) 1987. *The social and cultural construction of risk: essays on risk selection and perception*. Dordrecht: Kluwer Academic Publications.

Kasperson, R. E. & P. J. M. Stallen (eds) 1990. *Communicating risks to the public*. Dordrecht: Kluwer.

Kates, R. W. 1978. *Risk assessment of environmental hazard*. New York: John Wiley.

Kinston, W. & R. Rosser 1974. Disaster: effects on mental and physical state. *Journal of Psychosomatic Research* **18**, 437–56.

Select bibliography

Kohn, R. & I. Levav 1990. Bereavement in disaster: an overview of the research. *International Journal of Mental Health* **19**, 61–76.

Kreps, G. A. 1984. Sociological inquiry and disaster research. *Annual Reviews of Sociology* **10**, 309–30.

Kreps, G. A. 1989. *Social structure and disaster.* Newark, Delaware: University of Delaware Press.

Krimsky S. & A. Plough 1988. *Environmental hazards: communicating risk as a social process.* Dover, Mass.: Auburn House.

Kunreuther, H. 1968. The case for comprehensive disaster insurance. *Journal of Law and Economics* **11**, 133–63.

Lima, B. R. & M. Gittelman (eds) 1990. Coping with disasters: the mental health component; Parts 1 & 2. *International Journal of Mental Health* **19**(1 & 2).

Lindell, M. K. & R. W. Perry 1991. *Behavioural foundations of community emergency planning.* New York: Hemisphere Press.

Lystad, M. (ed.) 1988. *Mental health response to mass emergencies: theory and practice.* New York: Brunner Mazel.

McPherson, H. J. & T. F. Saarinen 1977. Floodplain dwellers' perception of the flood hazard at Tucson, Arizona. *Annals of Regional Science* **11**, 25–40.

Mather, A. S. 1982. The changing perception of soil erosion in New Zealand. *Geographical Journal* **148**, 207–18.

Mileti, D. S. & J. M. Nigg 1984. Earthquakes and human behaviour. *Earthquake Spectra* **1**, 89–106.

Palm, R. 1981. Public response to earthquake hazard information. *Annals of the Association of American Geographers* **71**, 389–99.

Palm, R. 1987. Pre-disaster planning: the response of residential real estate developers to special studies zones. *International Journal of Mass Emergencies and Disasters* **5**, 95–102.

Parad, H. J., H. L. P. Resnik, L. G. Parad (eds) 1976. *Emergency and disaster management: a mental health source book.* Bowie, Maryland: Charles Press.

Perry, R. W. & M. K. Lindell 1978. The psychological consequences of natural disaster: a review of research on American communities. *Mass Emergencies* **3**, 105–15.

Quarantelli, E. L. 1957. The behaviour of panic participants. *Sociology and Social Research* **41**, 187–94.

Quarantelli, E. L. (ed.) 1978. *Disasters: theory and research.* Beverly Hills, California: Sage.

Saarinen, T. F. 1966. *Perception of drought hazard on the Great Plains.* Department of Geography, University of Chicago.

Saarinen, T. F. 1976. *Environmental planning: perception and behavior.* Boston: Houghton Mifflin.

Scanlon, J. 1988. Winners and losers: some thoughts about the political economy of disaster. *International Journal of Mass Emergencies and Disasters* **6**, 47–64.

Sheets, P. D. 1980. *Archaeological studies of disaster: their range and value.* Boulder, Colorado: Natural Hazards Research and Applications Information Center.

Simpson-Housley, P. & P. Bradshaw 1978. Personality and the perception of earthquake hazard. *Australian Geographical Studies* **16**, 65–72.

Slovik, P., B. Fishhoff, S. Lichtenstein 1979. Rating the risks. *Environment* **21**, 14–39.

Smolka, A. & G. Berz 1989. The Mexico earthquake of September 19, 1985: an analysis of the insured loss and implications for risk assessment. *Earthquake Spectra* **5**, 223–48.

Sowder, B. J. (ed.) 1985. *Disasters and mental health: selected contemporary perspectives*. Rockville, Maryland: Center for Mental Health Studies of Emergencies, National Institute for Mental Health.

Starr, C., R. Rudman, C. Whipple 1976. Philosophical basis for risk analysis. *Annual Reviews of Energy* **1**, 629–62.

Tobriner, S. 1984. A history of reinforced masonry construction designed to resist earthquakes, 1755–1907. *Earthquake Spectra* **1**, 125–50.

Torry, W. I. 1978. Natural disasters, social structure and change in traditional societies. *Journal of African and Asian Studies* **13**, 167–83.

Turner, R. H., J. M. Nigg, D. Heller Paz 1986. *Waiting for disaster: earthquake watch in California*. Berkeley & Los Angeles, California: University of California Press.

US National Research Council 1989. *Improving risk communication*. Washington, DC: National Academy Press.

Vinso, J. D. 1977. Financial implications of natural disasters: some preliminary indications. *Mass Emergencies* **2**, 205–18.

Wang, S.-W. & A.-C. Zhao 1981. Droughts and floods in China, 1440–1979. In *Climate and history: studies in past climates and their impact on man*, T. M. L. Wigley, M. J. Ingram, G. Farmer (eds), 271–88. Cambridge: Cambridge University Press.

White, G. F. 1988. Paths to risk analysis. *Risk Analysis* **8**, 171–5.

Whyte, A. V. T. & I. Burton (eds) 1980. *Environmental risk assessment*. New York: John Wiley.

Conclusion

CHAPTER TEN

Towards an international strategy against disasters

The lessons of disaster

The human response to natural hazards seems to be generating or at least permitting an increase in property losses, especially in countries where economic growth is rapid and modern technology is spreading fast. Some hazards are created by persistent inhabitance of dangerous areas or by alteration of the land or water, while others are exacerbated by efforts to reduce the risk.

The record of disasters teaches the following important lessons. First, the high cost of destruction of life and property engendered by natural catastrophes makes it imperative that experience, skills and research capacity are shared at the international level. Secondly, in natural hazards studies the emphasis has been placed firmly on technological perspectives such as the design of monitoring and warning systems. A social component is required: in fact, the success or failure of mitigation programmes will be strongly influenced by people's perceptions of the threat of disaster and how to adjust to it, and by the organization and cultural make-up of society.

Thirdly, in developed countries, heavy reliance on civil and structural engineering works for protection against disasters has led to increases in both productivity and vulnerability. But to rely on only one strategy is to court socio-economic disaster, environmental degradation and an increased burden of debt and relief obligations. Moreover, there has been a widely held belief that hazard problems and disaster impacts can be solved using one or other of the specialized solutions: relief operations after avalanches, dam building to mitigate floods, anti-seismic construction against earthquakes, and so on. But a better approach would involve integrating all the various structural and non-structural methods in a comprehensive programme of hazard reduction. Thus the aim should be one of reducing the "hazardousness of place", that is, of mitigating all hazards affecting a particular zone. The creation of the United Nations Office of the Disaster Relief Co-ordinator was one positive move in this direction. Other international agencies can give comprehensive assistance on particular topics: the Food and Agriculture Organization on crop insurance, UNESCO on anti-seismic construction, the World Meteorological Office on weather hazards, and so on. But hazard mitigation planners should aim to evaluate all risks and institute strategies to reduce them to roughly comparable, and it is to be hoped low, levels.

Lastly, although they are great socio-economic levellers at the outset, disasters tend to reinforce the pre-existing divisions in society: the rich become richer and more powerful, while the poor become marginalized and increasingly disadvantaged. Equitable planning against natural catastrophe should ensure that control of the resources for reconstruction and economic recovery is not vested entirely in the hands of the most wealthy and influential members or groups of society. This, of course, is easier said than done!

But to some extent these lessons have been learnt by those who have to contend with disasters, for, throughout the world, governments are increasingly coming to realize that problems can be tackled and mitigation achieved only by comprehensive planning at the national and international level.

National policy and disasters

National policies on natural hazards tend to involve four possible approaches, which vary from minimum intervention to full-scale mitigation and emergency planning.

First, **disaster relief** is acceptable as a limited policy where the threat of major losses from a particular hazard is low. This does not represent mitigation, but merely a subsidy to the survivors of a disastrous event; hence, where there are reasons for a national hazard reduction policy to be lacking, government policy can be limited to disaster relief. But serious attention should be given to controlling secondary effects such as inflation and profiteering. Attention must also be given to groups of people who are peripheral to the disaster, especially those who are not granted relief, but suffer the effects of inflation, high prices, opportunism and profiteering during the aftermath of the crisis. Some social and political control must be exercised over those who use the disaster as an attempt to profit by the disadvantages of the victims (for example, by obtaining control of vital resources which have been rendered scarce by the disaster impact). Conflict of interest may occur between rival groups, such as environmentalists who oppose reservoir construction and hydrologists who wish to dam rivers as a means of controlling flood flow.

Hazards that have low disaster potential may in the end receive little attention from government and be allotted low priority during the apportionment of funds. For instance, Canada decided to have an explicit national policy of not mitigating floods at the federal level, leaving the problem instead to provinces and towns, whose resources and sphere of action are more limited. In any event, developed countries should usually aim to combine disaster relief with hazard insurance schemes and a social welfare policy to safeguard disadvantaged groups.

Secondly, **technological adjustments** offer limited prospects for control of natural events and their effects. This sort of policy often stems from the realization that disaster relief on its own is insufficient to mitigate the hazard. Such policies often place an excessive reliance on technology, at the expense of flexibility and a diversity of approaches. In the worst cases, reliance on the technological answer may exacerbate the very problems that it is designed to ameliorate, as in the case of the engineering solution to coastal erosion (Ch. 4). This method commonly fails to take social reactions into account, although they are vital factors, given that the functionality of technology depends to a large extent on how it is perceived by its users. Another potential drawback of technological approaches is that they may supplant traditional, "tried and tested" solutions to the hazard in question, yet their own constancy and efficacy may be dubious (as in the case of novel remedies to the post-disaster shelter problem, as outlined in Ch. 6).

Among the research community, opinions are divided as to whether technological change will reduce the death toll in disasters (Burton et al. 1978: 223) or whether its effects will be more than offset by population growth and above all the growth of the poorer sections of society. Coburn et al. (1989) argued that the current annual world population growth of 1.6 per cent (leading to doubling of total numbers in only 40 years) means that if death tolls in disasters are to be kept stable the average vulnerability of the world's building stock needs to be reduced at a reciprocal rate, or in other words halved in 40 years. In this respect, present trends point to steady progress in industrialized countries and little change in the developing world. Given the concentration of disaster impacts in the latter, this bodes ill for future global tolls of destruction and injury. The total per capita cost of seismically retrofitting a concrete framed building averages $5,000–20,000, but 80 per cent of the world's population lives in housing that cost no more than $500 to build (using local materials and family labour). In order to provide safe housing, investment would need to be increased by a factor of two or three.

It is paradoxical that Western society, which is so rich in information, should be so superficial in its treatment of multifaceted problems (Lemons 1989). At the same time, environmental hazards are an issue that defies adequate analysis from a sectoral, or partial, rather than holistic, viewpoint. At the root of the problem is society's fascination with technology, which, however, it has not yet learnt to use rationally and in moderation. Now technology is inherently separate from the value systems which employ it: it can solve or create problems. But in its relentless pursuit of technocracy society has tended to rely excessively on data and quantification. When facts and numbers are treated as explanation in their own right, and when mankind attempts to assign numerical values to intangible factors, that is when ethics, objectives and value systems become purely intrinsic and susceptible to being ignored or taken for granted (Miller 1985). *Tactics* replace *strategy*. Instead it should be recognized that the technological approach is only one of several forms of rationale, including the economic, social, legal, political and ecological ones. Problems should not be over-compartmentalized (Caldwell 1987).

Thirdly, **comprehensive damage reduction** may be the aim of policies, which are thus directed towards mapping hazards, regulating activities, devising emergency plans, designing engineering works, instituting financial and insurance schemes against the hazard, collecting and disseminating information, prediction and warning. An integrated approach of this kind is to be encouraged, but must be tailored to the resources available to carry it out. Moreover, such a strategy requires considerable sophistication and hence must be scaled down where the organization that can be mustered is likely to be rudimentary.

Finally, **multiple hazard management** can be achieved by combining strategies into a comprehensive policy. This involves equalizing risks in developed countries and selecting particular risks for attention in developing countries. There is a wide diversity in the acceptability of risks for each type of natural or man-made hazard. Hence, there are few plans for multiple hazard reduction and few, even, for comprehensive single hazard reduction. In developed nations, hazard management needs to be linked to environmental protection; in developing countries it must also be linked to socio-economic development.

The belief that national policies of hazard mitigation are needed has been very slow to develop (Davis & Seitz 1982). In the face of competing demands for scarce resources, the common response has been to provide help only when an overwhelmingly strong case can be made – for example, during the immediate aftermath of a major national disaster. In fact, most permanent agencies or large-scale hazard reduction works post-date major damaging events that stimulated their creation at a time when official attention was focused on natural disasters and public opinion was calling for something to be done. Hence, disaster tends to reassert and strengthen support for both national policies of hazard mitigation and particular strategies for controlling hazardous events or human vul-nerability to them. In the time between disasters, attention tends to switch to other matters, which may be fatal in terms of hazard policy formulation.

There is no universal blueprint for a national policy against disasters. It depends upon the particular configuration of hazards and their impacts, and upon conflicting demands for scarce national resources. Nevertheless, at the world scale, national policies *have* succeeded in reducing potential levels of death and destruction. If the levels have continued to increase, it is the result of a tendency towards increased vulnerability, and intervention by governments and agencies has slowed down the rate. Decision-makers must weigh the cost of sustaining the losses against the expense of loss-reduction measures. They must also bear in mind that economic loss is only a fraction of the true toll of disasters, which also produce death, injury, disruption of daily life and hardship.

Public policy and government action are often insufficient to reduce the very large and increasing social and economic costs of hazards: is there a widening of the gap between hazard mitigation on the one hand and impacts and vulnerability on the other? Cost appears to be the major influence on decision-making in this sphere, rather than practicability. But organization may be the key to successful implementation of decisions once they have been made. Experts and government officials alike need a structure of mutual involvement. In this respect it appears that the regional level is the right one for most hazard reduction plans: it provides the interface between national policy (which is, of course, still vital) and local initiatives, and it appears to be an ideal level for co-ordination of efforts among smaller bodies or for distribution of resources from central government to local administrations.

The International Decade for Natural Disaster Reduction

Whatever the operative rôle of the regional level of government, there is also an increasing need for worldwide initiatives. Global interconnectivity is increasing by leaps and bounds, and a major disaster no longer has purely domestic implications. The importance of Third World countries as sources of raw materials and primary production, and their increasing debt burdens make it imperative that the developed world interest itself in solutions to their disaster problems, by supplying relief, aid and expertise and by sharing technology. The existence under United Nations auspices of several agencies concerned with natural disasters, and the increasing fund of experience held by the main international charities are encouraging signs for the promotion and diffusion of hazard mitigation around the world.

Dr Frank Press, the President of the US National Academy of Sciences, first proposed the IDNDR in 1984 at the Eighth World Congress on Earthquake Engineering. Eventually, the idea was adopted by the United Nations, under Resolution 42/169 of 1987. The United States continued to take the lead by issuing two prospectuses on American efforts connected with the Decade (US National Research Council 1987, 1989). Several proponents and practitioners have published their ideas on how the Decade should be put into operation (Holland 1989, Housner 1989, Lechat 1990, Oaks & Bender 1990, Bates et al. 1991).

It seems that the principal thrust of the Decade will be to promote the international sharing of ideas and data (through conferences and other co-operative efforts), and the development of national and worldwide networks for monitoring the agents which produce disasters. It is thus hoped that by the end of the twentieth century all countries will have made national assessments of natural disaster risk, drawn up preparedness plans at the national and local levels and set up or gained access to warning systems at all appropriate scales from the global to the local (IDNDR Committee 1991).

Laudable as these aims are, the initiative has not escaped criticism, calling into question the American prediction that it will be possible to halve the impact of natural disasters by the year 2000. Mitchell (1988) offered a powerful critique of the IDNDR in general and the US Decade for Natural Disaster Reduction in particular. He argued that new research, such as that proposed under the auspices of the Decade, is not easily justified when much current knowledge and expertise remain unutilized. Moreover, co-operative science has a mixed record of success in reducing the toll of natural hazards. Its efficacy should be evaluated before any large new projects are proposed, especially those requiring expensive and sophisticated monitoring, experimentation and management initiatives which are unlikely to work in poor countries.

According to Mitchell, broad, societal processes are not easy to change, yet are fundamental to human vulnerability. They are not considered explicitly in the prospectus for the USDNDR, and similarly, non-structural approaches and the applications of alternative, low-level technology are not receiving adequate consideration. Finally, as initially specified, the Decade lacks a strong focus on a few relatively simple issues. Resources will undoubtably be insufficient to solve, or even tackle, all major problems of disasters around the world.

Hence, in the eyes of some practitioners, the IDNDR has assumed the status of a "technofix", in which the proponents of technology and hard science use it as a justification for generating yet more of the same. Sociologists (Dynes 1990), Third World development specialists (Merani 1991) and ecologists (Bunin 1989) have been quick to point out the lack of reference to their own specialities. An extreme interpretation would be that the Decade represents an attempt by engineers and physical and natural scientists to concentrate academic power and funding opportunities into their own hands in the name of applying their sciences. This, however, should not negate the fact that science and technology have very important rôles to play in monitoring and mitigation efforts.

As usual, one of the most pressing needs for the future is for information. But new structures for collecting data are not necessarily required. Rather, what is needed is some form of standardized procedure for gathering information, assembling the data and guaranteeing worldwide accessibility to them. Injury statistics, for example, should be internationally notifiable in the same way as communicable diseases. Under the aegis of the IDNDR, the United Nations is the obvious choice for such a task, but has so far not been supplied with adequate resources to carry it out. However, there are some positive signs of innovation. As I write, the international community, led by the seven most industrialized nations, has begun to debate the possibility of appointing a UN Commissioner for Disaster Relief. At last there is some consensus that relief efforts have been too haphazard and too disparate, and that duplication or lack of effort have all too often replaced collaboration and sharing. Providing that the social and perceptual problems of disaster receive the same priority as the technical ones, there is every likelihood that at the conclusion of the Decade a better, more unified strategy for tackling natural disaster will come into being.

References

Bates, F. L., R. R. Dynes, E. L. Quarantelli 1991. The importance of the social sciences to the International Decade for Natural Disaster Reduction. *Disasters* **15**, 288–9.

Bunin, J. 1989. Incorporating ecological concerns into the IDNDR. *Natural Hazards Observer* **14**, 4–5.

Burton, I., R. W. Kates, G. F. White 1978. *The environment as hazard*. New York: Oxford University Press.

Caldwell, L. 1987. The contextual basis for environmental decision-making: assumptions are predeterminants of choice. *Environmental Professional* **9**, 302–8.

Coburn, A. W., A. Pomonis, S. Sakai 1989. Assessing strategies to reduce fatalities in earthquakes. *International Workshop on Earthquake Injury Epidemiology for Mitigation and Response*. Baltimore: The Johns Hopkins University Press.

Davis, M. & S. T. Seitz 1982. Disasters and governments. *Journal of Conflict Resolution* **26**, 547–68.

Dynes, R. R. 1990. Social concerns and the IDNDR. *Natural Hazards Observer* **14**, 7.

Holland, G. L. 1989. Observations on the International Decade for Natural Disaster Reduction. *Natural Hazards* **2**, 77–82.

Housner, G. W. 1989. An International Decade for Natural Disaster Reduction, 1990–2000. *Natural Hazards* **2**, 45–75.

IDNDR Committee 1991. Report on the first session of the IDNDR Scientific and Technical Committee, Bonn, 4–8 March.

Lechat, M. F. 1990. The International Decade for Natural Disaster Reduction: background and objectives. *Disasters–* **14**, 1–6.

Lemons, J. 1989. The need to integrate values into environmental curricula. *Environmental Management* **13**, 133–47.

Merani, N. S. 1991. The International Decade for Natural Disaster Reduction. In *Managing natural disasters and the environment*, A. Kreimer & M. Munasinghe (eds), 36–8. Washington, DC: Environment Department, World Bank.

Miller, A. 1985, Technological thinking: its impact on environmental management. *Environmental Management* **9**, 179–90.

Mitchell, J. K. 1988. Confronting natural disasters: an International Decade for Natural Hazard Reduction. *Environment* **30**, 25–9.

Oaks, S. D. & S. O. Bender 1990. Hazard reduction and everyday life: opportunities for integration during the Decade for Natural Disaster Integration. *Natural Hazards* **3**, 87–90.

US National Research Council 1987. *Confronting natural disasters: an International Decade for Natural Hazard Reduction*. Washington, DC: National Academy Press.

US National Research Council 1989. *Reducing disasters' toll: the United States Decade for Natural Disaster Reduction*. Washington, DC: National Academy Press.

Select bibliography

Aguirre, B. E. & D. Bush 1992. Disaster programs as technology transfers. *International Journal of Mass Emergencies and Disasters* **10**, 161–78.

Bates, T. F. 1980. *Transferring earth science information to decision-makers: problems and opportunities as experienced by the US Geological Survey*. US Geological Survey Circular 813.

Boulle, P. L. 1990. Will the 1990s be a decade of increasingly destructive natural disasters? *Natural Hazards* **3**, 419–22.

Faupel, C. 1985. *The ecology of disaster*. New York: Irvington.

Hays, W. W., B. M. Pouhban 1991. Technology transfer. *Episodes* **14**, 66–72.

Journal of Geography 1988. The teaching of natural hazards (special issue). *Journal of Geography* **87**.

Olson, R. S. & D. C. Nilson 1982. Public policy analysis and hazards research: natural complements. *Social Science Journal* **19**, 89–103.

Waddell, E. 1977. The hazards of scientism: a review article. *Human Ecology* **5**, 69–76.

Yin, R. K. & G. B. Moore 1985. *The utilization of research: lessons from the natural hazards field*. Washington, DC: Cosmos Corporation.

Index

Pages numbers in **bold** refer to figures; numbers in *italics* refer to tables.